ENVIRONMENTAL IMPACT OF SHIPS

Shipping is responsible for transporting 90 per cent of the world's trade. This book provides a comprehensive review of the impact shipping has on the environment. Topics covered include pollutant discharges, such as atmospheric emissions, oil, chemical waste, sewage and biocides, as well as non-pollutant impacts, including invasive species, wildlife collisions, noise, physical damage and the environmental effects associated with shipwrecks and shipbreaking. The history of relevant international legislation is also covered. With chapters written by eminent international authors, this book provides a global perspective on the environmental impact of ships, making it a useful reference for advanced students and researchers of environmental science, as well as practitioners of maritime law and policy and marine business.

STEPHEN DE MORA recently retired as the chief executive of the Plymouth Marine Laboratory (PML) and PML Applications Ltd. Previously, he taught chemistry, environmental science and oceanography at universities in the UK (University of Lancaster), New Zealand (University of Auckland) and Canada (University of Quebec at Rimouski), and he worked at the International Atomic Energy Agency (IAEA) Marine Environment Laboratory in Monaco. Through the IAEA, he worked on several Regional Seas Programme and Global Environment Facility projects around the world. He was presented with a Distinguished Service Award in 2005, the same year in which the IAEA was a co-recipient of the Nobel Prize. He is the only *ad hominem* member of the UK Marine Science Coordination Committee. He serves as a Sargasso Sea Commissioner and on a number of other national and international committees.

TIMOTHY FILEMAN is a marine scientist, centre manager for the Ballast Water Centre and technical manager at PML Applications Ltd. He has more than 30 years' experience in marine science and was initially trained as an analytical chemist working on marine pollution-related issues for the UK government. He later moved into research in order to follow his interests in organic contaminant behaviour through estuaries and marine biogeochemistry before developing commercial services for PML. He developed the Ballast Water Centre, which grew from his experience in working with the shipping industry to deliver environmental services. He is a chartered scientist and marine scientist, as well as a member of the Institute of Marine Engineering, Science and Technology (IMarEST).

THOMAS VANCE is Centre Manager of the Centre for Marine Biofouling and Corrosion. He specializes in marine fouling community ecology and biofouling control. He has experience in designing, conducting and interpreting field- and laboratory-based experiments on marine invertebrate and algal assemblages, both in the UK and internationally. His practical experience includes diving surveys, field-based manipulative experimentation, marine invertebrate taxonomy, advanced image analysis, physiological assessments of fouling species and molecular analysis of biofilms, together with multivariate statistics and reporting.

CAMBRIDGE ENVIRONMENTAL CHEMISTRY SERIES

This wide-ranging series covers all areas of environmental chemistry, placing emphasis on both basic scientific and pollution-orientated aspects. It comprises a central core of textbooks, suitable for those taking courses in environmental sciences, ecology and chemistry, as well as more advanced texts (authored or edited) presenting current research topics of interest to graduate students, researchers and professional scientists. Books cover atmospheric chemistry; chemical sedimentology; freshwater chemistry; marine chemistry; and soil chemistry.

Latest books published in the series:

T. Nakajima et al., *Environmental Contamination from the Fukushima Nuclear Disaster*

W. Davison, *Diffusive Gradients in Thin-Films for Environmental Measurements*

P. G. Coble et al., *Aquatic Organic Matter Fluorescence*

S. Roy et al., *Phytoplankton Pigments: Characterization, Chemotaxonomy and Applications in Oceanography*

E. Tipping,*Cation Binding by Humic Substances*

D. Wright and P. Welbourn, *Environmental Toxicology*

ENVIRONMENTAL IMPACT OF SHIPS

Edited by

STEPHEN DE MORA
Plymouth Marine Laboratory

TIMOTHY FILEMAN
PML Applications Ltd

THOMAS VANCE
PML Applications Ltd

CAMBRIDGE
UNIVERSITY PRESS

University Printing House, Cambridge CB2 8BS, United Kingdom

One Liberty Plaza, 20th Floor, New York, NY 10006, USA

477 Williamstown Road, Port Melbourne, VIC 3207, Australia

314–321, 3rd Floor, Plot 3, Splendor Forum, Jasola District Centre, New Delhi – 110025, India

79 Anson Road, #06–04/06, Singapore 079906

Cambridge University Press is part of the University of Cambridge.

It furthers the University's mission by disseminating knowledge in the pursuit of education, learning, and research at the highest international levels of excellence.

www.cambridge.org
Information on this title: www.cambridge.org/9781108422376
DOI: 10.1017/9781108381598

First published 2020

A catalogue record for this publication is available from the British Library.

ISBN 978-1-108-42237-6 Hardback

Contents

Contributors

Thomas G. Bell
Plymouth Marine Laboratory, Plymouth, United Kingdom

Matej David
David Consult D.O.O., Izola, Slovenia

Caroline H. Fox
Dalhousie University, Halifax, Nova Scotia, Canada

Stephan Gollasch
Gollasch Consulting, Hamburg, Germany

C. Michael Hall
University of Cantebury, Christchurch, New Zealand

Peter Hinchliffe
International Chamber of Shipping, London, United Kingdom

M. Maruf Hossain
University of Chittagong, Chittagong, Bangladesh

Katherine Langford
3 Moorfields Street, Fig Tree Pocket, Australia

John A. Lewis
ES Link Services Pty Ltd, Castlemaine, Australia

Nikola Mandic
University of Split, Split, Croatia

Samantha Eslava Martins
Norwegian Institute for Water Research (NIVA), Oslo, Norway; Federal University of Rio Grande, Rio Grande do Sul, Brazil

Gorana Jelic Mrcelic
University of Split, Split, Croatia

Isabel Oliveira
University of Aveiro, Aveiro, Portugal

Richard John (Jan) Pentreath
Ropewalk, Penpol, Camelot House, Truro, Cornwall, United Kingdom

Ranka Petrinovic
University of Split, Split, Croatia

Roger C. Prince
Roger Prince, Pittstown, NJ, USA

Virginia Sciacca
Laboratori Nazionali del Sud (INFN), Catania, Italy

Christopher T. Taggart
Dalhousie University, Halifax, Nova Scotia, Canada

Kevin Thomas
University of Queensland, Brisbane, Australia

Simon J. Ussher
Plymouth University, Plymouth, United Kingdom

Salvatore Viola
Laboratori Nazionali del Sud (INFN), Catania, Italy

Mingxi Yang
Plymouth Marine Laboratory, Plymouth, United Kingdom

Preface

Environmental impacts of shipping arise through port and channel development, and ships, of which only the latter are considered in this book. Shipping is a vital industry supporting global trade, with over 90 per cent of goods transported by sea. Most of the worldwide fleet comprises 90,000 cargo vessels of one type or another, but ships are diverse in type, function and region of operation. Given that cruise liners, fishing vessels, research ships and naval vessels are not necessarily confined to the major shipping lanes, but rather travel throughout the world's oceans, the impact of ships on the marine environment is present everywhere that ships can go. Moreover, these effects can be felt in inland waters that are navigable and/or are connected directly or indirectly to the seas, such as the Great Lakes and the Caspian Sea.

This edited book provides a comprehensive review of the multifarious effects that ships can have on the environment. Currently, no such authoritative text exists. Whilst the emphasis is on pollutant discharges (air, oil, waste, sewage and biocides) from normal operations, other effects are considered (invasive species, wildlife collisions, noise and physical damage). With respect to end of life, chapters are devoted to the environmental effects of both shipwrecks and shipbreaking. Finally, the history of relevant international legislation is covered, together with a perspective from the shipping industry. This holistic approach recognizes the need in the community to understand fully the environmental consequences of shipping.

The drivers, pressures, state, impact and response (DPSIR) framework provides a useful conceptual model for assessing and managing the problems arising from the interactions between ships and the environment. Research into marine pollution (including noise) and other deleterious consequences, such as wildlife collisions, has led to numerous requirements to alter behaviour and practices at sea. The most important responses involve legal instruments to mitigate, remediate and/or prevent the impact of ships on the environment. Mechanisms encompass local by-laws, national regulations and international conventions, notably those of the International Maritime Organization. Thereafter, appropriate and responsible implementation by the shipping industry helps to protect marine and coastal environments.

1
Shipping, Ships and the Environment

THOMAS VANCE, TIMOTHY FILEMAN AND STEPHEN DE MORA

1.1 Introduction

Many people around the globe rely on the low-cost transport of goods and commodities that commercial shipping provides. Indeed, about 90 per cent of the world's traded goods are transported by sea, with more than 70 per cent of this being containerized cargo (United Nations Conference on Trade and Development, 2017). Shipping densities are illustrated in Figure 1.1, demonstrating the great concentration of traffic along key routes.

In most cases, the beneficiaries of this method of trade are unaware of the environmental impacts ships cause. The diverse nature of the environmental impacts of modern ships in terms of the sources of impact, their magnitude and their persistence is shown in Figure 1.2.

1.2 Shipping

1.2.1 International Demand for Shipping

Never before has the demand for global shipping been greater. However, there have been multiple and significant changes in recent decades relating to the technologies and operational procedures used by the shipping sector, with many of these changing the environmental impact of the industry. Data showing the growth of various categories of cargo are presented in Table 1.1.

Consequently, understanding the cumulative environmental impact of ships has never been more important. Research across the environmental disciplines is increasingly uncovering the complexity and incredible levels of interconnectivity between environmental partitions, as well as the fragility of many of the life-support systems provided by our planet. This is perhaps especially true in the field of marine environmental research, where recent discoveries regarding the intricacies of cell-to-cell communication between microbes and higher organisms, the sophisticated navigation behaviour of fish, reptiles and cetaceans and the susceptibility of marine organisms to synergistic environmental stressors are just a few of many possible examples of how our understanding of organism behaviour is improving all of the time.

As we learn more about the way our oceans function and the goods and services they provide us, it is even more pressing that we measure how our use of the oceans affects these

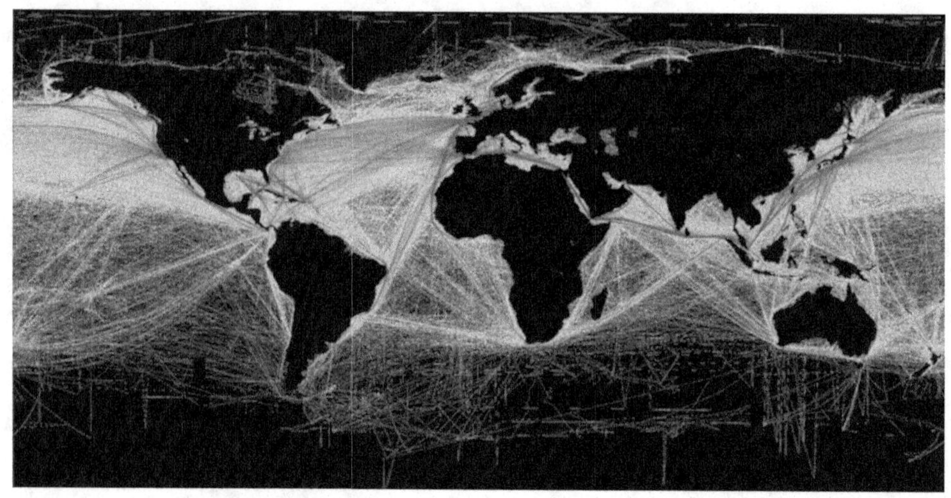

Figure 1.1 A global map of human impacts on marine ecosystems, showing relative commercial shipping densities.
Source: Halpern *et al.* (2008)

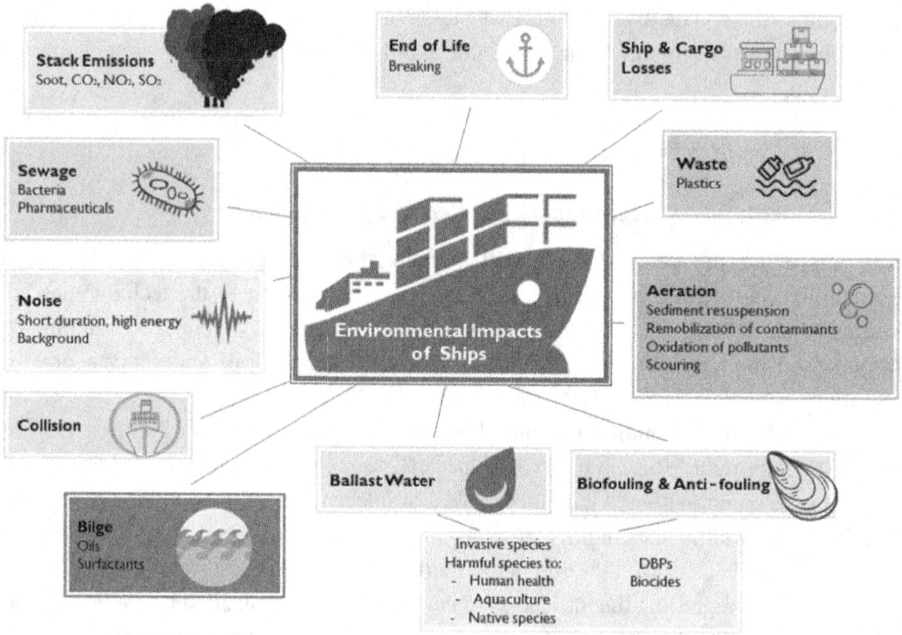

Figure 1.2 The diverse nature of the environmental impacts of modern ships. DBP = disinfection by-product.

processes and manage our use accordingly. This is particularly the case within the shipping industry, with rapid changes taking place in relation to the way ships are designed, the fuel sources they use and how fast they move. Even the very seascape of our planet where ships travel is subject to change (Figure 1.3), with new shipping routes opening up in previously

Table 1.1. *Growth in international seaborne trade over selected years (millions of tons loaded)*

Year	Oil and gas	Main bulks[a]	Dry cargo other than main bulks	Total (all cargos)
1970	1440	448	717	2605
1980	1871	608	1225	3704
1990	1755	988	1265	4008
2000	2163	1295	2526	5984
2005	2422	1709	2978	7109
2006	2698	1814	3188	7700
2007	2747	1953	3334	8034
2008	2742	2065	3422	8229
2009	2642	2085	3131	7858
2010	2772	2335	3302	8409
2011	2794	2486	3505	8785
2012	2841	2742	3614	9197
2013	2829	2923	3762	9514
2014	2825	2985	4033	9843
2015	2932	3121	3971	10,023
2016	3055	3172	4059	10,287

[a] Iron ore, grain, coal, bauxite, alumina and phosphate rock.
Source: Compiled by the United Nations Conference on Trade and Development secretariat based on data supplied by reporting countries and as published on government and port industry websites and by specialist sources. Data for 2006 onwards have been revised and updated to reflect improved reporting, including more recent figures and better information regarding the breakdown by cargo type. Figures for 2016 are estimates based on preliminary data or on the last year for which data were available

unpassable waters, such as the Northwest Passage, due to the consistent decline in sea ice (Brigham *et al.*, 2009). It has never been more pertinent to measure and understand the impact that the global shipping industry has on our environment and our long-term well-being.

1.2.2 Global Dependence on Commercial Shipping

The ability to move goods, commodities and people on vessels has long been associated with wealth and power. Notable examples include the clinker built vessels predominantly used by the Vikings to great effect for trade, commerce and warfare during AD 793–1066. Before this, the beginnings of the spice trade that fuelled the global economy during the Middle Ages were largely dependent on commercial vessels as far back as 3000 BC.

In modern times, approximately 10 billion tonnes of cargo are moved globally by commercial vessels (United Nations Conference on Trade and Development, 2017). Approximately 78 per cent of surface European Union freight is moved by sea shipping in Europe compared with 17 per cent by road and 3 per cent by rail (ACEA, 2011). Shipping is the global link that connects commodity suppliers with production and

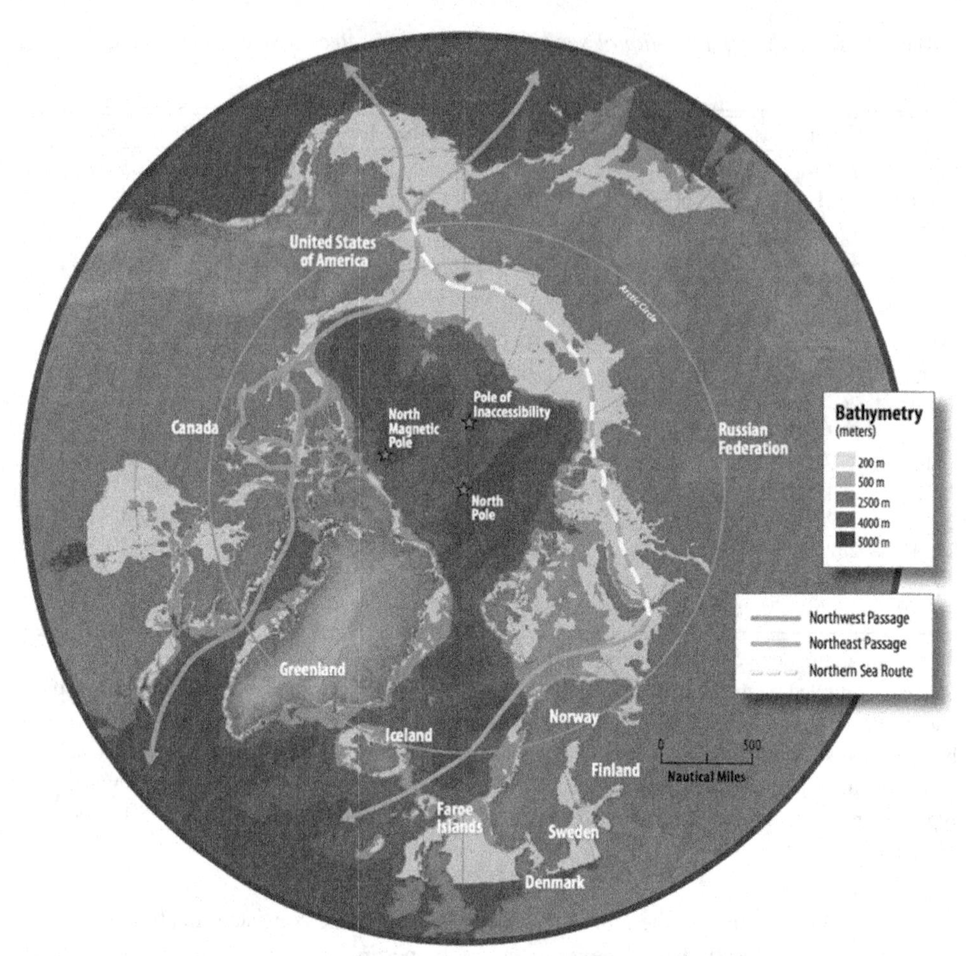

Figure 1.3 Map of the Arctic region, showing the Northeast Passage, Northern Sea Route and Northwest Passage shipping routes and bathymetry.

manufacture processes. On completion of manufacture, market-ready products are then distributed for sale across the globe. Increasingly, the by-products of our vast levels of global consumerism are also transported by shipping, such as waste and reuseable materials that can only be economically recycled in bulk form. Plastic waste generated from products manufactured and consumed in Western Europe is now a standard cargo destined for waste treatment plants in China. Recycled materials are then transported in bulk via ships back to manufacturing plants, so completing the link between product life cycles and shipping.

1.2.3 Transportation of Fuels

Of all the commodities, the transport of fuels such as coal and petrochemicals has historically been closely linked with shipping. Until the recent and rapid development of renewable energy capability, the transport of crude fuels by sea following its production and subsequent distribution after refining has been a lifeline that has directly underpinned a significant proportion of the global economy and allowed much of the recent rapid development of new technologies to take place. There is little doubt regarding our dependence and reliance on shipping. However, our understanding of the environmental cost of this dependence is less clear.

1.3 Ships

1.3.1 Vessels and Scale of the Industry

The scale of the shipping industry is vast, despite not being visually obvious to many who depend on it. Some 90,000 commercial vessels of all classes were reported in 2017, with a total global tonnage of 1.86 trillion deadweight tonnage (dwt) (United Nations Conference on Trade and Development, 2017). Of these, approximately 43 per cent of vessels are bulk carriers. A typical bulk carrier can be in the region of 290 m in length overall and 45 m in breadth extreme, with a hull area of approximately 19,000 m^2.

Some forms of environmental impact produced by these ships have been characterized by current research, such as greenhouse gas emissions. In contrast, other forms of input, such as scrubber emissions, noise and pollution during shipbreaking, have not yet been fully addressed in terms of measuring either the global scale of the input or the severity and persistence of the impact that is produced.

Regarding leisure vessels, determining the numbers of such vessels on a global scale is challenging, with many vessels being unregistered. Approximately 500,000 leisure vessels were recorded in the UK in 2012, with between 80,000 and 100,000 of these being motorboats (Royal Yachting Association, 2012). The number of vessels actually in service, however, is less clear. Although leisure vessels are subject to less regulation and record-keeping, they may play an important role in perpetuating some of the environmental impacts that originate from commercial vessels, with the spread of hull-fouling species being a prime example.

The fuel consumption of large commercial vessels is immense, and a typical car carrier of 200 m in length overall can burn between 50 and 60 tonnes of fuel per day when steaming at around 20 knots (Bialystocki & Konovessis, 2016). Combustion gases from heavy fuel oil produce far higher levels of greenhouse gas emissions and other pollutants that contribute to human health issues, such as particulates, compared to combustion gases from most other transport fuels. For example, typical heavy fuel oil contains several thousand times more sulfur than standard road diesel. To provide context on just one fraction of atmospheric inputs, the huge volume of fuel consumed by the international shipping industry currently produces approximately 2–3 per cent of global anthropogenic CO_2 emissions. This ranks CO_2 emissions from commercial shipping vessels somewhere between the emission ratings of Germany and Japan.

1.3.2 Changes in Technology and Operational Practice

There has been much change in terms of technology and operational practice in recent years, which has affected the environmental impact of ships. Although subject to short-term variation, the size of general commercial vessels has increased over recent decades, affording greater economies of scale to the trader and consumer (Malchow, 2017). Advances in the scale and efficiency of ship construction techniques have led in part to this increase in size. Increasing vessel size has also driven the expansion of ports, together with the deepening and widening of shipping channels, all of which result in environmental impacts at multiple scales that should be considered in assessing the overall impact of shipping.

The steaming speeds of vessels continue to change as a result of multiple drivers, including demand, fuel oil prices and propulsion technology. Vessel speed has direct implications for fuel consumption, exhaust emissions, marine noise levels, biocidal release rates from hull coatings and collision risks with cetaceans and reptiles. As commercial vessel speeds vary with bunker prices, the environmental impacts also vary in relative severity and importance. For example, when slow steaming, hull coatings designed for faster speeds become less effective, increasing fuel consumption and greenhouse gas emissions.

Ship speeds can also result in highly significant impacts on vulnerable species such as the North Atlantic right whale, for which approximately 35 per cent of deaths are attributed to collisions with commercial ships (Knowlton & Kraus, 2001).

Changes in hull design also track bunker prices, with substantial retrofitted hull modifications, such as the removal or replacement of bulbous bows, being carried out in attempts to optimize hull performance in order to match required steaming speeds.

Propulsion methods and engine designs are also changing, with resulting consequences for the environmental impact of ships. Additional retrofitted sections to standard fixed propellers are available that can improve the efficiency of the unit under certain conditions. Fixed propellers are often replaced by multidirectional pods and thrusters, which can influence levels of efficiency and manoeuvrability and the generation of marine noise. Modern propulsion systems can also influence levels of physical scour and erosion from wash-in ports, harbours and shallow-water anchorages, as is discussed in Chapter 8.

Engine design and operation have been the focus of much research and development, with modern engine configurations competing to offer increasing power-to-weight ratios and to achieve fuel savings. Advanced engine management systems that allow the engine to run at different speeds under different loads are also helping to save fuel and to reduce environmental impacts, along with fuel additives and blends that reduce harmful emissions from the exhaust stack.

Advances in engine design, propulsion systems and the ability to closely monitor a vessel's hull and fuel efficiency are driving the shipping industry to demand manufacturers of hull coatings to improve hull efficiency by increasing anti-fouling performance. Fuel efficiency and, to a lesser extent, the translocation of non-native marine organisms have moved higher up the agenda in global shipping circles, requiring coating manufacturers to innovate and to provide new products, despite the tightening up of permissible biocides for used in anti-fouling coatings.

Effective coatings can increase fuel efficiency and reduce exhaust emissions, yet between 80 and 90 per cent of the global fleet are operating with biocidal coatings, with much of the biocidal content leaching out and disassociating from the coating into the marine environment over the life cycle of the coating. Modern biocide-free anti-fouling coatings are becoming more widely used, but they do not always offer the same performance as biocidal systems, especially after extended static periods or at slow steaming speeds. Although the environmental benefits of biocide-free or low-leach-rate coating systems seem obvious, the environmental implications of the wide-scale use of these technologies are still not fully understood and are the subject of current research programmes.

Alternatives to heavy fuel oil as a fuel source are available, although the numbers of vessels powered by cleaner fuels such as liquified natural gas are relatively low, reaching 200 vessels in 2017 (Riviera, 2017). Recent advances in battery technology have enabled the use of electric engines for specialist vessels operating in particular routes. The year 2017 saw the introduction of one of the first electric battery-powered 2000-dwt coal carriers operating with a 40-nautical-mile range.

Trials of renewable energy use on commercial cargo ships have also been conducted since the 1980s in the form of wind-powered kites. Modern renewable energy options such as turbines, wings, rotors, kites and photovoltaics are all commercially available to supplement the primary propulsion systems of vessels or to provide power to auxiliary systems at sea and in dock.

Despite these advances, the majority of the global fleet (between 80 and 90 per cent) still operate on heavy fuel oil, which results in high levels of greenhouse gases being emitted by ships stacks. In order to reduce sulfur emissions from marine fuels, the International Maritime Organization (IMO) has introduced Emission Control Areas, which control emissions of SO_x, NO_x and particulate matter in coastal areas around North America and Europe, and the global sulfur cap for marine fuels was reduced from the current 3.5 per cent to 0.5 per cent in 2020. Far from being only an offshore problem, the common practice of running combustion engines in ports to provide power for electrical generators results in direct exposure of coastal communities to highly toxic exhaust gases.

1.4 The Environment

1.4.1 Environmental Impacts

Many of the environmental impacts of ships are felt in coastal communities and habitats. However, the impacts of ships can also be felt on a much wider scale, which makes them difficult to measure and quantify. For example, vessel exhaust stack emissions, which originate from a marine point source, can quickly spread across terrestrial environmental compartments, making it challenging to track them and to measure their full impact.

Specific characteristics of the receiving environment also affect the severity of the environmental impact. For example, many major shipping routes, harbours and anchorages are located in estuaries and waterbodies where environmental salinity drops well below typical marine levels. Environmental salinity can have significant implications for the environmental impacts of shipping in relation to the efficacy of the electro-chlorination systems of ballast water treatment systems, the efficacy of biocidal anti-fouling coatings and the fate of pollutants from vessel exhaust stack scrubbers. As a consequence, freshwater bodies that receive high levels of shipping activity may be uniquely susceptible to the environmental impact of ships.

1.4.2 Out of Sight, Out of Mind

Despite many citizens of our planet depending on shipping to underpin their modern way of life, it is likely that many of them will not be familiar with the full range of environmental impacts that vessels can produce, resulting in a phenomenon known as 'sea blindness'. There are many examples of the environmental impacts of shipping having escaped attention, perhaps simply because ships predominately operate on the world's oceans, where the impacts are less obvious than equivalent terrestrial industries. Examples of the incongruous environmental situation concerning shipping include exhaust scrubbers that remove many of the pollutants from the atmosphere only to deposit them in the oceans or the fact that shipping was exempt from the United Nations' Paris Agreement.

1.4.3 Environmental Legislation

Globally, we are becoming more aware of the impacts of shipping, and legislation is being applied to control and limit environmental degradation. Examples include the Energy Efficiency Design Index (EEDI), which seeks to non-prescriptively promote the use of more efficient equipment and engines and in doing so stimulate technological development and innovation.

The IMO's International Convention for the Control and Management of Ships' Ballast Water and Sediments (BWM Convention) and the IMO MARPOL low-sulfur fuels regulation seek to address other environmental inputs in order to reduce their impacts. Even secondary levels of environmental and human health protection, such as limits on volatile compound concentrations in marine coatings, are all contributing to increased environmental awareness and in many cases are driving technology development forwards to meet new environmental compliance standards.

However, despite these advances, substantial gaps remain between the implementation, compliance and enforcement of these rules and regulations on the high seas. Recent examples have demonstrated that, for large-scale operators, the fines and penalties associated with breach of environmental compliance are financially favourable compared to the levels of investment required to meet environmental standards.

Additionally, whilst relatively high-margin fractions of the shipping sector, such as the cruise industry, might be able to invest in meeting these new environmental standards, the procurement and operation of environmental compliance technology, such as ballast water treatment systems, are likely to be beyond the financial reach of the lower-margin fractions of the sector, such as the European short sea shipping industry. The economic cost of environmental compliance is likely to produce longer-term environmental and socio-economic impacts.

Unsurprisingly, new environmental legislation may be met with resistance from the commercial shipping industry. This is understandable to a certain extent, as the industry can justifiably claim that 99.98 per cent of oil reaches its destination port without issue (IMO, 2001). In addition, it is generally acknowledged that terrestrial sources of pollution in the sea are far greater than marine sources, yet these terrestrial sources receive less attention in terms of regulation. Despite this, estimates that marine transportation sources contribute approximately 12 per cent of total marine pollution suggest that scrutiny should not be relinquished (GESAMP, 1990).

1.5 The Future

Penalties for breach of environmental regulations need to be realistic in order to provide a sufficient deterrent, whilst still retaining flexibility to allow smaller organizations, with less capacity to invest, to comply. It is also clear that, when pushed, technology development continues to deliver alternatives with the potential to reduce the environmental impacts of shipping, with ballast water treatment systems being one recent example.

As the shipping industry develops and technology and operational standards change, it is imperative that we remain vigilant and open-minded about the source, fate and impacts of outputs from ships. Before global shipping can be managed in an informed way, a balanced, impartial and holistic view of the environmental impacts of ships is required. Crucially, we need to consider the long-term fate, severity and persistence of cumulative impacts and the full range of environmental compartments across which they transcend. It is exactly in this capacity that this book intends to contribute.

With concerted effort and awareness, we have the potential to develop the most efficient source of mass transport on the globe, allowing for the sustainable development of our planet. If we are not aware of the environmental impacts of shipping, or if we fail to mitigate them, there is grave potential for shipping to measurably accelerate the ongoing degradation of the oceanic environment upon which we all depend.

References

ACEA (2011). *Discussion Paper – 16th ACEA SAG Meeting – Global Trends in Transport Routes and Goods Transport: Influence on Future International Loading Units*. Aachen: RWTH Aachen University.

Bialystocki, N. and Konovessis, D. (2016). On the estimation of ship's fuel consumption and speed curve: a statistical approach. *Journal of Ocean Engineering and Science*, **1**, 157–166.

Brigham, L., McCalla, R., Cunningham, E. *et al.* (2009). Arctic marine shipping assessment (AMSA) Norway: Protection of the Arctic Marine Environment (PAME), Arctic Council. Archived from the original on 1 November 2014. https://cil.nus.edu.sg/wp-content/uploads/2014/06/2009-Arctic-Marine-Shipping.pdf

GESAMP (1990). The State of the Marine Environment. Rep. Stud. GESAMP No. 39 at 88. www.gesamp.org/publications/the-state-of-the-marine-environment

IMO (2001). Secretary General's comment at the IMO Council's 86th Session. Council Doc. C/86/10/2001. https://webaccounts.imo.org/Common/WebLogin.aspx?App=IMODOCS&ReturnUrl=https%3A%2F%2Fdocs.imo.org%2FCategory.aspx%3Fcid%3D30

Knowlton, A. R. and Kraus, S. D. (2001). Mortality and serious injury of Northern right whales (*Eubalaena glacialis*) in the western North Atlantic Ocean. *Journal of Cetacean Research and Management*, **2**, 193–208.

Halper, B. S., Walbridge, S., Selkoe, K. A. *et al.* (2008). A global map of human impact on marine ecosystems. *Science*, **319**, 948–952.

Malchow, L. (2017). Growth in containership sizes to be stopped? *Maritime Business Review*, **2**(3), 199–210.

Riviera (2017). The world's LNG-fuelled fleet in service in 2017. www.rivieramm.com/news-content-hub/the-worlds-lng-fuelled-fleet-in-service-in-2017-29232

Royal Yachting Association (2012). Vessels in the UK. Economic Contribution of the Recreational Boater. www.rya.org.uk/SiteCollectionDocuments/legal/Web%20Documents/Environment/Economic_Contribution_From_Recreational_Boating.pdf

United Nations Conference on Trade and Development (2017). Review of Maritime Transport, UNCTAD/RMT/2017 Sales No. E.17.II.D.10 ISBN 978-92-1-112922-9. https://unctad.org/en/PublicationsLibrary/rmt2017_en.pdf

2

Atmospheric Emissions from Ships

THOMAS G. BELL, MINGXI YANG AND SIMON J. USSHER

2.1 Introduction

Atmospheric emissions from ships have not been subject to the same regulations as those on land until very recently. Carbon emissions from the shipping industry are low (per tonne of transported goods) relative to other areas of the transport sector, namely road traffic and aviation. Regulatory controls of atmospheric pollutants such as sulfur dioxide (SO_2), nitrogen oxides (NO_x) and particulate matter (PM) were imposed on land-based anthropogenic emissions, but not applied to ships.

The shipping industry has grown as the global population has increased. The shipment of goods by sea dominates global trade: 90 per cent of non-bulk cargo such as food and clothes are transported by sea (Wan *et al.*, 2016). The trend in shipping activity seems unstoppable, continuously increasing as consumer demand rises. In 100 years, the number of ships larger than 100 gigatonnes has tripled and carbon dioxide (CO_2) emissions have risen over fourfold (Eyring *et al.*, 2005b). The tight link to global trade and population has led to projections that future shipping activity will keep on increasing (quadrupling by 2050 relative to 1990 levels; Eyring *et al.*, 2005a).

Shipping activity is significant enough to be observable from space (Tournadre, 2014), and the gaseous emissions of ships can be detected and tracked in highly trafficked regions (de Wildt *et al.*, 2012; Marbach *et al.*, 2009; Richter *et al.*, 2004). Ships have also become bigger, enabling them to carry greater loads with lower fuel consumption per tonne. However, the rise in shipping activity has led to greater scrutiny of ships' emissions and their impacts.

The annual emissions of the pollutants NO_x, SO_2 and PM from ship traffic greatly exceed those from aviation and are comparable to road traffic (Eyring *et al.*, 2005b). In 2013, shipping emissions in Europe contributed 18, 18 and 11 per cent to the total emissions of NO_x, SO_2 and $PM_{2.5}$ (PM smaller than 2.5 μm in diameter), respectively (Wan *et al.*, 2016). Many of the gaseous emissions influence important atmospheric processes that determine the lifetime of other pollutants and greenhouse gases in the atmosphere (Law *et al.*, 2013). Ship emissions also contain heavy metals and nutrients (Agrawal *et al.*, 2008a), which, when deposited onto the surface ocean, may have substantial impacts on ocean biogeochemistry (Paytan *et al.*, 2009).

Shipping activity is not equally distributed around the globe. An estimated 70 per cent of shipping activity and emissions are within 400 km of the coast and in or near to major ports (Dalsoren *et al.*, 2009; Endresen *et al.*, 2003; Eyring *et al.*, 2005b). The atmospheric emissions from ships are significant sources of air pollution in certain regions, with a recent global study suggesting that 400,000 people die each year directly as a result of ship emissions (Corbett *et al.*, 2007; Sofiev *et al.*, 2018). In the last decade, the International Maritime Organization has begun to impose tighter emission controls on the industry.

The initial regulation of atmospheric emissions has mainly focused on sulfur, limiting the emission of SO_2 in certain highly trafficked parts of the world (Johansson *et al.*, 2013). The industry has adjusted to these regulations by either switching to lower-sulfur-content fuels or by installing emission scrubbers (see discussion in Section 2.4). In the year 2020, more stringent emissions regulations will come into force, which represents a challenge for the industry and will have substantial implications for the environment.

In this chapter, we discuss the various factors that determine the impact of ship emissions on the atmosphere. We cover the composition of the fuel that ships typically burn (Section 2.2) and the composition of the resultant gaseous and particulate emissions (Section 2.3), including the influence of the combustion conditions and exhaust treatment (Section 2.4). We also describe the processes that influence the lifetime and evolution of ship atmospheric emissions (Section 2.5) and discuss their environmental implications (Section 2.6). Finally, we consider how the situation may change in the future (Section 2.7).

2.2 Fuel Composition

In the early twentieth century, unprocessed heavy fuel oil (HFO) came into general use in shipping engines, replacing coal as the fuel in steamships. In 2005, ships consumed approximately 350 Mt/year of fuel, and usage is continuously increasing (Paxian *et al.*, 2010). HFO is a residual fuel from the first stage of the oil refinery process of crude oil (Figure 2.1). HFO is characterized as a black, opaque and viscous fuel (viscosity = 180 mm^2/s at 50°C) with a high density, similar to that of water (>900 kg/m^3). HFO can be further processed or blended with distillate fuels (marine gasoil and marine diesel oil) to make intermediate fuel oils (IFOs) for shipping. Common fuels are IFO 380 (where 380 is the viscosity in mm^2/s) or lower-viscosity IFOs, which are blended with lighter distillates. It has become more common to have IFOs made completely from secondary distillates/fractions of residual oil by processes such as vacuum distillation and visbreaking (thermal cracking). Secondary processing produces a cleaner fuel and makes greater use of the residual fraction from the first stage of the oil refinery process.

International standards are used to classify the categories of marine bunker fuel oils. ISO 8217:2017 categorizes seven distillate fuels and six residual fuels and specifies the fuel requirements for marine diesel engines. The ISO 8217 standard has recently been updated to include regulations for use of biofuels (i.e., fatty acid methyl ester content) and oil sands (International Organization for Standardization 2017, ISO 8217:2017; www.iso.org/standard/64247.html).

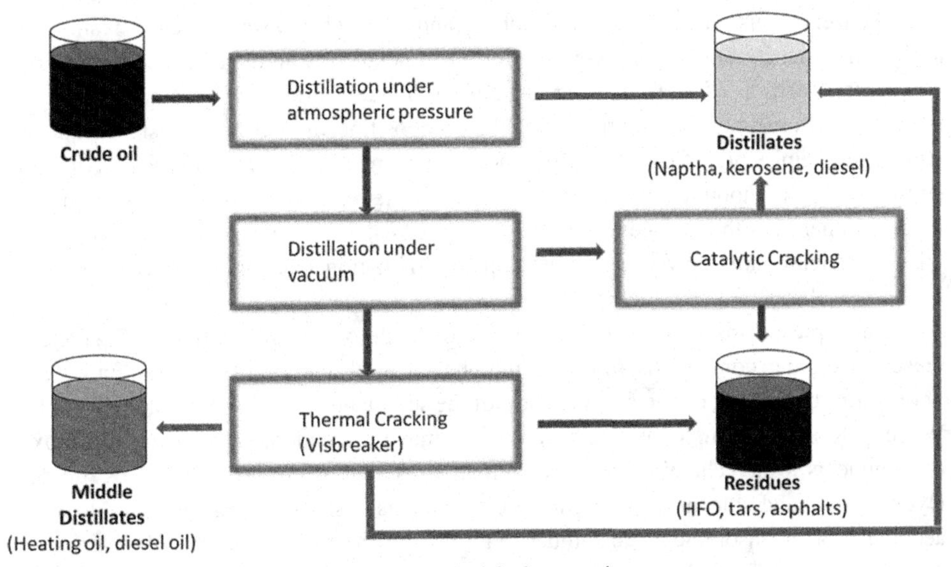

Figure 2.1 Simplified schematic of the typical fuel processing process.

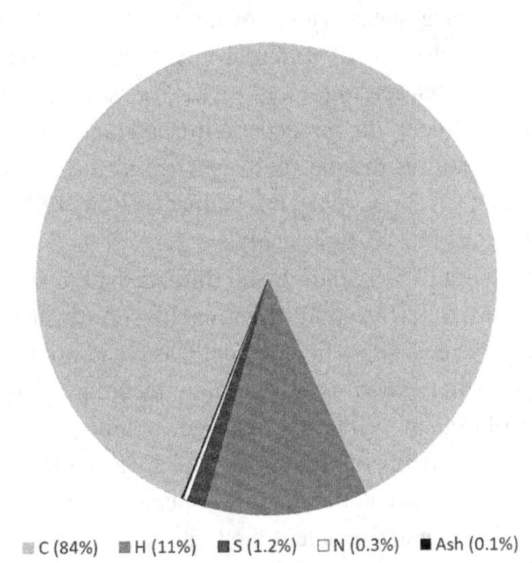

Figure 2.2 Pie chart of the typical elemental composition of heavy fuel oil.
Source: Data from Lyyränen *et al.* (1999)

A typical HFO elemental composition (Figure 2.2) shows that the main elements present are C, H, O, N and S, with the hydrocarbon component making up the vast majority of the fuel by mass (as expected for any biogenic organic substrate). The bulk of fuel oil is made up of a highly complex mixture of straight and branched saturated hydrocarbons along with polycyclic hydrocarbons. The complexity of oil and petroleum chemistry is beyond the scope of this chapter and readers are referred elsewhere (e.g., Tissot & Welte, 1984). Briefly, the major crude oil molecules can be characterized as: (1) saturated hydrocarbons (e.g., straight chain, branched alkanes and cycloalkanes); (2) aromatic hydrocarbons (e.g., benzene, napthenoaromatic molecules); and (3) high-molecular-weight resins and asphaltenes that contain a major contribution from polar S, N and O atoms.

The composition of HFO and the large mass of fuel burnt each year has significant implications for atmospheric emissions. The composition of fuel is most significant for atmospheric emissions of carbon, sulfur and trace elements rather than nitrogen. Nitrogen emissions depend upon the oxidation of nitrogen in the reacting air during the combustion process rather than the nitrogen in the fuel. The consistent percentage of hydrocarbons in oil allows reasonably low-uncertainty estimates of national and global CO_2 emissions based on well-documented fuel production rates.

Sulfur typically makes up 0.1–3.0 per cent of crude oil by weight (Figure 2.2) and is present as organic compounds, mostly as thiophenes, thiols and sulfides. The sulfur content of oil varies depending on the geochemistry of the sediments during oil formation (Tissot & Welte, 1984). For example, sedimentary material may have undergone *diagenesis* in anoxic environments, which chemically reduces sulfate, uses up oxygen and accumulates sulfides. Hydrothermal and volcanic activities also act as sources of sulfur and can be linked to metal accumulation, as insoluble metal sulfides (e.g., FeS or ZnS) are produced in these environments (Tissot & Welte, 1984).

There is less control on trace metal content of marine fuels, but it is possible to control sulfur content through the selection of different crude oils and during fuel processing. Unprocessed crude oils are often referred to as 'sweet' or 'sour', referring either to low or high sulfur content, respectively. In general, sulfur content increases as fuel density increases. ISO 8217 also requires marine fuel suppliers to differentiate sulfur content as high-sulfur fuel oil (HSFO; <3.5 per cent), low-sulfur fuel oil (LSFO; <1.0 per cent) and ultralow-sulfur fuel oil (ULSFO; <0.1 per cent).

Sulfur content can be tailored by blending and diluting with differing crude oils and by adjusting refinery processes. Selective sulfur removal (desulfurization) must be efficient in terms of energy and resource costs, low in chemical waste and not significantly perturb the fuel chemistry. Hydrodesulfurization is a common technique for desulfurization and involves the conversion of organic sulfur into hydrogen sulfide (H_2S) using metal catalysts. This technique requires high temperatures (300–400°C) and pressures of hydrogen gas (e.g., 10–100 atm) and does not remove all sulfur species. Considerable research is now underway to find more effective and efficient methods (e.g., oxidative desulfurization and bio-desulfurization; Zhao & Baker, 2015). The environmental implications of fuel sulfur content will be discussed in Section 2.6. Fuel sulfur content also affects engine design and maintenance because sulfur combustion products and residues are corrosive to piston liners in ships engines. Basic (high-pH) cylinder lubricant is often used to cope with the acidity.

Fuel contains trace elements and contaminants (<1 per cent; Figure 2.2) and the inorganic component of this that remains after combustion is termed the *ash content*. The trace elements are important, despite their low abundance. At certain levels in the environment, trace elements can be toxic (see discussion in Section 2.6.3). Trace elements become more concentrated during fuel processing, with more in the residue than the oil and distillates. Greater amounts of lower-volatility metals are left behind by the distillation process. Ash content by mass is typically 0.05 per cent (up to 0.15 per cent w/w) of any commercial fuel, while emulsified water constitutes ~0.05–0.20 per cent (up to 1 per cent w/w; International Bunker Industry Association; http://ibia.net/wp-content/uploads/2014/06/Vanadium-Sulphur-in-Marine-Fuels.pdf). The metal elements in oil ash are (in common order of abundance) vanadium (V), nickel (Ni), iron (Fe), potassium (K) and sodium (Na), often present in oil at concentrations of up to hundreds of μg/g (Ball *et al.*, 1960). The alkali (Na and K) and alkaline earth metals (magnesium, Mg, and calcium, Ca) are also relatively abundant, usually in combination with charged naphthenic acid or hydrated inorganic salts (e.g., chlorides and carbonates) in droplets of emulsified water. The ratios and concentrations of other trace metals such as aluminium (Al), silicon (Si) and many other metals can vary significantly depending on the crude oil source.

The principal classes of metal species in crude oil include: (1) metalloporphyrins of Ni, V, Fe or Cu; (2) poorly characterized non-porphyrin species; and (3) naphthenic acid salts that principally bind to hard cations such as Ca and Mg. Metal concentrations in different fuel sources are surprisingly variable, with some of the highest reported concentrations in fuels from Venezuela and Mexico (International Bunker Industry Association; http://ibia.net/wp-content/uploads/2014/06/Vanadium-Sulphur-in-Marine-Fuels.pdf). Metal content variability is strongly related to the asphaltene content in the oil. Asphaltenes are heavy and a highly aromatic fraction of molecules in oil. They are operationally defined as the fraction of oil that is not soluble in non-polar, light alkanes (e.g., n-heptane) but is soluble in toluene. Asphaltenes also contain a high percentage of sulfur and nitrogen. A significant fraction of organically bound vanadium and nickel found in crude oil (e.g., 39–80 per cent w/w for V in asphaltenes) exists as porphyrin complexes, although other non-porphyrin-rich fractions have also been characterized (Fish *et al.*, 1984).

Secondary contamination during the fuel refinery and transport processes is a significant source of trace elements within marine fuels. Examples include iron contamination from storage tank and pipeline corrosion and increased aluminium and silicon in blended fuels due to the release of particles during catalytic cracking. Metal contents are important as they can affect the processes within the refinery process and the long-term stability of the fuel. Metals can also be used to fingerprint fuels and identify their source region.

As discussed in Section 2.1 for sulfur, much of the variability of elements in oil is related to the conditions and geology of the regions in which they were formed. The high abundance of vanadium and nickel in fuels (~50–200 μg/g) is remarkable and contrasts with the low relative abundance in the Earth's crust and in biological tissue. Chemical processing controlled by palaeo-redox conditions (pH and Eh) in the sedimentary environment during diagenesis results in vanadium and nickel in the sediment exchanging with other more common biogenic metal complexes (e.g., Mg-, Fe-porphyrins; Lo Mónaco *et al.*, 2002).

2.3 Composition of Shipping Emissions to the Atmosphere

A multitude of gases and particles of differing compositions are emitted from a ship's exhaust stack. The in-stack composition of shipping emissions predominantly consists of aerosol particles (PM), gases (CO_2, NO_x and SO_2) and low-molecular-weight volatile organics (e.g., acetaldehyde, acetone and naphthalene). The concentrations and loadings of these are a function of the fuel type, the conditions within the combustion chamber and the nature of any exhaust treatment or 'scrubbing'. The proceeding subsections discuss these processes and our current understanding of the effects they have on the nature of the gases and particles that enter the atmosphere.

Equation (2.1) summarizes the basic equation for the chemical oxidation of marine fuels (i.e., combustion of a theoretical pure diesel fuel to generate propulsion and power in a ship's engine).

$$4C_{12}H_{23} + 71\ O_2 \rightarrow 48CO_2 + 46H_2O + \text{Energy} \tag{2.1}$$

Fuel is carried from the fuel tank to the engine room, injected into an engine cylinder as small liquid particles and then vaporized and combusted in hot compressed air in a controlled explosion. If combustion extends beyond the expected time due to poor fuel grade or poorly tuned engines it is called 'afterburning'. Afterburning can reduce engine efficiency, cause higher exhaust temperatures and reduce the firing pressure of the cylinder.

An important parameter for characterizing shipping emissions is the emission factor (EF), which is commonly used to upscale flux estimations or to compare certain chemicals or gases. The EF can be normalized to units of energy, EF_{Energy} (Eq. (2.2), units are g/kWh) or mass of fuel used, EF_{Mass} (Eq. (2.3), units are g/kg).

$$EF_{Energy} = \text{Mass of chemical constituent/Energy produced} \tag{2.2}$$

$$EF_{Mass} = \text{Mass of chemical constituent/Mass of fuel} \tag{2.3}$$

2.3.1 Ship Emissions of Gases

The major gases in a ship plume are the different forms of carbon, sulfur and nitrogen. Carbon can exist in multiple forms, the majority being fully oxidized CO_2. Partially oxidized carbon is typically found as carbon monoxide (CO). Other carbon compounds emitted have multiple carbon atoms in different structural forms that are bound up with additional elements such as hydrogen, nitrogen and sulfur (collectively referred to as volatile organic compounds (VOCs)). The freshly emitted sulfidic gases exist almost exclusively as SO_2, while nitrogen gases are a mix of nitric oxide (NO) and nitrogen dioxide (NO_2), collectively referred to as NO_x. Ozone (O_3) is not formed directly in the combustion chamber of the ship's engine, but can be produced downwind of ship emissions by a set of complex interactions with NO_x and hydroxyl radicals.

Data collected from the Penlee Point Atmospheric Observatory (on the south-west coast of the UK) provide a good example of an easily observed ship plume compared to the ambient atmosphere (Figure 2.3). SO_2, CO_2 and NO_x levels all increase as ship plumes pass over the observatory. At the same time, O_3 levels immediately drop due to reactions with NO within the plume. Note that as ship plumes age, O_3 is typically regenerated via cyclic reactions involving NO_x (see Section 2.5).

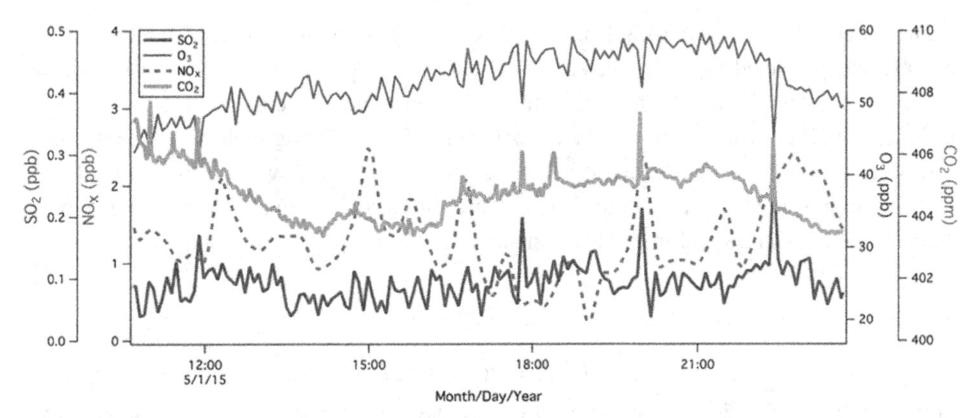

Figure 2.3 Time series of gaseous SO_2, O_3, CO_2 and NO_x levels in air coming from the English Channel. A ship plume 'event' is clearly observed at ~20:00 hours. Data were collected at the Penlee Point Atmospheric Observatory, which is located on the south-west coastline of the UK (Yang *et al.*, 2016). NO_x data are courtesy of Rebecca Cordell, University of Leicester.

2.3.1.1 Carbon Dioxide and Carbon Monoxide

CO_2 is the dominant product of fossil fuel combustion and is well known for its important heat-trapping (greenhouse) effect. The burning of fossil fuels by humans has substantially modified the natural cycle of CO_2, and the current climate impacts and potential future global effects are well documented (Ciais *et al.*, 2013). CO_2 is soluble in seawater and forms a weak acid (carbonic acid, H_2CO_3), which dissociates as carbonate ions and protons (acidity). The dissolution of anthropogenic CO_2 into the ocean has caused a detectable change in the seawater acidity (pH) (Santana-Casiano *et al.*, 2007).

CO is commonly produced by the incomplete combustion of fossil fuels. It is typically associated with urban pollution and is a useful tracer for terrestrial anthropogenic emissions (Rivier *et al.*, 2006). CO reacts with hydroxyl radicals (OH) in the atmosphere, and the products undergo cycling with NO_x to produce O_3, a major air pollutant. The simplified net reaction is:

$$CO + 2O_2 + hv \rightarrow CO_2 + O_3$$

where hv refers to the light absorbed by NO_x during the process.

Most emission inventories for CO_2 and CO from ships are based on bottom-up estimates that couple the annual tonnage of fuel used by the industry with discrete studies of engine EFs (e.g., Eyring *et al.*, 2005a, 2005b). Eyring *et al.* (2005a, 2005b) also make emission predictions using ship traffic demand and technology scenarios linked to assumptions about future economic growth. They estimate that 187 teragrams (Tg) of CO_2 were emitted by ships in 1950. In 2001, emissions had risen to 813 Tg CO_2 per year. Emission predictions for 2050 range anywhere between 1108 and 2001 Tg CO_2 per year.

Of the total oxides of carbon (CO_2 and CO) that are emitted in ship plumes, ~0.15 per cent (by mass) is in the form of CO (Eyring *et al.*, 2005a, 2005b). Assuming this proportion has remained constant and will not change in the future, estimated emissions of CO follow a similar trend to CO_2, increasing from 0.3 Tg CO per year in 1950 to 1.31 Tg CO per year in 2001 and projected to reach between 1.50 and 3.39 Tg CO per year by 2050 (Eyring *et al.*, 2005a). An additional (and large) uncertainty in future CO emissions is the variability in engine efficiency. EFs for CO range between 1.5 and 22 g/kg fuel, with at least some of this variability driven by different fuel types and engine type/power (Betha *et al.*, 2017). As the shipping industry adjusts to different fuel types and air quality regulations whilst striving to reduce costs and increase efficiency, the CO/CO_2 ratio and total CO emitted in ship plumes may decrease.

2.3.1.2 Sulfur Dioxide

Organic sulfur in fuel is readily oxidized during the combustion process to gaseous SO_2. Some SO_2 is further oxidized in the marine atmosphere in the presence of water vapour to form sulfuric acid. Sulfuric acid readily deposits onto particles (aerosols) or can even form new particles. Almost all (~95 per cent) of the fuel sulfur is liberated during the combustion process, such that the SO_2 EF (ratio of gaseous SO_2 to total carbon emitted by the source) is often assumed to be representative of the sulfur content of the fuel that has been burnt (e.g., Corbett & Koehler, 2003). There are relatively few data sets where the EF has been directly compared with the fuel sulfur content of the ship. Most analyses have observed levels of SO_2 and total carbon that reflect approximately what would be expected based on typical ship fuel sulfur content (Hobbs *et al.*, 2000; Schlager *et al.*, 2006; Sinha *et al.*, 2003). However, Chen *et al.* (2005) did observe SO_2 EFs that were a factor of 1.6–1.9 lower than previously published values.

A significant proportion of emissions are concentrated in or close to ports. Dalsoren *et al.* (2009) estimate that ships contribute greater than 10 per cent of the SO_2 in these areas. This is particularly relevant when considering the impact that these emissions may have on human health (see Section 2.6.2). Ports are often co-located with relatively large residential areas that have a high population density. The most intense shipping activity occurs around Asia. Nine of the ten largest container ports can be found along the east coast of the Asian continent (Wan *et al.*, 2016), contributing as much as 20 per cent of the atmospheric loading of SO_2 (Streets *et al.*, 2000). Since the work of Streets *et al.* (2000), shipping emissions have grown more rapidly in East Asia than anywhere else in the world. A recent estimate suggests that shipping activity in East Asia accounted for 16 per cent of global shipping emissions in 2013, compared to 4–7 per cent in 2002–2005 (Liu *et al.*, 2016). Sulfur emissions from ocean-going vessels in East Asia have not been regulated, so SO_2 emissions are likely to have grown at the same rate as shipping activity in the region.

Ship exhaust emissions of sulfur were unregulated (unlike terrestrial sulfur emissions) until 2005. Ships have typically burnt cheap, lower-grade fuels (HFOs) with high sulfur contents, which resulted in substantial sulfur emissions. It is estimated that in 2004 alone, ships emitted 16 Tg SO_2 to the atmosphere over the global ocean (Dalsoren *et al.*, 2009). Following the introduction of sulfur emission controls in 2015 (see Section 2.7.2), emissions have reduced to 10 Tg SO_2 (Johansson *et al.*, 2017), but the increase in global trade activity means that total sulfur emissions continue to increase.

Ship SO_2 emissions are not distributed equally over the oceans, with emissions concentrated in the northern hemisphere and in regions of intense shipping activity (e.g., the Gibraltar straits). Yang *et al.* (2016) observe that ambient SO_2 levels at a coastal site in the south-west of the UK were profoundly influenced by the combined effects of intense shipping activity in the English Channel and the prevailing wind direction. When winds blew from the south-west off the North Atlantic Ocean, concentrations of SO_2 were significantly lower than when the wind blew from the south-east over the English Channel. The difference in SO_2 concentrations in air masses from the south-west and south-east reduced when new limits on sulfur emissions came into force (Yang *et al.*, 2016).

2.3.1.3 Nitrogen Species

Nitrous Oxide (N_2O) N_2O is a minor product of high-temperature fossil fuel combustion. It is produced when nitrogen from the air or fuel is partially oxidized. The extent of emissions from marine vessel engines is unknown, but is typically considered to be small: an EF of 0.01–0.09 g/kWh for N_2O compared to 2–12 g/kWh for NO_x (Cooper, 2001). This is in part due to rapid thermal decomposition of N_2O, which is thought to proceed on a catalytic surface (M) as follows:

$$N_2O + M \rightarrow N_2 + O + M$$

$$N_2O + O \rightarrow N_2 + O_2$$

$$N_2O + O \rightarrow 2NO$$

The efficiency of this reaction in part determines the levels of more oxidized nitrogen species (NO and NO_2, referred to collectively as NO_x). As discussed in the next subsection, combustion conditions directly impact the NO_x and N_2O levels in the ship's exhaust.

Nitrogen Oxides (NO_x) NO_x (= NO + NO_2), major air pollutants, are formed from a mix of nitrogen (N_2) and oxygen (O_2) at high temperatures (e.g., in internal combustion engines). NO_x constitute some of the main pollutants from ship exhausts along with SO_2 and PM. Global ship NO_x emissions were estimated at 6.7 Tg N/year for the year 2015 (Johansson *et al.*, 2017), which represents ~40 per cent of the NO_x emissions from the transport industry and 15 per cent of global NO_x emissions. NO_x emissions do not scale simply with fuel consumption, with different ships emitting different amounts. Observations from the Gulf of Mexico and Texas by Williams *et al.* (2009) showed that bulk freight carriers and tankers have the highest NO_x emission (~80 g NO_x/kg fuel) of ships underway. Container carriers, passenger ships and tugs emit at a lower level (~60 g NO_x/kg fuel). The variability in NO_x emissions is likely due to a combination of engine type/size, loading (rpm) and age. Ships built at different times have to comply with different IMO regulations (Tier I–III; see Table 2.1).

The majority of NO_x immediately coming out of a ship stack is in the form of NO (96 per cent; EPA, 2000). The highly reactive NO participates in a series of chain reactions

Table 2.1. *MARPOL NO_x emission limits (taken from www.dieselnet.com/standards/inter/imo.php)*

		NO_x limit (g/kWh)		
Tier	Date	rpm < 130	130 ≤ rpm < 2000	rpm ≥ 2000
I	2000	17.0	$45 \times rpm^{-0.20}$	9.8
II	2011	14.4	$44 \times rpm^{-0.23}$	7.7
II	2016[a]	3.4	$9 \times rpm^{-0.20}$	1.96

[a] In NO_x Emission Control Areas (see Section 2.7.2). Tier II standards apply outside Emission Control Areas.

that generate NO_2. In coastal marine environments, these reactions can also result in the production of O_3, a secondary pollutant, irritant and greenhouse gas (see Section 2.5 for details).

Impact of NO_x Emissions on Atmospheric Chemistry NO_x emissions from ships are predicted to have a major influence on the marine atmosphere. An early modelling study suggested 100-fold greater NO_x levels over busy shipping lanes compared to a case where ship NO_x emissions are omitted (Lawrence & Crutzen, 1999). These authors also predicted ship-driven increases in the concentrations of O_3 by a factor of two and increases in the hydroxyl radical (OH) by up to a factor of five. However, such high concentrations of NO_x in marine environments and such a large impact of ship NO_x emissions are not supported by observations (Davis *et al.*, 2001; Kasibhatla *et al.*, 2000). More recent work suggests that NO_x are removed from the system more quickly within a plume (~2 hours) than outside of the plume (several hours); this is mostly because the high concentration of OH inside of a ship plume results in rapid NO_2 + OH reactions, removing NO_x from the atmosphere (Chen *et al.*, 2005; Song *et al.*, 2003b). Accounting for this 'ageing' of ship plumes reduces the impact of NO_x emissions on marine atmospheric chemistry, especially over pristine marine areas (Holmes *et al.*, 2014; Vinken *et al.*, 2011). Accounting for plume dilution on NO_x, O_3 and OH budgets is discussed in more detail in Section 2.5.

Satellite Monitoring of NO_x Based on the technique of differential optical absorption spectroscopy, ship emissions of NO_x are visible from space in the form of NO_2. Many studies have used the SCIAMACHY (SCanning Imaging Absorption spectroMeter for Atmospheric CHartographY) instrument on board the ENVISAT satellite (e.g., Richter *et al.*, 2004). In October 2017, Sentinel-5P was launched, carrying the TROPOspheric Monitoring Instrument. High-spatiotemporal resolution data products of atmospheric NO_x became available in July 2018. Satellite measurements have been useful for establishing and monitoring the spatial distribution and temporal changes in ship NO_x emissions (e.g., Boersma *et al.*, 2015). Over major shipping lanes in the Mediterranean Sea, the Red Sea, the Indian Ocean and the South China Sea, shipping activity significantly increased between 2003 and 2008, but sharply decreased from 2008 to 2009 as a result of the global economic recession. These changes in shipping activity are reflected in the satellite observations of NO_x levels (de Wildt *et al.*, 2012).

2.3.1.4 Volatile Organic Compounds and Methane

VOCs arise from ship engines due to incomplete combustion of fuel. The shipping contribution to the global VOC burden is thought to be relatively small. Agrawal *et al.* (2008b) made comprehensive measurements of a large array of volatile organic gases in emissions from a crude oil tanker. They determined EFs for an array of carbonyl compounds, C_4–C_8 hydrocarbons, polycyclic aromatic hydrocarbons (PAHs) and higher-carbon alkanes. Amongst the VOCs emitted, formaldehyde was by far the most abundant, followed by other low-molecular-weight aldehydes (e.g., acetaldehyde) and acetone. Simple aromatic compounds including benzene, toluene and xylene and small PAHs such as naphthalene, fluorine and phenanthrene were also found to be fairly abundant (Cooper, 2003). Cooper *et al.* (1996) reported that PAHs make up about 1 per cent of the total VOC emissions from ships. The total VOC emissions themselves are only ~1 per cent of the NO_x emissions from ships, such that O_3 production from ships is much more sensitive to the emissions of NO_x than VOCs.

Reports of methane (CH_4) emissions from ships are relatively rare. From a study of two diesel-powered ships in the southern Atlantic Ocean, Sinha *et al.* (2003) found that CH_4 emissions were roughly 10 per cent of the NO_x emissions. The relatively low CH_4 emission rates from ships and the ubiquitously high background level of CH_4 mean that elevated CH_4 is often not detectable in ship plumes. Sinha *et al.* (2003) estimate that global ship CH_4 emissions are 0.37–0.47 Tg/year, which is a few orders of magnitude lower than the major CH_4 sources (fossil fuel, natural vegetation, etc.) and is an insignificant contribution to the current global CH_4 budget.

2.3.2 Ship Emissions of Aerosols

Ships produce plumes containing a high density of particulates (or aerosols), which can be transported very long distances across the globe in relatively short timescales (days to weeks). Aerosols are defined as very fine particles of liquid or solid suspended in a gas (in this case, marine air) and are usually found in the nanometre to micrometre diameter size range.

The EF of aerosols >5 nm in diameter ranges from 0.5 to 5.0×10^{16} aerosols/kg fuel (Alföldy *et al.*, 2013; Chen *et al.*, 2005; Hobbs *et al.*, 2000; Kasper *et al.*, 2007; Petzold *et al.*, 2008; Sinha *et al.*, 2003) . Aircraft measurements of freshly emitted aerosols from ship fuel combustion indicate a typical diameter of 50 nm (e.g., Hobbs *et al.*, 2000; Petzold *et al.*, 2008), as well as a more volatile mode that is even smaller (diameter $\leq$20 nm). These nanoparticulates dominate the particle number concentration. For example, measurements of container ship emissions show that only ~10 per cent of aerosol particles had diameters >100 nm (Murphy *et al.*, 2009). A strong correlation is observed between particle number concentration and gaseous SO_2 in ship plumes (Alföldy *et al.*, 2013). From a coastal site in Denmark, ship emissions were estimated to contribute 11–19 per cent of the particle number concentration (12–490 nm in diameter) and 9–18 per cent of the particle mass concentration (12–150 nm in diameter) when air arrived from over a shipping lane (Kivekäs *et al.*, 2014).

The combustion process in a ship engine completely alters the elemental composition of the resulting aerosols compared to the original fuel. The bulk of the material is volatilized and oxidized into the gaseous phase (see Section 2.4). The black colour of soot makes it the most visually arresting part of ship plumes, but soot is generally a fairly small fraction of the ship-emitted aerosols. Sinha *et al.* (2003) present data on aerosol composition from the emissions of two ships in the South Atlantic Ocean. They observed ~4 per cent black carbon by mass and used this to estimate global soot emissions from ships: 19–26 Gg/year (~0.2 per cent of total anthropogenic soot emissions).

Aerosol mass is dominated by sulfate and organic carbon, although the relative contributions vary. For example, aircraft measurements of ship tracks found the relative contributions of soot, sulfate and organic carbon to aerosol mass were approximately 3, 40 and 21 per cent, respectively (Petzold *et al.*, 2008). Ash (including trace metals) also constituted a small fraction (~4 per cent) and water contributed the remaining mass. However, a different study in the Gulf of Mexico observed that soot accounted for 15 per cent of the total aerosol mass emitted from ships (Lack *et al.*, 2009). The rest of the aerosol mass was primarily made up of sulfate (46 per cent) and organic carbon (39 per cent) (see Figure 2.4). The components of ship emission aerosols are discussed in more detail in the subsections that follow.

2.3.2.1 Black Carbon/Soot and Organic Carbon

Soot carbon, a class of carbonaceous aerosols, is a product of incomplete combustion. A hotter, more oxygen-rich combustion environment should result in less soot generation. Soot emissions from ships are highest at low engine loads (Lack *et al.*, 2009). Soot carbon absorbs strongly over the entire visible wavelength and does not volatilize below ~400°C,

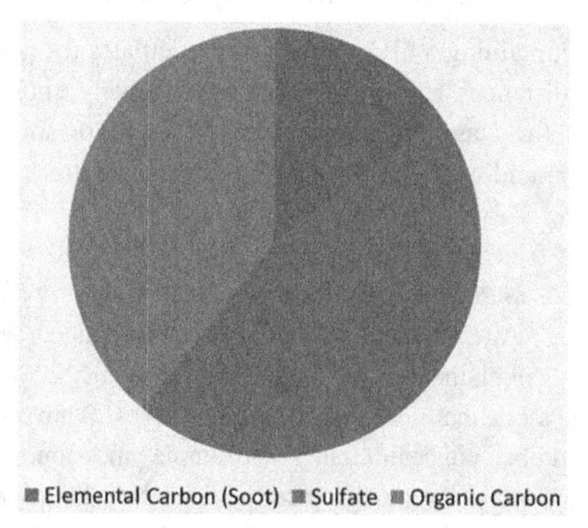

Figure 2.4 A typical breakdown of the major components of ship-emitted aerosols.
Source: Example data from Lack *et al.* (2009)

so it is often operationally defined as black carbon or elemental carbon. Freshly emitted soot particles consist of almost pure carbon and form graphitic agglomerates (often called spherules) with a typical diameter of a few tens of nanometres (Bond & Bergstrom, 2006). Primary soot aerosols coagulate and mix with volatile components downstream of the emission source (see Figure 2.5). Soot aerosols grow slowly and the median diameter is generally submicron. The emission of soot exerts positive climate forcing due to its strong light-absorbing properties. Global soot emissions due to anthropogenic fossil and biofuel emissions cause +0.4 W/m^2 warming (Ciais *et al.*, 2013). The capacity for black carbon to warm the Earth's surface is especially important when soot particles are emitted over an area of high reflectivity (e.g., ice, snow). Soot is also thought to be a cause of human morbidity and premature mortality (Smith *et al.*, 2009).

Soot is quite inert (i.e., there is minimal loss to chemical reactions in the atmosphere) and the particles are small, such that the dry deposition rate is slow (see Section 2.5 for more on ship plume evolution). Wet deposition is an important removal pathway for black carbon in the atmosphere, but soot is poorly soluble in water, so this is still relatively inefficient. Soot removal from the atmosphere is thus fairly slow and, once emitted from one region, can be transported long distances. The typical atmospheric lifetime for black carbon/soot is thus approximately 1 week (Cape *et al.*, 2012).

Organic carbon aerosols are also emitted as a result of incomplete combustion, but they volatilize at a much lower temperatures than soot. Organic carbon contains more hydrogen and oxygen and has a different absorption signature from soot (e.g., Yang *et al.*, 2009). Organic carbon tends to absorb much more in the ultraviolet than at longer wavelengths, which is why it is sometimes referred to as *brown carbon*. A fraction of particulate organic carbon is emitted directly from combustion along with soot (primary aerosols), while condensation of organic vapours also leads to the formation of organic carbon particles (secondary aerosols).

2.3.2.2 *Inorganic Ions (Sulfate, Nitrate, Ammonium and Phosphate) and Trace Metals*

Aerosols produced due to fuel combustion at sea inevitably contain oxidized and charged forms of the main groups of acidic elements. These can be a direct product of the oxide gases (discussed in Sections 2.3.1.1–2.3.1.3) or created by scavenging and condensation reactions in the atmosphere (e.g., HNO_3, H_3PO_4). In typical-pH environments near the sea surface (i.e., close to neutral acidity), ship-emitted aerosols will form salts (e.g., ammonium sulfate) with positively charged ions either in the core or the exterior of the particle. Aerosol deposition acts as a loss term for the main group elements from the atmosphere into the ocean.

Size-fractionated sampling of aerosols from container ship stacks burning HFOs reveal two predominant fractions: small, highly sulfidic nanoparticles and a relatively low percentage of larger (>320-nm) black, carbonaceous particles (Murphy *et al.*, 2009). The larger particles are aggregates (clusters or strands) that include graphitic soot aggregates and carbon-rich char particles, which contain a broader variety of chemical species. The smaller nanoparticles are predominantly hydrated sulfate and dominate the PM mass and

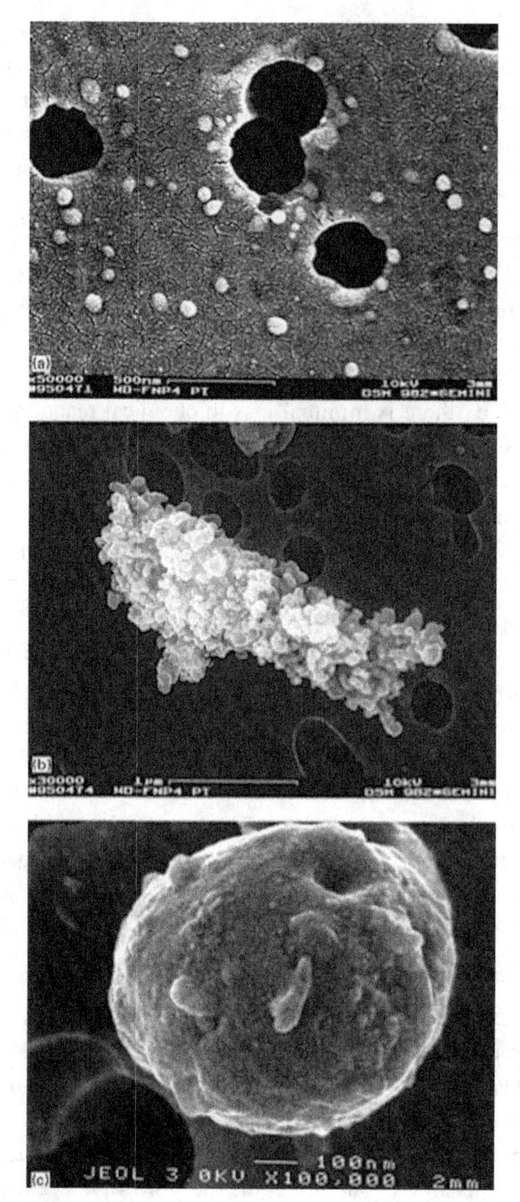

Figure 2.5 Scanning electron microscopy images of particles emitted by a ship burning heavy fuel oil: (a) nucleated primary spherical particles; (b) agglomerated particle consisting of the primary particles; and (c) fuel oil droplet residue particle.
Source: Images taken from Lyyränen *et al.* (1999), their figure 5

number concentrations (Murphy *et al.*, 2009). The most striking chemical change from the original fuel chemistry is the dominance of sulfate and water-bound sulfate aerosols after combustion, which can be up to 95 per cent of the total mass. A study of a container ship plume showed that for 1 hour after emission the sulfate mass and organic mass in the particles remained at 70 and 28 per cent, respectively (Murphy *et al.*, 2009). However, it is important to note that the aerosol chemical composition, including the percentage sulfate (and hydrated sulfate), can change significantly depending on the engine load (e.g., Agrawal *et al.*, 2008a). Kasper *et al.* (2007) observe that soot and sulfate are minor components of the aerosols emitted from a low-speed marine diesel engine, with other organic material dominating the aerosol mass.

The sulfate fraction of freshly emitted aerosols typically scales directly with the sulfur content of the fuel that is burnt. For ships burning HFOs, the emitted aerosol mass is dominated by sulfate (and associated water), followed by organics (Agrawal *et al.*, 2008a). The organic carbon content of aerosol emissions from ships appears to be related to the sulfur content of the fuel that is burnt (Agrawal *et al.*, 2008b; Lack *et al.*, 2009). Lack *et al.* (2009) also observe that aerosols from a low-sulfur fuel absorb more light (i.e., proportionally more soot) and are less hygroscopic (due to lower sulfate). These changes in aerosol composition and properties could increase their lifetime and decrease their effectiveness as cloud condensation nuclei.

Nitrogen and phosphorus are abundant elements in the composition of living cellular material, but this has little bearing on the composition of ship-emitted aerosols. Phosphorus is negligible in combustion emissions compared to oxidized nitrogen and sulfur. The dominant source of phosphorus aerosols in the marine environment are land-based terrigenous sources (i.e., rock, dust, soil and sediment; Graham & Duce, 1979). Similarly, gas-phase ammonia and aerosol ammonium are not produced by combustion, such that the shipping contribution to atmospheric reduced nitrogen is negligible.

NO_x are the major nitrogen-containing gas-phase species arising from fuel combustion, but these are generated from ambient air within the combustion chamber. NO_x cycling is complex (see Section 2.5), but a net product of NO_x emissions is gas-phase nitric acid (HNO_3), which readily reacts to generate various forms of nitrate aerosol (e.g., sodium nitrate, ammonium nitrate). Ammonia also reacts with ship-derived sulfate aerosols. The above processes either contribute to secondary aerosol formation (different from the direct formation of combustion particles, which are termed 'primary aerosols') or aerosol growth.

The majority (~75 per cent) of ship-emitted SO_2 is oxidized to sulfate in the marine atmosphere (Yang *et al.*, 2016), with the rest being lost due to dry deposition onto the ocean surface. SO_2 oxidation mainly takes place in cloud water (e.g., Yang *et al.*, 2011). In the aerosol phase, acidic sulfate is primarily titrated (i.e., balanced) by basic ammonium (NH_4^+), forming ammonium (bi)sulfate. Models predict that reduced sulfur emissions will result in more NH_4^+ reacting with HNO_3, enhancing the formation of ammonium nitrate aerosols (e.g., Ansari & Pandis, 1998). Matthias *et al.* (2010) model the impacts of ship emissions in the atmosphere over the North Sea. Their results highlight the influence that ship emissions have upon atmospheric chemistry. The shipping lanes in the English Channel contain elevated SO_2 and NO_x levels. In contrast, elevated sulfate and nitrate aerosols due to ship emissions occur further downwind, near northern Germany and Denmark. A source of reduced nitrogen (ammonia or ammonium) is important for the formation of nitrate aerosol. Ammonia and ammonium emissions are predominantly from terrestrial sources, meaning that enhanced concentrations of nitrate aerosols derived from ship NO_x emissions are predicted close to land (Matthias *et al.*, 2010).

Aerosol trace metal levels in the lower atmosphere are typically below 1 $\mu g/m^3$ unless measured closed to pollution sources or high-emission sites. Trace metals in aerosols from ships are derived from the ash component of the marine fuel. Most metals are non-volatile (with the notable exception of low-melting-point metals such as mercury), so the aerosol trace metal content is expected to be in the same ratio as the original fuel. However, there are some additional contributions that result from engine and exhaust manifold wear. Ship-emitted particles thus contain clear enrichments in vanadium and other trace metals compared to the original fuel (Murphy *et al.*, 2009). Switching to a different fuel type can influence the trace metal composition in aerosols emitted by ships. For example, aerosol vanadium levels were reduced by ~60 per cent in the San Francisco Bay area when ships switched to lower-sulfur fuels to comply with the 2009 sulfur emission regulations along the west coast of the USA (Tao *et al.*, 2013). The variability of trace metal concentrations in different fuel types also provides the opportunity to use metal ratios to fingerprint emissions and to trace ship plumes (e.g., Kulkarni *et al.*, 2007; Moreno *et al.*, 2008; Viana *et al.*, 2009).

2.4 Combustion Conditions and Exhaust Treatments

Following combustion and any further reactions in the exhaust flue, ship emissions are released into the atmosphere. Ship emissions of any element (e.g., carbon, sulfur, iron) to the atmosphere can be approximated by Eq. (2.4):

$$\text{Emission} = E_{fuel} \times M_{fuel} \times K_{eff} \tag{2.4}$$

where E_{fuel} is the elemental mass per kg of fuel, M_{fuel} is the mass of fuel combusted and K_{eff} represents an efficiency factor for the conversion of the element into the gas phase. It is important to note that this cannot be used for nitrogen, as nitrogen emissions are the result of the oxidation of nitrogen gas ($N_{2(g)}$) from ambient air that is mixed with the fuel during combustion. No consideration of fuel type and quality is required for nitrogen. A first approximation is to use the engine-specific emission estimates from the manufacturer. The methods for sampling and analysis of emissions from engines (gases and PM) must conform to ISO 8178-1.

The efficiency of the oxidation and emission of all chemical elements in fuel into the gas phase is confounded by a range of variables that impact K_{eff}. Global fuel consumption by ships is known reasonably precisely, but models also require a detailed knowledge of the EFs in order to get realistic emission estimates. EFs are related to the individual characteristics and design of the engine used and its operating conditions. EFs depend strongly on several variables, including: (1) vessel size and engine type; (2) vessel activity and engine load; (3) fuel and lubricating oil; and (4) adaptations to atmospheric sulfur emission regulations.

2.4.1 Vessel Size and Engine Type

Modelled EFs for NO_x and SO_2 are predicted to be higher for larger vessels such as container ships (Dalsoren *et al.*, 2009), and this prediction is supported by field measurements. Williams *et al.* (2009) test a broad selection of ships and find that bulk freight carriers and tanker ships have the highest emissions of NO_x (75–90 g NO_x/kg fuel) and SO_2 (20–30 g SO_2/kg fuel). Engine size is an important determinant of aerosol particle number and size distribution. A large engine generates higher levels of metals in the exhaust than a small engine burning similar fuel (Lyyränen *et al.*, 1999). Celo *et al.* (2015) observe large differences in the emissions of trace metals (e.g., zinc, iron, aluminium, nickel, vanadium) coming from different vessels burning the same type of fuel.

Poor combustion efficiency (due to engine ageing, poor maintenance) results in the production of higher levels of CO. Lack *et al.* (2008) observe a strong positive correlation between black carbon and CO, suggesting that lower combustion efficiency encourages black carbon emissions. Medium-speed engines are common on smaller vessels such as tugboats, fishing vessels and ferries. Medium-speed engines generally burn distillate fuels and emit more than twice as much black carbon as vessels with slow- and high-speed diesel engines. In contrast, medium-speed diesel engines typically produce lower NO_x emissions when stationary than container ship engines and tugs. Regulation of ship speed has been used as a policy tool for reducing NO_x emissions (see discussion in Section 2.7.2).

2.4.2 Vessel Activity and Engine Load

Increased fuel consumption due to elevated engine speed/load inevitably increases atmospheric emissions from ships. Emission estimates are thus typically adjusted to engine load by weighting them (i.e., mass emitted per kWh; see Eq. (2.2)). Comparing weighted EFs highlights whether certain types of ship or the activities of a ship (e.g., speed) influence emissions. For example, positive correlations have also been observed between engine load and the weighted emissions of NO_x, SO_2 and PM (Agrawal *et al.*, 2008a). This is partly because engine loading influences the efficiency and temperature of combustion. It is worth noting that these trends are not completely predictable, as the authors found that the weighted NO_x emissions from this container ship were also elevated at very low engine loads (8 per cent).

The chemical composition of aerosols emitted in ship exhausts is also likely to be influenced by vessel activity. Low engine loads tend to increase black carbon emissions significantly (Lack & Corbett, 2012). Ships moving through Arctic waters tend to have highly variable engine loads that are not necessarily coupled to ship speed when they move through sea ice. Breaking sea ice requires high engine loads even though these ships are not moving very fast. Changing engine speed and load thus has ramifications for black carbon emissions and for subsequent radiative impacts (see Section 2.6.1).

2.4.3 Fuel and Lubricating Oil

Different fuels are used by shipping companies/vessels and fuel-type usage depends upon ship location. The choice of fuel is determined by cost (low-sulfur fuels are cleaner but more expensive) and by emission regulations, which are region specific (see Section 2.7.2). Combustion of low-sulfur fuel inevitably produces low-sulfur emissions, with reduced sulfate aerosol concentrations in the ship stack emissions. Combustion of low-sulfur fuels also produces less particulate mass than high-sulfur fuel (>0.5 per cent) emissions (Lack *et al.*, 2009). Winnes and Fridell (2009) compared emissions from the combustion of HFO (higher sulfur content) with emissions from marine gas oil (lower sulfur content). Except for SO_2, there was no significant change in the exhaust gas emissions of CO_2, CO, NO_x and total hydrocarbons when normalized to energy released (g/kWh). The authors also observed a reduction in the particle mass emissions (0.33 and 1.34 g/kWh for low- and high-sulfur fuels, respectively) and changes in the particle size distribution. It is worth noting that the number of very small particles (300–400 nm) actually increased in the emissions from low-sulfur fuel combustion. This may mean that the perceived health benefits of switching to low-sulfur fuel (see Section 2.6.2) will not be as pronounced as predicted.

In addition to the choice of fuel, lubricating oils (additives) are used by the industry to help prolong the life of the engine. The trace element emissions are most influenced, as lubricating oil is enriched in elements such as calcium and zinc, which are then found in the exhaust aerosols (e.g., Lyyränen *et al.*, 1999).

2.4.4 Adaptations to Atmospheric Sulfur Emission Regulations

The atmospheric emissions from many ships are controlled by recent regulations (Sulfur Emission Control Areas (SECAs) were established in 2015), and more stringent global regulations will come into force in 2020 (see Section 2.7.2 for further detail). To adapt to these regulations, the industry has chosen to either:

(1) Switch to low-sulfur-content fuels (marine gas oil or alternatives such as liquefied natural gas (LNG)) or
(2) Install exhaust gas treatment units (commonly referred to as scrubbers), which scrub the sulfur from the exhaust fumes.

A scrubber enables a ship to conform to low-sulfur emission regulations while continuing to use high-sulfur-content fuels. Scrubbers react with the sulfur in the flue to remove it before the exhaust enters the atmosphere. There are three main types of scrubbers: dry scrubbers, freshwater scrubbers (closed loop) and seawater scrubbers (open loop). Dry scrubbers clean the exhaust gases with solids such as hydrated lime-treated granules. Open- and closed-loop scrubbers are wet scrubbers that use a uniform high-pressure spray of water into the ship exhaust to remove the sulfur. A closed-loop scrubber cleans the exhaust gases with freshwater that is rich in caustic soda. An open-loop seawater scrubber is a popular solution for the shipping industry because it uses untreated seawater to remove the sulfur:

$$2SO_{2(g)} + 2H_2O_{(l)} + O_2 \rightarrow 2SO_4{}^{2-}{}_{(aq)} + 4H^+{}_{(aq)}$$

The system is particularly efficient because it takes advantage of the natural alkalinity of seawater to neutralize the acidity (H^+) produced by converting gaseous SO_2 to aqueous sulfate ($SO_4{}^{2-}{}_{(aq)}$):

$$HCO_3{}^-{}_{(aq)} + 2H^+{}_{(aq)} \rightarrow CO_{2(g)} + H_2O_{(l)}$$

Wet scrubbers typically increase the sulfate content of the scrubber water and reduce the pH (make it more acidic). Wet scrubbers also inadvertently remove other elements from the ship exhaust. Fridell and Salo (2016) observe low particle numbers and complete removal of ultra-fine particles when high-sulfur fuel emissions are scrubbed. The emissions of PM by mass reduce by 75 per cent and the total number of particles reduces by ~90 per cent.

Water from an open-loop scrubber undergoes a crude treatment process before returning to the ocean. Typically, a cyclonic (centrifugal) separator removes the small particles and a settling tank traps the heavier particles for sludge disposal. However, Fridell and Salo (2016) observed that the scrubber-induced reduction in solid-phase exhaust particles (~50 per cent) was much less than the reduction in total particle number (~90 per cent). These data suggest that the scrubber preferentially removed the volatile component of the aerosols. The scrubber also reduced the levels of polycyclic aromatic compounds in the exhaust. Turner *et al.* (2017) also suggest that scrubbers may efficiently remove PAHs (organic pollutants that are typically bound to aerosols) and trace metals from ship exhaust into the scrubber water. Lieke *et al.* (2013) show that scrubbers may even influence the micro- and nano-structural characteristics of the particles that do not get removed into the scrubber water.

Scrubbers may help ships conform to sulfur emission regulations. The results above also indicate that scrubbers modify atmospheric emissions and seawater composition in unforeseen ways. The potential consequences of open-loop seawater scrubber systems are discussed in Section 2.6.3.

2.5 Evolution of Ship Plumes in the Atmosphere

2.5.1 Entrainment of Background Air

The gases and particles emitted from a ship stack immediately begin to mix with background air in the lower atmosphere. The plume expands and entrains background air in both the vertical and horizontal. The vertical dilution of ship emissions is constrained by the mixing height of the atmospheric marine boundary layer (MBL; in the order of 500–1000 m). This layer is highly turbulent, and its behaviour is directly influenced by interactions between the atmosphere and ocean. The height of the MBL is determined by its stability, which is primarily influenced by horizontal wind speed and the direction/magnitude of the ocean–atmosphere heat gradient. Plume expansion can be described using an empirically determined coefficient to model the nature of the expansion (von Glasow *et al.*, 2003). The parameters needed to estimate plume expansion are influenced by atmospheric stability. von Glasow *et al.* (2003) describe the horizontal and vertical expansion of a ship plume as in Eqs (2.5) and (2.6):

$$h_{t1} = h_{t0}(t_1/t_0)^{0.6} \tag{2.5}$$

$$w_{t1} = w_{t0}(t_1/t_0)^{0.75} \tag{2.6}$$

where w and h refer to the height and width of the plume, respectively. Subscripts refer to the plume at time t and an initial reference time (t_0). The authors suggest that at t_0 = 1 s, h_{t0} and w_{t0} are approximately 5.5 and 10.0 m, respectively. Plume size at a fixed distance from source is thus a function of time (as is shown in Eqs (2.5) and (2.6)) and the transit speed (horizontal wind speed). In a clean marine atmosphere, the contrast between gas/aerosol concentrations within and outside of a fresh ship plume is extremely large. von Glasow *et al.* (2003) use a model to demonstrate that the background aerosols (sulfate and sea salt) can significantly impact the gas-phase chemistry (especially NO_x) within the ship plume. Ship plume dispersion is discussed further in Chosson *et al.* (2008).

2.5.2 Wet and Dry Deposition

Removal of gases and particles out of the atmosphere to a surface occurs via wet or dry deposition. Wet deposition is the transfer of airborne species to the surface in aqueous form (rain, snow or fog). Wet deposition includes: (1) the dissolution of soluble gases into water/cloud droplets; (2) the removal of particles that serve as nuclei for water droplets; and (3) the removal of particles that collide with droplets within and below clouds. The relative importance of dry versus wet deposition is thus determined by whether the species of interest is in particulate or gaseous form, its solubility in water and the amount and intensity of precipitation.

Dry deposition is the direct transfer of gases or particles to the surface without the aid of precipitation. The dry deposition flux of gases and particles is a function of the local concentration of the depositing species and the deposition velocity. For particles, the major factor influencing the deposition velocity is the particle size, with larger particles increasingly influenced by gravitational settling, as well as near-surface turbulence and diffusion (for more detail on particle deposition onto a water surface, see Slinn & Slinn, 1980; Slinn *et al.*, 1978). For gases, the deposition velocity is controlled by factors such as wind speed and the stability of the atmospheric boundary layer, as well as the roughness of the surface and its reactivity with the species that is being deposited (for more detail on gas exchange across the air–sea interface, see Liss, 1973; Liss & Slater, 1974).

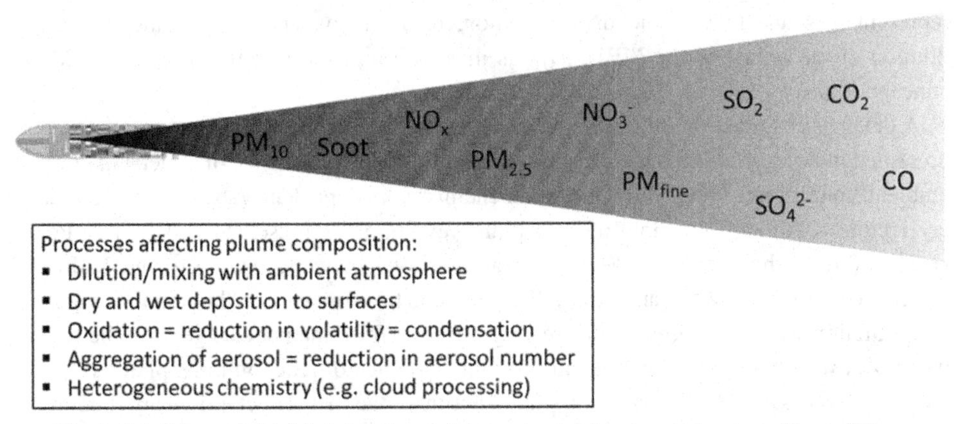

Figure 2.6 Schematic highlighting the evolving nature of the chemical composition within a ship plume. Multiple processes determine the change in concentration of gases and aerosols within the plume.

2.5.3 Chemical Processing

A ship plume is modified by *in situ* chemical reactions and interactions between gas and particle phases (Figure 2.6). Entrainment of background air with high relative humidity influences particle composition. Aerosols can have a solid and an aqueous phase, and the interactions with the surrounding gas-phase constituents are influenced by what is in the aqueous phase. Some chemical reactions also result in new products – for example, the oxidation reaction of SO_2 with water vapour forms sulfuric acid and, ultimately, sulfate aerosol. Other reactions modify the concentrations in the ship plume and/or in the background air. Coagulation and ageing of the air mass will likely reduce the ratio between aerosol number and a conservative tracer such as CO_2 or CO. Petzold *et al.* (2008) observed that the non-volatile mode (e.g., soot) was conserved as ship plumes dispersed and aged, while the number of volatile aerosols (e.g., sulfate, organic carbon) decreased with increasing plume age, presumably due to rapid coagulation. The measurement of the aerosol number emitted from ships is likely to be quite sensitive to the distance downwind of the emission source.

The vertical dilution of a ship plume (Eq. (2.5)) has a very large impact on the plume concentration (C, moles m^{-3}) in a relatively short period of time. The vertical mixing time of a ship plume (t) is in the order of minutes (Chosson *et al.*, 2008). We present plume concentration changes (ΔC) in a simplistic manner by considering the column-integrated concentration change (i.e., $\Delta C = \delta \langle C \rangle / \delta t$ (moles/m^2/min). This is determined by the balance between production and loss terms, expressed here as fluxes (F, moles/m^2/min; Eq. (2.7)):

$$\Delta C = F_{PROD} - (F_{DEST} + F_{WET} + F_{DRY} + F_{MIX}) \tag{2.7}$$

where chemical reactions with plume air and the background MBL (m) air can result in both production (F_{PROD}) and destruction (F_{DEST}) within the plume, F_{WET} and F_{DRY}

represent loss due to wet and dry deposition, respectively, and F_{MIX} is the horizontal dilution of the plume in the MBL. F_{MIX} is thus also a product of the background MBL concentrations.

A detailed examination of how sulfur emissions in a ship plume can evolve over time is presented in Song *et al.* (2003a). The authors highlight the importance of understanding the ambient conditions, specifically the photochemistry and presence/absence of boundary layer clouds. Another example of ship plume evolution is presented in Charlton-Perez *et al.* (2009), who highlight the non-linear reactions between NO_x, O_3 and hydroxyl radicals (see Sections 2.5.4 and 2.5.5). These interactions mean that the initial changes in concentration within the ship plume are very different from the net change once the plume has mixed with background marine air. The relative rates of reaction time, plume dilution and loss processes (e.g., deposition) all determine the net effect on many important chemical species. The difference in behaviour of gases with contrasting chemical reactivity is evidenced by the observations of Schlager *et al.* (2006). The authors used an aircraft to intersect the plume of a large container ship off the west coast of France on multiple occasions. Measurements of CO_2 and total NO_x ($NO_y = HNO_3 + NO_x$) were made in the initial plume (<60 s old) and after the plume had aged by 18 min. NO_y is much more reactive than CO_2, and the increase in the NO_y/CO_2 ratio could be used to estimate the net production of NO_y in the plume.

2.5.4 Photochemical Production and Loss Processes

In a fresh ship plume, NO rapidly reacts with available O_3, forming NO_2:

$$NO + O_3 \rightarrow NO_2 + O_2$$

Thus, downstream of a ship plume, most of the NO_x is in the form of NO_2. During the day, NO_2 is photolysed by sunlight, giving back NO and an oxygen atom:

$$NO_2 + h\nu \rightarrow NO + O$$

The oxygen atom combines rapidly with O_2 to reform O_3:

$$O + O_2 + M \rightarrow O_3 + M$$

Note that this is a 'null' cycle in that no new O_3 is formed. However, NO can also react with the hydroperoxyl radical (HO_2) and organic radicals, leading to new O_3 formation.

2.5.5 Removal of NO_x from the Atmosphere

During the day, NO_x is removed from the atmosphere mainly by reaction with the OH radical:

$$OH + NO_2 + M \rightarrow HNO_3 + M$$

The reaction of NO_2 with OH is one of the main pathways of aerosol nitrate formation, because HNO_3 reacts with available ammonium (NH_4^+) to form ammonium nitrate. Photolysis and further oxidation by OH, precipitation (e.g., rain) and deposition to surfaces are the other important sinks for HNO_3.

At night, without photolysis and reactions with OH, NO_2 reacts with O_3:

$$NO_2 + O_3 \rightarrow NO_3 + O_2$$

The resultant nitrate radical (NO_3) can react with another molecule of NO_2 to form N_2O_5. In the atmosphere, NO_3 and N_2O_5 exist in a close thermal equilibrium:

$$NO_3 + NO_2 \leftrightarrow N_2O_5$$

Near the plume origin, NO_3 can also react with NO, reforming NO_2:

$$NO_3 + NO \rightarrow 2NO_2$$

Similar to HNO_3, N_2O_5 is highly soluble and is removed from the atmosphere by wet and dry deposition. N_2O_5 also reacts on aerosol surfaces in the marine atmosphere (e.g., sea salt) to form HNO_3. Both NO_3 and N_2O_5 have very short lifetimes during the daytime and so only have appreciable concentrations at night.

There is a major distinction between NO_x–O_3 chemistry in a semi-polluted marine environment and in a heavily polluted urban region. In marine environments where NO_x levels are relatively low (once the ship plumes have dispersed), additional NO_x emissions from ships generally lead to a net production of O_3 (e.g., Holmes *et al.*, 2014). In contrast, in polluted regions with already very high NO_x levels, additional NO_x emissions usually result in a reduction of O_3. Reduced O_3 is due to the increased importance of the OH + NO_2 cycle termination reaction, which removes both NO_x and OH + HO_2 from the system. This is one reason why O_3 levels in cities are often lower than in surrounding rural areas. In such high-NO_x environments, the production of O_3 is more sensitive to the availability of VOCs.

2.6 Environmental Implications of Shipping Emissions

2.6.1 Climate Impacts from Ship Emissions

2.6.1.1 Greenhouse Gases

It is widely documented that the Earth's climate is changing due to anthropogenic activity (Ciais *et al.*, 2013). Short-wave solar radiation is absorbed in the upper atmosphere, but a proportion gets through the atmosphere and is absorbed by the Earth's surface. A fraction of this energy is re-emitted into the atmosphere as long-wave radiation (heat). Greenhouse gases do not particularly interact with short-wave radiation, but trap long-wave radiation and prevent it from escaping into the upper atmosphere. CO_2 is the best-known example of a greenhouse gas because its emissions due to anthropogenic activity have increased substantially since the Industrial Revolution. In 1960, atmospheric CO_2 was around

315 parts per million (ppm), whereas levels today exceed 400 ppm (Ciais *et al.*, 2013). Shipping activity has of course contributed to the rise in atmospheric CO_2 levels, and it is predicted to contribute significantly more if activity continues to increase unchecked. Current projections suggest that as much as 2 gigatonnes of CO_2 will be emitted by ships into the marine atmosphere every year by 2050 (Eyring *et al.*, 2005a). The IMO has pledged to reduce the global CO_2 emissions from ships.

Greenhouse gases can be compared by how efficiently they trap heat. This is called their 'global warming potential' (GWP), which is referenced to CO_2 (GWP = 1). Despite being present at much lower levels in the atmosphere, gases such as nitrous oxide (N_2O; GWP = 268), CH_4 (GWP = 86) and sulfur hexafluoride (SF_6; GWP = 16,300) have a profound impact on our climate (GWPs quoted here refer to an impact over a 20-year timescale). Ships do not emit much SF_6, but they do emit low ppm levels of N_2O (Cooper, 2001) and ppb levels of CH_4 (Sinha *et al.*, 2003).

As discussed in Section 2.5, NO_x emissions play an important role in determining O_3 levels in the MBL. O_3 is critical to the Earth's radiative balance. O_3 in the stratosphere absorbs much of the incoming ultraviolet radiation, simultaneously cooling the underlying lower atmosphere (troposphere) while warming the stratosphere. In the lower atmosphere, O_3 has a radiative effect that is similar in nature to CO_2 and other greenhouse gases. Stevenson *et al.* (2013) calculate that global changes in tropospheric O_3 have contributed 410 mW/m^2 since the year 1750. The lifetime of O_3 is short, such that it cannot be compared to other greenhouse gases using the GWP metric. In addition to its direct radiative forcing effect, O_3 is highly reactive and is a source of OH radical. The OH radical plays a major role in determining the abundance and distribution of other gases that are climatically important. A good example of this effect is the influence of O_3/OH on CH_4.

Early calculations by Lawrence and Crutzen (1999) suggested that ship emissions of NO_x could have significant impacts on radiative forcing by causing changes in O_3 levels and altering the lifetime of atmospheric CH_4. However, these calculations did not account for the interplay between the non-linear production of O_3 and the rapid concentration changes in a dynamically evolving ship plume. More recent analyses (Charlton-Perez *et al.*, 2009; Holmes *et al.*, 2014) have considered the scale at which ship plume chemistry occurs – typically hundreds of kilometres (less than the grid cell of global models). By incorporating an improved representation of ship plumes, NO_x perturbations are estimated to be smaller than previously suggested. A net radiative forcing due to a 5 per cent increase in ship NO_x emissions (1 Tg N/year) is estimated to cause changes of +3.4 mW/m^2 (due to short-lived O_3 increase), –5.7 mW/m^2 (due to CH_4 decrease) and –1.7 mW/m^2 (due to the long-lived O_3 decrease associated with the reduction in CH_4). This gives a net radiative forcing of –4 mW/m^2 and is at the low end of previous estimates (Holmes *et al.*, 2014 and references therein). Over a relative short timescale (years to a few decades), the warming effect due to NO_x-driven O_3 production and the cooling effect due to OH-driven CH_4 destruction are opposite in sign and similar in magnitude. Thus, the combined effect from these short-lived forcers is thought to be fairly small (<10 mW/m^2; see Figure 2.7).

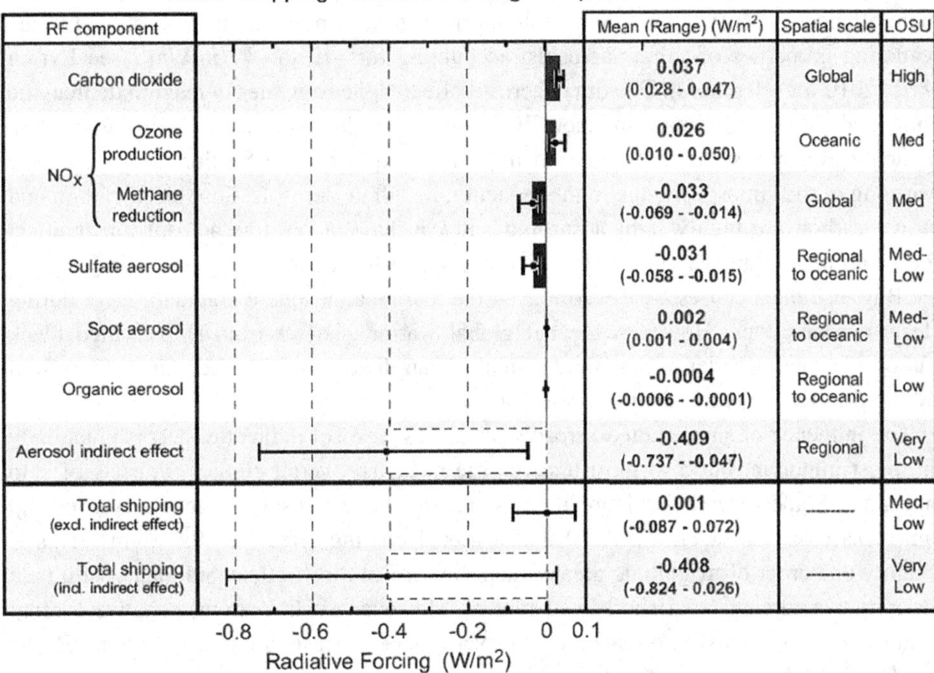

Figure 2.7 Global average annual mean radiative forcing (RF) and literature ranges due to emissions from shipping in W/m^2 for 2005. The boxes show the mean of the lower and upper estimates reported in the literature and the whiskers show the range of literature values given by the highest and lowest estimates. The typical geographical extent (spatial scale) of the RF and the level of scientific understanding (LOSU) are also indicated. RF contributions with very low LOSUs are displayed with dashed lines. This figure does not include the positive RF that could possibly occur from the interaction of black carbon with snow, which has so far not been investigated for ships.
Source: Taken from Eyring *et al.* (2010), their figure 14

2.6.1.2 Aerosol Radiative Effects

The largest influence of ship emissions on the global climate is likely to be through the effects of aerosols on radiation and clouds (see Figure 2.6). Aerosols affect the Earth's radiative balance directly by scattering/absorbing incoming radiation (termed the 'aerosol direct effect') and indirectly via their influence on clouds (termed the 'aerosol indirect effect'). As discussed in Section 2.3.2, aerosols from ship emissions roughly fall into two categories: those that are directly emitted as particles (primary aerosols; e.g., soot carbon) and those that are converted from gas-phase emissions (secondary aerosols; e.g., sulfate from SO_2 oxidation). A large fraction of the aerosol number in ship emissions is associated with primary aerosols. In contrast, most of the aerosol volume (as well as mass) in ship emissions is within secondary aerosols. Ship emissions are estimated to account for roughly half of the total atmospheric SO_2 mass in busy shipping corridors such as the North Atlantic (Capaldo *et al.*, 1999).

All aerosols scatter incoming radiation, and larger aerosols scatter incoming radiation more efficiently. Most of the direct cooling effect from ship emissions is attributed to the scattering properties of sulfate aerosols, accounting for –12 to –47 mW/m^2 (see Eyring *et al.*, 2010 for references). The direct aerosol effect alone is greater in magnitude than the combined radiative forcing from short-lived forcers. Fuglestvedt *et al.* (2009) suggest that reducing SO_2 emissions from ships to improve air quality (see Section 2.7.2) causes a warming of the atmosphere due to the reduction in sulfate aerosols. Soot/black carbon and brown carbon are highly light absorbing and can have a positive aerosol direct effect (i.e., warming). Ramana and Devi (2016) show that black carbon emitted from ships in the Bay of Bengal causes a net warming of the marine atmosphere boundary layer during clear-sky conditions. However, the net global warming effect from ship-emitted black carbon is about an order of magnitude smaller than the direct aerosol cooling effect from sulfate (see Eyring *et al.*, 2010 for references).

The influence of ship-emitted aerosols on clouds (aerosol indirect effect) is potentially the most important but also most uncertain term in the overall climatic impacts of ship emissions. Some general circulation models predict a large aerosol indirect effect from ship emissions (e.g., Lauer *et al.*, 2007) and a global cooling effect of –0.5 W/m^2. This is roughly one order of magnitude greater than the aerosol direct effect and greater still than the radiative forcing from short-lived forcers. However, such a strong negative forcing is not evident in analyses of satellite observations near dense shipping corridors (Peters *et al.*, 2011), highlighting the large uncertainty in our understanding of aerosol–cloud interactions.

In marine regions with consistent stratus clouds and high ship traffic, highly reflective ‘ship tracks’ (or streaks of clouds) several kilometres wide and several hundred kilometres long are often observed. Ship plumes cause cloud droplet numbers to increase, and the cloud droplet radius tends to decrease (assuming that the atmospheric water column content does not change; Radke *et al.*, 1989). As a result, the optical depth and reflectivity (albedo) of the clouds increase, resulting in ‘whiter’ clouds that reflect more incoming light back to space. In previously cloud-free regions, aerosols from ships can also ‘activate’ and become cloud droplets, initiating cloud formation. Visible ship tracks provide irrefutable evidence that aerosols strongly enhance the radiative forcing at the local cloud scale (Christensen & Stephens, 2011; Coakley *et al.*, 1987). Devasthale *et al.* (2006) analyse satellite observations of cloud properties over land, coastal areas and the open sea of Europe between 1997 and 2002. Mean cloud albedo was higher and cloud top temperature was lower over coastal areas and over the English Channel compared to the open waters of the North Sea and North-East Atlantic.

However, individual ship plumes do not always produce ship tracks, or at least not tracks with albedos that are distinctly different from the background (Frick & Hoppel, 2000). Environmental conditions and the size distribution/composition of ship-emitted aerosols, as well as those of background aerosols, all seem to influence the formation of ship tracks (e.g., Hobbs *et al.*, 2000). For example, the impact of ship emissions on clouds seems to be more pronounced in regions of low ambient droplet concentration (Durkee *et al.*, 2000), likely because clouds in such clean conditions are more susceptible to ship perturbations than clouds in already polluted regions. Increases in cloud droplet number from ship emissions can also affect the liquid water content, coverage and lifetime of clouds through precipitation suppression (Albrecht, 1989; Ferek *et al.*, 2000). However, the overall magnitude and even the sign of this effect are not well known.

In summary, individual ship tracks can be highly effective at scattering radiation back to space, but the integrated effect of ship emissions on cloud properties and the global radiation budget is highly uncertain. Previous model assessments of this effect range from globally important (Lauer *et al.*, 2007; Righi *et al.*, 2015) to nearly negligible (Unger *et al.*, 2010), largely due to the aforementioned uncertainties.

2.6.2 Air Pollution

Atmospheric emissions from ships are a source of particles to the atmosphere and thus have the potential to impact air quality and human health. Small particles (diameters less than 2.5 μm, commonly referred to as $PM_{2.5}$) easily penetrate the lungs and have been associated with a wide range of health effects. Air pollution causes asthma and early mortality due to cardiopulmonary and lung cancer (e.g., Pope *et al.*, 2002). The majority of shipping activity occurs within a short distance of the coast (70 per cent within 400 km; Eyring *et al.*, 2010). Humans often reside in coastal regions, with a lot of shipping activity and ports in close proximity to residential dwellings. In 2010, 80 per cent of the world's megacities (defined as having >10 million inhabitants) had a coastal influence (von Glasow *et al.*, 2013).

The atmospheric lifetime of $PM_{2.5}$ is long enough that ship emissions in regions with high shipping activity have been estimated to substantially impact human mortality rates. Corbett *et al.* (2007) model shipping-related PM emissions and estimate that they were responsible for approximately 60,000 cardiopulmonary and lung cancer deaths annually, with most occurring near the coast in Europe and Asia. The authors also project that annual mortalities would increase by 40 per cent by 2012.

Much of the $PM_{2.5}$ from ships is associated with the high sulfur content of the fuel. Sulfur emissions have been regulated by the IMO in select coastal areas in recent years, and global regulation will come into force in 2020 (see Section 2.7.2). The potential impact of sulfur emission controls on human health was investigated by Winebrake *et al.* (2009). The authors estimated approximately 87,000 premature deaths annually in 2012 in the absence of sulfur emission control. Simulation of stringent sulfur emission controls in coastal areas reduced premature deaths by 43,500 per year.

The number of deaths calculated in the work of Corbett *et al.* (2007) and Winebrake *et al.* (2009) are likely underestimates. This is discussed in a more recent study (Sofiev *et al.*, 2018), which considered ship emissions and air pollution human health impacts in much greater detail. The authors used the Automatic Identification System ship traffic inventory (Jalkanen *et al.*, 2016) and global chemical transport models to estimate the spatial distribution of emissions and the atmospheric dispersion of ship-emitted PM. Importantly, the authors also applied a more accurate and sensitive concentration–response function to evaluate the public health impacts of ship emissions. In an unregulated sulfur

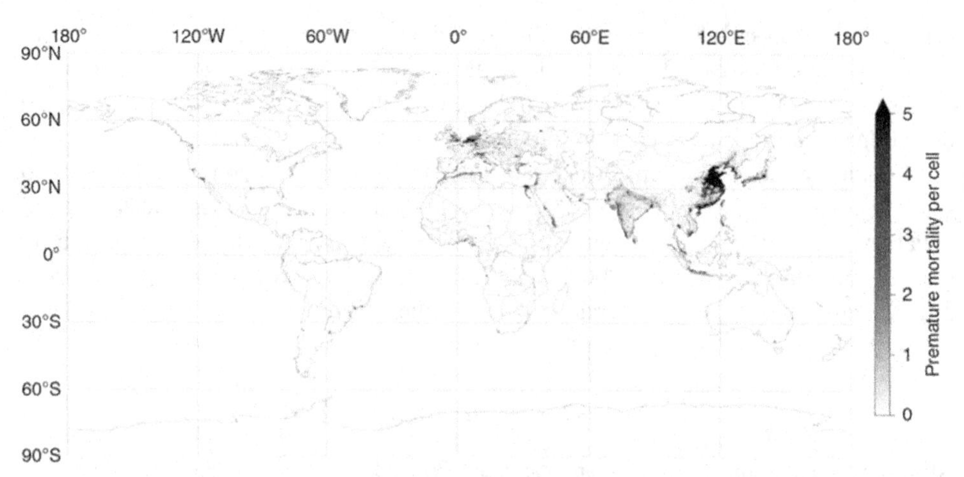

Figure 2.8 Map of combined mortality (from cardiovascular disease and lung cancer) due to $PM_{2.5}$ emissions from ships under a business-as-usual case for 2020. The in-grid-cell minimum and maximum mortality estimates are 0 and 2550, respectively.
Source: Taken from Sofiev *et al.* (2018), their figure 3

emissions scenario, Sofiev *et al.* (2018) estimate that in 2020 there will be approximately 400,000 premature deaths from lung cancer and cardiovascular disease (Figure 2.8) and approximately 14 million childhood asthma cases. The introduction of emission controls would reduce these impacts to approximately 250,000 premature deaths and 6.4 million cases of asthma.

2.6.3 Ocean Ecosystem Impacts

2.6.3.1 Contribution of Ship Emissions to Ocean Acidification

Surface ocean acidity (pH) is buffered by the dissolved inorganic carbon (DIC) pool (carbonic acid, H_2CO_3, and the bicarbonate, HCO_3^-, and carbonate, CO_3^{2-}, ions). Increased CO_2 levels in the atmosphere are gradually altering the carbonate system. Dissolved CO_2 is a weak acid, which means that increased concentrations increase the equilibrium concentrations of H^+ according to Eq. (2.8):

$$CO_{2(aq)} + H_2O_{(l)} \leftrightarrow H_2CO_{3(aq)} \leftrightarrow H^+_{(aq)} + HCO^-_{3(aq)} \leftrightarrow H^+_{(aq)} + CO^-_{3(aq)} \quad (2.8)$$

The term given to this process is ‘ocean acidification’ because increased CO_2 results in more free protons (H^+ ions) in seawater. Ocean acidification is predicted to be a major change (0.3–0.4 pH units over the twenty-first century), severely impacting organisms across the globe (Orr *et al.*, 2005). This is likely to have a long-lasting and catastrophic effect on marine ecosystems, as they depend on the biodiversity of organisms with calcium carbonate skeletons, shells and corals.

Shipping-derived aerosols and gases contain a lot of acidic carbon, nitrogen and sulfur species. The deposition of acidic species from combustion sources has the potential to change seawater acidity. Initial modelling work suggested that absorbed acidic gases from ship emissions are almost completely buffered by the seawater DIC pool (Doney *et al.*, 2007; Hunter *et al.*, 2011). However, a more recent study using a higher-resolution model suggests that the impact of ship emissions on seawater pH may be significant in certain regions (Hassellov *et al.*, 2013). The authors showed that ship emissions in areas with high ship traffic may cause regional pH reductions of the same order of magnitude as CO_2-driven ocean acidification.

2.6.3.2 Impacts of Ship Emissions on Plankton Growth and Atmospheric Carbon Removal

The ocean biological carbon pump is a biogeochemically important pathway in the ocean (Emerson *et al.*, 1997). The removal of organic matter from the sea surface is an important CO_2 removal pathway. The oceans have taken up approximately 25 per cent of the atmospheric CO_2 that has been produced due to human activities, limiting its greenhouse effects (Bakker *et al.*, 2016). Primary production of microscopic, photosynthetic plankton in the sea surface is defined at the uptake of CO_2 in the presence of sunlight to create new organic matter. Plankton growth also requires the harvesting of dissolved nutrients from the sea surface. The organic material enters the food chain and a certain percentage of it is transported to the ocean interior via settling detritus and faecal material. The efficiency of this 'carbon pump' is influenced by the type of plankton that is growing and sinking – larger plankton sink faster and are more likely to reach the ocean floor. Any material leaving the surface ocean is thus removing carbon from the surface ocean and acting as an overall sink for CO_2.

The deposition of trace metals in aerosols is important for marine biogeochemistry. Plankton require the major nutrients (nitrate, phosphate and silicate) to grow, but they also need trace elements such as iron (Fe), zinc (Zn) and nickel (Ni). Iron is a particularly interesting trace element because its availability restricts plankton growth in a lot of the world's oceans (Jickells *et al.*, 2005). Various degrees of co-limitation with macronutrients have also been observed, as well as the limitation of tiny cyanobacteria by cobalt (Dixon, 2008). Any restriction on plankton growth has the potential to influence the biological carbon pump. Although it is ubiquitous in the Earth's crust, iron is insoluble in seawater and is quickly removed via aggregation and precipitation. This results in very low iron concentrations (<50 ng/L) in the water column. Regular and consistent supplies of iron from external sources are required to sustain the growth of phytoplankton to support marine ecosystems. For example, Saharan dust aerosols are enriched in Fe, and the positive effects of their deposition upon plankton growth and health are well documented (e.g., Moore *et al.*, 2009).

One potential positive response to shipping emissions is that the deposition of ship-emitted aerosols is a transport mechanism for soluble forms of essential trace metals (e.g., Fe) to the surface ocean. Trace metals associated with anthropogenic emissions are often more soluble and bioavailable than natural dust aerosols. A global model study has suggested that combustion-derived Fe could contribute half of the soluble Fe supply to Northern Hemisphere oceans (Ito, 2013). Experiments with ambient seawater plankton assemblages have observed a response to the addition of aerosols containing elevated trace metal concentrations (Mackey *et al.*, 2012). The authors show that open ocean and coastal communities responded differently to the aerosol additions, and that certain plankton populations responded more than others. Nickel is required by plankton and has been shown to limit nitrogen uptake and thus growth in certain regions of the ocean (e.g., Dupont *et al.*, 2010). Similar relationships have been found for other metals such as zinc, manganese and cobalt (Mahowald *et al.*, 2018). A potential impact of aerosols derived from shipping emissions deposition could be to alleviate the limitation of trace metals in specific parts of the ocean. There have been no measurements to date that have looked at the impact of ship-derived aerosols on the marine plankton community.

2.6.3.3 *Impact of Ship-Emitted Toxic Heavy Metals and Pollutants*

Shipping and other combustion aerosols also include a suite of more toxic elements that can have a negative effect on phytoplankton and non-photosynthetic bacteria. The most notable elements are cadmium (Cd), lead (Pb), antimony (Sb), mercury (Hg), arsenic (As), tin (Sn), silver (Ag) and copper (Cu). These elements are toxic mainly because they are able to enter cells and replace other essential metals in vital reaction sites in enzymes and cell functional groups. Inoculation of microbial and phytoplankton cultures with low doses of toxic metals restricts growth and causes morbidity at very low levels ($<$0.05 mg/L; Wilson & Freeberg, 1980). In addition to direct toxicity, there are also antagonistic effects from imbalances in metals such as cadmium and manganese in seawater (Sunda & Huntsman, 1998).

Changes in the oxidation state (by losing or gaining electrons) of a metal can impact toxicity. For example, chromium(VI) is highly toxic compared to chromium(III), which acts as a micronutrient (Marieschi *et al.*, 2015). Metals are also often bound to organic molecules, to more common ions (e.g., chloride) or to smaller molecules (e.g., water), and their toxicity can be influenced by the chemical speciation. For example, elemental mercury (Hg^0) can be oxidized (to Hg^{2+}), which is bioavailable and can be converted to the highly toxic monomethylmercury (Soerensen *et al.*, 2010).

Most atmospheric trace metals are found in water droplets or in the particulate phase. The exceptions are arsenic, selenium and mercury, which also exist in the gas phase. Metal solubility in marine aerosols is highly variable and depends on chemical speciation. Water-soluble trace metals have been observed in ship aerosol emissions, containing a lot of variability (often over an order of magnitude) in metal concentrations (Kuang *et al.*, 2017). The bulk concentrations of airborne toxic trace metals are mostly focused around industrialized nations in the Northern Hemisphere. However, fine submicron aerosols are transported around the globe and can deposit thousands of miles away in pristine environments. Atmospheric deposition of toxic heavy metals from combustion sources is a particular problem in polar regions. The transport cycle of heavy metals such as mercury, cadmium and lead results in high concentrations in polar environments. Antarctic phytoplankton find µg/L levels of these metals highly toxic (Echeveste *et al.*, 2014). Copper deposition from natural and anthropogenic sources may also affect phytoplankton (Paytan *et al.*, 2009). These authors used models of atmospheric deposition and toxicity effects to show that phytoplankton in large areas of the coastal and open ocean may be impacted by aerosol

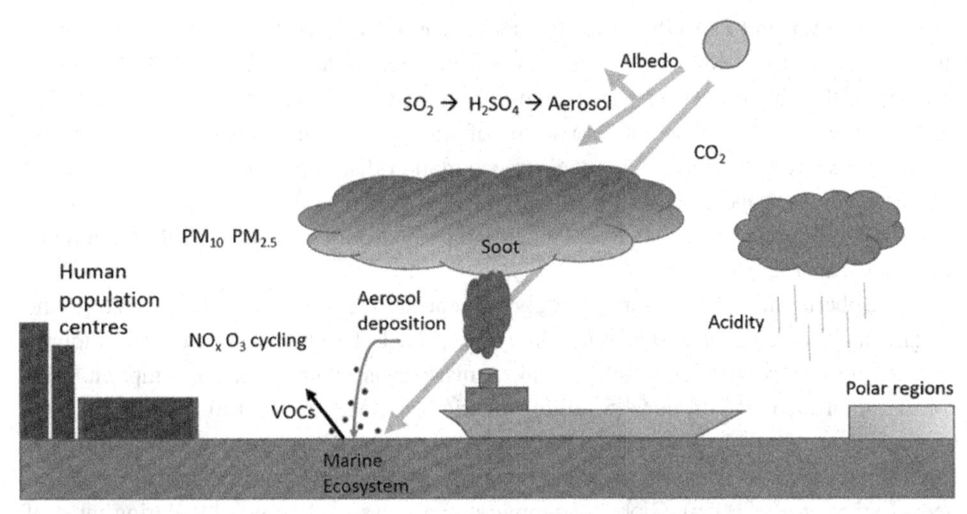

Figure 2.9 Schematic summarizing gas and aerosol ship emissions to the atmosphere and their impacts on marine and coastal environments.

copper deposition. The conclusions of this study have been corroborated by more recent observations (Jordi *et al.*, 2012; Stuart *et al.*, 2009).

Ship-emitted pollutants such as heavy metals may have a large impact on the ocean, even though the industry is a relatively minor source compared to other land-based emissions. The major impact from ship-emitted aerosols will be concentrated in regions that experience a lot of ship traffic (Figure 2.9). This local effect is likely to be even more focused where open-loop scrubbers are used by ships to conform to sulfur emission controls. Open-loop scrubber systems could influence atmospheric aerosol processing because they remove acidic sulfur species from the gas phase and thus potentially change trace metal solubility (Baker & Jickells, 2017). However, these scrubber systems are also likely to enhance and localize the input of trace metals and other organic pollutants into seawater in ports and the major shipping lanes. Very few studies have looked at the composition of scrubber water. Measurements by Turner *et al.* (2017) show trace metal concentrations that are sufficiently elevated to cause concern. Vanadium was ~50 times higher than in seawater. Copper and zinc levels were also elevated, but it is unclear whether they originated from exhaust emissions or from metal fittings within the scrubber system (Turner *et al.*, 2017). We are aware of no research that examines the impacts of ship-emitted organic pollutants and soot on the marine ecosystem.

2.7 Shipping Emissions in a Changing World

2.7.1 Changes in Shipping Traffic

In approximately the last 100 years, the total number of motor ships in the global fleet is estimated to have increased from 72,000 to 88,000 (Endresen *et al.*, 2007). However, a typical cargo ship is now larger, meaning that more goods can be transported. Estimates of global fleet transport capacity typically use an internal volume that adjusts for the non-linear relationship between ship size and capacity (gross tonnage, unitless). The gross tonnage of the global fleet increased from 22.4 to 558.0 million between 1900 and 2000 (Endresen *et al.*, 2007). There are a number of ways to estimate global shipping activity, including the weight of goods and their transportation distance. A recent review showed that international seaborne trade has approximately tripled over the last 20 years – in excess of 10,000 million tons of goods were loaded in 2016, leading to >50,000 billion ton miles in global seaborne trade (UNCTAD, 2017).

Atmospheric ship emissions are strongly influenced by the number and size of ships, the weight of goods and the distance they are transported. The global measure of shipping activity most relevant to atmospheric ship emissions is annual fuel consumption. Fuel consumption increased from 64.5 million metric tons (Mt) in 1950 to 280 Mt in 2001 (Eyring *et al.*, 2005b). Ship traffic demands are governed by global economic growth resulting in increased seaborne trade and more (number and size/power) ships in the global fleet (Eyring *et al.*, 2005a). Global economic growth was relatively stable during most of the 2000s (3.2 per cent/year during 2001–2008), but was lower and more variable in recent years, fluctuating between 2.2 and 2.6 per cent during 2015–2017 (UNCTAD, 2017). Seaborne trade is not divided equally between different sizes of economies or geographic regions. World seaborne trade activity is mainly in the developing economies (~60 per cent) and developed economies (~35 per cent), with only a small contribution from transition economies (~5 per cent). This leads to a strong skew in the regional distribution of trade towards Asia (~50 per cent), with the Americas and Europe also making substantial (~20 per cent each) contributions (UNCTAD, 2017). Future seaborne trade intensity and distribution will vary not just as a function of global economic growth, but also as a result of the balance in different countries' economic development.

Continued growth of the global population and global economy will drive an increase in shipping activity, and this will have profound effects on ship emissions. Ship-emitted particles are projected to substantially alter future atmospheric composition (sulfate and nitrate aerosols), as well as aerosol–cloud impacts on the Earth's radiation budget (Righi *et al.*, 2015). The relative distribution of these impacts depends upon future trade and the strength of economic development in different regions of the world. For example, Jonson *et al.* (2015) use a regional model to predict changes in air pollution in the Baltic Sea and North Sea. These authors predict that NO_x pollution from shipping will substantially increase by 2030 unless additional nitrogen emission controls are introduced (see Section 2.7.2).

Technological developments have enabled shipping activity to be measured more accurately. The Automatic Identification System (AIS; www.marinetraffic.com) is an automatic vessel position reporting system that is installed on all vessels that exceed 300 tonnes. AIS data have been used within a Ship Traffic Emission Assessment Model (STEAM) to estimate ship emissions with unparalleled resolution and geographical distribution (Jalkanen *et al.*, 2016; see Figure 2.10). Shipping activity has also been measured using altimeters on satellites, and the data agree well with AIS density maps (Tournadre, 2014). Elevated atmospheric NO_2 levels driven by intense emissions in major shipping lanes have also been observed from space (de Wildt *et al.*, 2012; Richter *et al.*, 2004).

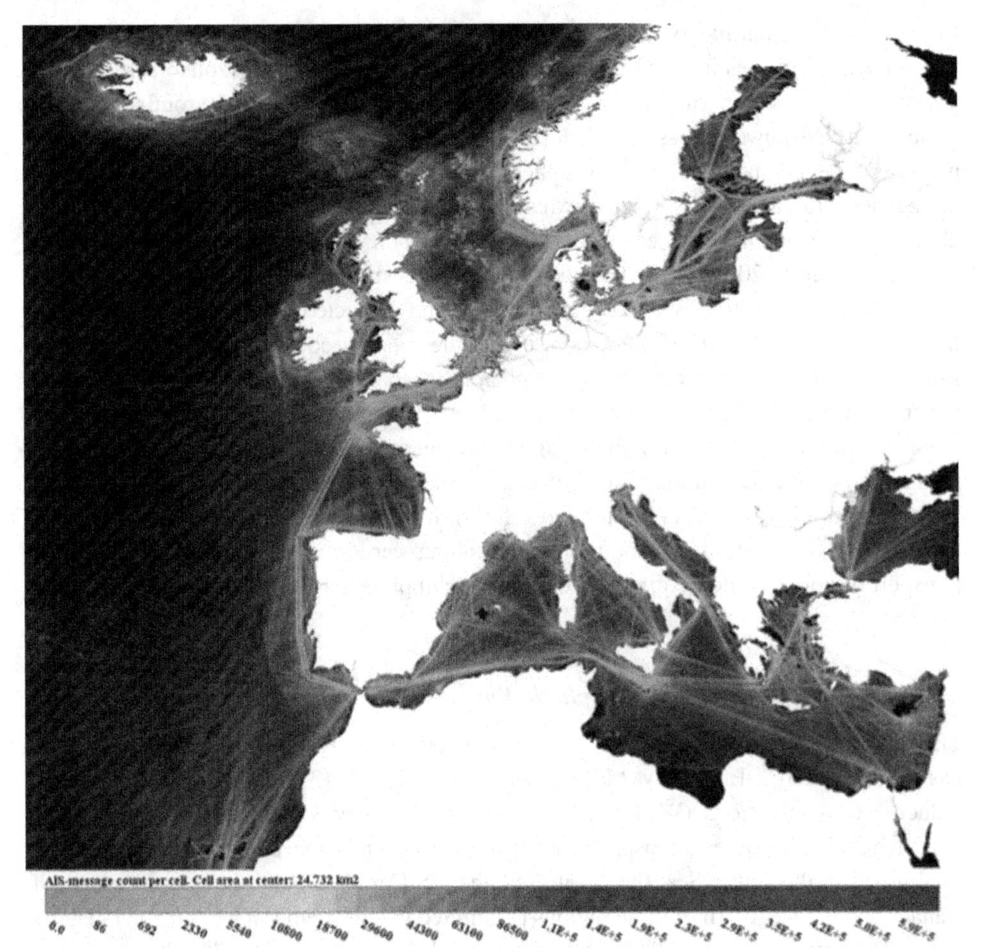

Figure 2.10 The geographical coverage of the terrestrial Automatic Identification System (AIS) network in Europe. The shade scale illustrates the number of position reports per unit area received in the European Union sea areas in 2011. Note the contrast between the intense activity in the shipping lanes close to the coast compared to the open ocean.
Source: Taken from Jalkanen *et al.* (2016), their figure 1

AIS and satellite tools offer the ability to monitor how the shipping industry responds to changes in socio-economics. For example, trends in atmospheric NO_2 detected by satellite over major shipping lanes in the Mediterranean Sea, Red Sea, Indian Ocean and South Chinese Sea correspond to variations in international trade volumes (de Wildt *et al.*, 2012).

The shipping industry may also respond to current and future changes in environmental conditions. For example, climate change projections suggest that Arctic sea ice will continue to retreat to the extent that ships will be able to routinely transit through the Northwest Passage (Melia *et al.*, 2016). The Northwest Passage is a potential route between the Pacific Ocean and Atlantic Ocean through the Canadian Arctic Archipelago. This passage would cut thousands of miles from the journeys of ships travelling between

Europe and Asia, but has typically been impassable. Declining sea ice extent has already been linked to changes in shipping activity in the Canadian Arctic (Pizzolato *et al.*, 2016), and the industry is preparing for low enough ice cover to be able to make routine journeys through the Northwest Passage. Adjustments to shipping routes into areas that have previously been unaffected by atmospheric emissions from ships will have important implications for air quality (e.g., Eckhardt *et al.*, 2013). Winther *et al.* (2014) model 2012 shipping activity and estimate that black carbon deposition around Iceland was enhanced by up to 20 per cent. The authors also suggest that by 2050 significantly higher black carbon deposition will be found along the expected Arctic diversion routes. In contrast, Browse *et al.* (2013) suggest that the relative contribution of black carbon from ships is small compared to the black carbon transported from non-shipping sources at lower latitudes.

A changing climate is also influencing the incidence of extreme weather. For example, rising sea surface temperatures are driving an increase in the frequency, duration and intensity of tropical storms (Webster *et al.*, 2005). If these storms become frequent enough to influence the cost of transporting goods along certain shipping routes, the future transport of goods could be diverted to different shipping routes or even over land.

2.7.2 Regulation of Atmospheric Emissions and Industrial Adaptation

The shipping industry is constantly evolving to be more energy and cost efficient. The IMO has had an Energy Efficiency Management Plan since 2013, with the explicit aim of reducing CO_2 emissions (Wan *et al.*, 2016). The industry is also adapting to specific IMO regulation of other atmospheric emissions. Four emission control areas have been established in the Baltic Sea, the North Sea, the US Caribbean and the coastal waters of Canada and the USA. All of these regions are limited to sulfur emissions of 0.1 per cent by mass. The rest of the world's oceans are limited to 3.5 per cent sulfur by mass, but additional emission controls will come into force in 2020 with a global 0.5 per cent cap (Turner *et al.*, 2017). The aim of current and future sulfur regulations is to reduce the impact of ship emissions on air quality. For example, modelling by Johansson *et al.* (2013) suggests that the 2015 regulations within the European SECA would lead to large reductions in sulfur and PM emissions (87 per cent and 48 per cent, respectively).

Multiple studies have looked at the level of industrial compliance and the effects upon atmospheric composition since the regional SECAs came into force in 2015. Ship plume observations have been made using observing platforms at sea, on land and in the air. Regulatory compliance has been good and has coincided with dramatic reductions in sulfur (SO_2 and sulfate) and PM (Beecken *et al.*, 2014, 2015; Kattner *et al.*, 2015; Lack *et al.*, 2011; Seyler *et al.*, 2017; Yang *et al.*, 2016). The impact of the 2020 global regulation of sulfur emissions is more difficult to predict. This regulation will affect regions of the world such as Asia, which conducts nearly half of global seaborne trade. Nine of the ten largest container ports in the world are in Asia, and they contribute up to 20 per cent of atmospheric ship sulfur emissions (Wan *et al.*, 2016). These ports and the intense shipping traffic in this region are not yet subject to any sulfur emission controls. We do not know how easy is will be to regulate, monitor and enforce compliance in these areas of the world.

The reduction in ship sulfur emissions should lead to a proportional reduction in sulfate aerosol mass concentration, and it is on this basis that a climatic warming effect as a result of the 2020 regulation change has been predicted (e.g., Sofiev *et al.*, 2018). The uncertainty in this prediction is large because the dominant term in the climatic effect of ship emissions – the aerosol indirect effect (i.e., aerosol–cloud interactions) – is more sensitive to the number concentration and size distribution of aerosols, rather than to the total aerosol mass concentration. It is also worth noting that there are competing climate impacts from shipping that occur at different timescales (i.e., short-lived forcers (SO_2, NO_X, O_3, aerosols: hours to days) versus long-lived forcers (CH_4: 10 years, CO_2: ~100 years)). Fuglestvedt *et al.* (2009) argue that the difference in short-lived and long-lived forcer lifetimes makes the timescale extremely important when determining whether reduced sulfur emissions from ships cause net cooling or warming. A reduction in ship CO_2 emissions coincidental with a sulfur emissions reduction will likely be needed to prevent further warming of the atmosphere.

Additional uncertainty is introduced when considering the degree of compliance to the regulation. Sulfur emission control will be expensive and is a major challenge for the shipping industry. Shipping companies can choose to adapt their engines to burn fuel with a lower sulfur content or to install emission scrubbers. Both options are expensive and have ramifications for industry (see further discussion in Wang *et al.*, 2007) and for other atmospheric emissions such as PM (e.g., Alföldy *et al.*, 2013; Fridell & Salo, 2016). It is also unclear how well emission limits can be enforced. Remote monitoring techniques have been well tested in areas in close proximity to ports and major shipping lanes (e.g., Balzani Lööv *et al.*, 2014; Berg *et al.*, 2012), but would be challenging and prohibitively expensive to implement on the high seas. The impact of the 2020 regulation on atmospheric emissions is thus not easy to predict.

Switching to open-loop seawater scrubbers enables ships to comply with limits on atmospheric sulfur emissions (see Section 2.4.4), but may redirect many of the atmospheric outputs directly into the seawater (Fridell & Salo, 2016; Turner *et al.*, 2017). Wet scrubbers will influence the aerosols emitted to the atmosphere, as they are very efficient at removing soluble aerosols but less efficient at removing the insoluble fraction (Fridell & Salo, 2016). Ships burning low-sulfur fuels are likely to emit smaller aerosols (Hobbs *et al.*, 2000) that are less hygroscopic (Lack *et al.*, 2009) and less effective at forming clouds. The inevitable large-scale changes in how the industry chooses to treat its exhaust emissions to address regulation aimed at reducing air pollution (see Section 2.6.2) has potentially profound implications for the climatic impact of ship emissions (see Section 2.6.1), as well as for surface ocean ecosystems (see Section 2.6.3). To our knowledge, there is no planned regulation of the emission to air or water of organic pollutants or trace metals from open-loop systems.

There has been some regulation of atmospheric NO_x emissions along the North American coastline and in the US Caribbean. Control of ship NO_x emissions is more difficult to achieve than control of ship sulfur emissions. In the coastal waters of California, a Vessel Speed Reduction programme is in place (Lack *et al.*, 2011), as there is evidence that lower ship speeds and engine loads may reduce NO_x emissions (e.g., Agrawal *et al.*, 2008a). Satellite measurements of NO_x suggest that vessels are now travelling 30 per cent slower through the Suez Canal, and that this has contributed to a 45 per cent reduction in NO_x emissions (Boersma *et al.*, 2015). Adoption of LNG as an alternative ship fuel would also contribute to lower NO_x emissions. The shipping industry is more likely to adapt and use LNG if NO_x emission limits are introduced. There has been some discussion by Baltic Sea and North Sea countries about implementing NO_x emission limits in their regional waters, but the debate is ongoing and political agreement is ultimately required in order for regulation to be implemented (Jonson *et al.*, 2015).

2.8 Summary and Concluding Thoughts

This chapter considered the gas and particle composition of ship emissions to the atmosphere and the factors that control these emissions (Sections 2.2–2.5). Emissions depend on vessel size and activity, fuel type and engine age, type and load. Emissions are rapidly processed in the atmosphere, and the downwind products are determined by the environmental conditions and the complex reactions with other gases and particles. Section 2.6 outlined the known and potential impacts of ship emissions on atmospheric chemistry, ocean biogeochemistry, climate and human health. Section 2.7 discussed the big changes on the horizon for the shipping industry, which include greater trade volumes, changes in shipping routes such as through the Arctic passage and the regulation of atmospheric emissions. The imminent changes for the industry are due to the 2020 regulation of worldwide ship sulfur emissions, and by the time this book is published, these will likely be in force. It is difficult to predict what the impacts will be on shipping activity or the response from industry to the regulations. It is clear that the changes are likely to have profound environmental and human health implications.

Acknowledgements

We thank Tim Smyth (Plymouth Marine Laboratory) for encouraging our interest in ship emissions to the atmosphere. The enthusiasm and effort from him and Frances Hopkins (Plymouth Marine Laboratory) enabled us to set up the Penlee Point Atmospheric Observatory. We will forever be grateful to the late Roland von Glasow for encouraging us to persevere with our research on this topic. T. G. Bell and M. Yang received additional support from the UK Natural Environment Research Council (grants NE/N015932/1, NE/N018044/1 and NE/S005390/1). S. J. Ussher was supported by a Marie Curie Career Integration Grant from the European Commission (PCIG-GA-2012-333143 DISCOSAT) and ATLANToS (EU Horizon 2020 grant agreement no. 633211). This is contribution no. 7 from the Penlee Point Atmospheric Observatory.

References

Agrawal, H., Malloy, Q. G. J., Welch, W. A., Wayne Miller, J. & Cocker III, D. R. (2008a). In-use gaseous and particulate matter emissions from a modern ocean going container vessel. *Atmospheric Environment*, **42**(21), 5504–5510.

Agrawal, H., Welch, W. A., Miller, J. W. & Cocker, D. R. (2008b). Emission measurements from a crude oil tanker at sea. *Environmental Science & Technology*, **42**(19), 7098–7103.

Albrecht, B. A. (1989). Aerosols, cloud microphysics, and fractional cloudiness. *Science*, **245**(4923), 1227.

Alföldy, B., Lööv, J. B., Lagler, F. *et al.* (2013). Measurements of air pollution emission factors for marine transportation in SECA. *Atmospheric Measurement Techniques*, **6**(7), 1777–1791.

Ansari, A. S. & Pandis, S. N. (1998). Response of inorganic PM to precursor concentrations. *Environmental Science & Technology*, **32**(18), 2706–2714.

Baker, A. R. & Jickells, T. D. (2017). Atmospheric deposition of soluble trace elements along the Atlantic Meridional Transect (AMT). *Progress in Oceanography*, **158**, 41–51.

Bakker, D. C. E., Pfeil, B., Landa, C. S. *et al.* (2016). A multi-decade record of high-quality fCO_2 data in version 3 of the Surface Ocean CO_2 Atlas (SOCAT). *Earth System Science Data*, **8**(2), 383–413.

Ball, J. S., Wenger, W. J., Hyden, H. J., Horr, C. A. & Myers, A. T. (1960). Metal content of twenty-four petroleums. *Journal of Chemical & Engineering Data*, **5**(4), 553–557.

Balzani Lööv, J. M., Alfoldy, B., Gast, L. F. L. *et al.* (2014). Field test of available methods to measure remotely SO_x and NO_x emissions from ships. *Atmospheric Measurement Techniques*, **7**(8), 2597–2613.

Beecken, J., Mellqvist, J., Salo, K., Ekholm, J. & Jalkanen, J. P. (2014). Airborne emission measurements of SO_2, NO_x and particles from individual ships using a sniffer technique. *Atmospheric Measurement Techniques*, **7**(7), 1957–1968.

Beecken, J., Mellqvist, J., Salo, K., *et al.* (2015). Emission factors of SO_2, NO_x and particles from ships in Neva Bay from ground-based and helicopter-borne measurements and AIS-based modeling. *Atmospheric Chemistry and Physics*, **15**(9), 5229–5241.

Berg, N., Mellqvist, J., Jalkanen, J. P. & Balzani, J. (2012). Ship emissions of SO_2 and NO_2: DOAS measurements from airborne platforms. *Atmospheric Measurement Techniques*, **5**(5), 1085–1098.

Betha, R., Russell, L. M., Sanchez, K. J. *et al.* (2017). Lower NO_x but higher particle and black carbon emissions from renewable diesel compared to ultra low sulfur diesel in at-sea operations of a research vessel. *Aerosol Science and Technology*, **51**(2), 123–134.

Boersma, K. F., Vinken, G. C. M. & Tournadre, J. (2015). Ships going slow in reducing their NO_X emissions: changes in 2005–2012 ship exhaust inferred from satellite measurements over Europe. *Environmental Research Letters*, **10**(7), 074007.

Bond, T. C. & Bergstrom, R. W. (2006). Light absorption by carbonaceous particles: an investigative review. *Aerosol Science and Technology*, **40**(1), 27–67.

Browse, J., Carslaw, K. S., Schmidt, A. & Corbett, J. J. (2013). Impact of future Arctic shipping on high-latitude black carbon deposition. *Geophysical Research Letters*, **40**(16), 4459–4463.

Capaldo, K., Corbett, J. J., Kasibhatla, P., Fischbeck, P. & Pandis, S. N. (1999). Effects of ship emissions on sulphur cycling and radiative climate forcing over the ocean. *Nature*, **400**, 743.

Cape, J. N., Coyle, M. & Dumitrean, P. (2012). The atmospheric lifetime of black carbon. *Atmospheric Environment*, **59**, 256–263.

Celo, V., Dabek-Zlotorzynska, E. & McCurdy, M. (2015). Chemical characterization of exhaust emissions from selected Canadian marine vessels: the case of trace metals and lanthanoids. *Environmental Science & Technology*, **49**(8), 5220–5226.

Charlton-Perez, C. L., Evans, M. J., Marsham, J. H. & Esler, J. G. (2009). The impact of resolution on ship plume simulations with NO_x chemistry. *Atmospheric Chemistry and Physics*, **9**(19), 7505–7518.

Chen, G., Huey, L. G., Trainer, M. *et al.* (2005). An investigation of the chemistry of ship emission plumes during ITCT 2002. *Journal of Geophysical Research: Atmospheres*, **110**(D10), D10S90.

Chosson, F., Paoli, R. & Cuenot, B. (2008). Ship plume dispersion rates in convective boundary layers for chemistry models. *Atmospheric Chemistry and Physics*, **8**(16), 4841–4853.

Christensen, M. W. & Stephens, G. L. (2011). Microphysical and macrophysical responses of marine stratocumulus polluted by underlying ships: evidence of cloud deepening. *Journal of Geophysical Research: Atmospheres*, **116**(D3), n.p.

Ciais, P., Sabine, C., Bala, G. *et al.* (2013). Carbon and other biogeochemical cycles. In T. F. Stocker, D. Qin, G.-K. Plattner, eds., *Climate Change 2013: The Physical Science Basis. Contribution of Working Group I to the Fifth Assessment Report of the Intergovernmental Panel on Climate Change*. Cambridge: Cambridge University Press.

Coakley, J. A., Bernstein, R. L. & Durkee, P. A. (1987). Effect of ship-stack effluents on cloud reflectivity. *Science*, **237**(4818), 1020.

Cooper, D. A. (2001). Exhaust emissions from high speed passenger ferries. *Atmospheric Environment*, **35**(24), 4189–4200.

Cooper, D. A. (2003). Exhaust emissions from ships at berth. *Atmospheric Environment*, **37**(27), 3817–3830.

Cooper, D. A., Peterson, K. & Simpson, D. (1996). Hydrocarbon, PAH and PCB emissions from ferries: a case study in the Skagerak–Kattegatt–Öresund region. *Atmospheric Environment*, **30**(14), 2463–2473.

Corbett, J. J. & Koehler, H. W. (2003). Updated emissions from ocean shipping. *Journal of Geophysical Research: Atmospheres*, **108**(D20), 4650.

Corbett, J. J., Winebrake, J. J., Green, E. H. *et al.* (2007). Mortality from ship emissions: a global assessment. *Environmental Science & Technology*, **41**(24), 8512–8518.

Dalsoren, S. B., Eide, M. S., Endresen, O. *et al.* (2009). Update on emissions and environmental impacts from the international fleet of ships: the contribution from major ship types and ports. *Atmospheric Chemistry and Physics*, **9**(6), 2171–2194.

Davis, D. D., Grodzinsky, G., Kasibhatla, P. *et al.* (2001). Impact of ship emissions on marine boundary layer NO_x and SO_2 distributions over the Pacific Basin. *Geophysical Research Letters*, **28**(2), 235–238.

de Wildt, M. D., Eskes, H. & Boersma, K. F. (2012). The global economic cycle and satellite-derived NO_2 trends over shipping lanes. *Geophysical Research Letters*, **39**, 1802.

Devasthale, A., Krüger, O. & Graßl, H. (2006). Impact of ship emissions on cloud properties over coastal areas. *Geophysical Research Letters*, **33**(2), n.p.

Dixon, J. L. (2008). Macro and micro nutrient limitation of microbial productivity in oligotrophic subtropical Atlantic waters. *Environmental Chemistry*, **5**(2), 135–142.

Doney, S. C., Mahowald, N., Lima, I. *et al.* (2007). Impact of anthropogenic atmospheric nitrogen and sulfur deposition on ocean acidification and the inorganic carbon system. *Proceedings of the National Academy of Sciences of the United States of America*, **104**(37), 14580–14585.

Dupont, C. L., Buck, K. N., Palenik, B. & Barbeau, K. (2010). Nickel utilization in phytoplankton assemblages from contrasting oceanic regimes. *Deep Sea Research Part I: Oceanographic Research Papers*, **57**(4), 553–566.

Durkee, P. A., Chartier, R. E., Brown, A. *et al.* (2000). Composite ship track characteristics. *Journal of the Atmospheric Sciences*, **57**(16), 2542–2553.

Echeveste, P., Tovar-Sánchez, A. & Agustí, S. (2014). Tolerance of polar phytoplankton communities to metals. *Environmental Pollution*, **185**, 188–195.

Eckhardt, S., Hermansen, O., Grythe, H. *et al.* (2013). The influence of cruise ship emissions on air pollution in Svalbard – a harbinger of a more polluted Arctic? *Atmospheric Chemistry and Physics*, **13**(16), 8401–8409.

Emerson, S., Quay, P., Karl, D. *et al.* (1997). Experimental determination of the organic carbon flux from open-ocean surface waters. *Nature*, **389**(6654), 951–954.

Endresen, Ø., Sørgård, E., Behrens, H. L., Brett, P. O. & Isaksen, I. S. A. (2007). A historical reconstruction of ships' fuel consumption and emissions. *Journal of Geophysical Research: Atmospheres*, **112**(D12), n.p.

Endresen, Ø., Sørgård, E., Sundet, J. K. *et al.* (2003). Emission from international sea transportation and environmental impact. *Journal of Geophysical Research: Atmospheres*, **108**(D17), 4560.

EPA (2000). *Analysis of Commercial Marine Vessels Emissions and Fuel Consumption Data*. Washington, DC: US Environmental Protection Agency.

Eyring, V., Isaksen, I. S. A., Berntsen, T. *et al.* (2010). Transport impacts on atmosphere and climate: shipping. *Atmospheric Environment*, **44**(37), 4735–4771.

Eyring, V., Kohler, H. W., Lauer, A. & Lemper, B. (2005a). Emissions from international shipping: 2. Impact of future technologies on scenarios until 2050. *Journal of Geophysical Research: Atmospheres*, **110**(D17), n.p.

Eyring, V., Kohler, H. W., van Aardenne, J. & Lauer, A. (2005b). Emissions from international shipping: 1. The last 50 years. *Journal of Geophysical Research: Atmospheres*, **110**(D17), n.p.

Ferek, R. J., Garrett, T., Hobbs, P. V. *et al.* (2000). Drizzle suppression in ship tracks. *Journal of the Atmospheric Sciences*, **57**(16), 2707–2728.

Fish, R. H., Komlenic, J. J. & Wines, B. K. (1984). Characterization and comparison of vanadyl and nickel compounds in heavy crude petroleums and asphaltenes by reverse-phase and size-exclusion liquid chromatography/graphite furnace atomic absorption spectrometry. *Analytical Chemistry*, **56**(13), 2452–2460.

Frick, G. M. & Hoppel, W. A. (2000). Airship measurements of ship's exhaust plumes and their effect on marine boundary layer clouds. *Journal of the Atmospheric Sciences*, **57**(16), 2625–2648.

Fridell, E. & Salo, K. (2016). Measurements of abatement of particles and exhaust gases in a marine gas scrubber. *Proceedings of the Institution of Mechanical Engineers, Part M: Journal of Engineering for the Maritime Environment*, **230**(1), 154–162.

Fuglestvedt, J., Berntsen, T., Eyring, V. *et al.* (2009). Shipping emissions: from cooling to warming of climate – and reducing impacts on health. *Environmental Science & Technology*, **43**(24), 9057–9062.

Graham, W. F. & Duce, R. A. (1979). Atmospheric pathways of the phosphorus cycle. *Geochimica et Cosmochimica Acta*, **43**(8), 1195–1208.

Hassellov, I. M., Turner, D. R., Lauer, A. & Corbett, J. J. (2013). Shipping contributes to ocean acidification. *Geophysical Research Letters*, **40**(11), 2731–2736.

Hobbs, P. V., Garrett, T. J., Ferek, R. J. *et al.* (2000). Emissions from ships with respect to their effects on clouds. *Journal of the Atmospheric Sciences*, **57**(16), 2570–2590.

Holmes, C. D., Prather, M. J. & Vinken, G. C. M. (2014). The climate impact of ship NO_x emissions: an improved estimate accounting for plume chemistry. *Atmospheric Chemistry and Physics*, **14**(13), 6801–6812.

Hunter, K. A., Liss, P. S., Surapipith, V. *et al.* (2011). Impacts of anthropogenic SO_x, NO_x and NH_3 on acidification of coastal waters and shipping lanes. *Geophysical Research Letters*, **38**, n.p.

Ito, A. (2013). Global modeling study of potentially bioavailable iron input from shipboard aerosol sources to the ocean. *Global Biogeochemical Cycles*, **27**(1), 1–10.

Jalkanen, J. P., Johansson, L. & Kukkonen, J. (2016). A comprehensive inventory of ship traffic exhaust emissions in the European sea areas in 2011. *Atmospheric Chemistry and Physics*, **16**(1), 71–84.

Jickells, T. D., An, Z. S., Andersen, K. K. *et al.* (2005). Global iron connections between desert dust, ocean biogeochemistry, and climate. *Science*, **308**(5718), 67–71.

Johansson, L., Jalkanen, J. P., Kalli, J. & Kukkonen, J. (2013). The evolution of shipping emissions and the costs of regulation changes in the northern EU area. *Atmospheric Chemistry and Physics*, **13**(22), 11375–11389.

Johansson, L., Jalkanen, J.-P. & Kukkonen, J. (2017). Global assessment of shipping emissions in 2015 on a high spatial and temporal resolution. *Atmospheric Environment*, **167**(Suppl. C), 403–415.

Jonson, J. E., Jalkanen, J. P., Johansson, L., Gauss, M. & Denier van der Gon, H. A. C. (2015). Model calculations of the effects of present and future emissions of air pollutants from shipping in the Baltic Sea and the North Sea. *Atmospheric Chemistry and Physics*, **15**(2), 783–798.

Jordi, A., Basterretxea, G., Tovar-Sánchez, A., Alastuey, A. & Querol, X. (2012). Copper aerosols inhibit phytoplankton growth in the Mediterranean Sea. *Proceedings of the National Academy of Sciences of the United States of America*, **109**(52), 21246–21249.

Kasibhatla, P., Levy, H., Moxim, W. J. *et al.* (2000). Do emissions from ships have a significant impact on concentrations of nitrogen oxides in the marine boundary layer? *Geophysical Research Letters*, **27**(15), 2229–2232.

Kasper, A., Aufdenblatten, S., Forss, A., Mohr, M. & Burtscher, H. (2007). Particulate emissions from a low-speed marine diesel engine. *Aerosol Science and Technology*, **41**(1), 24–32.

Kattner, L., Mathieu-Üffing, B., Burrows, J. P. *et al.* (2015). Monitoring compliance with sulfur content regulations of shipping fuel by *in situ* measurements of ship emissions. *Atmospheric Chemistry and Physics*, **15**(17), 10087–10092.

Kivekäs, N., Massling, A., Grythe, H. *et al.* (2014). Contribution of ship traffic to aerosol particle concentrations downwind of a major shipping lane. *Atmospheric Chemistry and Physics*, **14**(16), 8255–8267.

Kuang, X. M., Scott, J. A., da Rocha, G. O. *et al.* (2017). Hydroxyl radical formation and soluble trace metal content in particulate matter from renewable diesel and ultra low sulfur diesel in at-sea operations of a research vessel. *Aerosol Science and Technology*, **51**(2), 147–158.

Kulkarni, P., Chellam, S. & Fraser, M. P. (2007). Tracking petroleum refinery emission events using lanthanum and lanthanides as elemental markers for $PM_{2.5}$. *Environmental Science & Technology*, **41**(19), 6748–6754.

Lack, D., Lerner, B., Granier, C. *et al.* (2008). Light absorbing carbon emissions from commercial shipping. *Geophysical Research Letters*, **35**(13), n.p.

Lack, D. A. & Corbett, J. J. (2012). Black carbon from ships: a review of the effects of ship speed, fuel quality and exhaust gas scrubbing. *Atmospheric Chemistry and Physics*, **12**(9), 3985–4000.

Lack, D. A., Corbett, J. J., Onasch, T. *et al.* (2009). Particulate emissions from commercial shipping: chemical, physical, and optical properties. *Journal of Geophysical Research: Atmospheres*, **114**(D7), n.p.

Lack, D. A., Cappa, C. D., Langridge, J., *et al.* (2011). Impact of fuel quality regulation and speed reductions on shipping emissions: implications for climate and air quality. *Environmental Science & Technology*, **45**(20), 9052–9060.

Lauer, A., Eyring, V., Hendricks, J., Jöckel, P. & Lohmann, U. (2007). Global model simulations of the impact of ocean-going ships on aerosols, clouds, and the radiation budget. *Atmospheric Chemistry and Physics*, **7**(19), 5061–5079.

Law, C. S., Breviere, E., de Leeuw, G. *et al.* (2013). Evolving research directions in Surface Ocean–Lower Atmosphere (SOLAS) science. *Environmental Chemistry*, **10**(1), 1–16.

Lawrence, M. G. & Crutzen, P. J. (1999). Influence of NO_x emissions from ships on tropospheric photochemistry and climate. *Nature*, **402**(6758), 167–170.

Lieke, K. I., Rosenørn, T., Pedersen, J. *et al.* (2013). Micro- and nanostructural characteristics of particles before and after an exhaust gas recirculation system scrubber. *Aerosol Science and Technology*, **47**(9), 1038–1046.

Liss, P. S. (1973). Processes of gas exchange across an air-water interface. *Deep-Sea Research*, **20**(3), 221–238.

Liss, P. S. & Slater, P. G. (1974). Flux of gases across the air–sea interface. *Nature*, **247**(5438), 181–184.

Liu, H., Fu, M., Jin, X. *et al.* (2016). Health and climate impacts of ocean-going vessels in East Asia. *Nature Climate Change*, **6**(11), 1037–1041.

Lo Mónaco, S., López, L., Rojas, H. *et al.* (2002). Distribution of major and trace elements in La Luna Formation, Southwestern Venezuelan Basin. *Organic Geochemistry*, **33** (12), 1593–1608.

Lyyränen, J., Jokiniemi, J., Kauppinen, E. I. & Joutsensaari, J. (1999). Aerosol characterisation in medium-speed diesel engines operating with heavy fuel oils. *Journal of Aerosol Science*, **30**(6), 771–784.

Mackey, K. R. M., Buck, K. N., Casey, J. R. *et al.* (2012). Phytoplankton responses to atmospheric metal deposition in the coastal and open-ocean Sargasso Sea. *Frontiers in Microbiology*, **3**, 359.

Mahowald, N. M., Hamilton, D. S., Mackey, K. R. M. *et al.* (2018). Aerosol trace metal leaching and impacts on marine microorganisms. *Nature Communications*, **9**(1), 2614.

Marbach, T., Beirle, S., Platt, U. *et al.* (2009). Satellite measurements of formaldehyde linked to shipping emissions. *Atmospheric Chemistry and Physics*, **9**(21), 8223–8234.

Marieschi, M., Gorbi, G., Zanni, C., Sardella, A. & Torelli, A. (2015). Increase of chromium tolerance in *Scenedesmus acutus* after sulfur starvation: chromium uptake and compartmentalization in two strains with different sensitivities to Cr(VI). *Aquatic Toxicology*, **167**, 124–133.

Matthias, V., Bewersdorff, I., Aulinger, A. & Quante, M. (2010). The contribution of ship emissions to air pollution in the North Sea regions. *Environmental Pollution*, **158**(6), 2241–2250.

Melia, N., Haines, K. & Hawkins, E. (2016). Sea ice decline and 21st century trans-Arctic shipping routes. *Geophysical Research Letters*, **43**(18), 9720–9728.

Moore, C. M., Mills, M. M., Achterberg, E. P. *et al.* (2009). Large-scale distribution of Atlantic nitrogen fixation controlled by iron availability. *Nature Geoscience*, **2**(12), 867–871.

Moreno, T., Querol, X., Alastuey, A. & Gibbons, W. (2008). Identification of FCC refinery atmospheric pollution events using lanthanoid- and vanadium-bearing aerosols. *Atmospheric Environment*, **42**(34), 7851–7861.

Murphy, S. M., Agrawal, H., Sorooshian, A. *et al.* (2009). Comprehensive simultaneous shipboard and airborne characterization of exhaust from a modern container ship at sea. *Environmental Science & Technology*, **43**(13), 4626–4640.

Orr, J. C., Fabry, V. J., Aumont, O. *et al.* (2005). Anthropogenic ocean acidification over the twenty-first century and its impact on calcifying organisms. *Nature*, **437**(7059), 681–686.

Paxian, A., Eyring, V., Beer, W., Sausen, R. & Wright, C. (2010). Present-day and future global bottom-up ship emission inventories including polar routes. *Environmental Science & Technology*, **44**(4), 1333–1339.

Paytan, A., Mackey, K. R., Chen, Y. *et al.* (2009). Toxicity of atmospheric aerosols on marine phytoplankton. *Proceedings of the National Academy of Sciences of the United States of America*, **106**(12), 4601–4605.

Peters, K., Quaas, J. & Graßl, H. (2011). A search for large-scale effects of ship emissions on clouds and radiation in satellite data. *Journal of Geophysical Research: Atmospheres*, **116**(D24), n.p.

Petzold, A., Hasselbach, J., Lauer, P. *et al.* (2008). Experimental studies on particle emissions from cruising ship, their characteristic properties, transformation and atmospheric lifetime in the marine boundary layer. *Atmospheric Chemistry and Physics*, **8**(9), 2387–2403.

Pizzolato, L., Howell, S. E. L., Dawson, J., Laliberté, F. & Copland, L. (2016). The influence of declining sea ice on shipping activity in the Canadian Arctic. *Geophysical Research Letters*, **43**(23), 12146–12154.

Pope, I. C., Burnett, R. T., Thun, M. J. *et al.* (2002). Lung cancer, cardiopulmonary mortality, and long-term exposure to fine particulate air pollution. *JAMA*, **287**(9), 1132–1141.

Radke, L. F., Coakley, J. A. & King, M. D. (1989). Direct and remote sensing observations of the effects of ships on clouds. *Science*, **246**(4934), 1146.

Ramana, M. V. & Devi, A. (2016). CCN concentrations and BC warming influenced by maritime ship emitted aerosol plumes over southern Bay of Bengal. *Nature Scientific Reports*, **6**, 30416.

Richter, A., Eyring, V., Burrows, J. P. *et al.* (2004). Satellite measurements of NO_2 from international shipping emissions. *Geophysical Research Letters*, **31**(23), L23110.

Righi, M., Hendricks, J. & Sausen, R. (2015). The global impact of the transport sectors on atmospheric aerosol in 2030 – part 1: land transport and shipping. *Atmospheric Chemistry and Physics*, **15**(2), 633–651.

Rivier, L., Ciais, P., Hauglustaine, D. A. *et al.* (2006). Evaluation of SF_6, C_2Cl_4, and CO to approximate fossil fuel CO_2 in the Northern Hemisphere using a chemistry transport model. *Journal of Geophysical Research: Atmospheres*, **111**(D16), n.p.

Santana-Casiano, J. M., Gonzalez-Davila, M., Rueda, M. J., Llinas, O. & Gonzalez-Davila, E. F. (2007). The interannual variability of oceanic CO_2 parameters in the Northeast Atlantic subtropical gyre at the ESTOC site. *Global Biogeochemical Cycles*, **21**(1), 1015.

Schlager, H., Baumann, R., Lichtenstern, M. *et al.* (2006). Aircraft-based trace gas measurements in a primary European ship corridor. In R. Sausen, A. Blum & D. S.

Lee, eds., *Proceedings of an International Conference on Transport, Atmosphere and Climate (TAC)*. Oxford: Office for Official Publications of the European Communities, pp. 83–88.

Seyler, A., Wittrock, F., Kattner, L. *et al.* (2017). Monitoring shipping emissions in the German Bight using MAX-DOAS measurements. *Atmospheric Chemistry and Physics*, **17**(18), 10997–11023.

Sinha, P., Hobbs, P. V., Yokelson, R. J. *et al.* (2003). Emissions of trace gases and particles from two ships in the southern Atlantic Ocean. *Atmospheric Environment*, **37**(15), 2139–2148.

Slinn, S. A. & Slinn, W. G. N. (1980). Predictions for particle deposition on natural waters. *Atmospheric Environment*, **14**(9), 1013–1016.

Slinn, W. G. N., Hasse, L., Hicks, B. B. *et al.* (1978). Some aspects of the transfer of atmospheric trace constituents past the air–sea interface. *Atmospheric Environment (1967)*, **12**(11), 2055–2087.

Smith, K. R., Jerrett, M., Anderson, H. R. *et al.* (2009). Public health benefits of strategies to reduce greenhouse-gas emissions: health implications of short-lived greenhouse pollutants. *Lancet*, **374**(9707), 2091–2103.

Soerensen, A. L., Sunderland, E. M., Holmes, C. D. *et al.* (2010). An improved global model for air-sea exchange of mercury: high concentrations over the North Atlantic. *Environmental Science & Technology*, **44**(22), 8574–8580.

Sofiev, M., Winebrake, J. J., Johansson, L. *et al.* (2018). Cleaner fuels for ships provide public health benefits with climate tradeoffs. *Nature Communications*, **9**(1), 406.

Song, C. H., Chen, G. & Davis, D. D. (2003a). Chemical evolution and dispersion of ship plumes in the remote marine boundary layer: investigation of sulfur chemistry. *Atmospheric Environment*, **37**(19), 2663–2679.

Song, C. H., Chen, G., Hanna, S. R., Crawford, J. & Davis, D. D. (2003b). Dispersion and chemical evolution of ship plumes in the marine boundary layer: investigation of $O_3/NO_y/HO_x$ chemistry. *Journal of Geophysical Research: Atmospheres*, **108**(D4), 4143.

Stevenson, D. S., Young, P. J., Naik, V. *et al.* (2013). Tropospheric ozone changes, radiative forcing and attribution to emissions in the Atmospheric Chemistry and Climate Model Intercomparison Project (ACCMIP). *Atmospheric Chemistry and Physics*, **13**(6), 3063–3085.

Streets, D. G., Guttikunda, S. K. & Carmichael, G. R. (2000). The growing contribution of sulfur emissions from ships in Asian waters, 1988–1995. *Atmospheric Environment*, **34**(26), 4425–4439.

Stuart, R. K., Dupont, C. L., Johnson, D. A., Paulsen, I. T. & Palenik, B. (2009). Coastal strains of marine *Synechococcus* species exhibit increased tolerance to copper shock and a distinctive transcriptional response relative to those of open-ocean strains. *Applied and Environmental Microbiology*, **75**(15), 5047–5057.

Sunda, W. G. & Huntsman, S. A. (1998). Processes regulating cellular metal accumulation and physiological effects: phytoplankton as model systems. *Science of the Total Environment*, **219**(2), 165–181.

Tao, L., Fairley, D., Kleeman, M. J. & Harley, R. A. (2013). Effects of switching to lower sulfur marine fuel oil on air quality in the San Francisco Bay area. *Environmental Science & Technology*, **47**(18), 10171–10178.

Tissot, B. P. & Welte, D. H. (1984). *Petroleum Formation and Occurrence*. Berlin: Springer-Verlag.

Tournadre, J. (2014). Anthropogenic pressure on the open ocean: the growth of ship traffic revealed by altimeter data analysis. *Geophysical Research Letters*, **41**(22), 7924–7932.

Turner, D. R., Hassellov, I. M., Ytreberg, E. & Rutgersson, A. (2017). Shipping and the environment: smokestack emissions, scrubbers and unregulated oceanic consequences. *Elementa – Science of the Anthropocene*, **5**, 45.

UNCTAD (2017). *Review of Maritime Transport. United Nations Conference on Trade and Development (UNCTAD/RMT/2017)*. New York: United Nations.

Unger, N., Bond, T. C., Wang, J. S. *et al.* (2010). Attribution of climate forcing to economic sectors. *Proceedings of the National Academy of Sciences of the United States of America*, **107**(8), 3382–3387.

Viana, M., Amato, F., Alastuey, A. *et al.* (2009). Chemical tracers of particulate emissions from commercial shipping. *Environmental Science & Technology*, **43**(19), 7472–7477.

Vinken, G. C. M., Boersma, K. F., Jacob, D. J. & Meijer, E. W. (2011). Accounting for non-linear chemistry of ship plumes in the GEOS-Chem global chemistry transport model. *Atmospheric Chemistry and Physics*, **11**(22), 11707–11722.

von Glasow, R., Lawrence, M. G., Sander, R. & Crutzen, P. J. (2003). Modeling the chemical effects of ship exhaust in the cloud-free marine boundary layer. *Atmospheric Chemistry and Physics*, **3**, 233–250.

von Glasow, R., Jickells, T. D., Baklanov, A. *et al.* (2013). Megacities and large urban agglomerations in the coastal zone: Interactions between atmosphere, land, and marine ecosystems. *Ambio*, **42**(1), 13–28.

Wan, Z., Zhu, M., Chen, S. & Sperling, D. (2016). Pollution: three steps to a green shipping industry. *Nature*, **530**(7590), 275–277.

Wang, C., Corbett, J. J. & Winebrake, J. J. (2007). Cost-effectiveness of reducing sulfur emissions from ships. *Environmental Science & Technology*, **41**(24), 8233–8239.

Webster, P. J., Holland, G. J., Curry, J. A. & Chang, H. R. (2005). Changes in tropical cyclone number, duration, and intensity in a warming environment. *Science*, **309**(5742), 1844.

Williams, E. J., Lerner, B. M., Murphy, P. C., Herndon, S. C. & Zahniser, M. S. (2009). Emissions of NO_x, SO_2, CO, and HCHO from commercial marine shipping during Texas Air Quality Study (TexAQS) 2006. *Journal of Geophysical Research: Atmospheres*, **114**(D21), D21306.

Wilson, W. B. & Freeberg, L. R. (1980). *Toxicity of Metals to Marine Phytoplankton Cultures*. Washington, DC: Environmental Research Laboratory, Office of Research and Development, US Environmental Protection Agency.

Winebrake, J. J., Corbett, J. J., Green, E. H., Lauer, A. & Eyring, V. (2009). Mitigating the health impacts of pollution from oceangoing shipping: An assessment of low-sulfur fuel mandates. *Environmental Science & Technology*, **43**(13), 4776–4782.

Winnes, H. & Fridell, E. (2009). Particle emissions from ships: dependence on fuel type. *Journal of the Air & Waste Management Association*, **59**(12), 1391–1398.

Winther, M., Christensen, J. H., Plejdrup, M. S. *et al.* (2014). Emission inventories for ships in the arctic based on satellite sampled AIS data. *Atmospheric Environment*, **91**, 1–14.

Yang, M., Howell, S. G., Zhuang, J. & Huebert, B. J. (2009). Attribution of aerosol light absorption to black carbon, brown carbon, and dust in China – interpretations of atmospheric measurements during EAST-AIRE. *Atmospheric Chemistry and Physics*, **9**(6), 2035–2050.

Yang, M., Huebert, B. J., Blomquist, B. W. *et al.* (2011). Atmospheric sulfur cycling in the southeastern Pacific – longitudinal distribution, vertical profile, and diel variability observed during VOCALS-REx. *Atmospheric Chemistry and Physics*, **11**(10), 5079–5097.

Yang, M., Bell, T. G., Hopkins, F. E. & Smyth, T. J. (2016). Attribution of atmospheric sulfur dioxide over the English Channel to dimethyl sulfide and changing ship emissions. *Atmospheric Chemistry and Physics*, **16**(8), 4771–4783.

Zhao, H. & Baker, G. A. (2015). Oxidative desulfurization of fuels using ionic liquids: a review. *Frontiers of Chemical Science and Engineering*, **9**(3), 262–279.

3

Oil Pollution from Operations and Shipwrecks

ROGER C. PRINCE

3.1 Introduction

Ships are essential facilitators of modern economies – they are by far the most cost-effective mode of long-distance transport for both finished goods and raw materials, and over 90 percent of world trade is carried by sea (IMO, 2017a). There were more than 50,000 ships in the world's merchant fleets in 2015. Bulk carriers – carrying solids such as coal and grains – accounted for about a fifth of the fleet, with a combined capacity of around 705 million tons deadweight. General cargo carriers numbered about 17,000, crude oil tankers about 7000 and container ships about 5000 (Statista, 2018). The amounts moved are staggering; for example, the International Tanker Owners Pollution Federation (ITOPF, 2018) reported that seaborne oil trade has averaged 100 trillion barrel-miles per year since 2000. Financial efficiency pushes for ever-larger vessels, and concomitantly ever-larger canals (Chen *et al.*, 2016) and dock facilities (Mongelluzzo, 2016).

Most ships reach their destinations without mishap, but all have the potential to cause pollution during their routine operations. Catastrophic accidents are likely to release at least fuel oil, and in the case of oil tankers, potentially large amounts of cargo. Fortunately, both sources of pollution are declining steadily due to the focused efforts of regulatory bodies and industry. This chapter will review oil pollution from both routine operations and from catastrophic losses, discussing the chemical composition of crude and fuel oils, how they get into the marine environment and the various regulatory bodies aiming to minimize this input. It will also discuss their fate once spilled and how responders minimize their environmental impacts.

3.2 Petroleum

Petroleum, literally rock oil, is used on a staggering scale – more than 13 million tonnes per day in 2016 (EIA, 2020b). It is the geothermally processed product of algae that grew many millions of years ago (with some terrigenous input in certain cases). The average age of commercial crude oil is about 100 million years (Tissot & Welte, 1984), although some are Precambrian (more than a billion years old; Jackson *et al.*, 1986; McKirdy *et al.*, 1981), while other important reserves, such as the California Midway reservoirs, date from only 10–13 million years ago (Brooks, 1952). Petroleum is being generated today in the

Table 3.1. *Conversion factors for petroleum*

Tonnes	m^3	Barrels	Gallons (US)	Liters
1	1.16	7.3	306	1160

Obviously, tonnes are a measure of mass, while the others are measures of volume, so the precise conversion from tonnes will depend on the density of the petroleum. These factors are for a crude oil of API gravity 32.6° – a medium crude oil typical of those in commerce. Refined products are usually less dense, but the effect on these values is small.

Guaymas Basin (Simoneit & Kvenvolden, 1994) and the Kamchatka volcanic region (Bazhenova *et al.*, 1998; Varfolomeev *et al.*, 2011), where near-surface geothermal heat replaces deep geothermal energy to convert recent biomass to hydrocarbons. When we consider that the primeval atmosphere on Earth contained no oxygen (Kasting & Seifert, 2002) and that photosynthesis is the only known significant generator of oxygen on our planet, it is obvious that buried reduced carbon must complement all oxygen in the atmosphere. Thus, the total amount of buried biomass is staggering. How much of this has been converted to fossil fuel reserves remains a matter of debate, but it is unlikely that exploitation to date has significantly reduced the total (Smil, 2006). Thus, petroleum is likely to be a major energy source for some time unless political pressures intervene (Benes *et al.*, 2015; Kerschner & Hubacek, 2009).

Crude oils and refined fuels are very complex mixtures that vary significantly depending on their source (Speight, 2014; Tissot & Welte, 1984; USDOE, 2008). They contain many thousands – perhaps millions – of molecular species, many still uncharacterized, but they are conveniently grouped into four categories: saturated hydrocarbons, aromatic hydrocarbons, resins and asphaltenes. The first two are amenable to gas chromatography and have been intensely studied (Peters *et al.*, 2005). The last two contain heteroatoms such as oxygen, sulfur and nitrogen and are not usually amenable to gas chromatography. Until recently their identities were unknown, but recent developments in high-resolution mass spectrometry (Cho *et al.*, 2015; Hughey *et al.*, 2007; Kim *et al.*, 2005) are beginning to reveal their composition. A rule of thumb is that typical crude oil is about 30 percent linear and branched saturated hydrocarbons, about 30 percent cyclic saturated hydrocarbons, about 30 percent aromatic hydrocarbons and about 10 percent resins and asphaltenes (Tissot & Welte, 1984). Commercially, crude oils are classified by their density in units of American Petroleum Institute (API) gravity, which is defined as (141.5/[specific gravity]) – 131.5 and expressed as degrees (°). Thus, distilled water has an API gravity of 10° and denser fluids will have lower API gravities. The median API gravity of crude oil processed in the USA in 2015 was 31.68° (EIA, 2020a), and this is probably typical worldwide. Almost all oils in commerce float on seawater, although a few very heavy bitumens and bunker fuels may "float" beneath the surface of freshwater, as happened with the 1989 *Presidente Rivera* spill in the Delaware River, USA (Wiltshire & Corcoran, 1991). Table 3.1 offers approximate conversion factors for the various units used to measure crude oil; oil is typically priced in barrels (bbl) in US$.

Once oils enter refineries, they are desalted and distilled into fractions that will eventually end up as gasoline, jet fuel, diesel fuels, residual fuels, etc. The higher-molecular-weight fractions are then processed to produce additional gasoline, jet and diesel fuels, but eventually the refinery is left with a residuum that can either be converted to coke (using the hydrogen in catalytic conversion processes) or be converted to "bunker fuels." Most ships run on these "bunker fuels" – an etymological holdover from when coal was kept in bunkers on board. There are four principal types of bunker fuel: residual oil, also known as Bunker C or intermediate fuel oil (IFO) 380, which is typically about 98 percent residual oil from the initial crude distillation plus a few percent lighter diluent to attain a prescribed viscosity (the 380 is an indication of a maximum viscosity of 380 centipoise [cP]); IFO 180, with more diluent to attain a lower viscosity (180 cP); marine diesel oil, a distillate fraction with perhaps a small amount of residual oil; and marine gas oil, a straight distillate fraction (USEPA, 2008). It is important to recognize that these products are sold based on a number of properties, especially viscosity and energy content, and increasingly on (low) sulfur and nitrogen content to meet legislative requirements. Despite nomenclature that they are "heavy" fuels, they all float on water (especially seawater), but they may need warming to be most effective as a fuel. Worldwide demand for maritime fuels is about 550,000 tonnes per day, some 4 percent of global oil consumption (EIA, 2020c).

3.3 Oil in the Sea

3.3.1 Seeps

As mentioned in Section 3.2, most commercially valuable crude oils were once mainly algal biomass that got buried and heated to convert the biomass to hydrocarbons (Tissot & Welte, 1984). Once oil and gas have been generated, they are expelled from their source rock and, being lighter than water, they rise through the overlying sedimentary rocks until they either get trapped and become a valuable resource or eventually make their way to the surface of the Earth as a seep. Even if there is a reservoir, oil and gas may eventually overflow it, and again get to the Earth's surface. Terrestrial seeps have been sources of fascination for thousands of years (Etiope, 2015), and the first US oil well in Titusville, Pennsylvania, was drilled in an area with several seeps (Black & Ladson, 2010). Marine seeps are often less obvious, although their cumulative release probably accounts for about half of all the oil that gets into the world's oceans every year (NRC, 2003). One of the most dramatic – and certainly the most studied – is the Coal Oil Point seep area in the nearshore of Santa Barbara, Southern California (Hein, 2013), which releases some 8500 tonnes of oil a year (Kvenvolden & Cooper, 2003), more than 2.6 million US gallons. The ARCO and Mobil oil companies installed *Platform Holly* at one of the seeps in 1966, and large submarine "tents" installed by ARCO in 1982 capture some of the escaping gas and oil. Over the years, the production of oil has reduced the pressure in the reservoir, and gas and oil seepages have significantly decreased (Quigley *et al.*, 1999) and seem to have essentially ceased (Palminteri, 2014). Nevertheless, this area is still a very significant source of hydrocarbons in the Pacific Ocean, and although seabird mortalities do occur (Henkel *et al.*,

2014) and tar still contaminates nearby shores (Del Sontro *et al.*, 2007), much is consumed by microbes almost as soon as it is released (Farwell *et al.*, 2009; Mendes *et al.*, 2015).

The Gulf of Mexico, with a surface area of 1.5 million km^2, is also home to many submarine seeps; satellite synthetic aperture radar identified 914 seep sites releasing more than 50,000 tonnes of oil to the surface every year – some 16 million US gallons per year (MacDonald *et al.*, 2015). Presumably, as in the case of the California Coal Oil Point seeps, surface slicks are only a small fraction of the total oil and gas released; Kvenvolden and Cooper (2003) estimated a total release of about 140,000 tonnes, or 43 million US gallons, but this may well be an underestimate, since at that time only 350 seeps were known compared to the 914 noted by MacDonald *et al.* (2015). Seeps are found around the world: in the Adriatic (Etiope *et al.*, 2014), off the shore of Sulawesi (Camplin & Hall, 2014) and all around the coast of Africa (Selley & Van der Spuy, 2016).

3.3.2 Terrestrial Sources

If seeps are the largest individual category of oil releases into the marine environment in most years, they are closely followed by anthropogenic inputs caused by the general use of petroleum products (NRC, 2003). Such releases include those from operating vessels, but are mainly from terrestrial sources that escape from increasingly paved urban areas. In 2003, the National Research Council (NRC) estimated that such sources contributed nearly 70 percent of the petroleum introduced into the world's oceans from anthropogenic sources and nearly 85 percent of the total petroleum input from anthropogenic sources into North American waters.

3.3.3 Shipping

3.3.3.1 Routine Operations

Ships are responsible for only a small fraction of the petroleum that gets into the world's oceans every year. Some comes from routine operations, but this is diminishing substantially as international regulations become more stringent. Historically, two tanker operations were particularly damaging: ballasting and cargo tank washing. Ballasting occurs as a tanker discharges its cargo and seawater is loaded to compensate as ballast. Eventually, this (oily) water must be discharged to be replaced by crude oil. Cargo tank washings are exactly what they sound like: water washing of cargo space to remove oily sludges (Griffin, 1994). The problem is that when the water enters cargo tanks, it mixes with oil residues from the cargo, and if this is discharged untreated, it can be a major source of pollution. Initial attempts to at least keep discharges away from coastlines included the 1954 International Convention for the Prevention of Pollution of the Sea by Oil, which prohibited such discharges within 50 miles (80 km) of the shore (Admiralty and Maritime Law Guide, 2017). A substantial improvement came with the International Convention for the Prevention of Pollution from Ships, 1973, as modified by the Protocol of 1978, commonly known as MARPOL 73/78 (IMO, 2016). Annex I, Regulations for the Prevention of Pollution by

Oil, came into force in 1983. It dealt with both oil tank and engine room waste and mandated oil–water separators, oil content meters for any waste disposed at sea and port reception facilities for waste that was too oily for discharge.

Initial improvements included load on top, where ballast and washing waters were allowed to sit long enough for the oil to float to the surface of the tank so that relatively clean water could be discharged from the bottom. Crude oil washing in place of water washing was another improvement, using heated crude oil to mobilize sludge into the oil rather than into the environment, and finally segregated ballast tanks were filled with water when there was no oil in the vessel and emptied as oil was taken aboard (Griffin, 1994) – all but the oldest tankers use segregated ballast tanks today. At the same time, monitoring of discharge water was required for all vessels, with an upper limit for oil in discharged water of 15 ppm, a requirement for continuous monitoring of oil in bilgewater and oily water in cargo tanks and accurate record keeping (IMO, 2006). It is widely agreed that the ratification of MARPOL 73/78 led to a substantial decrease in oil released at sea, with a concomitant decrease in the number of tarballs washing up on shorelines (Butler *et al.*, 1998; Kucuksezgin *et al.*, 2013; Smith & Knap, 1985).

Some areas, such as the Baltic Sea, instigated even more stringent rules, and the 1992 Helsinki Convention requires essentially all ships in the Baltic to deliver bilgewater to reception facilities, at no cost to themselves (Fitzmaurice, 1993). Significant progress has been made, and discharges have decreased by more than 90 percent (HELCOM, 2016). Nevertheless, the Baltic remains a damaged environment, with eutrophication – not oil – representing the biggest threat (Carstensen *et al.*, 2014).

Other routine operational oil inputs come from lubrication. All vessels have stern tubes through which the propeller shaft passes from inside the ship, and most vessels use oil lubrication for the stern tube. The pressure inside the bearing must be higher than the local water pressure to prevent water ingress, so there are constant – hopefully very small – leaks. USEPA (2011) cites a review by Etkin that found that the average rate across vessel types was 2.6 L (0.7 US gallons) per day, and that the lubricants were often those formulated for engine crankcase use. Not surprisingly, there is nascent research supporting a move to water lubrication and polymer bearings, and commercial adoption is underway (USEPA, 2011), although confusion between "bio-based" and "biodegradable" is rampant (Hayes, 2017).

3.3.3.2 Lightering

The transfer of fuel or cargo from one vessel to another is known as "lightering," and it is a very important process because many of the larger oil tankers are too large to dock at refineries. In 1998, the NRC (1998) estimated that >25 percent of the 7.5 million barrels of crude oil imported into the USA each day was lightered, principally in the Gulf of Mexico. Yet spills were infrequent (62 reported between 1984 and 1996), and the average spill size was about 1000 gallons. Lightering is also very important in some oil spill response operations – for example, most of the cargo of the stranded *Exxon Valdez* was removed, without incident, to the *Exxon Baton Rouge*, *Exxon San Francisco* and *Exxon Baytown*

Table 3.2. *The three largest tanker spills to date*

Name	Year	Cause	Fatalities	Location
Atlantic Empress	1979	Collision	27	30–300 km east of Tobago
Castillo de Bellver	1983	Fire	3	120 km northwest of Cape Town, South Africa
ABT Summer	1991	Explosion	5	1100 km west of Angola

Table 3.3. *A list of some oil spills where there have been detailed scientific investigations and descriptions of the spill, its effects on the environment and cleanup efforts*

Name	Year	Cause	Tonnes of oil	Location
Torrey Canyon	1967	Grounding	119,000	Isles of Scilly, UK
Amoco Cadiz	1978	Grounding	220,000	Brittany, France
Exxon Valdez	1989	Grounding	37,000	Prince William Sound, USA
Braer	1993	Grounding	85,000	Shetland Isles, UK
Sea Empress	1996	Grounding	72,000	Milford Haven, UK
Nakhodka	1997	Hull failure	6200	Oki Islands, Japan
Erika	1999	Hull failure	15,000	Bay of Biscay, France
Prestige	2002	Hull failure	63,000	Galicia, Spain
Hebei Spirit	2007	Collision	11,000	Daesan, South Korea

Google Scholar provides easy access to this literature, which totals hundreds of papers.

(US Congress House Committee on Merchant Marine and Fisheries, Subcommittee on Coast Guard and Navigation, 1989). Similar operations occur at many groundings (Buie, 1999).

3.3.3.3 Shipwrecks

Catastrophic shipwrecks are fortunately rather rare events, but beyond the potential loss of human life, they can have far-reaching environmental impacts. Of course, all vessels contain fuels, but it is wrecks of oil tankers that have by far the largest potential for releasing large volumes of oil into the environment. The International Tanker Owners Pollution Federation (ITOPF) maintains statistics on oil spills from vessels (ITOPF, 2018), and the number of spills and amount of oil lost to the sea have dropped dramatically since the 1970s. The three largest spills, all greater than 250,000 tonnes, occurred far out at sea (Table 3.2) and little oil washed ashore (ITOPF, 2018). Alas, other spills – albeit far smaller – often deposited most of their spilled oil on shorelines, and these have a substantial – albeit transient – impact. Table 3.3 lists some of these, highlighting those with a substantial scientific literature on the spill and spill response, as well as the extent and effects of oil stranding on shorelines. Tankers do not only carry unrefined crude oil, and spills of refined products and vegetable oils also occur. Examples include 600 tonnes of diesel and 300 tonnes of bunker fuel lost from the *Jessica* off the Galapagos in 2001 (Edgar *et al.*, 2003) and 1500 tonnes of sunflower oil lost from the *Kimya* off Anglesey in 1991 (Mudge, 1997).

These were very large vessels, and each released approximately 260,000 tonnes of oil. This can be compared to estimates of oil entering the world's oceans every year (>1,300,000 tonnes; NRC, 2003). The blowout from the *Deepwater Horizon* well in the Gulf of Mexico released some 437,000 tonnes of oil (Barbier, 2015).

3.3.3.4 War and Terrorism

The amount of oil lost at sea during major wars is substantial – Monfils (2005) found records of >7800 sunken vessels from World War II, including >860 oil tankers. Most Allied tankers carried refined products to Europe and Asia, not crude oil, but Japanese tankers moved crude oil from what had been the Dutch East Indies (now Indonesia; Wolborsky, 2014). Several tankers were sunk during the Iran–Iraq war (El-Shazly, 2016), and the tanker *Limburg* was attacked by suicide bombers off the shore of Yemen in 2002, spilling about a quarter of her cargo – some 9100 tonnes (Savage, 2014).

3.3.3.5 Design Improvements

The International Convention for the Prevention of Pollution by Ships (MARPOL 73/78 discussed in Section 3.3.3.1 (IMO, 2016) included regulations to improve the stability of oil tankers after collisions or stranding. Initially, this included locating the tanker ballast tanks, which are empty when the ship is loaded with oil, so that they would protect the cargo tanks, and in 1983 amendments prohibited oil tanks in the bow area, which were most likely to be damaged in a collision. The 1989 grounding of the *Exxon Valdez* in Alaska led to the passing of The Oil Pollution Act of 1990 (US Coast Guard, 2016), which mandated double hulls for all tankers operating in US waters by 2015, and MARPOL followed suit in 1992, demanding the retirement of single-hulled tankers at age 30 (IMO, 2016). Single-hull tankers will soon be gone from the seas. Other vessels will still be single-hulled, however, and even minor collisions can cause spills – for example, the cargo ship *Cosco Busan* grazed the San Francisco–Oakland Bay Bridge in 2007, spilling 173 tonnes of bunker fuel into the San Francisco Bay (Lemkau *et al.*, 2010). This spill highlighted the potential penalties for oil spills in areas of high economic and environmental value – the owners eventually agreed to a penalty of US$44.4 million, more than US$800 per gallon spilled (US Department of Justice, 2011).

3.4 Effects of Floating Oil Slicks

Oil slicks are serious hazards for birds, mammals and reptiles, and indeed it was the killing of 3600 seabirds and several seals and dolphins in the 1969 Santa Barbara blowout (Clarke & Hemphill, 2002) and 30,000 guillemots and razorbills in the 1967 *Torrey Canyon* spill (Bourne *et al.*, 1967) that spurred the development of effective dispersants to get the oil off the sea surface (Canevari, 1969). Dispersants were not used following the 1989 *Exxon Valdez* spill in Prince William Sound, and animal mortality included thousands of sea otters and hundreds of thousands of birds (Helm *et al.*, 2014). The cause of death was most likely thermal stress caused by loss of insulation, although lightly oiled animals may also have suffered toxicity.

Marine oil spills do not usually cause fish kills, but nearshore spills of distillate fuels have led to substantial invertebrate mortality – the 1996 spill of 2700 tonnes of heating oil from the tank barge *North Cape* is reported to have killed over 9 million lobsters (Gunter, 2014). This can be attributed to the wreck being nearshore so that there was little volume for dilution – and there was a significant storm that drove soluble components into the water column. The cause of death would thus have been acute toxicity – probably narcosis (Redman *et al.*, 2012) – rather than physical smothering, and this would be unlikely to occur in deep water.

Floating and dispersed slicks likely kill surface planktonic organisms, including fish and crustacean larvae, but this has not been documented as a serious harm in the field. Indeed, Fodrie and Heck Jr. (2011) conclude that immediate, catastrophic losses of 2010 fish cohorts were largely avoided after the *Deepwater Horizon* tragedy, and that species composition was unchanged by the spill. Similar analyses suggest that even very large spills are unlikely to have a detectable population effect on Arctic cod populations in the Beaufort Sea (Galloway *et al.*, 2017). Beached oil likely smothers seaweeds and invertebrates, but again this effect is typically localized and unlikely to have population-wide effects.

3.5 Fate of Oil in the Sea

Oils and refined products that get into the oceans are subject to a range of physical, chemical and biological processes. Almost all oils in commerce float, allowing the prompt evaporation of molecules with fewer than about 15 carbon atoms (Fingas, 2013), and evaporation is thus the probable fate of most of a gasoline spill, three-quarters or more of a diesel spill and perhaps 20–40 percent of a typical crude oil spill. Heavy fuels such as marine bunker fuels do not contain a significant volatile fraction. Small aromatic molecules (especially the notorious BTEX – benzene, toluene, ethylbenzene and the xylenes) are reasonably water soluble and can dissolve out of a floating slick if they do not evaporate – for example, toluene has a solubility of about 380 ppm in seawater (Sutton *et al.*, 1975). Small polar molecules such as naphthenic acids may also leach out of floating slicks (Stanford *et al.*, 2007), but other hydrocarbons, especially saturated molecules such as linear, branched and cyclic alkanes, are essentially insoluble in water, and in general losses due to solubility amount to only a few percent of most oil products. A more important phenomenon is the reverse, where water "dissolves" into the spilled oil, forming water-in-oil emulsions known as "mousses" (Fingas & Fieldhouse, 2012) and eventually tarballs (Goodman, 2003; Warnock *et al.*, 2015) that can persist for decades. Fortunately, as we shall discuss in Section 3.6, chemical dispersants that break emulsions and stimulate the natural dispersion process are effective tools in preventing the formation of tarballs (Lessard & DeMarco, 2000).

A competing process, especially while oil is floating as a slick, is photochemical oxidation, which affects primarily aromatic hydrocarbons (and potentially alkenes and alkynes in refined products – these are absent from crude oils; Aeppli *et al.*, 2012; Garrett *et al.*, 1998). Of course, ultraviolet light cannot penetrate very far into a dark oil slick, so photooxidation has little effect on the bulk properties of spilled oil unless the oil is a thin layer. Nevertheless, the process continues after oil reaches a shoreline, and it is likely important in generating the polymerized "skin" that stabilizes "pavements" on beaches. The latter – layers of immobile, hardened oil and sediment – form when oil reaches a shoreline as a heavy, thick slick. Oil becomes trapped in the sediment, and the oil and the sediment become essentially saturated with each other (Owens *et al.*, 2008). Oil incorporated into such pavements is effectively preserved from weathering processes and biodegradation until this heavy, solidified material is physically disrupted, so another major goal of spill cleanup operations is to prevent the formation of pavements.

All of the processes listed above either change the location (evaporation, dissolution), polymerize (photooxidation) or emulsify the spilled oil, but what *removes* it from the environment? The simple answer is either combustion or biodegradation: some atmospheric photooxidation processes may eventually completely mineralize small hydrocarbons (Brigden *et al.*, 2001), but combustion and biodegradation are by far the major routes of removal of hydrocarbons from the environment – the conversion of hydrocarbons to CO_2 and H_2O. Some tanker accidents have caught fire – some 70 percent of the cargo of 145,500 tonnes of crude oil burned following the tragic 1991 *Haven* spill that killed six crew members off Genoa, Italy (Martinelli *et al.*, 1995). Efforts are sometimes made to set fire to wrecks – this was attempted with partial success at the 1967 wreck of the *Torrey Canyon* on the Isles of Scilly (Burrows *et al.*, 1974), and more recently on the wreck of the wood chip carrier *New Carissa* off the coast of Oregon in 1999 (Gallagher *et al.*, 2001). Perhaps surprisingly, floating oil must be several millimeters thick to sustain combustion, and in general this requires corralling the oil in a boom, preferably a fireproof one. Under optimal conditions, burning may consume >90 percent of the oil contained in a boom, but there is usually only a small window of opportunity for success before the oil absorbs so much water that it becomes impossible to ignite (Buist, 2003). Burning boomed oil was a significant response to the 2010 *Deepwater Horizon* blowout (Allen *et al.*, 2011), and it may prove to be a particularly promising response in ice-infested waters, especially with the use of "herding" chemicals in place of booms (Buist *et al.*, 2011).

But by far the most important process that removes oil from the environment is biodegradation by microorganisms (Hazen *et al.*, 2016), both prokaryotic (Prince *et al.*, 2010) and eukaryotic (Prince, 2010). This biodegradation occurs under aerobic and anaerobic conditions (Prince & Walters, 2016), in the tropics (Zahed *et al.*, 2010) and the Arctic (Garneau *et al.*, 2016; McFarlin *et al.*, 2014) and at depth (Prince *et al.*, 2016). It is remarkably effective, rapidly degrading the vast majority of hydrocarbons (Prince *et al.*, 2013), and it can be remarkably rapid, with a "half-life" when dispersed to the ppm range of 2 weeks or so (Prince *et al.*, 2017). (It must be noted that we do not have a clear understanding of the biodegradability of the asphaltenes and resins that make up about

10 percent of most crude oils [see Prince & Walters, 2016], but these molecules do not have the "oily" properties associated with petroleum.)

Nevertheless, the biodegradation of oil has to overcome two potentially limiting phenomena: The first is that oil is insoluble in water, and so biodegradation takes place at the oil–water interface. This means that the surface area available for microbial colonization limits biodegradation. The second limiting issue is that while oil is an energy-dense food for microbes, it is not a complete food – it contains no biologically available nitrogen, phosphorus, iron, etc. These two limitations define the opportunities for aiding the natural process of biodegradation in oil spill response – dispersants encourage the oil to break into tiny droplets that maximize the surface area available for microbial "attack," and fertilizers overcome nitrogen, phosphorus, etc., limitations. In fact, dispersed oil is sufficiently dilute that there are enough nutrients in seawater to allow microbial growth, so fertilizers are unnecessary at sea. In contrast, the biodegradation of significant amounts of oil on shorelines is usually nutrient limited, and fertilizers will stimulate biodegradation there. We will return to these options as we discuss oil spill response in the next section.

3.6 Oil Spill Response

Oil spill response organizations are poised to respond to accidents around the globe with fleets of vessels, planes and skimmers, as well as stockpiles of boom and dispersant (IMO, 2017b; The Global Response Network, 2017). Of course, no good can come from any significant spill, but responders and regulators aim to ameliorate the problem with minimal additional harm – the so-called Net Environmental Benefit Approach (API, 2013). Response plans typically rely on efforts to collect as much oil as possible with skimmers and booms (Fingas, 2012), but skimming is a very slow process, typically carried out at 1 knot, and oil may spread too quickly for skimmers to be effective. Furthermore, skimmers can only work under relatively calm conditions because waves allow oil to escape underneath the towed boom. Thus, it is unusual for skimming to be effective following a large spill – for example, McNutt *et al.* (2012) estimated that only 3 percent of the oil from the 2010 *Deepwater Horizon* blowout was collected by skimmers, despite enormous effort, and slightly more was burned.

When oil cannot be collected or burned, responders turn to the use of dispersants, preferably delivered from the air to maximize the application area. Dispersants are mixtures of surfactants and surface-tension modifiers in a solvent – for example, the components of Corexit 9500® are listed in Table 3.4. There has been a lot of concern around the use of dispersants, especially regarding their toxicity and efficacy (reviewed in Prince, 2015). The toxicity issue is a red herring – of course all chemicals have some measurable toxicity, but dispersants have to pass standard toxicity tests (usually with a shrimp and a fish; Word *et al.*, 2013) to be on the list of potential products maintained by different governments, and most are no more toxic than common household dishwashing products (Word *et al.*, 2013). Dispersants are applied at an aspirational dose of 5 gallons per acre (47 L/ha), so concentrations drop to substantially below any concentrations of concern for acute toxicity by

Table 3.4. *Components of Corexit 9500® (Word* et al., *2013)*

CAS no.	Trivial name	Role
577-11-7	Dioctyl sulfosuccinate	Anionic surfactant
1338-43-8	SPAN® 80	Nonionic surfactant
9005-65-6	TWEEN® 80	Nonionic surfactant
9005-70-3	TWEEN® 85	Nonionic surfactant
29911-28-1	Di(propylene glycol) butyl ether	Surface-tension minimizer
7732-18-5	Hydrocarbon	Solvent

All components are approved for food use or food contact.

mixing into the top meter of the sea. Dispersed oil is substantially more toxic, but again dilution soon reduces concentrations to below levels of acute concern. Indeed, a key issue to be borne in mind when evaluating research on dispersants is the enormous dilution that occurs upon successful dispersion (Lee *et al.*, 2013) – levels drop to a few ppm of oil within hours, and as an example, the concentration of oil in the dispersed deep "plume" from the *Deepwater Horizon* blowout was <1 ppm (Wade *et al.*, 2016). Misunderstanding also confuses the issue of efficacy. Regulators, such as the US Environmental Protection Agency (EPA), have promulgated a variety of tests to ensure that dispersant products are indeed effective, but in mandating simple laboratory tests with relatively low energy and low "passing grades" (which do a good job of discriminating between good and poor products), they have led many to believe that dispersants are only 40–50 percent effective in the field (discussed in Prince, 2016). In fact, tests with more realistic wave energy, such as in the OHMSETT facility (a wave tank in New Jersey that is 200 m long, 20 m wide and 2.5 m deep), routinely measure dispersant efficiencies of >95 percent, even at low temperatures with ice in the water (Belore *et al.*, 2009). Contrary to popular misconceptions, dispersants work reasonably well on even heavy crude oils and bunkers (Lessard & De Marco, 2000), albeit a little more slowly.

The remarkable but often underappreciated great benefit of dispersing crude oil at sea is how rapidly biodegradation then proceeds. Because dispersed oils dilute within a day to sub-ppm concentrations (Lee *et al.*, 2013; Prince *et al.*, 2017), the indigenous low levels of nutrients such as biologically available nitrogen, phosphorus and iron are adequate for substantial and rapid microbial growth and biodegradation, and disappearance has an apparent half-life of 7–14 days in temperate waters, and slightly slower (by less than a factor of two) in the Arctic (Garneau *et al.*, 2016; McFarlin *et al.*, 2014) or at 1500 m depth (Prince *et al.*, 2016). This contrasts with floating slicks that may persist for weeks to months and may then impact a shoreline where oil may persist for years.

It is important to remember that our current understanding is that most of the hydrocarbons of crude oils – typically 70–90 percent of a crude oil and close to 100 percent of a refined product – are readily biodegraded. The exceptions are polycyclic saturated molecules such as the hopanes and steranes, and these have the useful property of acting as fingerprints to distinguish different oil sources (Peters *et al.*, 2005; Prince & Walters, 2016). Most hydrocarbons, especially the alkanes, are significantly less dense than water, but as they are degraded, the residual oil – now enriched in the rather denser asphaltenes and resins – may eventually sink, being associated with "marine snow" (Vonk *et al.*, 2015).

Dispersants encourage dispersion by lowering the interfacial tension between oil and water, and it still takes energy from waves to actually break the oil into droplets. Storm waves may provide enough energy for oil to disperse without the need for added dispersants – as happened during the extreme weather of the 1993 *Braer* spill (Harris, 1995). If oil is not dispersed effectively, however, the other fates described above begin to dominate – photochemistry aids the formation of tarballs and emulsification incorporates water into oil to form mousses. Oil may well land on a shoreline. If this happens in calm weather on sandy shores it can be picked up with mechanical equipment, as happened in the 2007 *Sea Empress* spill (Colcomb *et al.*, 1997). But if oil gets onto cobble beaches and seeps between rocks it will be very difficult to collect. This is what happened following the 1989 *Exxon Valdez* spill in Prince William Sound, Alaska. In this case, the majority of the mobile oil was removed by washing it back into the sea and collecting it with skimmers (Nauman, 1991). The residual oil was too concentrated for the indigenous levels of nutrients in tidal water to facilitate substantial biodegradation, so shorelines were treated by carefully adding slow-release and oleophilic fertilizers: Biodegradation was stimulated several fold (Bragg *et al.*, 1994). But despite this success, it is very clear that encouraging biodegradation before oil gets to a shoreline will remove oil from the environment much more rapidly – in weeks or months rather than in several years.

Many people imagine that biodegradation would be stimulated by adding bacteria to oil slicks or oiled shorelines, but in fact, despite several tests, there is no evidence that this works. This is most probably because oil has been part of the biosphere for hundreds of millions of years, and so microbes that take advantage of this high-calorie food are ubiquitous and likely well adapted to their local environment. People sometimes imagine that genetically modified organisms are used, but this is not the case. It is true that the first patented microbe was designed to consume hydrocarbons (Chakrabarty, 1981), but none have ever been used at a spill.

3.7 Vegetable Oils

Large amounts of vegetable oils are shipped by sea, and their spills are becoming a problem (Bucas & Saliot, 2002). Vegetable oils are triglycerides – usually with C_{16} and C_{18} fatty acids esterified to the glycerol. The majority of the fatty acids have one to three unsaturated bonds, and they readily polymerize when exposed to sunlight and air. As such, they can be very persistent at sea (Mudge, 1997). Although they might seem more benign than petroleum, they have very similar effects on birds and other wildlife, and may even attract them to their peril. Dispersants have not been reported to work well on these products, so collection with booms is the only realistic spill response option at sea. Sorbent pads have been used on shorelines, but vegetable oils have been surprisingly persistent in areas where they have been left to weather without effective cleanup (Bucas & Saliot, 2002; Mudge, 1997).

3.8 Summary

As has been made clear in this chapter, shipping was once responsible for significant oil pollution during normal operations, and especially so following catastrophic accidents. The International Convention for the Prevention of Pollution from Ships, 1973 and its addenda have eliminated most of the operational discharges and have led to safer vessels, so much less oil is lost from ships today than 50 years ago. Occasional catastrophic spills still occur, but oil spill response organizations with substantial resources stand ready to respond around the world. Collecting large volumes of spilled oil is very difficult, and when this cannot be achieved, techniques that stimulate the biodegradation of spilled oil can significantly minimize environmental harm. This can be done by increasing the surface area for microbial "attack" by applying dispersants to facilitate the dispersion of oil as tiny droplets that diffuse apart. If this fails and oil reaches shorelines that cannot readily be physically cleaned, adding carefully formulated fertilizers can stimulate biodegradation. In both cases, approaches that work with natural phenomena rather than in ignorance of them are likely to be the most effective.

References

Admiralty and Maritime Law Guide (2017). International Convention for the Prevention of Pollution of the Sea by Oil, 1954. www.admiraltylawguide.com/conven/oilpol1954.html

Aeppli, C., Carmichael, C. A., Nelson, R. K. *et al.* (2012). Oil weathering after the *Deepwater Horizon* disaster led to the formation of oxygenated residues. *Environmental Science & Technology*, **46**, 8799–8807.

Allen, A. A., Jaeger, D., Mabile, N. J. & Costanzo, D. (2011). The use of controlled burning during the Gulf of Mexico Deepwater Horizon MC-252 oil spill response. In *Proceedings of the International Oil Spill Conference*. Washington, DC: American Petroleum Institute, Paper 194.

API (2013). Net Environmental Benefit Analysis for effective oil spill preparedness and response. www.api.org/~/media/Files/EHS/Clean_Water/Oil_Spill_Prevention/NEBA/NEBA-Net-Environmental-Benefit-Analysis-July-2013.pdf

Barbier, C. J. (2015). MDL-2179 oil spill by the oil rig "*Deepwater Horizon*." www.laed.uscourts.gov/sites/default/files/OilSpill/Orders/1152015FindingsPhaseTwo.pdf

Bazhenova, O. K., Arefiev, O. A. & Frolov, E. B. (1998). Oil of the volcano Uzon caldera, Kamchatka. *Organic Geochemistry*, **31**, 421–428.

Belore, R. C., Trudel, K., Mullin, J. V. & Guarino, A. (2009). Large-scale cold water dispersant effectiveness experiments with Alaskan crude oils and Corexit 9500 and 9527 dispersants. *Marine Pollution Bulletin*, **58**, 118–128.

Benes, J., Chauvet, M., Kamenik, O. *et al.* (2015). The future of oil: geology versus technology. *International Journal of Forecasting*, **31**, 207–221.

Black, B. & Ladson, M. (2010). Oil at 150: energy past and future in Pennsylvania. *Pennsylvania Legacies*, **10**, 6–13.

Bourne, W. R., Parrack, J. D. & Potts, G. R. (1967). Birds killed in the *Torrey Canyon* disaster. *Nature*, **215**, 1123–1125.

Bragg, J. R., Prince, R. C., Harner, E. J. & Atlas, R. M. (1994). Effectiveness of bioremediation for the *Exxon Valdez* oil spill. *Nature*, **368**, 413–418.

Brigden, C. T., Poulston, S., Twigg, M. V., Walker, A. P. & Wilkins, A. J. (2001). Photo-oxidation of short-chain hydrocarbons over titania. *Applied Catalysis B: Environmental*, **32**, 63–71.

Brooks, B. T. (1952). Evidence of catalytic action in petroleum formation. *Industrial & Engineering Chemistry*, **44**, 2570–2577.

Bucas, G. & Saliot, A. (2002). Sea transport of animal and vegetable oils and its environmental consequences. *Marine Pollution Bulletin*, **44**, 1388–1396.

Buie, G. W. (1999). Some thoughts on salvage operations during oil spills. In *Proceedings of the International Oil Spill Conference*. Washington, DC: American Petroleum Institute, pp. 1205–1209.

Buist, I. (2003). Window-of-opportunity for *in situ* burning. *Spill Science & Technology Bulletin*, **8**, 341–346.

Buist, I., Potter, S., Nedwed, T. & Mullin, J. (2011). Herding surfactants to contract and thicken oil spills in pack ice for in situ burning. *Cold Regions Science and Technology*, **67**, 3–23.

Burrows, P., Rowley, C. & Owen, D. (1974). Torrey Canyon: a case study in accidental pollution. *Scottish Journal of Political Economy*, **21**, 237–258.

Butler, J. N., Wells, P. G., Johnson, S. & Manock, J. J. (1998). Beach tar on Bermuda: recent observations and implications for global monitoring. *Marine Pollution Bulletin*, **36**, 458–463.

Camplin, D. J. & Hall, R. (2014). Neogene history of Bone Gulf, Sulawesi, Indonesia. *Marine & Petroleum Geology*, **57**, 88–108.

Canevari, G. P. (1969). General dispersant theory. In *Proceedings of the International Oil Spill Conference*. Washington, DC: American Petroleum Institute, pp. 171–177.

Carstensen, J., Andersen, J. H., Gustafsson, B. G. & Conley, D. J. (2014). Deoxygenation of the Baltic Sea during the last century. *Proceedings of the National Academy of Sciences of the United States of America*, **111**, 5628–5633.

Chakrabarty, A. N. (1981). *Microorganisms Having Multiple Compatible Degradative Energy-Generating Plasmids and Preparation Thereof*, US Patent 4, 259,444.

Chen, J., Zeng, X. & Deng, Y. (2016). Environmental pollution and shipping feasibility of the Nicaragua Canal. *Marine Pollution Bulletin*, **113**, 87–93.

Cho, Y., Ahmed, A., Islam, A. & Kim, S. (2015). Developments in FT-ICR MS instrumentation, ionization techniques, and data interpretation methods for petroleomics. *Mass Spectrometry Reviews*, **34**, 248–263.

Clarke, K. C. & Hemphill, J. J. (2002). The Santa Barbara oil spill: a retrospective. *Yearbook of the Association of Pacific Coast Geographers*, **64**, 157–162.

Colcomb, K., Bedborough, D., Lunel, T. *et al.* (1997). Shoreline cleanup and waste disposal issues during the Sea Empress incident. In *Proceedings of the International Oil Spill Conference*. Washington, DC: American Petroleum Institute, pp. 195–198.

Del Sontro, T. S., Leifer, I., Luyendyk, B. P., & Broitman, B. R. (2007). Beach tar accumulation, transport mechanisms, and sources of variability at Coal Oil Point, California. *Marine Pollution Bulletin*, **54**, 1461–1471.

Edgar, G. J., Snell, H. L. & Lougheed, L. W. (2003). Impacts of the *Jessica* oil spill: an introduction. *Marine Pollution Bulletin*, **47**, 273–275.

EIA (2020a). Crude oil input qualities. www.eia.gov/dnav/pet/pet_pnp_crq_a_EPC0_YCG_d_a.htm

EIA (2020b). Energy outlook. www.eia.gov/outlooks/steo/report/global_oil.cfm

EIA (2020c). Oil medium term market outlook. www.iea.org/search?q=Oil%20medium%20term%20market%20outlook

El-Shazly, N. E. (2016). *The Gulf Tanker War: Iran and Iraq's Maritime Swordplay*. Berlin: Springer.

Etiope, G. (2015). *Natural Gas Seepage: The Earth's Hydrocarbon Degassing*. Berlin: Springer.

Etiope, G., Panieri, G., Fattorini, D. *et al.* (2014). A thermogenic hydrocarbon seep in shallow Adriatic Sea (Italy): gas origin, sediment contamination and benthic foraminifera. *Marine & Petroleum Geology*, **57**, 283–293.

Farwell, C., Reddy, C. M., Peacock, E. *et al.* (2009). Weathering and the fallout plume of heavy oil from strong petroleum seeps near Coal Oil Point, CA. *Environmental Science & Technology*, **43**, 3542–3548.

Fingas, M. (2012). *The Basics of Oil Spill Cleanup*. Boca Raton, FL: CRC Press.

Fingas, M. (2013). Modeling oil and petroleum evaporation. *Journal of Petroleum Science Research*, **2**, 104–115.

Fingas, M. & Fieldhouse, B. (2012). Studies on water-in-oil products from crude oils and petroleum products. *Marine Pollution Bulletin*, **64**, 272–283.

Fitzmaurice, M. (1993). The new Helsinki convention on the protection of the marine environment of the Baltic Sea area. *Marine Pollution Bulletin*, **26**, 64–67.

Fodrie, F. J. & Heck Jr., K. L. (2011). Response of coastal fishes to the Gulf of Mexico oil disaster. *PLoS One*, **6**, e21609.

Gallagher, J. J., Hile, H. B. & Miller, J. A. (2001). The old New Carissa; a study in patience. In *Proceedings of the International Oil Spill Conference*. Washington, DC: American Petroleum Institute. pp. 85–90.

Gallaway, B. J., Konkel, W. J. & Norcross, B. L. (2017). Some thoughts on estimating change to Arctic cod populations from hypothetical oil spills in the eastern Alaska Beaufort Sea. *Arctic Science*, **3**, 716–729.

Garneau, M. È., Michel, C., Meisterhans, G. *et al.* (2016). Hydrocarbon biodegradation by Arctic sea-ice and sub-ice microbial communities during microcosm experiments, Northwest Passage (Nunavut, Canada). *FEMS Microbiology Ecology*, **92**, fiw130.

Garrett, R. M., Pickering, I. J., Haith, C. E. & Prince, R. C. (1998). Photooxidation of crude oils. *Environmental Science & Technology*, **32**, 3719–3723.

The Global Response Network (2017). The Global Response Network. https://globalresponsenetwork.org

Goodman, R. (2003). Tar balls: the end state. *Spill Science & Technology Bulletin*, **8**, 117–121.

Griffin, A. (1994). MARPOL 73/78 and vessel pollution: a glass half full or half empty? *Indiana Journal of Global Legal Studies*, **1**, 489–513.

Gunter, C. T. (2014). Potential impacts from a worst case discharge from an United States offshore wind farm. In *Proceedings of the International Oil Spill Conference*. Washington, DC: American Petroleum Institute, pp. 869–877.

Harris, C. (1995). The Braer incident: Shetland Islands, January 1993. In *Proceedings of the International Oil Spill Conference*. Washington, DC: American Petroleum Institute, pp. 813–819.

Hayes, D. G. (2017). Commentary: the relationship between "biobased," "biodegradability" and "environmentally-friendliness" (or the absence thereof). *Journal of the American Oil Chemists Society*, **94**, 1329–1331.

Hazen, T. C., Prince, R. C. & Mahmoudi, N. (2016). Marine oil biodegradation. *Environmental Science & Technology*, **50**, 2121–2129.

Hein, F. J. (2013). Overview of heavy oil, seeps, and oil (tar) sands, California. In F. J. Hein, D. Leckie, S. Larter & J. R. Suter, eds., *Heavy-Oil and Oil-Sand Petroleum*

Systems in Alberta and Beyond: AAPG Studies in Geology #64. Tulsa, OK: American Association of Petroleum Geologists, pp. 407–435.

HELCOM (2016). Illegal discharges of oil in the Baltic Sea. http://helcom.fi/baltic-sea-trends/environment-fact-sheets/hazardous-substances/illegal-discharges-of-oil-in-the-baltic-sea

Helm, R. C., Costa, D. P., DeBruyn, T. D. *et al.* (2014). Overview of effects of oil spills on Marine Mammals. In M. Fingas, ed., *Handbook of Oil Spill Science and Technology*. Hoboken, NJ: John Wiley, pp. 455–475.

Henkel, L. A., Nevins, H., Martin, M. *et al.* (2014). Chronic oiling of marine birds in California by natural petroleum seeps, shipwrecks, and other sources. *Marine Pollution Bulletin*, **79**, 155–163.

Hughey, C. A., Galasso, S. A. & Zumberge, J. E. (2007). Detailed compositional comparison of acidic NSO compounds in biodegraded reservoir and surface crude oils by negative ion electrospray Fourier transform ion cyclotron resonance mass spectrometry. *Fuel*, **86**, 758–768.

IMO (2006). Pollution prevention equipment required under MARPOL 73/78. www.amnautical.com/products/pollution-prevention-equipment-under-marpol-2006-edition#.V7pT3Y73rxA

IMO (2016). International Convention for the Prevention of Pollution from Ships (MARPOL). www.imo.org/en/About/Conventions/ListOfConventions/Pages/International-Convention-for-the-Prevention-of-Pollution-from-Ships-(MARPOL).aspx

IMO (2017a). IMO Profile. Overview. https://business.un.org/en/entities/13

IMO (2017b). Oil spill organizations and resource providers. www.imo.org/en/OurWork/Environment/PollutionResponse/OilPollutionResources/Pages/Oil%20Spill%20Organizations%20and%20Resource%20Providers.aspx

ITOPF (2018). Oil tanker spill statistics 2017. www.itopf.com/knowledge-resources/data-statistics/statistics

Jackson, M. J., Powell, T. G., Summons, R. E. & Sweet, I. P. (1986). Hydrocarbon shows and petroleum source rocks in sediments as old as 1.7×10^9 years. *Nature*, **322**, 727–729.

Kasting, J. F. & Siefert, J. L. (2002). Life and the evolution of Earth's atmosphere. *Science*, **296**, 1066–1068.

Kerschner, C. & Hubacek, K. (2009). Assessing the suitability of input–output analysis for enhancing our understanding of potential economic effects of peak oil. *Energy*, **34**, 284–290.

Kim, S., Stanford, L. A., Rodgers, R. P. *et al.* (2005). Microbial alteration of the acidic and neutral polar NSO compounds revealed by Fourier transform ion cyclotron resonance mass spectrometry. *Organic Geochemistry*, **36**, 1117–1134.

Kucuksezgin, F., Pazi, I., Gonul, L. T. & Duman M. (2013). Distribution and sources of polycyclic aromatic hydrocarbons in Cilician Basin shelf sediments (NE Mediterranean). *Marine Pollution Bulletin*, **71**, 330–335.

Kvenvolden, K. A. & Cooper, C. K. (2003). Natural seepage of crude oil into the marine environment. *Geo-Marine Letters*, **23**, 140–146.

Lee, K., Nedwed, T., Prince, R. C. & Palandro, D. (2013). Lab tests on the biodegradation of chemically dispersed oil should consider the rapid dilution that occurs at sea. *Marine Pollution Bulletin*, **73**, 314–318.

Lemkau, K. L., Peacock, E. E., Nelson, R. K. *et al.* (2010). The M/V *Cosco Busan* spill: source identification and short-term fate. *Marine Pollution Bulletin*, **60**, 2123–2129.

Lessard, R. R. & DeMarco, G. (2000). The significance of oil spill dispersants. *Spill Science & Technology Bulletin*, **6**, 59–68.

MacDonald, I. R., Garcia-Pineda, O., Beet, A. *et al.* (2015). Natural and unnatural oil slicks in the Gulf of Mexico. *Journal of Geophysical Research: Oceans*, **120**, 8364–8380.

Martinelli, M., Luise, A., Tromellini, E. *et al.* (1995). The M/C Haven oil spill: environmental assessment of exposure pathways and resource injury. In *Proceedings of the International Oil Spill Conference*. Washington, DC: American Petroleum Institute, pp. 679–685.

McFarlin, K. M., Prince, R. C., Perkins, R. & Leigh, M. B. (2014). Biodegradation of dispersed oil in arctic seawater at –1°C. *PLoS One*, **9**, e84297.

McKirdy, D. M., Aldridge, A. K. & Ypma, P. J. (1983). A geochemical comparison of some crude oils from pre-Ordovician carbonate rocks. In M. Bjoroy, ed., *Advances in Organic Geochemistry 1981: International Conference Proceedings*. Hoboken, NJ: Wiley-Blackwell, pp. 99–107.

McNutt, M. K., Chu, S., Lubchenco, J. *et al.* (2012). Applications of science and engineering to quantify and control the *Deepwater Horizon* oil spill. *Proceedings of the National Academy of Sciences of the United States of America*, **109**, 20222–20228.

Mendes, S. D., Redmond, M. C., Voigritter, K. *et al.* (2015). Marine microbes rapidly adapt to consume ethane, propane, and butane within the dissolved hydrocarbon plume of a natural seep. *Journal of Geophysical Research: Oceans*, **120**, 1937–1953.

Monfils, R. (2005). The global risk of marine pollution from WWII shipwrecks: examples from the seven seas. In *Proceedings of the International Oil Spill Conference*. Washington, DC: American Petroleum Institute, pp. 1049–1054.

Mongelluzzo, W. (2016). No easy solutions to problems plaguing US ports. www.joc.com/port-news/us-ports/no-easy-solutions-problems-plaguing-us-ports_20160324.html

Mudge, S. (1997). Can vegetable oils outlast mineral oils in the marine environment? *Marine Pollution Bulletin*, **34**, 213.

Nauman, S. A. (1991). Shoreline clean-up: equipment and operations. In *Proceedings of the International Oil Spill Conference*. Washington, DC: American Petroleum Institute, pp. 141–148.

NRC (1998). *Oil Spill Risks from Tank Vessel Lightering*. Washington, DC: National Academies Press.

NRC (2003). *Oil in the Sea III*. Washington, DC: National Academies Press.

Owens, E. H., Taylor, E. & Humphrey, B. (2008). The persistence and character of stranded oil on coarse-sediment beaches. *Marine Pollution Bulletin*, **56**, 14–26.

Palminteri, J. (2014). Underwater seep tents dry up off Goleta. https://keyt.com/news/2014/07/23/underwater-seep-tents-dry-up-off-goleta

Peters, K. E., Walters, C. C. & Moldowan, J. M. (2005). *The Biomarker Guide*. Cambridge: Cambridge University Press.

Prince, R. C. (2010). Eukaryotic hydrocarbon degraders. In K. N. Timmis, ed., *Handbook of Hydrocarbon and Lipid Microbiology*. Berlin: Springer-Verlag, pp. 2066–2078.

Prince, R. C. (2015). Oil spill dispersants: boon or bane? *Environmental Science & Technology*, **49**, 6376–6384.

Prince, R. C. (2016). Biostimulation of marine crude oil spills using dispersants. In T. J. McGenity, K. N. Timmis & B. Nogales, eds., *Hydrocarbon and Lipid Microbiology Protocols*. Berlin: Springer, pp. 95–104.

Prince, R. C. & Walters, C. C. (2016). Biodegradation of oil and its implications for source identification. In S. A. Stout & Z. Wang, eds., *Standard Handbook Oil Spill Environmental Forensics*, 2nd ed. Cambridge, MA: Academic Press, pp. 869–916.

Prince, R. C., Gramain, A. & McGenity, T. J. (2010). Prokaryotic hydrocarbon degraders. In K. N. Timmis, ed., *Handbook of Hydrocarbon and Lipid Microbiology*. Berlin: Springer-Verlag, pp. 1672–1692.

Prince, R. C., McFarlin, K. M., Butler, J. D. *et al.* (2013). The primary biodegradation of dispersed crude oil in the sea. *Chemosphere*, **90**, 521–526.

Prince, R. C., Nash, G. W. & Hill, S. J. (2016). The biodegradation of crude oil in the deep ocean. *Marine Pollution Bulletin*, **111**, 354–357.

Prince, R. C., Butler, J. D. & Redman, A. D. (2017). The rate of crude oil biodegradation in the sea. *Environmental Science & Technology*, **51**, 1278–1284.

Quigley, D. C., Hornafius, J. S., Luyendyk, B. P. *et al.* (1999). Decrease in natural marine hydrocarbon seepage near Coal Oil Point, California, associated with offshore oil production. *Geology*, **27**, 1047–1050.

Redman, A. D., Parkerton, T. F., McGrath, J. A. & Di Toro, D. M. (2012). PETROTOX: an aquatic toxicity model for petroleum substances. *Environmental Toxicology & Chemistry*, **31**, 2498–2506.

Savage, C. (2014). Guantánamo detainee pleads guilty in 2002 attack on tanker off Yemen. *New York Times*. www.nytimes.com/2014/02/21/us/guantanamo-detainee-ahmed-muhammed-haza-al-darbi.html?_r=0

Selley, R. C. & Van der Spuy, D. (2016). The oil and gas basins of Africa. *Episodes* **39**, 429–445.

Simoneit, B. R. & Kvenvolden, K. A. (1994). Comparison of ^{14}C ages of hydrothermal petroleums. *Organic Geochemistry*, **21**, 525–529.

Smil, V. (2006). Peak oil: a catastrophist cult and complex realities. *World Watch*, **19**, 22–24.

Smith, S. R. & Knap, A. H. (1985). Significant decrease in the amount of tar stranding on Bermuda. *Marine Pollution Bulletin*, **16**, 19–21.

Speight, J. G. (2014). *The Chemistry and Technology of Petroleum*, 5th ed. Boca Raton, FL: CRC Press.

Stanford, L. A., Kim, S., Klein, G. C., *et al.* (2007). Identification of water-soluble heavy crude oil organic-acids, bases, and neutrals by electrospray ionization and field desorption ionization Fourier transform ion cyclotron resonance mass spectrometry. *Environmental Science & Technology*, **41**, 2696–2702.

Statista (2018). Number of ships in the world merchant fleet between January 1, 2008 and January 1, 2017, by type. www.statista.com/statistics/264024/number-of-merchant-ships-worldwide-by-type

Sutton, C. & Calder, J. A. (1975). Solubility of alkylbenzenes in distilled water and sea water at 25.0°. *Journal of Chemical and Engineering Data*, **20**, 320–322.

Tissot, B. P. & Welte, D. H. (1984). *Petroleum Formation and Occurrence*. Berlin: Springer-Verlag.

US Coast Guard (2016). Oil Pollution Act of 1990 (OPA). www.uscg.mil/Mariners/National-Pollution-Funds-Center/About_NPFC/OPA

US Congress House Committee on Merchant Marine and Fisheries, Subcommittee on Coast Guard and Navigation (1989). *Liability for Oil Pollution Damages*. Washington, DC: US Government Printing Office.

US Department of Justice (2011). Ship owners and operators to pay $44 million in damages and penalties for 2007 Bay Bridge crash and oil spill. https://casedocuments.darrp.noaa.gov/southwest/cosco/pdf/09-19-11_Cosco%20Busan_Overall_Consent_Decree_Press_Release%20FINAL.pdf

USDOE (2008). Strategic petroleum reserve crude oil assay manual. www.spr.doe.gov/reports/docs/CrudeOilAssayManual.pdf

USEPA (2008). Global trade and fuels assessment – future trends and effects of requiring clean fuels in the marine sector. https://nepis.epa.gov

USEPA (2011). Environmentally acceptable lubricants EPA 800-R-11-002. https://nepis.epa.gov

Varfolomeev, S. D., Karpov, G. A., Synal, H. A., Lomakin, S. M. & Nikolaev, E. N. (2011). The youngest natural oil on Earth. *Doklady Chemistry*, **438**, 144–147.

Vonk, S. M., Hollander, D. J. & Murk, A. J. (2015). Was the extreme and wide-spread marine oil-snow sedimentation and flocculent accumulation (MOSSFA) event during the *Deepwater Horizon* blow-out unique? *Marine Pollution Bulletin*, **100**, 5–12.

Wade, T. L., Sericano, J. L., Sweet, S. T., Knap, A. H. & Guinasso, N. L. (2016). Spatial and temporal distribution of water column total polycyclic aromatic hydrocarbons (PAH) and total petroleum hydrocarbons (TPH) from the *Deepwater Horizon* (Macondo) incident. *Marine Pollution Bulletin*, **103**, 286–293.

Warnock, A. M., Hagen, S. C. & Passeri, D. L. (2015). Marine tar residues: a review. *Water Air & Soil Pollution* **226**, 1–24.

Wiltshire, G. A. & Corcoran, L. (1991). Response to the Presidente Rivera major oil spill, Delaware River. In *Proceedings of the International Oil Spill Conference*. Washington, DC: American Petroleum Institute, pp. 253–258.

Wolborsky, S. L. (2014). *Choke Hold: The Attack on Japanese Oil in World War II*. Auckland: Pickle Partners Publishing.

Word, J. Q., Clark, J. R. & Word, L. S. (2013). Comparison of the acute toxicity of Corexit 9500 and household cleaning products. *Human & Ecological Risk Assessment*, **21**, 707–725.

Zahed, M. A., Aziz, H. A., Isa, M. H. & Mohajeri, L. (2010). Effect of initial oil concentration and dispersant on crude oil biodegradation in contaminated seawater. *Bulletin of Environmental Contamination & Toxicology*, **84**, 438–442.

4

Waste and Sewage

C. MICHAEL HALL

4.1 Introduction

Marine litter and waste has become one of the major environmental issues of the early twenty-first century. Around 6.4 million tonnes of litter are deposited into the oceans each year (UNEP, 2005), a figure that continues to grow as a result of a variety of social and economic factors, including consumerism and the purchase of single-use products, coastal urbanization, shipping, poor waste management and the use of plastics. Indeed, as Bergmann *et al.* (2015, p. x) noted, 'The ubiquity of litter in the open ocean is prominently illustrated by numerous images of floating debris from the ocean garbage patches and by the fact that the search for the missing Malaysia Airlines flight MH370 in March 2014 produced quite a few misidentifications caused by litter floating at the water surface.' Eriksen *et al.* (2014) estimated that there was a minimum of 5.25 trillion plastic particles weighing 268,940 tons afloat in the sea, but this figure does not include debris on beaches or on the sea floor. Galgani *et al.* (2015, p. 29) suggest that plastics 'typically constitute the most important part of marine litter sometimes accounting for up to 100 % of floating litter', while 90 per cent of litter caught in benthic trawls is also plastic (Galgani *et al.*, 2015). However, it is important to note that even though the amount of plastic that is produced is increasing, and is expected to continue to do so in the future (Taylor, 2017), the 'predominance of plastics in litter is not the result of relatively more plastics being littered compared to paper, paperboard or wood products reaching the oceans, but because of the exceptional durability or persistence of plastics in the environment' (Andrady, 2015, p. 58).

Growing awareness of the impacts of plastics on the marine environment (Bergmann *et al.*, 2015), together with concerns over discarded fishing nets and the environmental effects of the rapidly growing cruise ship sector (Klein, 2002), especially in relation to sewage discharge (Hall *et al.*, 2017), has meant that there is a high media profile and a strong public interest in shipping waste. Of course, much of the waste and litter that are in the ocean are from land rather than marine sources (National Research Council, 2008). Nevertheless, government and consumer concern over the problem means that the issue of the production and sustainable disposal of waste from the world's shipping fleet is now receiving more attention than ever before.

Marine litter, also known as marine debris, is defined by the United Nations Environment Programme (UNEP) as 'any persistent, manufactured or processed solid material

Table 4.1. *Sources of marine debris, litter, waste and sewage*

Main sea-based sources	Main land-based sources
Merchant shipping, ferries and cruise liners	Municipal landfills (waste dumps) located on the coast
Fishing vessels	Riverine transport of waste from landfills and litter along inland waterways
Government vessels: military fleets and research vessels	Discharges of untreated sewage and storm water (including overflows)
Private pleasure craft	Discharges from industrial facilities (accidental spills, solid waste from landfills and untreated waste water)
Offshore oil and gas platforms	Riverine transport of agricultural waste and run-off along inland waterways
Aquaculture facilities	Tourism, recreation and other leisure activities at the coast

Sources: Derraik (2002), Galgani *et al.* (2015) and Strand *et al.* (2015)

discarded, disposed of or abandoned in the marine and coastal environment' (UNEP, 2005, p. 3), and it includes a range of materials, including plastics, metals, sanitary waste, paper, cloth, wood, glass, rubber and pottery (Galgani *et al.*, 2013). Additional related sources of marine pollution, although not usually defined as marine litter but still posing a significant marine waste management problem, include semi-solid remains of, for example, vegetable and mineral oils, sewage and chemicals. Marine litter includes any anthropogenic item deliberately discarded or unintentionally lost in the sea and along the coastline or transported into the marine environment by rivers, drainage, sewage systems, coastal processes or winds (García-Rivera *et al.*, 2017). The Office for the London Convention/Protocol and Ocean Affairs (2016) argue that marine litter can be broadly categorized according to its source, being either land-based or sea-based. Land-based sources include urban areas, domestic and industrial activities, tourism, ports and harbours and agriculture. These enter the marine environment through rivers, ephemeral streams, wastewater and sewage inputs, as well as from wave action on the coast. Sea-based sources include fisheries, recreational boats, commercial shipping, ferries, energy production, aquaculture and legal and illegal dumping activities (Table 4.1). Sea-based sources can be further subdivided into operational waste (e.g., fishing equipment), non-operational waste (e.g., galley waste), public littering (e.g., from pleasure craft) and sewage. At a global scale, land-based sources are estimated to contribute approximately 80 per cent of maritime litter, with the remaining 20 per cent coming from sea-based sources (Strand *et al.*, 2015; UNEP, 2005). However, a complicating matter in differentiating between ship-based waste and other waste sources is that many ships deposit waste in ports, from which it may then go to landfill sites or other locations where, if poorly managed, it may then enter the marine environment.

Marine waste and litter constitute significant direct and indirect threats to human health and safety, as well as to the wider environment (Bakir *et al.*, 2014; Bergmann *et al.*, 2015; Laist, 1987; Ryan, 2015; Setälä *et al.*, 2014). Significantly, marine litter also poses a direct economic threat to marine-dependent industries and fishing, shipping and tourism (Mouat *et al.*, 2010; Newman *et al.*, 2015).

Commercial fisheries are affected by fish stock becoming by-catch in discarded fishing nets ('ghost catch') (Anderson & Alford, 2014), damage to fishing vessels (Jones, 1995) and the generation of negative perceptions of seafood products. Mouat *et al.* (2010) examined the annual direct economic impact of marine litter on Scottish fishing vessels (i.e., the costs of repairs and direct losses to earnings) and estimated that, on average, marine litter costs each fishing vessel €17,000–€19,000 per year. Two-thirds of this cost (€12,000) were incurred in the time spent clearing litter from nets (calculated using the average value of 1 hour's fishing time as estimated by participating vessels in the project). Overall, the direct costs to the Scottish fishing industry were therefore between €11.7 and €13.0 million every year, equivalent to approximately 5 per cent of the fleet's annual revenue (Mouat *et al.*, 2010). Significantly, these economic estimates did not include the potential economic costs of the ghost catch. The shipping industry is threatened by the hazard that marine litter poses to navigation and the damage it can cause to the water intakes on ships. Approximately 3.5 per cent of all call-outs by the Royal National Lifeboat Institution (RNLI) in the UK are for fouled propellers/impellers (RNLI, 2017).

Tourism is impacted by the aesthetic damage that marine waste can create in coastal areas (Polasek *et al.*, 2017; Rangel-Buitrago *et al.*, 2017), which can affect destination images and branding, as well as by its effect on tourism resources such as beaches, coral reefs and marine wildlife, and charismatic fauna in particular (Eagle *et al.*, 2016). The cost of removing beach litter may be substantial: in the UK, the cost to coastal municipalities was estimated to be in the region of €18–19 million per annum, and €10.4 million per annum in Belgium and The Netherlands (Mouat *et al.*, 2010). Costs are also incurred by harbours and marinas in clearing marine litter, although no systematic work is available. Mouat *et al.* (2010) reported that one UK marina had an annual bill of €39,000 for litter removal.

Impacts to fisheries and tourism also represent some of the potential effects of marine litter on ecosystem services (McIlgorm *et al.*, 2011). For the 21 economies of the Asia–Pacific Rim, McIlgorm *et al.* (2011) estimated that marine debris-related damage to marine industries cost US$1.26 billion per annum in 2008 terms, equivalent to 0.3 per cent of the gross domestic product for the marine sector of the region. Some forms of marine-based tourism, and especially cruise ships, are also potentially impacted by the reputational costs of marine waste if they are also perceived as being a potential contributor to the marine litter problem (Hall *et al.*, 2017). Although floating macroalgae, wood and volcanic pumice have been part of the natural marine flotsam assemblage for millions of years, marine litter has some significantly different characteristics: (1) it tends to remain afloat for longer; (2) it is more widespread; (3) it is increasingly present in oceanic regions where natural floating substrata occur less frequently; and (4) anthropogenic litter is of no nutritional value to most organisms (Kiessling *et al.*, 2015).

One of the most difficult issues in managing and costing the environmental impacts of ship-based waste and sewage is being able to determine just how much waste shipping is responsible for. Substantial spatial variations may exist in terms of the relative significance of land and marine sources because of their different industrial activities and waste management cultures, while different reporting regimes will also affect the extent of the information collected. Waste generation statistics may not be very accurate in some jurisdictions because of poor record keeping, and illegal dumping and incineration may mean that some waste contributions are not properly accounted for. Seas At Risk (2011) notes that around half of all ships calling at major ports use incinerators. However, in addition to the serious concerns over the emission of dangerous pollutants from incinerators, they argue that 'the practice of incineration makes it difficult for authorities to assess whether or not a ship has complied with regulations as incinerator ashes are both difficult to measure and associate to types and quantities of garbage items' (Seas At Risk, 2011, p. 4). Sheavly and Register (2007) suggest that, at a global level, only about 27 per cent of all ship waste is delivered to reception facilities, while the majority is dumped or incinerated.

In an early analysis, Horsman (1982) estimated that merchant ships dumped 639,000 plastic containers a day around the world and therefore represented a major source of plastic debris. In a report for the European Parliament, TRT Trasporti e Territorio Srl (2007) provided an estimate of the amount of waste generated by the European Union (EU) and global shipping fleets of over 100 GT (gross tonnage). The study used a sample of ship garbage management plans and garbage record books to develop a solid waste balance of a typical 20-crew international cargo ship, with certain material types being parameterized for other ship types (roll-on/roll-off (Ro-Ro) and only passenger ships) in proportion to the number of passengers transported. TRT Trasporti e Territorio Srl (2007) estimated that 10.5 million m^3 of solid waste was produced by the world fleet in 2006, of which 30 per cent (3.1 million m^3) was by the EU fleet. A total of 41 per cent of solid waste belongs to MARPOL Annex V, Category 4 (other garbage including comminuted paper, glass, metals, etc., in order to save holding spaces), 29 per cent is food waste and 14 per cent is plastics. However, at a global level, TRT Trasporti e Territorio Srl (2007; see also Sheavly & Register, 2007) report that only approximately 27 per cent of ship waste production appeared to have been given to reception facilities, with the majority being discharged or incinerated (Table 4.2). This figure changed substantially if ships were operating in Annex V special areas (e.g., the Mediterranean, North Sea, Baltic), and particularly the EU fleet Ro-Ro and passenger ships, because of different waste management requirements, leading to 51 per cent of solid waste in the EU fleet going to reception facilities instead of being incinerated or discharged at sea (Table 4.3).

A major issue with respect to the assessment of the waste contribution of shipping to the oceans is not only whether desired standards and reporting measures have been followed, but also whether the contributions of many fishing and recreational vessels have been included. For example, in the early 1990s, recreational leisure and fishing craft were identified as being responsible for approximately 52 per cent of all rubbish dumped in US waters (Derraik, 2002). Lost fishing equipment is consistently reported as the main component of marine litter in those areas that experience a high degree of fisheries pressure – examples include Blanes Canyon in the Catalan Sea in the north-west Mediterranean (Ramirez-Llodra *et al.*, 2013), California (Moore & Allen, 2000), the Celtic Sea (Galgani *et al.*, 2000), the East China Sea (Lee *et al.*, 2006) and the Nazaré Canyon in the West Iberian Margin of the Atlantic off Portugal (Mordecai *et al.*, 2011). In their study of the composition, spatial distribution and source of marine litter recovered by commercial

Table 4.2. *Solid waste of the world fleet produced on board (MARPOL Annex V categories) by category and disposal (ships >100 GT)*

Impact type	Total ship-generated solid waste (before disposal) (m^3/year)	Category 1, plastics (m^3/year)	Category 2, floating packaging covering materials (m^3/year)	Category 3, paper, rags, glass, metals, bottles and other similar residues (m^3/year)	Category 4, triturated paper, rags, glass, metals, bottles, etc. (m^3/year)	Category 5, food waste (m^3/year)	Category 6, other waste mixed with dangerous substances (m^3/year)	Total solid waste discharged overboard or incinerated (m^3/year)	Total solid waste to reception facilities (m^3/year)
Tanker (oil, chemicals, liquified gas, others)	957,541	108,716	85,177	51,106	323,672	229,977	158,893	719,201	238,340
Bulk carriers	526,379	59,764	46,823	28,094	177,928	126,423	87,347	395,359	131,020
General and specialized cargo	1,450,574	164,694	129,034	77,420	490,328	348,391	240,707	1,089,514	361,060
Containers and reefers	391,469	44,446	34,823	20,894	132,326	94,021	64,960	294,029	97,440
Ro-Ro passenger and Ro-Ro cargo	5,587,794	840,206	25,752	394,969	2,501,468	1,777,359	48,040	4,479,537	1,108,257

Table 4.2. (*cont.*)

Impact type	Total ship-generated solid waste (before disposal) (m^3/year)	Category 1, plastics (m^3/year)	Category 2, floating packaging covering materials (m^3/year)	Category 3, paper, rags, glass, metals, bottles and other similar residues (m^3/year)	Category 4, triturated paper, rags, glass, metals, bottles, etc. (m^3/year)	Category 5, food waste (m^3/year)	Category 6, other waste mixed with dangerous substances (m^3/year)	Total solid waste discharged overboard or incinerated (m^3/year)	Total solid waste to reception facilities (m^3/year)
Total cargo and cargo/passenger ships	8,913,758	1,217,826	321,609	572,482	3,625,722	2,576,171	599,947	6,977,641	1,936,117
Cruise and passenger ships	1,576,481	230,545	22,157	108,376	686,380	487,691	41,333	632,191	944,291
Total	10,490,239	1,448,371	343,766	680,858	4,312,102	3,063,862	641,280	7,609,832	2,880,408

Sources: EMAS-Ship (2006) in TRT Trasporti e Territorio Srl (2007)

Table 4.3. *Solid waste of the EU fleet produced on board (MARPOL Annex V categories) by category and disposal (ships >100 GT)*

Impact type	Total ship-generated solid waste (before disposal) (m^3/year)	Category 1, plastics (m^3/year)	Category 2, floating packaging covering materials (m^3/year)	Category 3, paper, rags, glass, metals, bottles and other similar residues (m^3/year)	Category 4, triturated paper, rags, glass, metals, bottles, etc. (m^3/year)	Category 5, food waste (m^3/year)	Category 6, other waste mixed with dangerous substances (m^3/year)	Total solid waste discharged overboard or incinerated (m^3/year)	Total solid waste to reception facilities (m^3/year)
Tanker (oil, chemicals, LG, others)	166,085	18,857	14,774	8864	56,141	39,889,	27,560	124,745	41,340
Bulk carriers	102,769	11,668	9142	5485	34,738	24,682	17,053	77,189	25,580
General and specialized cargo	188,262	21,375	16,747	10,048	63,637	45,216	31,240	141,402	46,860
Containers and reefers	83,806	9515	7455	4473	28,328	20,128	13,907	62,946	20,860
Ro-Ro passener and Ro-Ro cargo	2,241,011	336,969	10,328	158,404	1,003,225	712,818	19,267	988,679	1,252,332
Total cargo and cargo/ pax ships	2,781,933	398,383	58,445	187,274	1,186,070	842,734	109,027	1,394,961	1,386,972
Cruise and passenger ships	360,335	52,223	6147	24,549	155,478	110,471	11,467	144,691	215,644
Total	3,142,268	450,606	64,592	211,823	1,341,548	953,205	120,493	1,539,652	1,602,616

Sources: EMAS-Ship (2006) in TRT Trasporti e Territorio Srl (2007)

trawlers in the Spanish Southeast Mediterranean, García-Rivera *et al.* (2017) reported that fishing activity was the source of 29.16 per cent of macro-marine litter, almost 68.10 per cent of plastic litter and 25.10 per cent of metal litter. The sources of the remaining litter could not be directly identified, revealing the high degree of uncertainty that usually exists regarding the specific origins of marine litter. However, following a qualitative analysis of marine traffic, these authors suggested that the likely sources in open waters were merchant ships and recreational and fishing vessels in coastal waters (García-Rivera *et al.*, 2017). Although the exact contribution of ship waste to the overall extent of waste found in the oceans is difficult to ascertain, it is clearly substantial, while some insights are available from the analysis of marine litter.

Shipping has long been recognized as a significant factor in marine littering in the North Sea/German Bight given that it is one of the world's busiest shipping areas (Galgani *et al.*, 2000; Vauk & Schrey, 1987). Seas At Risk, an umbrella organization of European environmental non-governmental organizations that promotes policies for marine protection at European and international levels, suggests that in the North Sea up to 40 per cent of marine litter comes from the maritime sector, with as much as 90 per cent of the plastic found on beaches in The Netherlands originating from shipping and fisheries (Seas At Risk, 2011).

Research on the Norwegian and Barents seas found that the largest densities of marine litter occurred close to the coast in areas with high maritime activity (e.g., shipping and fisheries), where 5 tons/km^2 was not uncommon (Buhl-Mortensen & Buhl-Mortensen, 2017). Fisheries-related litter dominated the findings in the Norwegian Sea, although ship traffic, aquaculture and oil production were also identified as contributors. The locations with a high density of litter (2000 items or 1500 kg/km^2 or more) are in areas of high fishing intensity or in canyons and troughs. The highest density was >6000 items/km^2, which is 30 times the background value of 200 items/km^2. The Norwegian findings are also similar to the results from other areas with high fishing activities, such as on oceanic ridges and seamounts (Pham *et al.*, 2014; Woodall *et al.*, 2015).

Vieira *et al.* (2015) examined the distribution, type and abundance of marine litter in Ormonde and Gettysburg, the two seamounts of Gorringe Bank, located in the North Atlantic Ocean approximately 125 km south-west of Portugal. The Gorringe Bank is characterized by a high level of fishing activities, as well as being located on Atlantic and Mediterranean shipping routes. The origin of litter on the Gorringe Bank was mostly from maritime activities. Lost or discarded fishing gear accounted for the majority of the observed items (56 per cent), while other frequent types of litter included dumped glass bottles (15 per cent), metal (e.g., ship artefacts, tins and cans; 10 per cent) and plastics (10 per cent); a chair was also identified at 2340 m water depth. There were also spatial differences in the distribution of litter. Discarded and lost fishing gear was observed mostly on the summit and upper flank (~0–500 m depth) of the Gettysburg and Ormonde seamounts. Fishing gear, while still present, was gradually replaced by heavier items such as glass bottles and then metal at greater depths. Glass bottles were concentrated in the fished areas of the gentle southern slope of the Ormonde Seamount (500–1000 m depth). The predominant maritime origin of litter on the Gorringe Bank observed by Viera *et al.*

(2015) is consistent with the results obtained by Pham *et al.* (2014) on the Condor Seamount in the Azores, where lost fishing gear accounted for 73 per cent of the debris found near the summit (185–265 m depth). The economic value of fish stocks on seamounts tends to be high, and Ressurreição and Giacomello (2013) demonstrated that spatial accessibility to such locations becomes a significant factor in marine littering, given that the pressure over these oceanic features increases with their proximity to fishing harbours. Nevertheless, the economic effects of marine litter on fishing fleets are also high.

4.2 Sewage and Grey Water

Under international maritime law, sewage (black water) is covered under Annex IV of the International Convention for the Prevention of Pollution from Ships, 1973, as modified by the Protocol of 1978 (MARPOL). Because of the large numbers of people they carry, sewage management is a major problem for the rapidly growing cruise ship sector (Hall *et al.*, 2017). Butt (2007) estimated that approximately 50 tonnes of sewage per day are produced by an average cruise ship, equating to between 20 and 40 L/person/day. In addition, cruise ships also discharge considerable quantities of grey water, which is other waste water from sources such as kitchens, laundries and showers (Gössling *et al.*, 2015). It has been estimated that a cruise ship of 3000 passengers can generate 340–960 m^3 of grey water daily (Guilbaud *et al.*, 2012). Like sewage, grey water also contains organic matter, which, because of the often coastal nature of much cruise traffic, can have considerable impacts on algal growth and eutrophication, as well as being a potential vector for biological invasion (Hall *et al.*, 2010). TRT Trasporti e Territorio Srl (2007) arrived at an overall estimate of more than 250 million tonnes of grey and black waters being discharged by global shipping in 2006, of which 25 per cent was due to the EU fleet. Based on assumptions regarding the extent to which maritime shipping met regulatory requirements and were fitted with treatment systems, TRT Trasporti e Territorio Srl (2007) estimated discharges of 46,000 tonnes/year of organic matter (biological oxygen demand) and of about 9000 tonnes of nitrogen substances and phosphorous legal discharges by international shipping of grey and black waters, 31 per cent of which is from the EU fleet. Nevertheless, assessments of liquid (non-oil) waste disposal (sewage sludge sent to reception facilities or burnt on board by incinerators or power plants) were limited to only two categories of vessel: (1) Ro-pax and Ro-Ro cargo ships and (2) cruise and passenger ships (Table 4.4) (TRT Trasporti e Territorio Srl, 2007).

Table 4.4. *Estimated level of sewage sludge of the world and EU fleets (ships >100 GT) sent to reception facilities or burnt on board by incinerators or power plants*

Vessel type	World fleet (tonnes/year)	EU fleet (tonnes/year)
Ro-pax and Ro-Ro cargo ships	460,498	184,685
Cruise and passenger ships	126,356	28,622

Sources: Danish Shipowners' Association (2000) in TRT Trasporti e Territorio Srl (2007)

4.3 Environmental Impacts of Waste

The entanglement of wildlife in marine litter, often discarded or lost fishing nets and ropes (Baulch & Perry, 2014), is regarded as the most visible effect of plastic pollution on marine organisms (Kühn *et al.*, 2015). Entanglement in discarded fishing equipment can affect the mobility of animals and their capacity to eat and breathe (Baulch & Perry, 2014). Although they are likely primarily derived more from land-based sources than shipping, plastic debris can also be ingested by animals (fish, invertebrates, marine birds, marine mammals, turtles) and affect individual fitness, with subsequent consequences for reproduction and survival, even if they do not directly cause mortality. The number of species known to have been affected by either entanglement with or ingestion of plastic debris has doubled between the assessments conducted by Laist in 1997 and Kühn *et al.* in 2015, from 267 to 557 species among all groups of wildlife. For marine turtles, the number of affected species increased from 86 to 100 per cent (7 of 7 species), for marine mammals from 43 to 66 per cent (81 of 123 species) and for seabirds from 44 to 50 per cent (203 of 406 species) (Kühn *et al.*, 2015). In addition, marine litter, and especially plastic and other synthetic materials, represent a long-lived substrate that can transport hitch-hiking 'alien' species horizontally to ecosystems elsewhere (Kiessling *et al.*, 2015) or vertically from the sea surface to the sea floor (Kühn *et al.*, 2015), and therefore present significant potential for biological invasion and marine ecosystem change (Molnar *et al.*, 2008). For example, Kiessling *et al.* (2015) report that 387 taxa, including prokaryotic and eukaryotic microorganisms, seaweeds and invertebrates, have been found rafting on floating litter in all major oceanic regions. Most of these taxa (335) were associated with plastic substrata, such as domestic waste, plastic fragments or buoys made of plastic. Significantly for the purposes of the present chapter, over a third of these taxa (132) were recorded from items that previously served maritime purposes, mainly buoys and fishing gear (Kiessling *et al.*, 2015).

Globally, hundreds of thousands of marine birds and mammals perish in active fishing gear (Read *et al.*, 2006; Žydelis *et al.*, 2013), along with non-target species, but there are no accurate estimates available for the actual number of animals that become entangled in fisheries' debris and litter. However, on the basis of species records, Kühn *et al.* (2015) suggest that the percentage of species recorded as entangled is high and includes 100 per cent of marine turtles (7 of 7 species), 67 per cent of seals (22 of 33 species), 31 per cent of whales (25 of 80 species) and 25 per cent of seabirds (103 of 406 species). Assessments of entanglement for fish, invertebrates and small species of reptiles are not available as they are not recorded for the many thousands of species that exist. In comparison to the listings by Laist (1997), the number of marine bird, turtle and mammal species with known entanglements increased from 89 (21 per cent) to 161 (30 per cent) between 1997 and 2015 (Kühn *et al.*, 2015), although entanglement trends for individual species are difficult to assess.

The estimated time over which discarded and lost fishing gear entangles and kills organisms varies substantially and is gear and site specific (Erzini *et al.*, 2008; Kühn *et al.*, 2015; Newman *et al.*, 2011). In a review of ghost fishing, Matsuoka *et al.* (2005) estimated catch durations of derelict gill and trammel nets of between 30 and 568 days, although ghost fishing efficiency appears to sometimes decrease exponentially (Ayaz *et al.*,

2006; Baeta *et al.*, 2009; Tschernij & Larsson, 2003). For example, in an experiment on 24 fleets of normal commercial gill nets deployed in a typical gill net fishing ground in the southern Baltic Sea, Tschernij and Larsson (2003) found 80 per cent of the catch in the experimental bottom gill nets during the first 3 months, thereafter stabilizing at around 5–6 per cent of the initial level, although they suggest that there was strong evidence that the nets would have continued to catch fish beyond the observed 27 months of the experiment. Lost fishing gear can carry on trapping until it is heavily colonized by marine organisms (biofouling), thereby altering the weight, mesh size and visibility of nets (Humborstad *et al.*, 2003; Sancho *et al.*, 2003). In deeper water (>200 m deep), catches from lost nets are expected to stabilize to approximately 20 per cent of the catch from actively fished nets after 45 days (Humborstad *et al.*, 2003). However, Large *et al.* (2009) report that such nets may continue to 'fish' for periods of at least 2–3 years, and potentially longer, primarily as a result of lesser rates of biofouling and tidal scouring occurring in deep water.

4.4 Responses to Ship Waste, Litter and Sewage

There is a range of regulatory and behavioural interventions available to supranational, national and regional actors to enable reductions in ship waste. These are also closely interconnected to changes in technology to reduce the amount of ship waste. Table 4.5 outlines with examples some of the regulatory and behavioural responses to waste reduction. Educational programmes, while significant, may only have a limited impact on behaviours as, while seafarers and shipping companies and owners may be aware of waste management issues, knowledge alone may not overcome other incentives to retain existing behaviours, such as increased financial returns by dumping waste at sea. Therefore, there is increased interest in the use of social marketing and behavioural economic approaches to change embedded social and economic practices (Chen, 2015; Newman *et al.*, 2015). Such measures are also significant as the development of educational and promotional campaigns to change waste management behaviours may act to limit governmental interventions and provide a basis for collaborative action between different stakeholders (Hall, 2014, 2016). Nevertheless, legislative and regulatory actions are also hugely important in the shipping sector because of the inherent difficulties of policing the marine environment, especially because of transboundary issues and the problems of managing transnational space (Goldfarb *et al.*, 1998; Hastings & Potts, 2013). As Seas At Risk (2011, p. 5) observed, '[S]hipping sets itself apart from other industries due to its geographic bearing and that international legislation does not involve a closed system of full waste accountability.'

International regulations under MARPOL, the International Convention for the Prevention of Pollution from Ships, govern the what and the where of waste that can be discharged overboard (Table 4.6). MARPOL prohibits the disposal of plastics at sea, as well as most other ship garbage, and requires signatories to the convention to ensure that adequate reception facilities for ship-generated waste are available (Newman *et al.*, 2015). MARPOL therefore reflects the main focus of international, governmental and non-governmental

Table 4.5. *A continuum of behavioural interventions for altering ship-based marine littering behaviour*

Intervention	Education	Social Marketing	Legislation & Regulation
Behavioural state of target audience	Unaware of impact of littering/considering behavioural change/maintaining change behaviour	Aware of need to change/not considering change	Entrenched littering behaviours/no desire to change
Rights	Individual rights of shipping companies and owners stronger		State/societal rights stronger
Locus of power	Resides in individual seafarers and ship owners and managers	Relatively balanced between change agency and individual seafarers and shipping companies	Resides in change agency (although overall extent of power depends on capacity of change agency to implement)
Behavioural influences from government	Lessons	Hugs/nudges	Shoves/smacks
International interventions	2006 IMO Marine Environment Protection Committee Action Plan on Tackling the Inadequacy of Port Reception Facilities; UNEP Regional Sea Programme & Global Programme of Action; UNEP Guidelines on the Use of Market-based and Economic Instruments; UNEP/IOC Guidelines on Surveying and Monitoring of Marine Litter; UNEP Global Partnership of Marine Litter, Clean Seas Campaign	Seas At Risk campaigns; Friends of the Earth Cruise Report Campaign	UN Convention on the Law of the Sea (UNCLOS); Annex IV (with respect to sewage) and Annex V (with respect to garbage) of MARPOL 73/78; London Protocol
Supranational interventions	Commission for the Conservation of Antarctic Marine Resources (CCAMLR) Marine Debris Program; KIMO (Kommunenes Internasjonale Miljøorganisasjon) educational initiatives	Fishing for Litter (FFL) projects in the OSPAR area; KIMO campaigns	1992 Helsinki Convention and HELCOM (Helsinki Commission) Strategy for Port Reception Facilities (Baltic Strategy); EU Port Reception Facility Directive; EU Marine Strategy Framework Directive
National/ regional interventions	Virginia Department of Environmental Quality's Virginia Eastern Shorekeeper programme; GhostNets Australia	Fishing for Litter Scotland; Net Recycling Scotland; Gestes Propres ('Clean Habits') 'I Sail, I Sort' campaign	US Marine Debris Research, Prevention, and Reduction Act; UK Merchant Shipping (Prevention of Pollution by Sewage and Garbage from Ships) Regulation 2008

Sources: After Chen (2015), Hall (2014) and Newman *et al.* (2015)

organization actors in ship waste management policy in the creation of incentives to encourage ships to use port waste facilities rather than dump waste at sea. This has been the approach, for example, favoured by both the International Maritime Organization (IMO) and the EU (Pérez *et al.*, 2017).

In reflecting the polluter-pays principle, the costs of waste collection and disposal are usually recovered through the collection of port fees. In Europe, this is done under European Directive 2000/59/EC, which establishes that all ships that stop over in EU ports must deliver their waste in port, except when they can prove that they can store it until the following stopover port. In order to reduce the economic incentive of throwing waste into the sea, each ship calling at an EU port therefore pays a fee to contribute to the costs of port reception facilities for ship-generated waste. Nevertheless, a number of issues have emerged from such an approach. High fees for waste collection can act as a disincentive for ships to discharge their waste at port when they could discard their waste overboard for free (if not caught). Therefore, a balance is required between cost recovery of waste handling and not discouraging disposal at port (Newman *et al.*, 2015). Within Europe, a diversity of tariffs has led to a lack of transparency in their development and application, as well as distorting competition between ports (Pérez *et al.*, 2017). Significantly with respect to the relationship between waste management practices and behavioural change, if the fee is not proportional to the amount of waste produced, it may act as a financial incentive to encourage waste reduction on board vessels (Ikonen, 2013). It has also been suggested that the diversity of tariffs in the EU makes it very difficult to determine whether the objective of reducing the discharge of waste into the sea has been achieved (Pérez *et al.*, 2017).

In response to this issue, Pérez *et al.* (2017) argue that a differential tariff should be developed for different types of waste given that the main factors that affect the volume of delivery of oily waste (primarily ship size) are not the same as those that determine the delivery of garbage (primarily number of people), while they also suggest that technology, the type of ship and the characteristics of the route should also be considered. In doing so, the payback period of investments in green shipping technologies could be shortened, but although 'the environmental charge may not change the performance of ship owners in the short term ... the incentive may be effective when it comes to the development of new ships' (Pérez *et al.*, 2017, p. 410). Another approach is that of the HELCOM Convention in the Baltic, which provides a number of recommendations regarding the introduction of a 'no special fee' or 'indirect fee' in Baltic ports. An indirect fee means that the cost of delivering garbage waste and sewage to port reception facilities is included in the fee paid by all ships visiting the port, irrespective of the quantities discharged (Ikonen, 2013). Such an approach has been adopted by several Baltic Sea ports (Gothenburg, Copenhagen, Klaipeda, Helsinki and Stockholm; Øhlenschlæger *et al.*, 2013), with the 'no special fee' system being regarded as effectively preventing cost from becoming a disincentive for using port reception facilities for waste and sewage disposal (Newman *et al.*, 2015). Such measures are also significant given the rapid growth of the cruise industry in the Baltic region and the difficulties involved in effective nitrogen removal from cruise ship waste water systems (Köster *et al.*, 2016).

Table 4.6. *Summary of discharge provisions of the revised MARPOL Annex V*

Type of garbage	Ships within special areas: Mediterranean Sea, Baltic Sea, Black Sea, Red Sea, the Gulfs area, North Sea, Antarctica (south of 60°S) and Wider Caribbean Region including the Gulf of Mexico and the Caribbean Sea	Ships outside special areas	Offshore platforms and all ships within 500 m of such platforms
Food wastes comminuted or ground (capable of passing through a screen with openings ≤25 mm)	Discharge permitted ≥12 nautical miles from the nearest land and en route	Discharge permitted ≥3 nautical miles from the nearest land and en route	Discharge permitted ≥12 nautical miles from the nearest land
Food wastes not comminuted or ground	Discharge prohibited	Discharge permitted ≥12 nautical miles from the nearest land and en route	Discharge prohibited
Cargo residues (not harmful to the marine environment) not contained in wash water	Discharge prohibited	Discharge permitted ≥12 nautical miles from the nearest land and en route	Discharge prohibited
Cargo residues (not harmful to the marine environment) contained in wash water	Discharge only permitted in specific circumstances: (1) if both the port of departure and the next port of destination are within the special area and the ship will not transit outside the special area between these ports and (2) if no adequate reception facilities are available at those ports; and ≥12 nautical miles from the nearest land and en route	Discharge permitted ≥12 nautical miles from the nearest land and en route	Discharge prohibited
		Discharge permitted	Discharge prohibited

Cleaning agents and additives (not harmful to the marine environment) contained in cargo hold wash water	Discharge only permitted in specific circumstances: (1) if both the port of departure and the next port of destination are within the special area and the ship will not transit outside the special area between these ports and (2) if no adequate reception facilities are available at those ports; and $\geq$12 nautical miles from the nearest land and en route		
Cleaning agents and additives (not harmful to the marine environment) contained in deck and external surfaces wash water	Discharge permitted	Discharge permitted	Discharge prohibited
Animal carcasses	Discharge prohibited	Discharge permitted as far from the nearest land as possible and en route	Discharge prohibited
All other garbage including plastics, domestic wastes, cooking oil, incinerator ashes, operational wastes and fishing gear	Discharge prohibited	Discharge prohibited	Discharge prohibited
Mixed garbage	When garbage is mixed with or contaminated by other substances prohibited from discharge or having different discharge requirements, the more stringent requirements shall apply		

Sources: International Maritime Organization (2014) and Marine Environment Protection Committee and International Maritime Organization (2011)

4.5 Future

Writing over 40 years ago, Hees (1977) observed that ship sewage discharges in the open ocean were predicted to be completely harmless. Such predictions have been shown to be profoundly wrong but, unfortunately, they also reflect a common attitude with respect to the oceans as a dumping ground for human waste. Although shipping is not responsible for the majority of the marine litter that is now found in the oceans, it remains a major contributor despite the introduction of a range of international conventions and actions. Undoubtedly, new technologies will play a part in reducing waste, but they will take time to have a wider effect given the size of the global shipping fleet and the replacement rate of older ships. In the meantime, the shipping fleet continues to grow, and the cruise ship sector, which is a major producer of sewage and food waste because of the amount of passengers that are now carried, remains one of the fastest-growing tourism sectors (Hall *et al.*, 2017). The sheer size of many of the cruise vessels, at 3000–5000 passengers and crew, also means that they produce as much waste and sewage as are produced by a small town. The increasing size of cruise ships will only increase pressures for improved waste management and pollution control, particularly given growing opposition to cruise ships at some destinations (Giuffrida, 2017). For example, the *Harmony of the Seas*, owned by Royal Caribbean, carries 6780 passengers and a crew of 2100 (Vidal, 2016). However, arguably, this is a situation faced by the wider shipping industry, given growing awareness of marine litter and waste.

Undoubtedly, regulatory and economic measures to reduce the extent of ship-based marine littering have grown in recent years, but their application, along with research on ship waste and sewage, is spatially uneven and primarily focused on the fleets and regions of the developed world. Many areas, particularly in developing countries, lack analyses and provision of adequate port waste disposal facilities, as well as the incentives to use them and reduce dumping at sea. The situation therefore reflects a classic transboundary pollution problem, given that the impacts of dumping at sea can spread beyond the locations at which waste is deposited (Hastings & Potts, 2013).

Ship-based waste and sewage technologies are improving, although substantial challenges in nitrogen and micro-plastic removal in particular remain (Köster *et al.*, 2016). Adoption rates and the size of the global shipping fleet mean that the greatest possibility for substantial rapid waste reduction from shipping lies in behavioural change, whether as a result of regulatory developments and/or financial incentives. However, such initiatives require action and investment from governments and port authorities in the provision of adequate disposal facilities and enforcement. Successive neoliberal market policies have meant that often much of the cost of such actions has been borne by the public rather than by the polluter, although the shipping industry – particularly the fisheries sector – is increasingly aware of the economic impacts of marine litter on its own businesses. Nevertheless, the economic significance of shipping and its political influence likely mean that it will require a much wider global public backlash against the level of marine pollution to force governments to reduce both land- and ship-based contributions. Effective policies and accompanying reductions will therefore only occur when global monitoring systems are established together with the worldwide availability of waste disposal facilities. Without effective public pressure, the likelihood of this occurring remains low.

References

Anderson, J. A. & Alford, A. B. (2014). Ghost fishing activity in derelict blue crab traps in Louisiana. *Marine Pollution Bulletin*, **79**, 261–267.

Andrady, A. L. (2015). Persistence of plastic litter in the oceans. In M. Bergmann, L. Gutow & M. Klages, eds., *Marine Anthropogenic Littery*. Cham: Springer, pp. 57–74.

Ayaz, A., Acarli, D., Altinagac, U. *et al.* (2006). Ghost fishing by monofilament and multifilament gillnets in Izmir Bay, Turkey. *Fisheries Research*, **79**, 267–271.

Baeta, F., Costa, M. J. & Cabral, H. (2009). Trammel nets' ghost fishing off the Portuguese central coast. *Fisheries Research*, **98**, 33–39.

Bakir, A., Rowland, S. J. & Thompson, R. C. (2014). Enhanced desorption of persistent organic pollutants from microplastics under simulated physiological conditions. *Environmental Pollution*, **185**, 16–23.

Baulch, S. & Perry, C. (2014). Evaluating the impacts of marine debris on cetaceans. *Marine Pollution Bulletin*, **80**, 210–221.

Bergmann, M., Gutow, L. & Klages, M., eds. (2015). *Marine Anthropogenic Litter*. Cham: Springer.

Buhl-Mortensen, L. & Buhl-Mortensen, P. (2017). Marine litter in the Nordic Seas: distribution composition and abundance. *Marine Pollution Bulletin*, **125**, 260–270.

Butt, N. (2007). The impact of cruise ship generated waste on home ports and ports of call: a study of Southampton. *Marine Policy*, **31**, 591–598.

Chen, C.-L. (2015). Regulation and management of marine litter. In M. Bergmann, L. Gutow & M. Klages, eds., *Marine Anthropogenic Litter*. Cham: Springer, pp. 395–428.

Danish Shipowners' Association (2000). *Environmental Reports for Ship Operations*, Copenhagen: DSA.

Derraik, J. G. B. (2002). The pollution of the marine environment by plastic debris: a review. *Marine Pollution Bulletin*, **44**, 842–852.

Eagle, L., Hamann, M. & Low, D. R. (2016). The role of social marketing, marine turtles and sustainable tourism in reducing plastic pollution. *Marine Pollution Bulletin*, **107**, 324–332.

EMAS-Ship (2006). *Manuale metdologico per l'implementazione di EMAS nelle compagnie di navigazione Confitarma*. Rome: Confitarma.

Eriksen, M., Lebreton, L. C., Carson, H. S. *et al.* (2014). Plastic pollution in the world's oceans: more than 5 trillion plastic pieces weighing over 250,000 tons afloat at sea. *PLoS One*, **9**(12), e111913.

Erzini, K., Bentes, L., Coelho, R. *et al.* (2008). Catches in ghost-fishing octopus and fish traps in the northeastern Atlantic Ocean (Algarve, Portugal). *Fishery Bulletin*, **106**, 321–327.

Galgani, F., Leauté, J. P., Moguedet, P. *et al.* (2000). Litter on the sea floor along the European coasts. *Marine Pollution Bulletin*, **40**, 516–527.

Galgani, F., Hanke, G., Werner, S. *et al.* (2013). *European Marine Strategy Framework Directive Working Group on Good Environmental Status (WG-GES) Technical Subgroup on Marine Litter. Guidance on Monitoring of Marine Litter in European Seas*. Luxemburg: Publications Office of the European Union.

Galgani, F., Hanke, G. & Maes, T. (2015). Global distribution, composition and abundance of marine litter. In M. Bergmann, L. Gutow & M. Klages, eds., *Marine Anthropogenic Litter*. Cham: Springer, pp. 29–56.

García-Rivera, S., Lizaso, J. L. S. & Millán, J. M. B. (2017). Composition, spatial distribution and sources of macro-marine litter on the Gulf of Alicante seafloor (Spanish Mediterranean). *Marine Pollution Bulletin*, **121**, 249–259.

Giuffrida, A. (2017). 'Imagine living with this crap': tempers in Venice boil over in tourist high season. *The Guardian*, 23 July. www.theguardian.com/world/2017/jul/23/venice-tempers-boil-over-tourist-high-season

Goldfarb, W., Krogmann, U. & Hopkins, C. (1998). Unsafe sewage sludge of beneficial biosolids: liability, planning, and management issues regarding the land application of sewage treatment residuals. *Boston College Environmental Affairs Law Review*, **26**, 687–768.

Gössling, S., Hall, C. M. & Scott, D. (2015). *Tourism and Water*. Bristol: Channelview.

Guilbaud, J., Massé, A., Andrès, Y., Combe, F. & Jaouen, P. (2012). Influence of operating conditions on direct nanofiltration of greywaters: application to laundry water recycling aboard ships. *Resources, Conservation and Recycling*, **62**, 64–70.

Hall, C. M. (2014). *Tourism and Social Marketing*. Abingdon: Routledge.

Hall, C. M. (2016). Intervening in academic interventions: framing social marketing's potential for successful sustainable tourism behavioural change. *Journal of Sustainable Tourism*, **24**, 350–375.

Hall, C. M., James, M. & Wilson, S. (2010). Biodiversity, biosecurity, and cruising in the Arctic and sub-Arctic. *Journal of Heritage Tourism*, **5**, 351–364.

Hall, C. M., Wood, H. & Wilson, S. (2017). Environmental reporting in the cruise industry. In R. Dowling & C. Weeden, eds., *Cruise Ship Tourism*, 2nd ed. Wallingford: CABI, pp. 441–464.

Hastings, E. & Potts, T. (2013). Marine litter: progress in developing an integrated policy approach in Scotland. *Marine Policy*, **42**, 49–55.

Hees, W. (1977). Sewage discharges from ships transiting coastal salt waters. *Journal of the American Water Resources Association*, **13**, 215–230.

Horsman, P. V. (1982). The amount of garbage pollution from merchant ships. *Marine Pollution Bulletin*, **13**, 167–169.

Humborstad, O.-B., Løkkeborg, S., Hareide, N.-R. & Furevik, D. M. (2003). Catches of Greenland halibut (*Reinhardtius hippoglossoides*) in deepwater ghost-fishing gillnets on the Norwegian continental slope. *Fisheries Research*, **64**, 163–170.

Ikonen, M. (2013). No-special-fee system for ships in the Baltic Sea ports. Paper presented at International Conference on Prevention and Management of Marine Litter in European Seas, Berlin. www.marine-litter-conference-berlin.info/userfiles/file/online/No-special-fee%20system%20for%20ships%20in%20the%20Baltic%20Sea%20ports_Ikonen.pdf

International Maritime Organization (2014). Simplified overview of the discharge provisions of the revised MARPOL Annex V which entered into force on 1 January 2013. www.imo.org/en/OurWork/Environment/PollutionPrevention/Garbage/Documents/2014%20revision/Annex%20V%20discharge%20requirements%2007-2013.pdf

Jones, M. M. (1995). Fishing debris in the Australian marine environment. *Marine Pollution Bulletin*, **30**, 25–33.

Kiessling, T., Gutow, L. & Thiel, M. (2015). Marine litter as a habitat and dispersal vector. In M. Bergmann, L. Gutow & M. Klages, eds., *Marine Anthropogenic Litter*. Cham: Springer, pp. 141–181.

Klein, R. A. (2002). *Cruise Ship Blues: The Underside of the Cruise Ship Industry.* Gabriola Island, BC: New Society Publishers.

Köster, S., Westhof, L. & Keller, L. (2016). Stand der Technik der Abwasserreinigung an Bord von Kreuzfahrtschiffen [The state of the art of wastewater treatment on cruise ships]. *GWF, Wasser – Abwasser*, **157**, 528–537.

Kühn, S., Bravo Rebolledo, E. L. & van Franeker, J. A. (2015). Deleterious effects of litter on marine life. In M. Bergmann, L. Gutow & M. Klages, eds., *Marine Anthropogenic Litter*. Cham: Springer, pp. 75–116.

Laist, D. W. (1987). Overview of the biological effects of lost and discarded plastic debris in the marine environment. *Marine Pollution Bulletin*, **18**, 319–326.

Laist, D. W. (1997). *Impacts of Marine Debris: Entanglement of Marine Life in Marine Debris Including a Comprehensive List of Species with Entanglement and Ingestion Records*. New York: Springer.

Large, P. A., Graham, N. G., Hareide, N.-R. *et al.* (2009). Lost and abandoned nets in deep-water gillnet fisheries in the Northeast Atlantic: retrieval exercises and outcomes. *ICES Journal of Marine Science*, **66**, 323–333.

Lee, D.-I., Cho, H.-S. & Jeong, S.-B. (2006). Distribution characteristics of marine litter on the sea bed of the East China Sea and the South Sea of Korea. *Estuarine, Coastal and Shelf Science*, **70**, 187–194.

Marine Environment Protection Committee & International Maritime Organization (2011). *Resolution MEPC.201(62) Adopted on 15 July 2011. Amendments to the Annex of the Protocol of 1978 Relating to the International Convention for the Prevention of Pollution from Ships, 1973* (Revised MARPOL Annex V). London: International Maritime Organisation. Available from www.imo.org/en/KnowledgeCentre/IndexofIMOResolutions/Marine-Environment-Protection-Committee-(MEPC)/Documents/MEPC.201(62).pdf

Matsuoka, K., Nakashima, T. & Nagasawa, N. (2005). A review of ghost fishing: scientific approaches to evaluation and solutions. *Fisheries Science*, **71**, 691–702.

McIlgorm, A., Campbell, H. F. & Rule, M. J. (2011). The economic cost and control of marine debris damage in the Asia–Pacific region. *Ocean and Coastal Management*, **54**, 643–651.

Molnar, J. L., Gamboa, R. L., Revenga, C. & Spalding, M. D. (2008). Assessing the global threat of invasive species to marine biodiversity. *Frontiers in Ecology and the Environment*, **6**, 485–492.

Moore, S. L. & Allen, M. J. (2000). Distribution of anthropogenic and natural debris on the mainland shelf of the Southern California Bight. *Marine Pollution Bulletin*, **40**, 83–88.

Mordecai, G., Tyler, P. A., Masson, D. G. & Huvenne, V. A. I. (2011). Litter in submarine canyons off the west coast of Portugal. *Deep-Sea Research II: Topical Studies in Oceanography*, **58**, 2489–2496.

Mouat, J., Lopez Lozano, R. & Bateson, H. (2010). *Economic Impacts of Marine Litter*. Lerwick: Kommunenes Internasjonale Miljøorganisasjon (KIMO) International Secretariat.

National Research Council (2008). *Tackling Marine Debris in the 21st Century*. Washington, DC: National Academies Press.

Newman, S. J., Skepper, C. L., Mitsopoulos, G. E. A. *et al.* (2011). Assessment of the potential impacts of trap usage and ghost fishing on the northern demersal scalefish fishery. *Reviews in Fisheries Science*, **19**, 74–84.

Newman, S., Watkins, E., Farmer, A., ten Brink P. & Schweitzer, J.-P. (2015). The economics of marine litter. In M. Bergmann, L. Gutow & M. Klages, eds., *Marine Anthropogenic Litter*. Cham: Springer, pp. 367–394.

Office for the London Convention/Protocol and Ocean Affairs (2016). *1996 Protocol to the Convention on the Prevention of Marine Pollution by Dumping of Wastes and Other Matter, 1972*. London: Office for the London Convention/Protocol and Ocean Affairs, International Maritime Organization.

Øhlenschlæger, J. P., Newman, S. & Farmer, A. (2013). *Reducing Ship Generated Marine Litter – Recommendations to Improve the EU Port Reception Facilities Directive*. London: Institute for European Environmental Policy.

Pérez, I., González, M. M. & Jiménez, J. L. (2017). Size matters? Evaluating the drivers of waste from ships at ports in Europe. *Transportation Research Part D*, **57**, 403–412.

Pham, C. K., Ramirez-Llodra, E., Alt, C. H. *et al.* (2014). Marine litter distribution and density in European seas, from the shelves to deep basins. *PLoS One*, **9**(4), e95839.

Polasek, L., Bering, J., Kim, H. *et al.* (2017). Marine debris in five national parks in Alaska. *Marine Pollution Bulletin*, **117**, 371–379.

Ramirez-Llodra, E., De Mol, B., Company, J. B., Coll, M. & Sardà, F. (2013). Effects of natural and anthropogenic processes in the distribution of marine litter in the deep Mediterranean Sea. *Progress in Oceanography*, **118**, 273–287.

Rangel-Buitrago, N., Williams, A. & Anfuso, G. (2017). Killing the goose with the golden eggs: litter effects on scenic quality of the Caribbean coast of Colombia. *Marine Pollution Bulletin*, **125**, 22–38.

Read, A. J., Drinker, P. & Northridge, S. (2006). Bycatch of marine mammals in U.S. and global fisheries. *Conservation Biology*, **20**, 163–169.

Ressurreição, A. & Giacomello, E. (2013). Quantifying the direct use value of Condor seamount. *Deep-Sea Research II: Topical Studies in Oceanography*, **98**(Pt A), 209–217.

RNLI (2017). *RNLI 2016 Operational Statistics*. Poole: RNLI.

Ryan, P. (2015). A brief history of marine litter research. In M. Bergmann, L. Gutow & M. Klages, eds., *Marine Anthropogenic Litter*. Cham: Springer, pp. 1–25.

Sancho, G., Puente, E., Bilbao, A., Gomez, E. & Arregi, L. (2003). Catch rates of monkfish (*Lophius* spp.) by lost tangle nets in the Cantabrian Sea (northern Spain). *Fisheries Research*, **64**, 129–139.

Seas At Risk (2011). Seas At Risk position paper. Ship waste dumping and the clean ship concept. How an improved EU PRF Directive can play a key role in cleaning up the seas. www.seas-at-risk.org/images/pdf/Seas_At_Risk_Position_Paper160911.pdf

Setälä, O., Fleming-Lehtinen, V. & Lehtiniemi, M. (2014). Ingestion and transfer of microplastics in the planktonic food web. *Environmental Pollution*, **185**, 77–83.

Sheavly, S. B. & Register, K. M. (2007). Marine debris & plastics: environmental concerns, sources, impacts and solutions. *Journal of Polymers and the Environment*, **15**, 301–305.

Strand, J., Tairova, Z., Danielsen, J. *et al.* (2015). *Marine Litter in Nordic Waters*. Copenhagen: Nordic Council of Ministers.

Taylor, M. (2017). $180bn investment in plastic factories feeds global packaging binge. *The Guardian*, 26 December. www.theguardian.com/environment/2017/dec/26/180bn-investment-in-plastic-factories-feeds-global-packaging-binge

TRT Trasporti e Territorio Srl (2007). *External Costs of Maritime Transport*. Brussels: European Parliament.

Tschernij, V. & Larsson, P. O. (2003). Ghost fishing by lost cod gill nets in the Baltic Sea. *Fisheries Research*, **64**, 151–162.

UNEP (2005). *Marine Litter, an Analytical Overview*. Nairobi: UNEP.

Vauk, G. J. M. & Schrey, E. (1987). Litter pollution from ships in the German Bight. *Marine Pollution Bulletin*, **18**, 316–319.

Vidal, J. (2016). The world's largest cruise ship and its supersized pollution problem. *The Guardian*, 21 May. www.theguardian.com/environment/2016/may/21/the-worlds-largest-cruise-ship-and-its-supersized-pollution-problem

Vieira, R. P., Raposo, I. P., Sobral, P. *et al.* (2015). Lost fishing gear and litter at Gorringe Bank (NE Atlantic). *Journal of Sea Research*, **100**, 91–98.

Woodall, L. C., Robinson, L. F., Rogers, A. D., Narayanaswamy, B. E. & Paterson, G. L. J. (2015). Deep sea litter: a comparison of seamounts, banks and a ridge in the Atlantic and Indian Oceans reveals both environmental and anthropogenic factors impact accumulation and composition. *Frontiers of Marine Science*, **2**, 1–10.

Žydelis, R., Small, C. & French, G. (2013). The incidental catch of seabirds in gillnet fisheries: a global review. *Biological Conservation*, **162**, 76–88.

5

Ballast Water

STEPHAN GOLLASCH AND MATEJ DAVID

5.1 Background

One of the most recent threats to any type of water caused by humans is species introduction through ballast water and sediment releases, which may result in harmful effects on the natural environment, human health, property and resources globally. One of the key species introduction vectors is shipping predominantly through species transfers in ballast water and biofouling of vessels (David & Gollasch, 2015a; Davidson & Simkanin, 2012; Ojaveer *et al.*, 2017; WGITMO, 2015). The relative importance of vectors may regionally be very different, and in some regions biofouling may prevail (Carlton & Eldredge, 2009). However, this chapter is limited to ballast water species transfers.

5.2 What Is Ballast Water?

When a vessel is not fully laden, additional weight is required to provide the vessel's seaworthiness (e.g., to compensate for the increased buoyancy that can result in a lack of propeller immersion, inadequate transversal and longitudinal inclination and other stresses on the vessel's hull). The material used for adding weight to the vessel is referred to as ballast. In the past, ballast material was solid, in the form of stones, gravel or sand. After the introduction of iron in vessel building during the middle of the nineteenth century, loading of water as ballast was preferred. This water was stored in cargo holds or tanks and was much easier to use and more time efficient compared to solid ballast. In addition, even when a vessel is fully laden, it can require ballast water operations due to, for example, an unequal weight distribution on board, due to certain weather and sea conditions, to compensate for draft restrictions and due to the consumption of fuel during the voyage. As a result, vessels fundamentally rely on ballast water for safe operations as a function of their design and construction (David, 2015).

Vessels usually conduct ballast water operations in the port opposite to their cargo operations. This means that when a vessel loads cargo, ballast water is discharged and vice versa. Depending on the vessel type, weather and sea conditions, as well as vessel operations, the ballasting or deballasting may also be carried out during navigation or at an anchorage.

Water ballast is transported in tanks that are different in number, volume, design and location within the ship. Some older vessels, mainly tankers, also use cargo holds or cargo tanks to ballast. Today's vessels have tanks that are dedicated to ballasting (i.e., segregated ballast tanks). However, ballasting in cargo holds may also be used today for bigger bulk carriers. Ballast tanks are connected to the ballast water pump(s) of a vessel by a ballast water pipeline. Ballast water is pumped on board via sea chest(s) and strainer(s).

At the end of the ballast water discharge process, tanks are close to empty and ballast pumps start losing suction. This results in some unpumpable water remaining in the ballast tanks, which is in general between 5 and 10 per cent of the ballast water tank volume. This means that vessels are never fully empty of ballast water, such that even when all pumpable ballast water has been released, some water and organisms still remain in the ballast tanks (David, 2015).

To calculate the amount of ballast water to be discharged in a port, a new generic ballast water discharge assessment model was developed. This model is based on vessel cargo operations and vessel dimensions (David *et al.*, 2012). The model was developed further and used to calculate the global ballast water discharges from all vessels engaged in international trade. The results showed that an impressive amount of ballast water (3.1 billion tonnes) had been discharged worldwide in 2013 (David, 2015).

5.3 Ballast Water Biology

In the early 1900s, the new finding of a non-indigenous species was attributed to ships' ballast water discharges for the first time. This was suggested by Ostenfeld (1908), who documented the Asian phytoplankton species *Odontella* (*Biddulphia*) *sinensis* in the North Sea following its bloom in 1903. Thereafter, ballast water was suggested as a species introduction vector many times, but it took approximately 70 years before the first biological ballast water sampling study was conducted by Medcof (1975). This was followed by several others (e.g., Briski *et al.*, 2010, 2011; David *et al.*, 2007; Gollasch, 1996; Gollasch *et al.*, 2000a, 2000b, 2002; Hallegraeff & Bolch, 1992; Hamer *et al.*, 2000; Locke *et al.*, 1991; McCollin *et al.*, 2008; Murphy *et al.*, 2002; Williams *et al.*, 1988).

These and other ballast water sampling studies documented that various bacteria, plant and animal species, including fish, can survive a vessel voyage in a ballast tank either in the water or in the sediment at the tank's bottom (e.g., Briski *et al.*, 2010, 2011; Carlton, 1985; Carlton & Geller, 1993; David *et al.*, 2007; Gollasch, 1996; Gollasch *et al.*, 2000a, 2002; Hallegraeff & Bolch, 1991; Hamer *et al.*, 2001; Locke *et al.*, 1991; McCollin *et al.*, 2008; Medcof, 1975; Murphy *et al.*, 2002; Williams *et al.*, 1988). Some organisms stay viable in ballast tanks for several months (e.g., Gollasch, 1996; Gollasch *et al.*, 2000a, 2000b) or even longer (Hallegraeff & Bolch, 1991).

Regarding the world's shipping fleet, it is estimated that 3000–4000 (Carlton & Geller, 1993; Gollasch, 1996) and possibly even 7000 (Carlton, 2001) different species are in transit within ships on a daily basis. More than 850 species are known to have been successfully introduced to and established in new environments (Hayes & Sliwa, 2003) and Hewitt stated in his recent speech at the International Council for the Exploration of the Sea (ICES) Annual Science Conference 2016 in Riga, Latvia, that nearly 2700 marine or estuarine species have an invasion history.

It was concluded that each single arriving vessel has the potential to introduce a new species, and furthermore that any introduced and established species has the potential to result in a significant negative impact on the recipient environment (e.g., Gollasch, 1996). Therefore, loading ballast water and sediment in one port and then discharging this in another port represents a substantial risk of transferring species to new environments.

5.4 New Species Arrivals

The annual number of new species records worldwide has paralleled shipping and is increasing. For example, in ICES member countries, a new introduction forming a new population beyond its natural range occurs approximately every 9 weeks (Minchin *et al.*, 2005). As an example, the Greater North Sea region has ~274 non-indigenous or cryptogenic[1] species. The majority of these species arrived between 1950 and 1999 (142 species). Since the beginning of the twenty-first century (until 2015), 60 such species have been found, and of these are 21 species that are new to Europe and were first recorded in the Greater North Sea region. The annual introduction rate has increased (Figure 5.1), with several new species found every year (AquaNIS Editorial Board, 2015; WGITMO, 2015).

In general, there is ground to believe that the number of released individuals (i.e., propagule pressure) and the number of released species (i.e., colonization pressure) are the key determinants of the number of species that become successfully introduced into new habitats. In short, this means that the more organisms and species are released, the more species introductions may occur (Briski *et al.*, 2012; Cope *et al.*, 2015; Lo *et al.*,

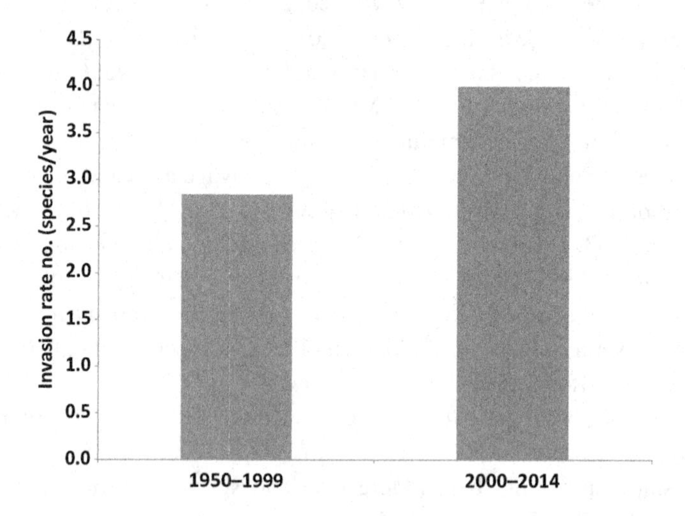

Figure 5.1 Annual rate of newly found non-indigenous or cryptogenic species in the Greater North Sea from 1950 to 1999 and from 2000 to 2014.

2012; Lockwood *et al.*, 2009). It was further stated by Simberloff (2009) that a continuing 'rain of propagules', especially when arriving from a variety of source regions, may reduce the expected genetic bottleneck for successful species introductions. This means that the more individuals of the same species that arrive, the greater the genetic diversity, which increases the likeliness of a wider tolerance of this species to the varying conditions of the new habitat (Simberloff, 2009).

In contrast, Ruiz *et al.* (2013) found no relationship between the quantity and frequency of ballast water discharges of foreign vessels and the number of introduced ballast water-mediated species in 16 large US bays. In addition, propagule supply is not a simple function of total ship arrivals (frequency and volume of ballast water discharges), frequency of vessel types and routes (source regions) among recipient ports or the transit survival of organisms, so invasion success has to be determined by the interaction of global shipping and local population dynamics (Seebens *et al.*, 2019).

Furthermore, the transport of a species alone does not necessarily result in the colonization of a new region; rather, there is a 'chain of events' that a species must endure to become established in a new environment (Carlton, 1985; Hayes, 1998). This starts with the ballast water uptake. Not all species reside year round in the water column to be pumped on board with ballast water. Some bottom-living species have only larval stages that reside in the water column during a certain period of the year. Thus, they can only be pumped on board at certain times. Although most species have a larval phase in their life cycle, some species lack such a stage (Grantham *et al.*, 2003; Shanks, 2009; Tardent, 1979; Thorson, 1946) and therefore cannot be taken on board within ballast water.

For organisms that survive the ballast water uptake process through a ship's pumps, the vessel's voyage must be endured, as well as the ballast water discharge pumping event. On arrival at the new environment, sufficient organism numbers will be needed to establish a founder population.

The minimum organism number required to develop a new population is generally unknown, but theoretically some species can generate new populations with low organism numbers (Bailey *et al.*, 2009). One example is *Melosira varians* (a phytoplanktonic diatom). In laboratory experiments, it was shown that in some cases *M. varians* was able to establish new populations with starting densities greater than 20 cells/mL (Aliff, 2015). This relatively low number of phytoplankton cells may occur in a single drop of water! An example of an intentional species introduction from which we know the number of individuals released is the gammarid *Gammarus tigrinus*. This North American species was introduced successfully to the Werra River (Germany) by importing 1000 individuals in a single event (Schmitz, 1960). Korsu and Huusko (2009) report that about 8000 individuals of the brook trout *Salvelinus fontinalis* are leading to highly successful establishment. More effort in this area was needed, and it took several trials during which more than 10,000 individuals were intentionally released until the large Kamchatka king crab *Paralithodes camtschatica* was successfully introduced to the Barents Sea (Orlov & Ivanov, 1978). From these examples, it seems that the larger an organism is, the greater the number of surviving individuals that are needed in order to successfully start a new population.

Survival of individuals depends on their tolerance of the conditions in the new environment. The degree of dispersal following discharge has an influence on the number of individuals needed for a successful founder population. In many cases, these windows of opportunity for establishment may depend on the precise location where the individuals are released (e.g., intentional releases or where the ballast discharge occurs). Colonization success may also depend on the season during which the arrival happens. For example, in cold-climate regions, a warm-water species might only survive in summer, and might not subsequently survive the winter. This is why tropical organisms in ballast water from, for example, the Port of Singapore (marine, tropical waters), when released in Hamburg (freshwater, cold-temperate climate), have a very low probability of surviving. Therefore, if the abiotic conditions of the recipient region match with species tolerance of these factors, this increases the invasion risk (David *et al.*, 2013; Drake *et al.*, 2015; Duncan *et al.*, 2014; Glibert, 2015; Smith *et al.*, 1999).

Without survival, no reproduction is possible, and without reproduction, colonization is impossible. Once a founder population is established (primary introduction), a species might spread further via a wide range of human activities and also via natural processes (secondary spread).

5.5 Species Impacts

The impacts of introduced species vary greatly and may cause considerable harm with consequent long-term implications. In general, the observed ecological impacts of introduced species include significant reductions in the abundance of native species, as well as changes to the physical and chemical composition of both sediments and the water column. Additional impacts include outcompeting native (commercial) species, fouling of aquaculture and fishing gear and fish kills through toxin production. However, some non-indigenous species are considered to be valuable resources, but in extreme cases a number of introduced species had almost catastrophic and seemingly irreversible impacts (e.g., Hayes & Sliwa, 2003).

It should be noted that not all introduced species are considered harmful. In some cases it is quite the reverse, because some species support important industries by providing employment and the (long-term) production of valued products. Examples include the many clam species, oysters and crabs that have been cultivated (Laruelle *et al.*, 1994; Miossec *et al.*, 2009; Sundet & Hoel, 2016). However, some species may become so prolific that they also create unwanted effects, as shown by the recent expansion and fouling of the Pacific oyster in the North Sea, which may be related to climate change (Reid & Valdés, 2011). This demonstrates that a latent threat to the environment, human health, property or resources will persist, as a currently non-impacting species may become invasive at a later stage.

Great harm can be caused by the introduction of a single harmful species. For instance, the zebra mussel *Dreissena polymorpha*, which is native to the Black Sea region, has expanded its range westwards in Europe and is now also extensively distributed in North

America. It has caused environmental changes to lakes and rivers, but of possibly greater importance is that the mussels have fouled the abstraction piping and pipework of power stations and municipal water supplies (Carlton & Geller, 1993; Hebert *et al.*, 1989; Johnson & Padilla, 1996; van der Velde *et al.*, 2010). Another example is the predatory sea star *Asterias amurensis*, which arrived in Australia from the north-west Pacific. It caused significant changes to bottom-dwelling communities, some of which are of economic importance (Buttermore *et al.*, 1994; Byrne & Morrice, 1997; Rossa *et al.*, 2003). A further predator, the comb jelly *Mnemiopsis leidyi*, was unintentionally introduced to the Black Sea from the eastern coast of the Americas. In its new environment, it developed into a massive organism concentration, resulting in heavy predation on zooplankton, including the larval stages of commercially important fishes (GESAMP, 1997; Vinogradov *et al.*, 2005). This newcomer declined in the Black Sea after a predatory comb jelly arrived, but the comb jelly continued to spread to the Caspian Sea, likely carried by shipping via the Volga–Don Canal (Ivanov *et al.*, 2000). Thereafter, it appeared in the Kiel Bight and has spread to several Baltic countries and to the southern North Sea (Javidpour *et al.*, 2006). It also expanded southwards to the Mediterranean (Shiganova *et al.*, 2001). However, in these locations, the impact was not as substantial as in the Black Sea.

These are only a few examples of an almost never-ending list of impacting non-indigenous species. All of these impacts sum to a considerable economic cost, but this cost is very difficult to quantify. More than 10 years ago, in the USA alone, a comprehensive study concluded that the estimated annual damage and/or control costs regarding introduced aquatic non-indigenous species was US$14.2 billion (Pimentel *et al.*, 2005). A more recent summary for Europe suggested costs of more than €12 billion annually, and this includes costs for the repair, management and mitigation of impacts (Shine *et al.*, 2010). It has to be noted that this cost summary includes terrestrial species impacts. For the aquatic sector, the costs are thought to be 10–15 per cent of the total amount.

Considering these negative impacts and the fact that every vessel transporting ballast water should be seen as representing a potential risk of introducing species, a precautionary approach would involve the regulation of species transportation vectors. Consequently, the likely most prominent vector – ballast water – was addressed first by a global convention.

5.6 The Ballast Water Management Convention

The significance of the ballast water issue was first addressed in 1973 by the International Maritime Organization (IMO), the United Nations body for the regulation of international maritime transport at the global scale. Dedicated work at the IMO resulted in the adoption of the International Convention for the Control and Management of Ships' Ballast Water and Sediments (BWM Convention) in February 2004 (IMO, 2004). The BWM Convention's main aim is to prevent, minimize and ultimately eliminate the risks to the environment, human health, property and resources that arise from the transfer of harmful aquatic organisms and pathogens (HAOPs) via ships' ballast waters and related sediments (David *et al.*, 2015a; Gollasch *et al.*, 2007). It should be noted that HAOPs in this context are not limited to non-indigenous species, but cover all aquatic species irrespective of their origin.

The BWM Convention consists of 22 articles followed by five sections with regulations. In addition, the 36-page document contains two appendices addressing standard formats and requirements regarding International Ballast Water Management Certificates, as well as documenting operations in a Ballast Water Record Book (IMO, 2004). In total, 15 guidelines support the uniform implementation of the BWM Convention by providing technical guidance.

As per the BWM Convention, certain obligations are to be met by all stakeholders, including the ship, the administrations (i.e., in their capacity as a flag state, port state and as the representative of a party) and the IMO (David *et al.*, 2015a).

5.6.1 D-1 and D-2

The standards regarding ballast water management are the Ballast Water Exchange (BWE) Standard (as stated in Regulation D-1) and the Ballast Water Performance Standard (see Regulation D-2). Regulation D-1 requires ships to exchange a minimum of 95 per cent of their ballast water volume. Regulation D-2 requires that discharged ballast water contains numbers of viable organisms below specified limits. The BWM Convention introduces these two different protective measures with a sequential implementation regime. From its entry into force in September 2017, the D-1 standard applies until ships have to meet D-2. Ships constructed before the entry into force of the BWM Convention will not be required to comply with Regulation D-2 until their first renewal survey following the date of entry into force of the BWM Convention (which means a relaxation of meeting Regulation D-2 up to a maximum of 5 years).

The reasoning behind BWE is that coastal organisms pumped on board during ballast water uptake are unlikely to survive when discharged at open sea. This is because of, for example, salinity issues and the lack of a hard substrate in the open sea to complete the life cycle of these coastal organisms. Similarly, high-sea organisms when pumped on board during the BWE at open sea are unlikely to survive when released in coastal waters with ballast water discharges due to possible salinity changes and the lack of suitable habitats. Furthermore, it is well known that organism concentrations are usually much lower in the open sea compared to coastal waters, which reduces the risk of species introductions. However, studies on board commercial vessels showed that it may be the case that organism concentrations in ballast tanks could be higher after a BWE (e.g., Macdonald & Davidson, 1998; McCollin *et al.*, 2001). This specifically may occur when the BWE is undertaken in shallower seas or during periods of high organism concentrations, such as algal blooms.

Therefore, the IMO requires that BWE shall, whenever possible, be conducted at least 200 nautical miles from the nearest land and in water depths of at least 200 m depth. Should this be impossible, then the BWE should be conducted as far from the nearest land as possible, and in all cases at least 50 nautical miles from the nearest land and in waters of at least 200 m depth (IMO, 2004). In any case, a ship shall not be required to deviate from its intended voyage or delay the voyage in order to conduct BWE. In sea areas where these BWE depth and distance requirements cannot be met, the port state may designate a ballast water exchange area (BWEA), which should be done in consultation with adjacent or other states, as applicable and according to the relevant IMO guidelines. In this case, a port state may require a ship to deviate from its intended route or delay its voyage in order to conduct BWE in a designated BWEA (David & Gollasch, 2008).

Due to the limitations of BWE, the IMO agreed on D-2 as a stronger protective regime. To meet the D-2 standard means that ships must discharge less than 10 viable organisms per cubic metre greater than or equal to 50 μm in minimum dimension and less than 10 viable organisms per millilitre less than 50 μm in minimum dimension and greater than or equal to 10 μm in minimum dimension, and the discharge of the indicator microbes shall not exceed the specified concentrations.

The D-2 standard for both organism groups greater than or equal to 10 μm in minimum dimension refers to all organisms, irrespective of their species (i.e., it is not only addressing non-indigenous or harmful organisms). As a result, taxonomic species identification is not required for the purposes of compliance testing with this standard.

To achieve the ballast water discharge standards set by the BWM Convention, different ballast water management systems (BWMSs) were developed (David & Gollasch, 2015a). The majority of BWMS use filtration as the pretreatment separation step. The dominating treatment process is to use an active substance, mostly generated on board by electrolysis/electrochlorination, closely followed by ultraviolet treatment as the second most frequently applied treatment. More than 60 BWMSs have received type approval according to the IMO guidelines (IMO, 2015, 2016), and the first BWMS that received United States Coast Guard type approval was announced in December 2016 (Coast Guard Maritime Commons, 2016).

When making use of active substances, BWMSs are evaluated for their environmental acceptability by an independent body (i.e., the Group of Experts on the Scientific Aspects of Marine Environmental Protection (GESAMP) Ballast Water Working Group). However, the long-term accumulating effects of the remaining active substances, disinfection by-products or applied neutralizers are unknown, but a recent analysis revealed that some disinfection by-products may reach levels of concern (David *et al.*, 2018; Dock *et al.*, 2019).

In this context, it is also interesting to note that unmanaged ballast water was documented to contain elevated levels of metals (Soleimani *et al.*, 2016), which may further contribute to these long-term effects.

5.6.2 Do All Vessels Need to Meet Ballast Water Management Requirements?

It was understood that no 'one size fits all' ballast water management approach is available because different states may have different geographical, environmental, socio-economic, organizational, political and other conditions, as well as different shipping patterns.

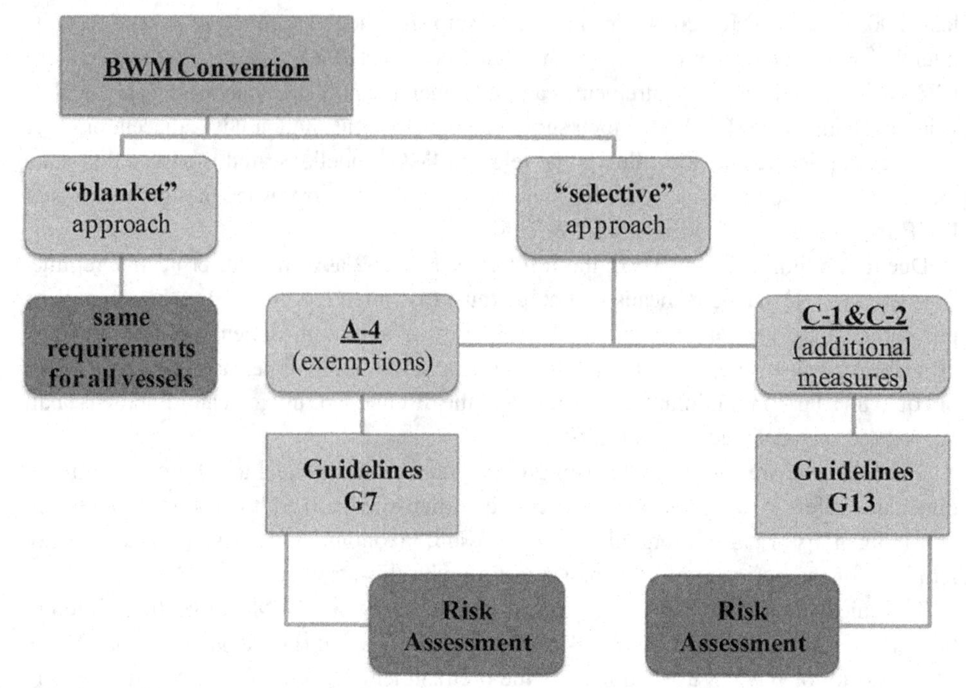

Figure 5.2 Blanket and selective ballast water management according to the BWM Convention.
Source: David and Gollasch (2015c)

The BWM Convention enables two different ballast water management approaches: 'blanket' and the 'selective' approaches. The blanket approach means that all ships intending to discharge ballast water in a port are required to conduct ballast water management measures. The selective approach does not apply the same ballast water management requirements to all vessels; the ballast water management measures required vary depending on the different levels of risk posed by the ballast water intended for discharge.

The BWM Convention also provides options for cases where vessels do not need to conduct BWM measures (i.e., Regulation A-3 Exceptions (see below) and Regulation A-4 Exemptions). Exemptions need to be based upon robust risk assessment. Such a risk assessment needs to be based upon the Guidelines on Risk Assessments under Regulation A-4 (G7) (IMO, 2007a). Should the risk be identified as acceptable, vessels may be exempted from ballast water management requirements. In other cases, when the risk is identified as (very) high, such ships may be required to take additional measures that may be introduced based on Regulation C-1 and based on the Guidelines for Additional Measures Including Emergency Situations (G13) (David & Gollasch, 2015c, 2018; David *et al.*, 2013a, 2015b; IMO, 2004, 2007b) (Figure 5.2).

In addition to exemptions, the BWM Convention also includes provisions for cases where vessels do not need to manage their ballast water at all. This applies to vessels that are in line with the Regulation A-3 Exceptions. These specific cases include the following (IMO, 2004):

(1) Ballast water uptake or discharge is needed for ensuring the safety of a ship in emergency situations.
(2) Accidental discharge results from damage to a ship or its equipment.
(3) Uptake or discharge of ballast water is used to avoid or minimize pollution incidents.
(4) Uptake and discharge of the same ballast water are conducted on the high seas.
(5) Uptake and discharge occur at the same location (David *et al.*, 2013b; Gollasch & David, 2012), provided no mixing occurs with other locations.

Granting an exemption or a permanent exception means that a vessel is not required to install a BWMS, with the clear benefit of avoiding capital and operational costs, as well as the burdens associated with documentation, certification and inspections (David *et al.*, 2015b).

5.7 Outlook

The more that species movements and their associated negative impacts are avoided, the better. This justifies all avoidance measures, including those that clearly have their limitations, such as BWE. We note that the BWM Convention entered into force in September 2017, and we hope that this will considerably reduce the new arrivals of species in our waters. The IMO understood that several uncertainties regarding the BWM Convention existed, and one of those was how to take a representative sample to prove compliance with the D-2 Standard. To promote research in this field, the IMO agreed on an experience-building phase, which was established after the entry into force of the BWM Convention in September 2017. During this phase, countries have been asked to conduct research and to share their results so that any open question can be answered. As a consequence, certain guidelines and/or the BWM Convention itself may be reopened for revision or amendment.

We recommend that coastal states should monitor their waters and document the number of newly found species that likely arrived by ballast water. The hopefully downwards trend of such species inventories may be used to justify any resources spent to address this species introduction vector. Other vectors, such as biofouling or species movements for aquaculture purposes, should also be regulated to reduce the risk of new species arriving even further and to reach the ultimate goal: to prevent, minimize and ultimately eliminate the risk of introducing species, especially those causing unwanted impacts.

References

Aliff, M. (2015). Evaluation of a Method for Ballast Water Risk-Release Assessment Using a Protist Surrogate. Retrieved from the University of Minnesota Digital Conservancy. http://hdl.handle.net/11299/174771

AquaNIS Editorial Board (2015). Information System on Aquatic Non-Indigenous and Cryptogenic Species. www.corpi.ku.lt/databases/aquanis

Bailey, S. A., Vélez-Espino, L. A., Johannsson, O. E., Koops, M. A. & Wiley, C. J. (2009). Estimating establishment probabilities of *Cladocera* introduced at low density: an evaluation of the proposed ballast water discharge standards. *Canadian Journal of Fisheries and Aquatic Sciences*, **66**(2), 261–276.

Briski, E., Bailey, S., Cristescu, M. E. & MacIsaac, H. J. (2010). Efficacy of 'saltwater flushing' in protecting the Great Lakes from biological invasions by invertebrate eggs in ships' ballast sediment. *Freshwater Biology*, **55**, 2414–2424.

Briski, E., Bailey, S. A. & MacIsaac, H. J. (2011). Invertebrates and their dormant eggs transported in ballast sediments of ships arriving to the Canadian coasts and the Laurentian Great Lakes. *Limnology and Oceanography*, **56**(5), 1929–1939.

Briski, E., Bailey, S. A., Casas-Monroy, O. *et al.* (2012). Relationship between propagule pressure and colonization pressure in invasion ecology: a test with ships' ballast. *Proceedings. Biological Sciences*, **279**(1740), 2990–2997.

Buttermore, R. E., Turner, E. & Morrice, M. G. (1994). The introduced northern Pacific seastar *Asterias amurensis* in Tasmania. *Memoirs of the Queensland Museum*, **36**(1), 21–25.

Byrne, M. & Morrice, M. G. (1997). Introduction of the northern Pacific asteroid *Asterias amurensis* to Tasmania: reproduction and current distribution. *Marine Biology*, **127** (4), 673–685.

Carlton, J. T. (1985). Transoceanic and interoceanic dispersal of coastal marine organisms: the biology of ballast water. *Oceanography and Marine Biology: An Annual Review*, **23**, 313–371.

Carlton, J. T. (2001). *Introduced Species in U.S. Coastal Waters: Environmental Impacts and Management Priorities*. Arlington, VA: Pew Oceans Commission.

Carlton, J. T. & Eldredge, L. G. (2009). *Marine Bioinvasions of Hawaii. Bishop Museum Bulletins in Cultural and Environmental Studies*. Honolulu, HI: Bishop Museum Press.

Carlton, J. T. & Geller, J. B. (1993). Ecological roulette: the global transport of nonindigenous marine organisms. *Science*, **261**, 78–82.

Coast Guard Maritime Commons (2016). Marine Safety Center issues Ballast Water Management System (BWMS) type-approval certificate to Optimarin AS. http://mariners.coastguard.dodlive.mil/2016/12/02/marine-safety-center-issues-ballast-water-management-system-bwms-type-approval-certificate-optimarin-as

Cope, R. C., Prowse, T. A. A., Ross, J. V., Wittmann, T. A. & Cassey, P. (2015). Temporal modelling of ballast water discharge and ship-mediated invasion risk to Australia. *Royal Society Open Science*, **2**, 150039.

David, M. (2015). Vessels and ballast water. In M. David & S. Gollasch, eds., *Global Maritime Transport and Ballast Water Management – Issues and Solutions. Invading Nature*. Dordrecht: Springer Science + Business Media, pp. 13–34.

David, M. & Gollasch, S. (2008). EU shipping in the dawn of managing the ballast water issue. *Marine Pollution Bulletin*, **56**(12), 1966–1972.

David, M. & Gollasch, S. (2015a). Introduction. In M. David & S. Gollasch, eds., *Global Maritime Transport and Ballast Water Management – Issues and Solutions. Invading Nature*. Dordrecht: Springer Science + Business Media, pp. 1–12.

David, M. & Gollasch, S. (2015b). Ballast water management systems for vessels. In M. David & S. Gollasch, eds., *Global Maritime Transport and Ballast Water Management – Issues and Solutions. Invading Nature*. Dordrecht: Springer Science + Business Media, pp. 109–132.

David, M. & Gollasch, S. (2015c). Ballast water management decision suport system. In M. David & S. Gollasch, eds., *Global Maritime Transport and Ballast Water Management – Issues and Solutions. Invading Nature*. Dordrecht: Springer Science + Business Media, pp. 225–260.

David, M. & Gollasch, S. (2018). How to approach ballast water management in European seas. *Coastal and Shelf Science*, **201**, 248–255.

David, M., Gollasch, S., Cabrini, M. *et al.* (2007). Results from the first ballast water sampling study in the Mediterranean Sea – the Port of Koper study. *Marine Pollution Bulletin*, **54**, 53–65.

David, M., Perkovič, W., Suban, V. & Gollasch, S. (2012). A generic ballast water discharge assessment model as a decision supporting tool in ballast water management. *Decision Support Systems*, **53**, 175–185.

David, M., Gollasch, S. & Leppäkoski, E. (2013a). Risk assessment for exemptions from ballast water management – the Baltic Sea case study. *Marine Pollution Bulletin*, **75**, 205–217.

David, M., Gollasch, S. & Pavliha, M. (2013b). Global ballast water management and the 'same location' concept: a clear term or a clear issue? *Ecological Applications*, **23**(2), 331–338.

David, M., Gollasch, S., Elliott, B. & Wiley, C. (2015a). Ballast water management under the Ballast Water Management Convention. In M. David & S. Gollasch, eds., *Global Maritime Transport and Ballast Water Management – Issues and Solutions. Invading Nature*. Dordrecht: Springer Science + Business Media, pp. 89–108.

David, M., Gollasch, S., Leppäkoski, E. & Hewitt, C. (2015b). Risk assessment in ballast water management. In M. David & S. Gollasch, eds., *Global Maritime Transport and Ballast Water Management – Issues and Solutions. Invading Nature*. Dordrecht: Springer Science + Business Media, pp. 133–170.

David, M., Linders, J., Gollasch, S. & David, J. (2018). Is the aquatic environment sufficiently protected from chemicals discharged with treated ballast water from vessels worldwide? A decadal environmental perspective and risk assessment. *Chemosphere*, **207**, 590–600.

Davidson, I. C. & Simkanin, C. (2012). The biology of ballast water 25 years later. *Biological Invasions*, **14**, 9–13.

Dock, A., Linders, J., David, M., Gollasch, S. & David, J. (2019). Is human health sufficiently protected from chemicals discharged with treated ballast water from vessels worldwide? – A decadal perspective and risk assessment. *Chemosphere*, **235**, 194–204.

Drake, D. A. R., Casas-Monroy, O., Koops, M. A. & Bailey, S. A. (2015). Propagule pressure in the presence of uncertainty: extending the utility of proxy variables with hierarchical models. *Methods in Ecology and Evolution*, **6**, 1363–1371.

Duncan, R. P., Blackburn, T. M., Rossinelli, S. & Bacher, S. (2014). Quantifying invasion risk: the relationship between establishment probability and founding population size. *Methods in Ecology and Evolution*, **5**, 1255–1263.

GESAMP (1997). Opportunistic settlers and the problem of the ctenophore *Mnemiopsis leidyi* invasion in the Black Sea. *GESAMP Reports and Studies*, **58**, 1–84.

Glibert, P. M. (2015). More than propagule pressure: successful invading algae have physiological adaptations suitable to anthropogenically changing nutrient environments. *Aquatic Ecosystem Health & Management*, **18**(3), 334–341.

Gollasch, S. (1996). Untersuchungen des Arteintrages durch den internationalen Schiffsverkehr unter besonderer Berücksichtigung nichtheimischer Arten. Doctoral dissertation (in German), University of Hamburg.

Gollasch, S. & David, M. (2012). A unique aspect of ballast water management requirements – the same location concept. *Marine Pollution Bulletin*, **64**, 1774–1775.
Gollasch, S., Lenz, J., Dammer, M. & Andres, H. G. (2000a). Survival of tropical ballast water organisms during a cruise from the Indian Ocean to the North Sea. *Journal of Plankton Research*, **22**(5), 923–937.
Gollasch, S., Rosenthal, H., Botnen, H. *et al.* (2000b). Fluctuations of zooplankton taxa in ballast water during short-term and long-term ocean-going voyages. *International Review of Hydrobiology*, **85**(5–6), 597–608.
Gollasch, S., Macdonald, E., Belson, S. *et al.* (2002). Life in ballast tanks. In: E. Leppäkoski, S. Gollasch & S. Olenin, eds., *Invasive Aquatic Species of Europe: Distribution, Impacts and Management*. Dordrecht: Kluwer Academic Publishers, pp. 217–231.
Gollasch, S., David, M., Voigt, M. *et al.* (2007). Critical review of the IMO International Convention on the Management of Ships' Ballast Water and Sediments. *Harmful Algae*, **6**, 585–600.
Grantham, B. A., Eckert, G. L. & Shanks, A. L. (2003). Dispersal potential of marine invertebrates in diverse habitats. *Ecological Applications*, **13**(Suppl. 1), 108–116.
Hallegraeff, G. M. & Bolch, C. J. (1991). Transport of toxic dinoflagellate cysts via ship's ballast water. *Marine Pollution Bulletin*, **22**, 27–30.
Hallegraeff, G. M. & Bolch, C. J. (1992). Transport of diatom and dinoflagellate resting spores in ships' ballast water: implications for plankton biogeography and aquaculture. *Journal of Plankton Research*, **14**(8), 1067–1084.
Hamer, J. P., McCollin, T. A. & Lucas, I. A. N. (2000). Dinofagellate cysts in ballast tank sediments: between tank variability. *Marine Pollution Bulletin*, **40**(9), 731–733.
Hamer, J. P., Lucas, I. A. N. & McCollin, T. A. (2001). Harmful dinoflagellate resting cysts in ships' ballast tank sediments; potential for introduction into English and Welsh waters. *Phycologia*, **40**, 246–255.
Hayes, K. R. (1998). Ecological risk assessment for ballast water introductions: a suggested approach. *ICES Journal of Marine Science*, **55**, 201–212.
Hayes, K. R. & Sliwa, C. (2003). Identifying potential marine pests – a deductive approach applied to Australia. *Marine Pollution Bulletin*, **46**, 91–98.
Hebert, P. D. N., Muncaster, B. W. & Mackie, G. L. (1989). Ecological and genetic studies on *Dreissena polymorpha* (Pallas): a new mollusc in the Great Lakes. *Canadian Journal of Fisheries and Aquatic Sciences*, **46**, 1587–1591.
IMO (2004). *International Convention for the Control and Management of Ships' Ballast Water and Sediments 2004*. London: International Maritime Organization.
IMO (2007a). *Guidelines for Risk Assessment under Regulation A-4 of the BWM Convention (G7). IMO, Marine Environment Protection Committee, Resolution MEPC.162(56), 13 July 2007*. London: International Maritime Organization.
IMO (2007b). *Guidelines for Additional Measures Regarding Ballast Water Management Including Emergency Situations (G13). Marine Environment Protection Committee, Resolution MEPC.161(56), 13 July 2007*. London: International Maritime Organization.
IMO (2015). *List of Ballast Water Management Systems That Make Use of Active Substances Which Received Basic or Final Approval. Note by the Secretariat. International Maritime Organization, BWM.2/Circ.34/Rev.4, May 2015*. London: International Maritime Organization.
IMO (2016). *Report of the MARINE ENVIRONMENT PROTECTION COMMITTEE on Its Seventieth Session. International Maritime Organization, MEPC 70/18, 11. November 2016*. London: International Maritime Organization.

Ivanov, V. P., Kamakin, A. M., Ushivtzev, V. B. *et al.* (2000). Invasion of the Caspian Sea by the comb-jellyfish *Mnemiopsis leidyi* (Ctenophora). *Biological Invasions*, **2**(3), 255–258.

Javidpour, J., Sommer, U. & Shiganova, T. (2006). First record of *Mnemiopsis leidyi* A. Agassiz 1865 in the Baltic Sea. *Aquatic Invasions*, **1**(4), 299–302.

Johnson, L. E. & Padilla, D. K. (1996). Geographic spread of exotic species: ecological lessons and opportunities from the invasion of the zebra mussel *Dreissena polymorpha*. *Biological Conservation*, **78**, 23–33.

Korsu, K. & Huusko, A. (2009). Propagule pressure and initial dispersal as determinants of establishment success of brook trout (*Salvelinus fontinalis* Mitchill 1814). *Aquatic Invasions*, **4**(4), 619–626.

Laruelle, F., Guillou, J. & Paulet, Y. M. (1994). Reproductive pattern of the clams, *Ruditapes decussatus* and *R. philippinarum* on intertidal flats in Brittany. *Journal of the Marine Biological Association of the United Kingdom*, **74**(2), 351–366.

Lo, V. B., Levings, C. D. & Chan, K. M. A. (2012). Quantifying potential propagule pressure of aquatic invasive species from the commercial shipping industry in Canada. *Marine Pollution Bulletin*, **64**, 295–302.

Locke, A., Reid, D. M., Sprules, W. G., Carlton, J. T. & van Leeuwen, H. C. (1991). Effectiveness of mid-ocean exchange in controlling freshwater and coastal zooplankton in ballast water. Canadian Technical Report. *Fisheries and Aquatic Sciences*, **1822**, 1–93.

Lockwood, J. L., Cassey, P. & Blackburn, T. M. (2009). The more you introduce the more you get: the role of colonization pressure and propagule pressure in invasion ecology. *Diversity and Distributions*, **15**, 904–910.

Macdonald, E. M. & Davidson, R. D. (1998). *Ballast Water Project. Fisheries Research Services Report 3/97*. Aberdeen: FRS Marine Laboratory.

McCollin, T., Macdonald, E. M., Dunn, J., Hall, C. & Ware, S. (2001). Investigations into ballast water exchange in European regional seas. In: *Proceedings of International Conference on Marine Bioinvasions, New Orleans, April 9–11 2001*. Dordrecht: Kluwer Academic Publishers, pp. 94–95.

McCollin, T. A., Shanks, A. M. & Dunn, J. (2008). Changes in zooplankton abundance and diversity after ballast water exchange in regional seas. *Marine Pollution Bulletin*, **56**, 834–844.

Medcof, J. C. (1975). Living marine animals in a ships' ballast water. *Proceedings of the National Shellfish Association*, **65**, 54–55.

Minchin, D., Gollasch, S. & Wallentinus, I. (2005). *Vector Pathways and the Spread of Exotic Species in the Sea. ICES Cooperative Research Report No. 271*. Copenhagen: International Council for the Exploration of the Sea.

Miossec, L., Le Deuff, R.-M. & Goulletquer, P. (2009). *Alien Species Alert: Crassostrea gigas (Pacific Oyster). ICES Cooperative Research Report No. 299*. Copenhagen: International Council for the Exploration of the Sea.

Murphy, K. R., Ritz, D. & Hewitt, C. L. (2002). Heterogeneous zooplankton distribution in a ship's ballast tanks. *Journal of Plankton Research*, **24**(7), 729–734.

Ojaveer, H., Olenin, S., Narscius, A. *et al.* (2017). Dynamics of biological invasions and pathways over time: a case study of a temperate coastal sea. *Biological Invasions*, **19**, 799–813.

Orlov, Y. I. & Ivanov, B. G. (1978). On the Introduction of the Kamchatka king crab *Paralithodes camtschatica* (Decapoda: Anomura: Lithodidae) into the Barents Sea. *Marine Biology*, **48**, 373–375.

Ostenfeld, C. H. (1908). On the immigration of *Biddulphia sinensis* Grev. and its occurrence in the North Sea during 1903–1907. *Medd Komm Havunders Ser Plankton*, **1**(6), 1–46.

Pimentel, D., Zuniga, R. & Morrison, D. (2005). Update on the environmental and economic costs associated with alien-invasive species in the United States. *Ecological Economics*, **52**, 273–288.

Reid, P. C. & Valdés, L. (2011). *ICES Status Report on Climate Change in the North Atlantic. ICES Co-operative Research Report No. 310*. Copenhagen: International Council for the Exploration of the Sea.

Rossa, D. J., Johnsona, C. R. & Hewitt, C. L. (2003). Variability in the impact of an introduced predator (*Asterias amurensis*: Asteroidea) on soft-sediment assemblages. *Journal of Experimental Marine Biology and Ecology*, **288**, 257–278.

Ruiz, G. M., Fofonnoff, P. W., Ashton, G., Minton, M. S. & Miller, A. W. (2013). Geographic variation in marine invasions among large estuaries: effects of ships and time. *Ecological Applications*, **23**(2), 311–320.

Schmitz, W. (1960). Die Einbürgerung von *Gammarus tigrinus* Sexton auf dem europäischen Kontinent. *Archives of Hydrobiology*, **57**, 223–225.

Seebens, H., Briski, E., Ghabooli, S. *et al.* (2019). Non-native species spread in a complex network: the interaction of global transport and local population dynamics determines invasion success. *Proceedings of the Royal Society B*, **286**, 20190036.

Shanks, A. L. (2009). Pelagic larval duration and dispersal distance revisited. *Biological Bulletin*, **216**, 373–385.

Shiganova, T., Mirzoyan, Z., Studenikina, E. *et al.* (2001). Population development of the invader ctenophore *Mnemiopsis leidyi*, in the Black Sea and in other seas of the Mediterranean basin. *Marine Biology*, **139**(3), 431–445,

Shine, C., Kettunen, M., Genovesi, P. *et al.* (2010). *Assessment to Support Continued Development of the EU Strategy to Combat Invasive Alien Species. Final Report for the European Commission*. Brussels: Institute for European Environmental Policy (IEEP).

Simberloff, D. (2009). The role of propagule pressure in biological invasions. *Annual Review of Ecology, Evolution, and Systematics*, **40**, 81–102.

Smith, L. D., Wonham, M. J., McCann, L. D. *et al.* (1999). Invasion pressure to a ballast-flooded estuary and an assessment of inoculant survival. *Biological Invasions*, **1**: 67–87.

Soleimani, F., Dobaradaran, S., Hayati, A., Khorsand, M. & Keshtkar, M. (2016). Data on metals (Zn, Al, Sr, and Co) and metalloid (As) concentration levels of ballast water in commercial ships entering Bushehr port, along the Persian Gulf. *Data in Brief*, **9**, 429–432.

Sundet, J. H. & Hoel, A. H. (2016). The Norwegian management of an introduced species: the Arctic red king crab fishery. *Marine Policy*, **72**, 278–284.

Tardent, P. (1979). *Meeresbiologie. Eine Einführung*. Stuttgart: Georg Thieme Verlag.

Thorson, G. (1946). Reproduction and development of Danish marine bottom invertebrates, with special reference to the planktonic larvae in the sound (Øresund). *Meddelelser fra Kommissionen for Danmarks Fiskeri-og Havundersøgelser, Serie Plankton*, **4**, 1–523.

Van der Velde, G., Rajagopal, S. & bij de Vaate, A. (2010). *The Zebra Mussel in Europe*. Leiden: Backhuys Publishers.

Vinogradov, M. E., Shushkina, E. A. & Lukasheva, T. A. (2005). Population dynamics of the ctenophores *Mnemiopsis leidyi* and *Beroe ovata* as a predator–prey system in the near-shore communities of the Black Sea. *Oceanology*, **45**(1), S161–S167.

WGITMO (2015). *Report of the Working Group on Introductions and Transfers of Marine Organisms (WGITMO). 18–20 March 2015, Bergen, Norway. ICES CM 2015/ SSGEPI:10*. Copenhagen: International Council for the Exploration of the Sea.

Williams, R. J., Griffiths, F. B., Van der Wal, E. J. & Kelly, J. (1988). Cargo vessel ballast water as a vector for the transport of non-indigenous marine species. *Estuarine, Coastal and Shelf Science*, **26**, 409–420.

Notes

1 Cryptogenic species are those where it is uncertain whether they are introduced or not.

6

Biocides from Marine Coatings

SAMANTHA ESLAVA MARTINS, ISABEL OLIVEIRA,
KATHERINE LANGFORD AND KEVIN THOMAS

6.1 Biofouling and a Brief History of the Different Approaches Used

Colonization by fouling organisms is a problem that has challenged operators of ships since humans first took to the seas. Reducing or preventing the fouling of a ship's hull is important to allow the vessel to pass efficiently through the water. For centuries, fouling has been controlled through the application of a coating that discourages fouling organisms. As early as the third century there are reports of the Greeks using tar and wax to coat ships' hulls, and as early as the sixteenth century there are reports of copper sheathing or mixtures containing arsenic being used as anti-fouling (AF) coatings on wooden ships. Many of these and the AF solutions that followed were based on the presence of a toxin in the paint to deter organisms from colonizing the painted surface. Cuprous oxide has been used as a biocide since the early nineteenth century and continues to be a common component of modern AF products. The twentieth century saw the introduction of contact leaching AF coatings designed to increase the efficacy and active lifetime of the coating. Typically, these paints contained copper and zinc as the biocidal additives and would be released through dissolution of the painted surface or leaching from an insoluble paint matrix. A major advancement in AF technology was the introduction of self-polishing paints where organotin (OT; typically tributyltin (TBT)) biocides that were incorporated into the paint polymer would allow for controlled release of the biocide as the polymer surface was hydrolyzed. Environmental concerns regarding the use of TBT as an AF biocide saw a ban on its use and the introduction of new, primarily organic biocides, often used alongside other biocides such as copper oxide. A common feature of these coatings has been the release of a biocide(s) into the environment. While modern coatings now aim to be biocide free, AF biocides continue to be developed and introduced onto the market. This chapter provides a comprehensive overview of our understanding of the potential harm caused by biocides released from AF paints applied to ships.

6.2 TBT and the Aftereffects

TBT is an OT compound that is well known for its worldwide application as a biocide, mostly in AF paints, as well as for its the reported adverse effects on the aquatic ecosystems (Fent, 2006; Sousa *et al.*, 2014; WHO, 1990) that led to its prohibition by the International

Maritime Organization (IMO) in 2008. This compound has three butyl groups covalently bonded to one tin (Sn^{4+}) atom. While inorganic tin is generally accepted as being nontoxic, OT's maximum biological activity occurs exactly in the trisubstituted form (Hoch, 2001) to which TBT belongs. Due to its high effectiveness at very low concentrations, TBT became particularly advantageous in terms of formulation and cost by permitting the production of extremely effective, versatile and economic AF paints during the 1960s (Omae, 2003).

At that time, TBT was considered less harmful than the preceding biocides due to its relatively short half-life in water (between 6 days and 2 weeks; Blunden & Chapman, 1982; Clark *et al.*, 1988; de Mora & Pelletier, 1997) and because it was believed to rapidly degrade into nontoxic substances. Nevertheless, TBT degradation rates based on laboratory studies were found to be longer in the environment than first expected (Burton *et al.*, 2006). TBT leaching from AF paints was also found to trigger well-documented adverse effects in nontarget organisms (Antizar-Ladislao, 2008; Sousa *et al.*, 2014) and to accumulate in sediments with half-lives varying between 1 year in surface aerobic sediments and up to decades in highly contaminated anoxic sediments (Langston & Pope, 1995; Stewart & de Mora, 1990). In addition, an evident relationship was described between TBT concentrations in water, sediments and biota and the proximity of areas with high naval activity such as harbors, marinas and shipyards (Barreiro *et al.*, 2001; Barroso & Moreira, 2002a; Fent, 2006; Stroben *et al.*, 1992), confirming AF paints as the main source of this compound in aquatic ecosystems. Furthermore, a number of transformation products, namely dibutyltin (DBT) and monobutyltin (MBT), were also found to be toxic, although to a lesser degree (Ferreira *et al.*, 2013; Gadd, 2000).

The first severe biological effects attributed to TBT were observed in Arcachon Bay (France), where oyster production almost ceased due to reported shell abnormalities, larvae mortality and low growth rates (Alzieu, 2000). Almost simultaneously, other effects on marine mollusks were reported, such as the reduction of gastropod densities and particularly the disruption of the endocrine and reproductive functions (Alzieu, 1991; Gibbs *et al.*, 1987; Matthiessen *et al.*, 1995; Oehlmann *et al.*, 1996) of more than 260 gastropod species worldwide (Titley-O'Neal *et al.*, 2011). This phenomenon is termed "imposex" and consists of the superimposition of male characteristics onto females (Smith, 1971). Imposex has been proven to be irreversible (Bryan *et al.*, 1986) and a reliable biomarker of TBT pollution. However, it has to be noted that triphenyltin (TPT) may also induce imposex in some gastropod species (Barroso *et al.*, 2002; Horiguchi *et al.*, 1997; Laranjeiro *et al.*, 2016; Schulte-Oehlmann *et al.*, 2000). Nevertheless, in Europe, the reported environmental concentrations of TPT are generally lower than those of TBT (Barreiro *et al.*, 2001; Barroso & Moreira, 2002b; Oliveira *et al.*, 2009; Ruiz *et al.*, 1998; Sousa *et al.*, 2009) and so imposex is generally accepted to be caused exclusively by TBT. On the other hand, special care has to be taken at locations where both OTs show similar concentrations (e.g., Japan; Eguchi *et al.*, 2009; Harino *et al.*, 2010). The effects of TBT have also been described in several other nontarget organisms such as bacteria, algae, fish, marine mammals and also humans (Antizar-Ladislao, 2008; Sousa *et al.*, 2014).

These undesirable effects drove several countries to limit TBT use during the 1980s. The first restrictive measures were taken in Europe, namely in France (1982) and the UK

(1987), followed by the USA, New Zealand (1988), Australia, Canada (1989), Japan (1990) and South Africa (1991) (Champ, 2000; Santillo *et al.*, 2001). Meanwhile, in Europe, through Directive 89/677/ECC, the European Union (EU) prohibited the use of TBT-based AF paints in boats less than 25 m in length and in submerged equipment intended to be used in fish and shellfish farming. However, the inefficacy of these measures, especially in areas close to TBT hotspots (i.e., harbors, marinas, dockyards), and the growing evidence of the global effects of TBT pollution (Alzieu, 1991; Barroso & Moreira, 2002b; Gibson & Wilson, 2003; Minchin *et al.*, 1997) prompted the adoption of the International Convention of the Control of Harmful Anti-fouling Systems in Ships (AFS Convention) supported by the IMO. The AFS Convention intended to forbid the reapplication of OT-based AF paints in all vessels from 2003 and to completely ban the use of these paints on ship hulls after 2008 (IMO, 2002). However, its entry into force only materialized in September 2008, after the ratification of 25 states representing 25 percent of the world's merchant shipping tonnage (IMO, 2002). Consequently, and on behalf of this measure, the EU anticipated the adoption of the AFS Convention through EU Regulation 782/2003/EEC (EC, 2003), compelling European vessels or others carrying the EU flag to implement the AFS Convention in European waters. Hence, in the EU, the most restrictive measure of the IMO ban effectively took place in 2008, while throughout much of the rest of world the ban was only implemented in 2013 (Langston *et al.*, 2015).

Almost a decade after the IMO OT ban, it is still possible to perceive the legacy of TBT pollution in aquatic ecosystems. The long-term fate of TBT in sediments, which have been acting as a major and persistent reservoir (Langston & Pope, 1995; Langston *et al.*, 2015), increases concern regarding this compound. Not only can TBT still be released into the environment by maintenance dredging, but it may also be desorbing from the sediments to the water column, possibly promoted by a chemical equilibrium due to the different chemical characteristics of both compartments (Ruiz *et al.*, 2015, 2008). Later studies still confirm recent TBT inputs and/or the existence of high TBT concentrations (Anastasiou *et al.*, 2016; Erdelez *et al.*, 2017; Lam *et al.*, 2017; Mattos *et al.*, 2017), although a decrease in the TBT environmental levels and a consequent imposex recovery have also been generally reported (Kim *et al.*, 2017; Laranjeiro *et al.*, 2018; Nicolaus & Barry, 2015; Wells *et al.*, 2017). However, the use of illegal paints cannot be ignored (Cacciatore *et al.*, 2018; Laranjeiro *et al.*, 2018; Nicolaus & Barry, 2015; Ruiz *et al.*, 2015). Furthermore, special attention should be given to countries that did not sign the treaty nor are regulated by national or regional laws and may still use TBT-based AF paints, including Chile (Batista *et al.*, 2016; Mattos *et al.*, 2017), Brazil (Artifon *et al.*, 2016; Petracco *et al.*, 2015), Venezuela (Paz-Villarraga *et al.*, 2015), Argentina (Commendatore *et al.*, 2015), South Africa (Okoro *et al.*, 2016) and Cape Verde (Lopes-Dos-Santos *et al.*, 2014), among others. Such countries may still be enticed to use TBT-based AF paints by companies that manufacture these paints in countries where TBT usage is banned, as was the case described by Turner and Glegg (2014) in 2014. It is important to continue to monitor TBT environmental levels to confirm whether the recovery trend is present in places that followed the OT ban and to alert countries with poor (or no) legislation on the environmental fate and effects of this biocide so that further deleterious effects can be prevented. The "not yet over" TBT story will certainly be used as an example in the future to raise awareness on the importance of the effective risk assessment of chemicals before they are released into the environment and the role of science in the protection and conservation of ecosystems.

6.3 Alternative Biocides in Marine Coatings

With the end of the legacy of OTs as anti-foulants, the use of metal-based marine coatings increased again, particularly copper-based paints. However, copper alone is not very efficient against biofouling, and new active substances and formulations have been developed to increase the efficiency of the paints. Being TBT-free, these co-biocides – often termed "booster" biocides – have been commonly employed along with copper in paint formulations.

The detrimental effects of OT-based biocides on nontarget organisms led the IMO not only to ban the use of OT compounds as anti-foulants, but also to encourage the development and use of tin-free AF coatings that meet the desired features for ideal biocides (IMO, 2002; Takahashi, 2009). Resolution A.895(21) adopted on November 1999 by the Diplomatic Conference of the Marine Environment Protection Committee (MEPC) of IMO "URGES ALSO Member Governments to encourage industries to continue to develop, test, and use as a high priority anti-fouling systems which do not adversely impact on non-target species and otherwise degrade the marine environment."

According to the IMO, the ideal biocides should have (1) broad-spectrum activity, (2) low toxicity to mammals, (3) low solubility in seawater, (4) low bioaccumulation in the food chain, (5) very low persistence in the environment, (6) compatibility with paint raw materials and (7) favorable price/performance (IMO, 2002). Initially, they were developed for use in small boats (<25 m), the first category of vessel for which TBT was banned from 1989 (Price & Readman, 2013). Currently, they are used in all kinds of vessels and submerged structures subject to biofouling. Many candidates have been marketed and used in the recent years, such as 4,5-dichloro-2-octyl-4-isothiazolin-3-one (DCOIT; Sea-Nine 211), Diuron, Irgarol 1051, chlorothalonil, dichlofluanid, zinc pyrithione (ZnPT) and triphenylborane pyridine (TPBP), among others (Dafforn *et al.*, 2011).

The increasing use of these alternative biocides has raised concern regarding the potential adverse effects that such compounds might pose to nontarget organisms, and effort has been made to gain knowledge of their behavior in the environment and their toxic effects on biota (Martins *et al.*, 2018). Currently, authorities in the environmental field make use of the outcomes of ecotoxicological tests, environmental risk assessments and other approaches in order to establish the environmental criteria of these substances. Many countries have taken legal measures to permit, restrict or ban the use of some alternative anti-foulants. Nevertheless, some alternative AF biocides have already been recognized in accordance with the AFS Convention, such as copper naphthenate, copper oxide, copper pyrithione (CuPT), copper thiocyanate, N-cyclopropyl-N'-(1,1-dimethylethyl)-6-(methylthio)-1,3,5-triazine-2,4-diamine, Diuron, dichlofluanid, DCOIT, medetomidine, 2-octyl-2H-isothiazol-3-one (octhilinone [OIT]), tetrachloroisophthalonitrile, N-(2,4,6-trichlorophenyl) maleimide, thiram, tolylfluanid, tralopyril, TPBP, zinc ethylene bisdithiocarbamate, ZnPT and ziram (www.cdlive.lr.org/information/Documents/Approvals/PaintsResinsReinforcements/paint15.pdf).

The alternative biocides applied more often as AF chemicals are mainly organic or organometallic compounds, and chemical properties such as the octanol/water partitioning coefficient (K_{ow}) and orgnic carbon/water partitioning coeffcient (K_{oc}) drive their partition behavior toward different environmental matrices, both biotic and abiotic. Overall, the chemical–physical properties of alternative biocides suggest that these compounds are neither persistent nor bioaccumulative, hence not being categorized as a persistent, bioaccumulative and toxic substance. Mostly, they have low water persistence, but some antifoulants such as Irgarol 1051 and Diuron may persist for several weeks in seawater (Thomas & Brooks, 2010). Photolysis and biodegradation are important processes for the removal of these compounds from the water column, reducing bioavailability (Table 6.1). However, accumulation in sediment may be an important sink for some biocides (Thomas & Brooks, 2010).

6.3.1 Irgarol 1051

The high toxicity of Irgarol 1051 (cybutryne) to primary producers has driven restrictions and bans at the regional scale. For example, its use has been banned in the UK and Denmark. Similarly, the EU did not approve Irgarol 1051 as an active substance for use in biocide products for AF control through Legal Act 107/2016.

However, this herbicide has been widely applied as an AF biocide after the TBT ban, and it has been regarded as the most frequently detected organic anti-foulant in environmental matrices (Castro *et al.*, 2011; Mohr *et al.*, 2009). In natural seawater, Irgarol 1051 has a half-life of 100–350 days (Thomas *et al.*, 2002), while it is very persistent under anaerobic conditions (Thomas *et al.*, 2003). The main metabolite M1 (2-methylthio-4-tert-butylamino-6-amino-*s*-triazine) can be produced through n-dealkylation of a cyclopropil group from the cyclopropylamino side chain at the six-position of the *s*-triazine ring during biodegradation (Yamada, 2007) (Figure 6.1). It also can be produced through chemical hydrolysis (Liu *et al.*, 1999, 1997) and photodegradation (Okamura *et al.*, 2002) stimulated by dissolved organic matter such as fulvic and humic substances (Okamura & Sugiyama, 2004). M1 is relatively stable in water ($t_{1/2}$ >200 days) and sediments ($t_{1/2}$ = 260 days) (Okamura, 2002; Thomas *et al.*, 2002). Other transformation products of Irgarol 1051 include M2 (3-[4-tertbutylamino-6-methylthiol-*s*-triazine-2-ylamino]-propionaldehyde) and M3 (N'-di-tert-butyl-6-methylthiol-*s*-triazine-2,4-diamine). There is a lack of information on the persistence of these compounds, but they have been reported to occur in the environment (Lam *et al.*, 2005).

6.3.2 Diuron

Diuron is a phenylurea compound that has been used since the 1950s, predominantly for weed control in nonagricultural applications (Thomas & Langford, 2009a), and its use as an

Table 6.1. *Physicochemical properties of TBT-free AF paints*

Substance, chemical name and chemical group	CAS number	Molecular weight	Maximum log K_{ow}	Log K_{oc}	Solubility (mg/L)	DT50 in water (days)	Elimination (DT50)	pKa	Vapor pressure (Pa)	Henry's law (Pa m^3/mol)	Bioaccumulation	Mode of action	Risk phrases
Irgarol 1051[a] Cybutryne 2-(Tert-butylamino)-4-cyclopropylamino)-6-(methylthio)-1,3,5-triazine Triazine	28159-98-0	253.37	3.2 at 10°C; 3.1 at 25°C	3.1	9 at 20°C, pH 7 8.8 at 20°C, pH 9	No hydrolysis; photolysis: 35.9–148 days at pH 7; not readily biodegradable	SW = 23 days Fish <3 days Algae = 9.2 days	4.12	$3.4E^{-5}$ Pa at 25°C	$4.1E^{-4}$	BCF marine fish = 250 L/kg BCF green macroalga = 5200 L/kg	Inhibitor of PS II electron transport	N R50/53
Diuron[a] 1-(3,4-Dichlorophenyl)-3,3-dimethylurea Phenylurea	330-54-1	233.10	2.82	3.4	35–36.4	31.4–365	14–27 days		$2.3E^{-7}$–$3.6E^{-2}$	$1.5E^{-6}$–0.2	Log BCF = 2.16 to *P. promelas* BCF <–3 to 74 to *C. carpio*	Inhibitor of PS II electron transport	
DCOIT 4,5-Dichloro-2-*n*-octyl-3-(2*H*)-isothiazolin-3-one Isothiazolinones	64359-81-5	282.00	2.85	4.19	2.26 at 10°C, pH 7 3.47 at 20°C, pH 7	Hydrolysis: 71 days at 25°C, pH 7; 178–201 days at 12°C, pH 7 148 days at 9°C, pH 8 Photolysis: 48–287 days at 9°C, SW; 38–225 days at 12°C, FW Biodegradation: 8.6–34.6 hours at 9°C; 6.8–27.8 hours at 12°C, estuaries	1.6 days at 12°C, FW <3.6 hours at 9°C, SW	NA	$9.8E^{-4}$ at 25°C	$3.3E^{-2}$ at 20°C, pH 7		Interruption of metabolic processes	N R50/53
OIT[a] Octyl-2H-isothiazol- 3-one Isothiazolinones	26530-20-1	213.34	3.62	3.32	525 at 20–25°C	Biodegradation at 20°C: 120 days SW 135 days sediment		14	$4.9E^{-3}$ at 20°C	$2.9E^{-3}$ at 20°C			R50/53

Table 6.1. (*cont.*)

Substance, chemical name and chemical group	CAS number	Molecular weight	Maximum log K_{ow}	Log K_{oc}	Solubility (mg/L)	DT50 in water (days)	Elimination (DT50)	pKa	Vapor pressure (Pa)	Henry's law (Pa m^3/mol)	Bioaccumulation	Mode of action	Risk phrases
TCMTB 2-(Thiocyanomethylthio) benzothiazole Thiazole	21564-17-0	238.40	2.7–3.3		10.4–45	31–36			$1.18E^{-3}$–607	$2.7E^{-2}$–1.2		Inhibitor of mitochondrial electron transport	
Chlorothalonil[a] 2,4,5,6-Tetrachloroisophthalonitrile Isophthalonitrile	1897-45-6	265.89	2.6–4.4	2.93–3.84	0.6–0.9	120°C, pH 5, 7, 9 = 16.1 days	Aerobic DT50 = 8.1–8.8 days SW Photolysis: 64.7 days, pH 5, 25°C	NA	$7.2E^{-5}$ at 25°C	$2.5E^{-2}$ at 25°C	BCF fish = 9.4–264	Inhibitor of mitochondrial electron transport	R50/53
Dichlofluanid[a] N'-Dimethyl-N-phenylsulfamide Organochlorine	1085-98-9	333.20	3.5 DMSA: 1.59 at 20°C	3.13	0.92 at 10°C 1.58 at 20°C DMSA: 1300 at 20°C	0.12–0.75		NA	$2.15E^{-5}$ at 20°C	$4.5E^{-3}$	BCF *L. macrochirus* = 72	Inhibitor of PS II electron transport	R50
Tolylfluanid N-(Dichlorofluoromethylthio)-N', N'-dimethyl-N-p-tolylsulfamide Organochlorine	731-27-1	347.30	3.9 DMST: 1.99	3.35	1.04 DMST: 677	Hydrolysis: 4.3 hours at pH 8, 20°C, SW 40 hours at pH 7, 20°C, FW	WSD: 0.3 days at 20°C; 0.6 days at 12°C	14	$2E^{-4}$ at 20°C	$7.7E^{-2}$	BCF fish = 74 L/kg		N R50
Zineb[a] Zinc ethylenebis (dithiocarbamate) (polymeric) Dithiocarbamate	12122-67-7	275.70	0.32	3	0.22 at 20°C, pH 7 5 at 20°C, pH 9	Hydrolysis: zineb <1 days Metabolites at pH 7, marine DIDT – 30.1%, 1 day ETU – 11.7%, 3 days	Zineb: Marine 10°C, 0.25 hours Brackish 20°C, 0.24 hours FW 20°C, 0.28 hours Metabolites	12.2	<$3.6E^{-5}$ Pa at 20°C	<0.046 at 20°C	BCF marine fish = 250 L/kg BCF green macroalga = 5200 L/kg	Inhibitor of PS II electron transport	N R50/53

						EU – 52.3%, 30 days Photolysis not expressive Not readily biodegradable	DIDT: Brackish 1.1 days, marine 3.2 days, FW 1 day ETU: brackish 8 days, marine 12 days, FW 33.4 days EU: brackish 28.6 days, FW 15.2 days, marine no degradation observed					
Mancozeb Zinc manganese ethylenebis dithiocarbamate Dithiocarbamate	8018-01-7	271.30	1.38	3	6.2 at 20°C	Hydrolysis: 0.67 days at 20°C Biodegradation: 0.5 days at 20°C, SW and sediment		10.2	$1.33E^{-5}$ at 20°C	$5.9E^{-4}$		
[[2-[(Dithiocarboxy) amino]ethyl]carbamodithioato]] (2-)-kS,kS']manganese Dithiocarbamate	12427-38-2	265.29	0.62	1.3	417 at 22–24°C	Hydrolysis: 1 day			7.5×10^{-8} mmHg at 25°C			
ETU (maneb, zineb, mancozeb metabolite) Ethylene thiourea	96-45-7	102.20	0.22	2.46	20,000	Photolysis: 1 day						
Thiram[a] Tetramethylthioperoxydicarbonic diamide Dithiocarbamate	137-26-8	240.42	1.7	3.98	16.5 at 20°C	Hydrolysis at 20°C: 3.5 Biodegradation at 20°C = 1.92, SW and sediment		8.19	$2.3E^{-3}$	$3.1E^{-2}$	Multisite inhibitor	H400 – aquatic acute I H410 – aquatic chronic I

Table 6.1. (*cont.*)

Substance, chemical name and chemical group	CAS number	Molecular weight	Maximum log K_{ow}	Log K_{oc}	Solubility (mg/L)	DT50 in water (days)	Elimination (DT50)	pKa	Vapor pressure (Pa)	Henry's law (Pa m^3/mol)	Bioaccumulation	Mode of action	Risk phrases
Ziram[a] Bis(dimethyldithiocarbamate) Dithiocarbamate	137-30-4	305.81	1.65	3.03	0.967 at 20°C	Hydrolysis at 20°C = 6.31 Photolysis at 20°C = 0.36 Biodegradation at 20°C = 0.3, SW and sediment		14	$1.8E^{-5}$ at 20°C	$5.7E^{-3}$ at 20°C			
ZnPT[a] Bis-(1-hydroxy-2(1H)-pyridinethionate-O,S) zinc Pyrithione	13463-41-7	317.70	0.9	4	6–8	<1			$1.0E^{-6}$–$1.3E^{-4}$	$5.0E^{-5}$		Multisite inhibitor	
CuPT[a] Bis(1-hydroxi-1H-pyridine-2-thionato-O,S)copper Pyrithione	14915-37-8	315.86	2.84	4	0,1 at 25°C, pH 7–9	Hydrolysis: 12.9-96 at pH 8.2, SW Biodegradation: 4.4–4.7 at 22°C, SW Photolysis: 0.5 hours, SW	3.4–14 days	4.67	$<10E^{-7}$ at 25°C		BCF fish = 7.7 L/kg BCF invertebrate = 8.0 L/kg		R50
TPBP	971-66-4	231.20	5.52[b]		1	1–34			$1.3E^{-4}$–< 0.133			Unknown	
TCMS-Pyridine 2,3,5,6-Tetrachloro-4-(methanesulfonyl)-pyridine	13108-52-6	294.90	NF		0.025	NF			$3.0E^{-5}$	NF		Inhibitor of mitochondrial electron transport	
Capsaicin 8-Methyl-N-vanillyl-6-nonenamide	404-86-4	305.46	3.04	3.04	10.3	13			$1.7E^{-6}$	$1.0E^{-13}$	BCF autotroph = 234 BCF invertebrate = 383 BCF fish = 543	Nervous system and metabolic disruptor	

Nonivamide *N*-[(4-hydroxy-3-methoxyphenyl) methyl]nonanamide	2444-46-4	293.41	4.16		NF	Aerobic biodegradation SW = 5.8			$3.3E^{-8}$	NF		Nervous system and metabolic disruptor	
Medetomidine[a] 4-[1-(2,3-Dimethylphenyl)ethyl]-1H-imidazole hydrochloride	86347-14-0	200.28	3.1		pH 7: 10°C = 353, 20°C = 425 pH 9 10°C = 834, 20°C = 153	51.3[c]	Sediment at 20°C = 50.2 days; at 9°C = 119.4 days	Dissociation constant = 7.1	$3.5E^{-6}$	$8.6E^{-6}$	BCF in fish = 1 L/kg	Activator of octopamine receptor	R50/53
TCPM[d] N-(2,4,6-trichlorophenyl) maleimide	13167-25-4	276.50	2.49	2.2	125.2 at 25°C		Water: 1440 hours Sediment: 13,000 hours		$1.5E^{-8}$ mmHg at 25°C	$1.33E^{-8}$ atm.m^3/mol	BCF = 16.51		
4-Bromo-2-(4-chlorophenyl)-5-(trifluoromethyl)-1H-pyrrole-3-carbonitrile	122454-29-9	349.50	3.47	3.66	0.168	Hydrolysis SW: 15 hours at 10°C, 3 hours at 25°C FW: pH 7, 10°C = 69 hours, pH 7, 25°C = 8 hours Photolysis at pH 7 = 8.9 hours	Aerobic biodegradation: FW, 21°C = 12.75 days SW, 21°C = 0.34 days Anaerobic degradation: FW, 21°C = 35.4 days; SW, 21°C = 0.68 days	7.08 at 26°C	$1.9E^{-8}$	$3.9E^{-5}$	BCF *C. carpio* < 3.2 mL/g	Inhibitor of mitochondrial electron transport	R50/53
Benzmethylamide N-methyl-2-[[2-(methylcarbamoyl) phenyl]disulfanyl]benzamide	2527-58-4	332.44											

Table 6.1. (*cont.*)

Substance, chemical name and chemical group	CAS number	Molecular weight	Maximum log K_{ow}	Log K_{oc}	Solubility (mg/L)	DT50 in water (days)	Elimination (DT50)	pKa	Vapor pressure (Pa)	Henry's law (Pa m^3/mol)	Bioaccumulation	Mode of action	Risk phrases
Fluorofolpet 2-{[Dichloro(fluoro) methyl]sulfanyl}-1H-isoindole-1,3(2H)-dione	719-96-0	280.10	2.5	2.75					0.0 ± 0.7 mmHg at 25°C				
IPBC 3-Iodoprop-2-yn-1-yl butylcarbamate	55406-53-6	281.10											

[a] In accordance with the AFS Convention (IMO, 2012).

[b] Log K_{ow} to tryphenylborane (TPB; Wendt *et al.*, 2013).

[c] Whole-system degradation DT50 (ECHA, 2014) (source: https://pubchem.ncbi.nlm.nih.gov/compound/maneb#section=Density).

[d] All properties were predicted by model EPISuite (www.chemspider.com/Chemical-Structure.75062.html).

FW = freshwater; NA = not applicable; NF = values not found; N R50 = very toxic to aquatic organisms; N R53 = may cause long-term effects in the aquatic environment; PS = photosystem; SW = seawater; WSD = water/sediment.

Sources: Mukherjee *et al.* (2009), Thomas and Brooks (2010), Yamada (2007), Dafforn *et al.* (2011), Wang *et al.* (2014), Castro *et al.* (2011), van Wezel and van Vlaardingen (2004), Thomas (2009), ECHA (2014), Liu *et al.* (2016), National Pesticide Information (http://npic.orst.edu/factsheets/archive/Capsaicintech.html#mode), Arrhenius *et al.* (2014), CBP (2014a, 2014b, 2014c), CBP (2015a, 2015b), KEMI (2014), US EPA (2005a, 2005b), www.chemicalland21.com/lifescience/phar/2-OCTYL-ISOTHIAZOLONE.htm, www.chemicalbook.com/ChemicalProductProperty_EN_CB3221648.htm, WFD (2012), New Zealand Environmental Protection Authority (2012), www.chemspider.com/Chemical-Structure.62954.html

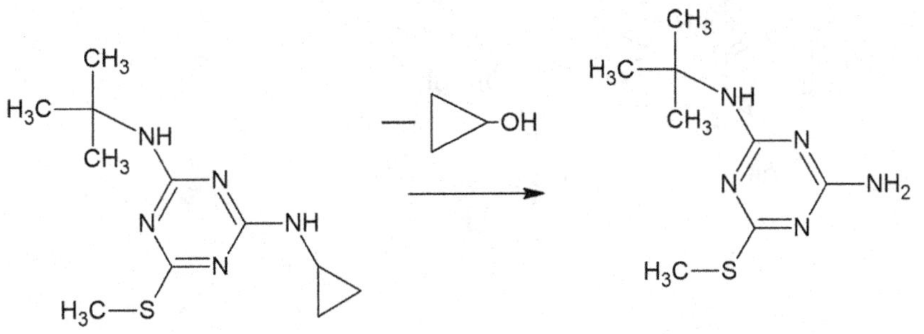

Irgarol 1051 M1

Figure 6.1 Degradation pathway of Irgarol 1051 to M1.
Source: Adapted from Thomas and Brooks (2010)

AF compound started in the 1980s (Ferrer *et al.*, 1997). As an herbicide, Diuron disrupts photosynthesis by blocking electron transport and the transfer of light energy (Bureau of Land Management, 2005), and it is regarded as very toxic to aquatic plants (Bureau of Land Management, 2005; Martins *et al.*, 2018).

Diuron is a fairly stable compound that is resistant to biodegradation and sunlight (Harino & Langston, 2009), stable to hydrolysis and unlikely to volatilize from aquatic systems (Bureau of Land Management, 2005) (Table 6.1). It therefore tends to be persistent in seawater and is less persistent in sediments (Thomas *et al.*, 2002). Diuron has a low tendency to bioaccumulate in aquatic organisms in spite of its persistence (Bureau of Land Management, 2005). Its main metabolites under aerobic conditions are 1-(3,4-dichlorophenyl)-3-methylurea (DCPMU) and 1-(3,4-dichlorophenyl)urea (DCPU) (Figure 6.2).

6.3.3 DCOIT

DCOIT (Sea-Nine 211) is a broad-spectrum biocide that has been used in AF paints since 1986. It prevents the growth and settlement of soft fouling (bacteria, fungi, algae) and hard fouling (barnacles) organisms on submerged surfaces, exhibiting rapid inhibition of growth at very low levels. DCOIT application is restricted to commercial boats and superyachts (size >25 m), buoys, sluice doors and offshore structures submerged in marine and brackish water. DCOIT is used mainly as a co-biocide (concentration 1–3 percent) in copper-based paints, increasing their efficiency (CBP, 2014a). As its mode of action, DCOIT reacts with proteins of organisms that come into contact with the coating surface, resulting in interruption of the metabolic processes and disruption of the physiological processes involved in the attachment of the organisms to solid surfaces, thus preventing fouling.

Once DCOIT is leached to the surrounding environment, biodegradation is the main route of degradation. At 9°C, biological half-lives in water and sediments are 14.0 and 3.6 hours for coastal and estuary areas, respectively, and 42.0 and 10.8 hours for ocean waters, respectively (Table 6.1). Metabolism of DCOIT involves the cleavage of the isothiazolone ring and subsequent oxidation of the alkyl metabolites formed (Figure 6.3), resulting in the formation

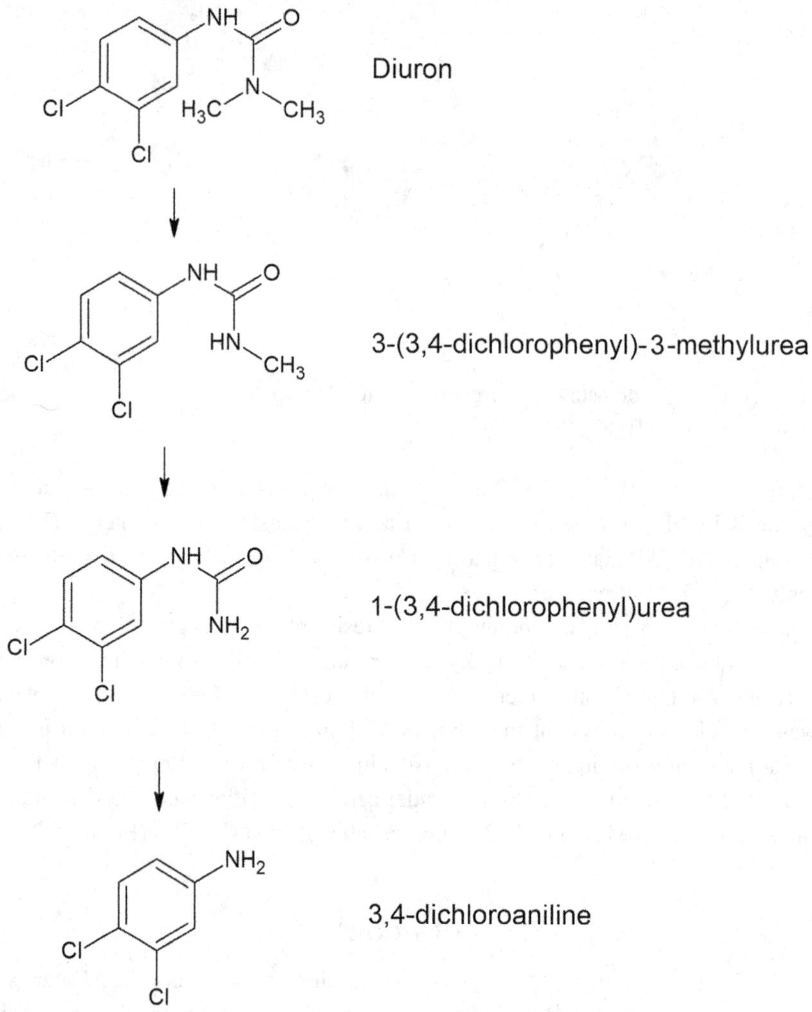

NH O
N
Cl H_3C CH_3
Cl

Diuron

NH O
HN CH_3
Cl
Cl

3-(3,4-dichlorophenyl)-3-methylurea

NH O
NH_2
Cl
Cl

1-(3,4-dichlorophenyl)urea

NH_2
Cl
Cl

3,4-dichloroaniline

Figure 6.2 Proposed degradation pathway for Diuron.
Source: Adapted from Thomas and Brooks (2010)

of its main metabolites N-(n-octyl) malonamic acid (NNOMA), N-(n-octyl) acetamide (NNOA), N-(n-octyl) oxamic acid (NNOOA) and 2-chloro-2-(n-octylcarbamoyl)-1-ethene sulfonic acid) (CBP, 2014a; Thomas & Brooks, 2010). The high K_{oc} for aquatic sediments (1.6×10^4 kg/L) suggests that DCOIT and its metabolites are rapidly partitioned into the sediment, binding there strongly and essentially irreversibly (Thomas, 2001).

6.3.4 Octhilinone

OIT is an isothiazolinone biocide with broad-spectrum antifungal activity. It has a similar structure to DCOIT, but without the chlorine atoms. It has different applications worldwide

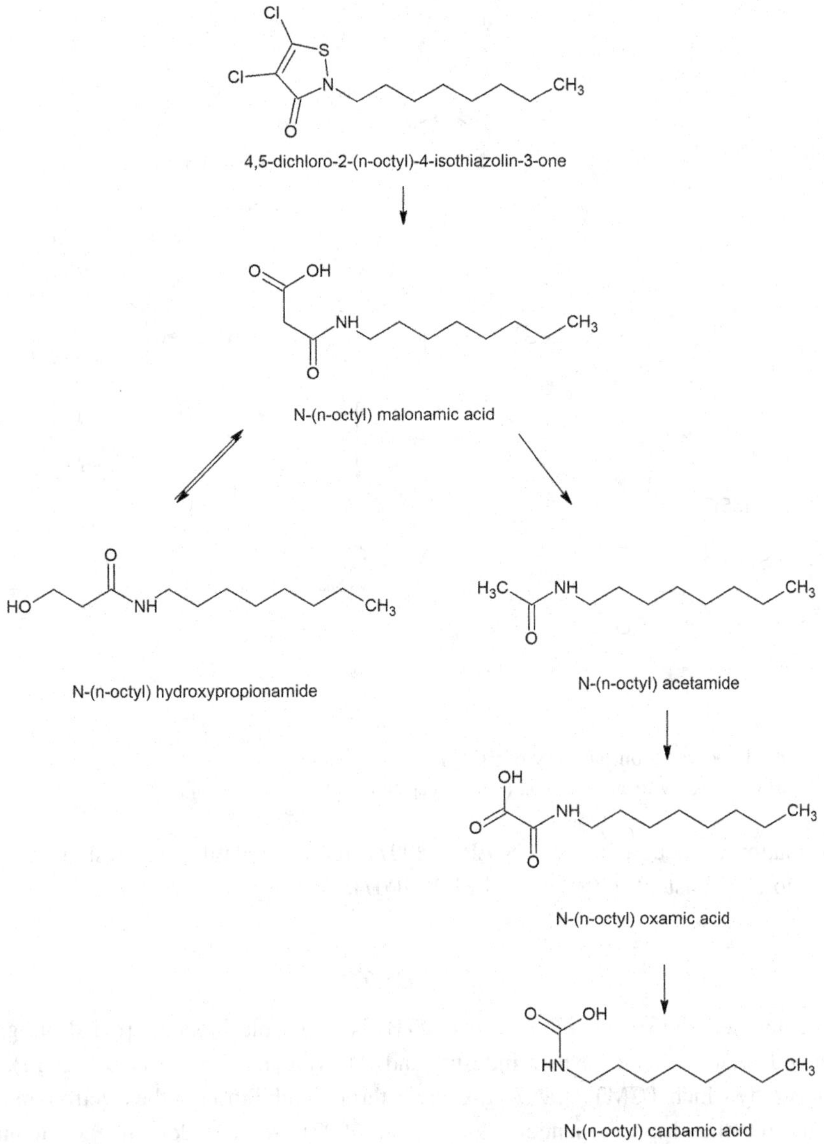

Cl
Cl
S
N
CH_3
O
4,5-dichloro-2-(n-octyl)-4-isothiazolin-3-one
O
OH
NH
CH_3
O
N-(n-octyl) malonamic acid
O
HO
NH
CH_3
N-(n-octyl) hydroxypropionamide
H_3C
NH
CH_3
O
N-(n-octyl) acetamide
OH
O
NH
CH_3
O
N-(n-octyl) oxamic acid
O
OH
HN
CH_3
N-(n-octyl) carbamic acid

Figure 6.3 Proposed degradation pathway for DCOIT under aerobic conditions.
Source: Adapted from Thomas and Brooks (2010)

and has been registered as an AF biocide in New Zealand (New Zealand Environmental Protection Authority, 2013).

With moderate solubility, it is expected that OIT will persist in water once introduced into the aquatic environment, with a moderate tendency to bind to sediments. OIT is hydrolytically stable and photolysis and biodegradation are the main processes of

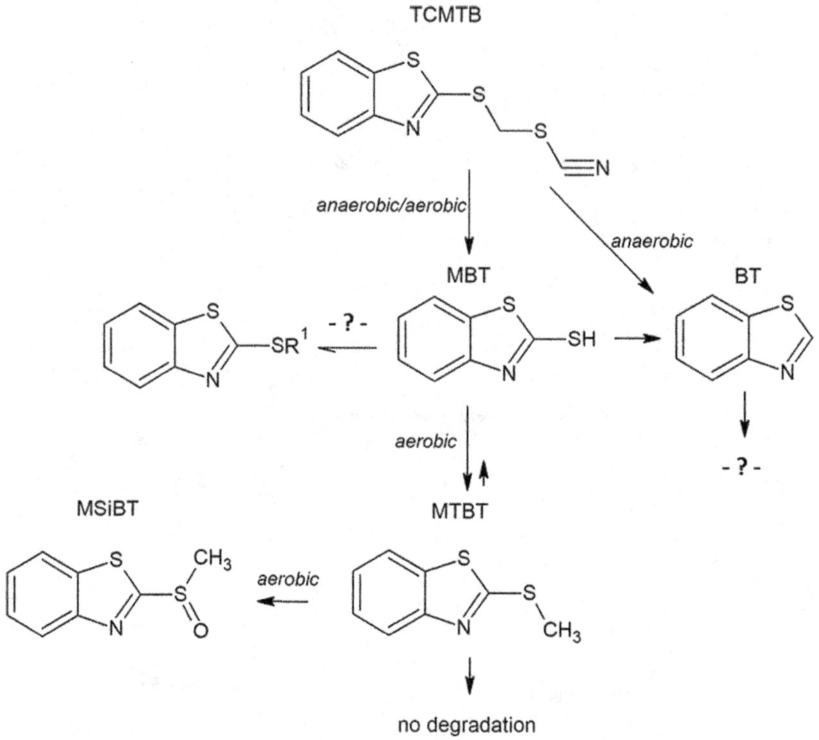

Figure 6.4 Degradation pathway of TCMTB.
Sources: Proposed by Reemtsma *et al.* (1995); adapted from Kurouani-Harani (2003)

transformation and degradation (US EPA, 2007). Its low volatility suggests that OIT is unlikely to be persistent in the air (US EPA, 2007).

6.3.5 TCMTB

2-(Thiocianomethylthio)-benzothiazol (TCMTB) is a thiazole biocide applied in agriculture, wood treatment, the leather industry and AF systems (Castro *et al.*, 2011). One mechanism by which TCMTB causes toxicity is through inhibition of the electron transport chain in mitochondria (Fernández-Alba *et al.*, 2002), so a wide range of nontarget organisms can be affected by TCMTB, which acts either as an herbicide or a fungicide.

Little is known about the TCMTB sorptive equilibrium in aquatic matrices, but it is known that TCMTB rapidly degrades via photolysis, contributing to its low environmental persistence. Photodegradation yields 2-mercaptobenzothiazole (MBT) as the main metabolite and traces of benzothiazole (BT) along unknown pathways (Figure 6.4).

6.3.6 Chlorothalonil

Chlorothalonil is an isophthalonitrile organochlorine fungicide that has been applied in agriculture worldwide for over 30 years, and its use as an AF biocide began after the ban on

OT-based paintings. Although being a polychlorinated aromatic compound, this substance does not have a high degree of persistence due to the two nitrile groups that activate the molecule (WFD, 2012). However, the presence of multiple reactive electrophilic centers makes chlorothalonil extremely toxic to aquatic organisms (Castro *et al.*, 2011; Martins *et al.*, 2018), hence the use of this biocide as an active substance in AF coatings has been gradually discontinued.

Once chlorothalonil reaches aquatic environments, degradation by photolysis in both fresh and estuarine waters occurs (Thomas & Brooks, 2010) by dechlorination of chlorothalonil (Van Scoy & Tjeerdema, 2014) (Figure 6.5). K_{oc} values indicate that chlorothalonil is expected to adsorb onto suspended solids and sediment, where a half-life of up to 8 days was reported (Walker *et al.*, 1988; WFD, 2012) (Table 6.1). Volatilization from water surfaces is not expected to be high, so this substance is unlikely to be persistent in the air compartment (WFD, 2012). The bioaccumulation potential of chlorothalonil is low to moderate, with a bioconcentration factor (BCF) up to 264 reported for fish (WFD, 2012).

6.3.7 Dichlofluanid

Dichlofluanid is an organochlorine fungicide that is unstable in aquatic systems, with a half-life of <24 hours in seawater, being rapidly hydrolyzed to N,N-dimethyl-N'-phenylsulfamide (DMSA) (Figure 6.6). Biodegradation and photolysis are also important processes of dichlofluanid degradation in water (CBP, 2006). Because of its rapid transformation, it is suggested that the metabolite DMSA be used for the risk assessment of dichlofluanid (van Wezel & van Vlaardingen, 2004). Dichlofluanid has a low tendency to volatilize; therefore, air has not been considered as a compartment of concern (CBP, 2006).

6.3.8 Tolylfluanid

This organochlorine biocide with antifungal activity has a very similar structure to dichlofluanid (Figure 6.7) and is intended to protect the hulls of pleasure craft and commercial ships as an active substance in AF paints (CBP, 2014b) registered in Europe, Japan and New Zealand (New Zealand Environmental Protection Authority, 2013; OPRF, 2010).

Tolylfluanid hydrolyzes rapidly under neutral and alkaline conditions, with a reported 50 per cent degradation time (DT50) of 40 hours in freshwater and 4.3 hours in seawater at 20°C (CBP, 2014b). The major metabolites formed are dimethylaminosulfotoluidid (DMST), DMST-benzoic acid and N,N-dimethylsulfamide (DMS) (Figure 6.7), which are hydrolytically stable (CBP, 2014b).

6.3.9 Zineb, Maneb and Mancozeb

Maneb and zineb are organic Mn and Zn complexes in which the metal ions are chelated to organic dithiocarbamate ligands (Harino & Langston, 2009). These and other

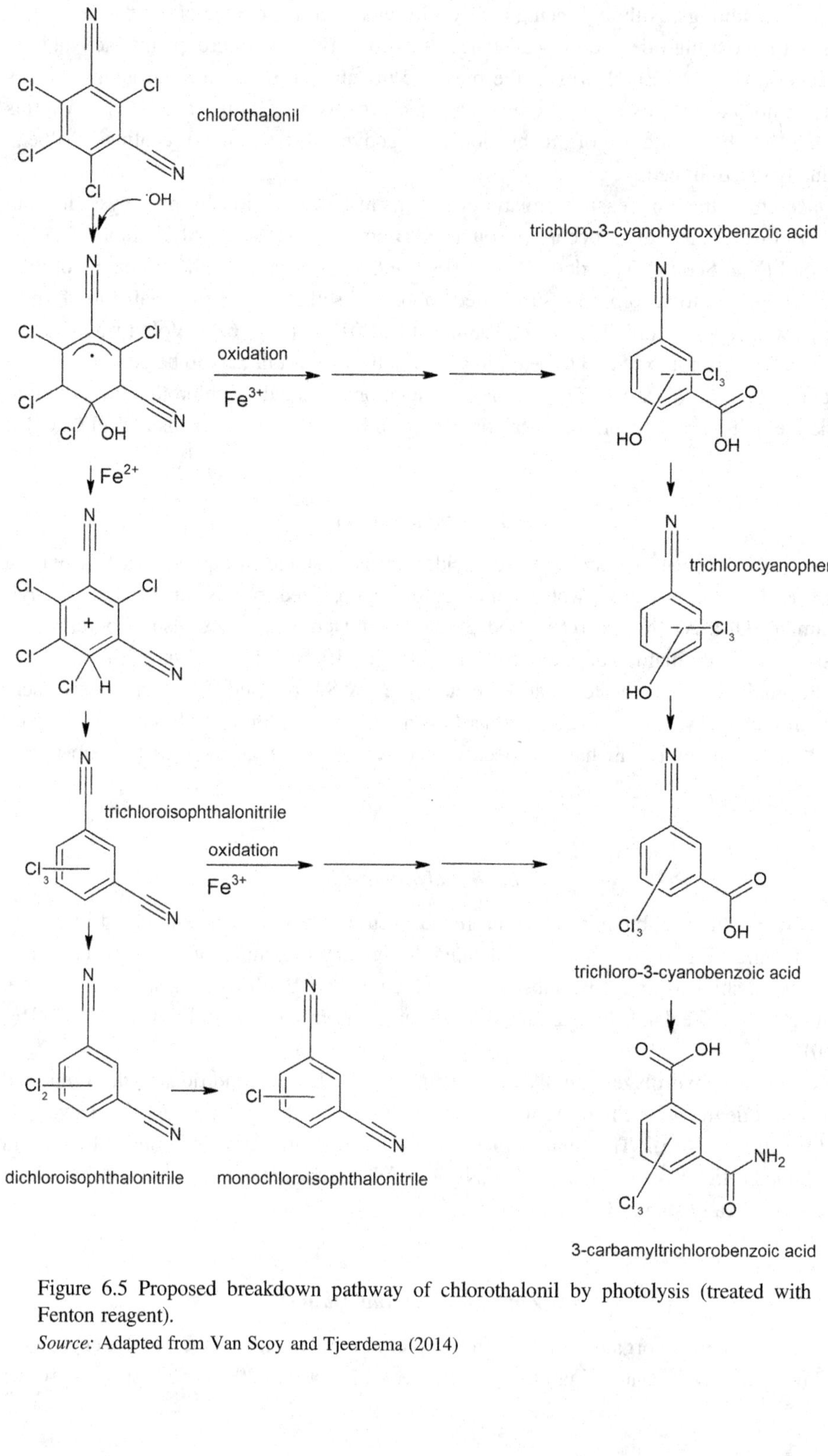

chlorothalonil

$\cdot OH$

oxidation

Fe^{3+}

trichloro-3-cyanohydroxybenzoic acid

Fe^{2+}

trichlorocyanophenol

trichloroisophthalonitrile

oxidation

Fe^{3+}

trichloro-3-cyanobenzoic acid

dichloroisophthalonitrile

monochloroisophthalonitrile

3-carbamyltrichlorobenzoic acid

Figure 6.5 Proposed breakdown pathway of chlorothalonil by photolysis (treated with Fenton reagent).

Source: Adapted from Van Scoy and Tjeerdema (2014)

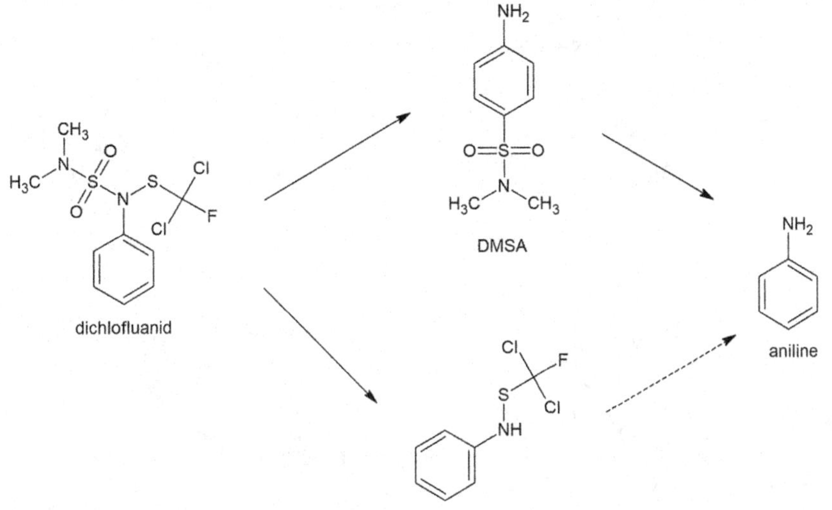

Figure 6.6 Degradation pathway for dichlofluanid.
Source: Adapted from Thomas and Brooks (2010)

dithiocarbamate biocides such as ziram and thiram (see Section 6.3.10) have been used in many AF paint formulations since the development of alternative organic co-biocides to replace TBT (Thomas, 2001).

Zineb is frequently used together with copper compounds in AF paints and is intended for use on commercial ships and pleasure craft (CBP, 2013) to protect the hull from a range of soft and hard fouling organisms, including red and green algae, diatoms (slimes) and invertebrate fouling organisms, such as mussels and barnacles (CBP, 2013).

There is a lack of data on environmental occurrence for zineb, which may be explained by its rapid hydrolysis (<96 hours) to its metabolites ethylenethiourea (ETU) (Figure 6.8), 5,6-dihydro-3H-imidazo (2,1-c)-1,2,4-dithiazole-3-thione (DIDT) and ethelyne diisothiocyanate (EDI) (Thomas *et al.*, 2002). The subsequent removal of ETU from natural waters is rapid ($t_{1/2}$ = 24–96 hours; Xu, 2000) because it is rapidly photolyzed in the presence of photosensitizers such as dissolved organic matter. However, photodegradation is not always relevant as a removal pathway as it requires clear water for sunlight to penetrate (NIVA, 2012).

Maneb also undergoes rapid hydrolysis to produce ETU, ethyleneura, ethylenbis (isothiocyanate) sulfide and carbimide. Photolysis is also important and it appears to form the same degradation products as by hydrolysis. The low K_{ow} suggests that maneb would not be significantly bioconcentrated by aquatic organisms (US EPA, 2005b).

As well as maneb and zineb, mancozeb is a member of the ethylene bisdithiocarbamate (EBDC) group of fungicides, and the toxicological properties are expected to be similar among all compounds. Mancozeb is a high-molecular-weight polymer composed of repeating single units containing manganese and zinc ions, with low water solubility and a low tendency to bioaccumulate. In the environment, mancozeb is expected to rapidly hydrolyze, resulting in a suite of compounds, including ETU (US EPA, 2005a). Because of

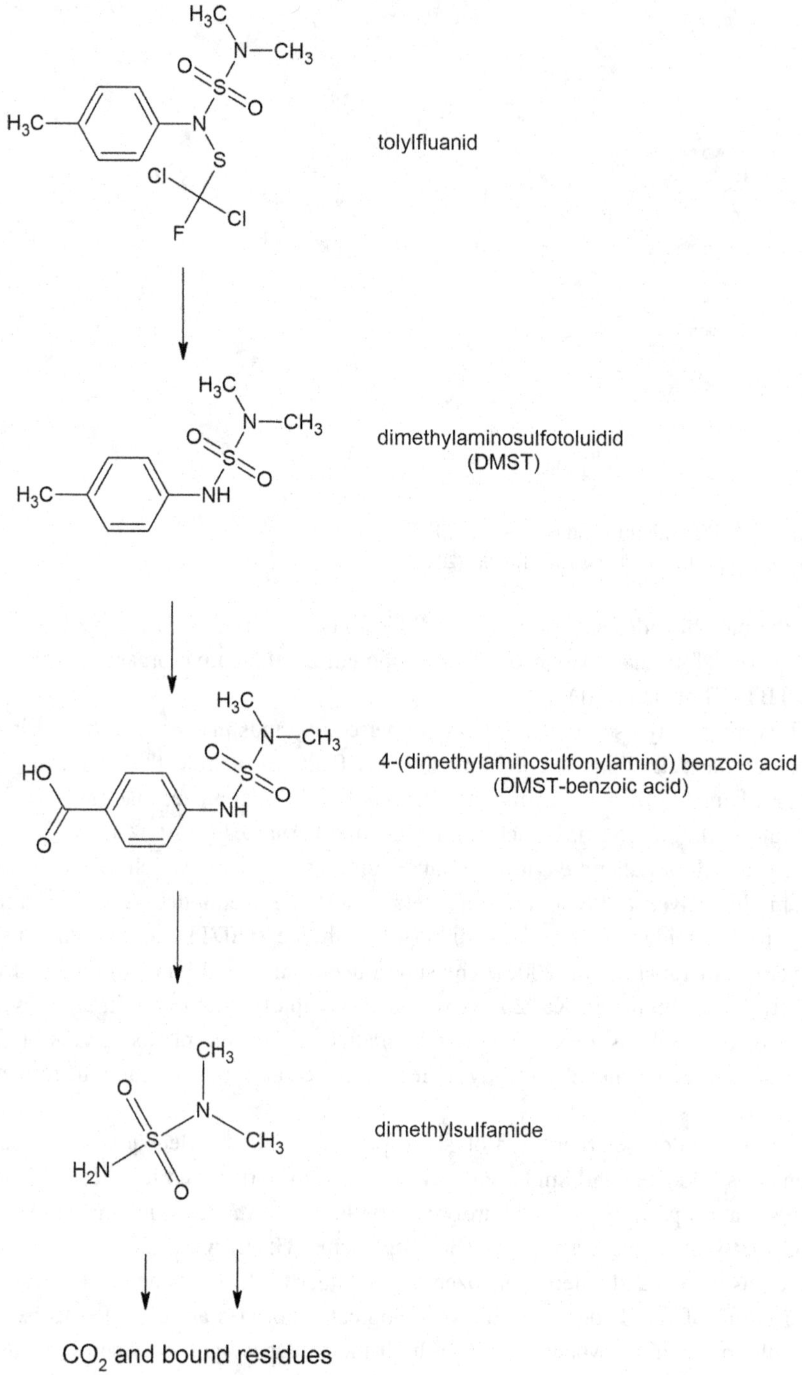

tolylfluanid

dimethylaminosulfotoluidid (DMST)

4-(dimethylaminosulfonylamino) benzoic acid (DMST-benzoic acid)

dimethylsulfamide

CO_2 and bound residues

Figure 6.7 The main degradation pathway of tolylfluanid in water and soil.
Source: Adapted from CBP (2009)

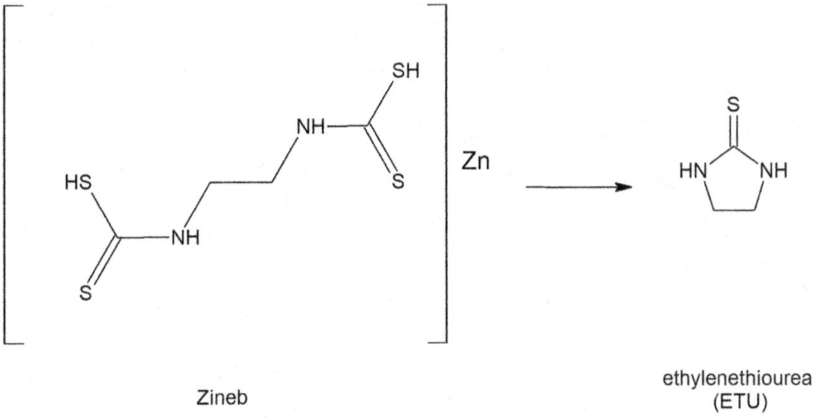

Figure 6.8 Transformation of zineb to ETU.
Source: Adapted from NIVA (2012)

its low volatility, long-range environmental transport is not expected to occur for zineb, maneb or mancozeb (CBP, 2013; US EPA, 2005a, 2005b).

ETU has been detected in the particulate phase of the water column (NIVA, 2012). This may be explained by the high affinity of zineb for particulate material, which may facilitate its rapid transformation to ETU. It is therefore recommended that the monitoring of zineb be performed through determining the concentration of ETU instead (NIVA, 2012).

6.3.10 Ziram and Thiram

Ziram and thiram are dithiocarbamate compounds mainly applied in agriculture as fungicides to protect fruits and seeds, and as vulcanization accelerators and antioxidants in the rubber industry (Castro *et al.*, 2011). The low K_{ow} and moderate water solubility suggest that ziram is not persistent in the environment, and studies have shown a maximum half-life of 18 days in marine ecosystems (van Wezel & van Vlaardingen, 2004). Hydrolysis, photodegradation and aerobic metabolism are the major routes of degradation of ziram (US EPA, 2003). Volatilization is not expected, taking into account the low vapor pressure of the substance (Table 6.1). Many metabolites can result from the degradation of ziram, such as CO_2, COS, CS_2, dimethyldithiocarbamic acid (DDC), N,N-dimethylformamide (DMF) and N,N-dimethylthioformamide (DMTF).

Thiram also undergoes rapid hydrolysis and photolysis, influenced mainly by pH. The degradation rate of thiram is higher in freshwater than in seawater (Castro *et al.* 2011; Harino & Langston, 2009). Its registration as an anti-foulant is restricted to Australia, Japan and New Zealand. In the latter, thiram's registration will expire in 2023.

6.3.11 Zinc Pyrithione and Copper Pyrithione

The pyrithione salts, such as ZnPT and CuPT, were introduced onto the market in the 1990s and have been used in marine AF paints since then as replacements for TBT because of

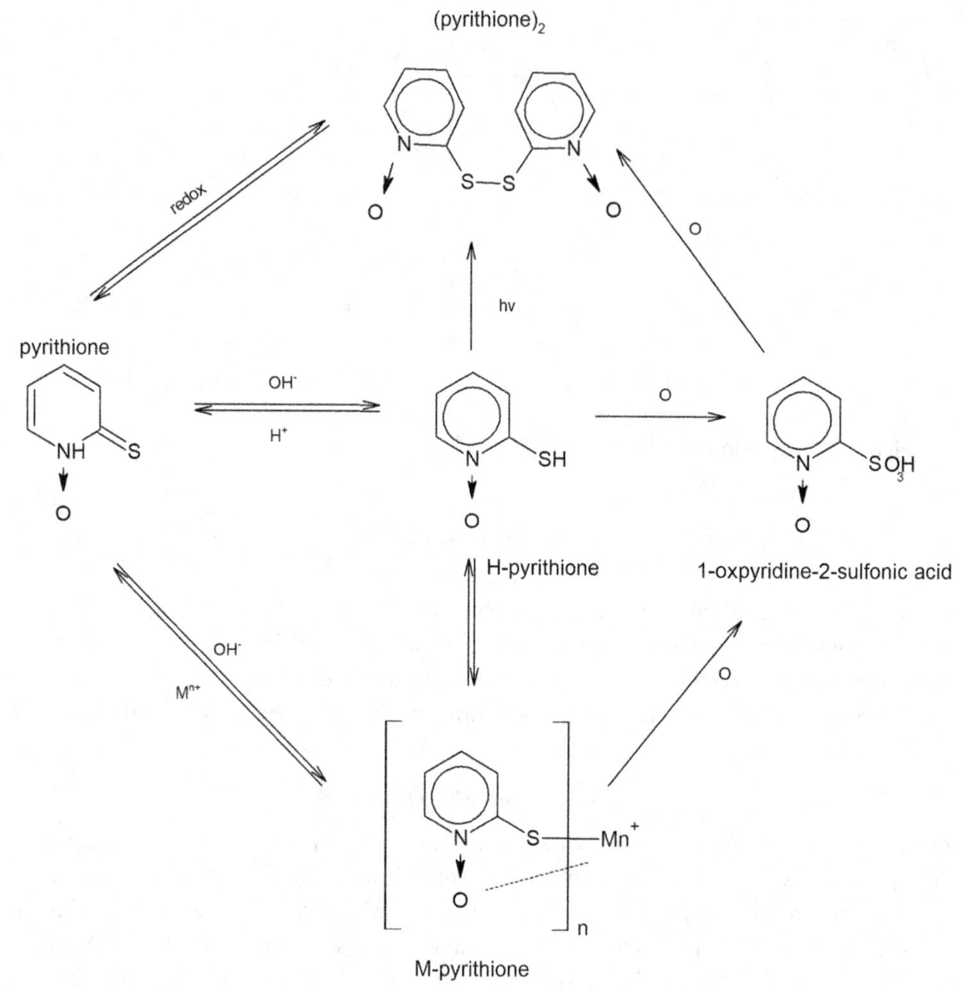

Figure 6.9 Pyrithione transformation in aquatic solution.
Source: Adapted from Seymour and Bailey (1981)

their broad antimicrobial activity, low water solubility and high degradability (Mochida *et al.*, 2006).

Pyrithione salts differ in their composition and may be used in different AF paints. However, they ionize to form similar species in the environment. Furthermore, ZnPT is reported to be readily converted to CuPT in the presence of copper (Nakajima & Yasuda, 1990).

ZnPT is unstable in seawater and is rapidly converted to its degradation products. The photodegradation half-life of ZnPT is reported to be <24 hours with the formation of 2-pyridine sulfonic acid (PSA). In the process, Zn^{2+} is released into the aquatic system (Figure 6.9). Due to the high sensitivity of pyrithiones to natural sunlight, these compounds may locally accumulate in deep waters or shallow sheltered waters where limited penetration of sunlight occurs (Thomas & Brooks, 2010). The half-life for the biodegradation processes is 4 days, and hydrolysis is slower, with a reported half-life of 96–120 days in seawater (Konstantinou & Albanis, 2004).

CuPT is a co-biocide used to prevent the settlement of soft fouling organisms such as marine algae and slimes (CAR, 2014). Its use in concentrations of up to 5 percent of the paint is approved in professional and leisure craft, as well as other objects intended to be submerged into water, such as aquaculture fish nets, pillings and drilling rigs (CAR, 2014; CBP, 2015a).

Once in the environment, CuPT exhibits low water solubility independently of pH (Table 6.1). The low Henry's law constant indicates that volatilization is not an important process and CuPT is unlikely to dissipate into the air compartment. CuPT is very sensitive to sunlight, showing rapid degradation ($t_{1/2} < 1$ hour). The biodegradation DT50 is approximately 4 days in seawater (Thomas & Brooks, 2010) and chemical hydrolysis is slower, with half-lives ranging from 96 to 120 days (CBP, 2015a).

A few studies have reported that pyrithiones disrupt the proton motive force in target organisms. Pyrithione acts by catalyzing the electroneutral exchange of H^+ and other ions with K^+ across cell membranes, resulting in the collapse of ion gradients important to cell functioning. This process may inhibit the membrane transport of nutrients and lead to the starvation and eventual death of the organisms (CAR, 2014), including nontarget organisms for which CuPT has been demonstrated to cause toxicity.

6.3.12 Triphenylborane Pyridine

TPBP has been the predominant AF biocide used in Japan since 1995 in commercial AF coatings for both ship hulls and fishnet applications (Amey & Waldron, 2004). TPBP has shown enhanced effectiveness when used together with metal pyrithiones and presents long-lasting AF effects (Mochida *et al.*, 2012).

There are only a few studies reporting the environmental fate of TPBP, with evidence of biodegradation, hydrolysis and photolysis in seawater. TPBP is highly sensitive to light, with a photodegradation half-life of 6.6 hours in natural seawater (Zhou *et al.*, 2007).

Among the degradation products, diphenylborane, monophenylborane, benzene, phenol and boric acid were detected under hydrolysis, thermolysis and photolysis in freshwater (Amey & Waldron, 2004), whereas phenol and biphenyl were the main degradation products produced by hydrolysis and photolysis in natural seawater (Figure 6.10).

6.3.13 TCMS-Pyridine

The biocide TCMS-pyridine (formula: 2,3,3,6-tetrachloro-4-methylsulfonyl pyridine) is little used as an active substance in AF paintings (Thomas & Langford, 2009b), and information on its physicochemical properties is scarce (Table 6.1). Its mode of action consists of inhibiting mitochondrial electron transport (Guardiola *et al.*, 2012), leading to cell death.

triphenylborane pyridine

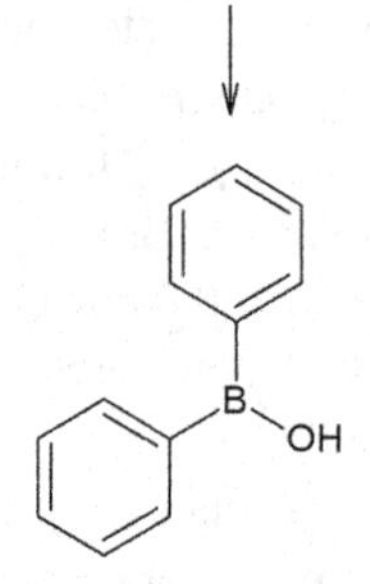

diphenylboron hydroxide

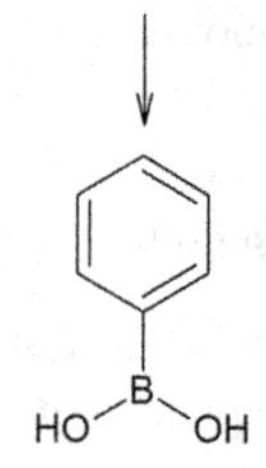

monophenylboron hydroxide

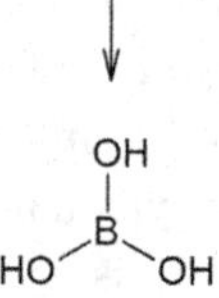

boron trihydroxide

Figure 6.10 Degradation pathway of TPBP.
Source: Adapted from Amey and Waldron (2004)

6.3.14 Capsaicin and Nonivamide

Capsaicin is considered an eco-friendly biocide because it consists of a natural extract from chili peppers. It is commonly used as an animal repellent and more recently has been introduced in the formulation of AF paints in China, in accordance with the National Environmental Protection Standard of the People's Republic of China (Liu *et al.*, 2016; Oliveira *et al.*, 2014). Nonivamide is another naturally occurring capsaicinoid found in chili peppers, although it is more commonly manufactured synthetically and is currently a candidate AF biocide, taking into account that it also exhibits AF properties and appears to be better suited for large-scale use (Liu *et al.*, 2016).

Information on the environmental behavior of capsaicin and nonivamide is scarce. However, environmental risk assessments have been performed for both substances, and it was concluded that they pose low risk to marine ecosystems and might be suitable for use as an eco-friendly active substance for AF coatings (Liu *et al.*, 2016; Wang *et al.*, 2014). For capsaicin, experiments showed that this compound is readily biodegradable and has a low tendency to bioaccumulate (Wang *et al.*, 2014). For nonivamide, natural degradation after sunlight exposure was rapid, resulting in low-toxicity or nontoxic products unlikely to pose a risk to marine ecosystems (Liu *et al.*, 2016).

6.3.15 Medetomidine

AF products containing medetomidine can be applied on the hulls of commercial and government ships, superyachts and pleasure craft, as well on structures and objects subject to immersion in marine waters. Medetomidine is designed to protect submerged structures from fouling by hard fouling marine organisms, acting via the activation of analogous octamine receptors in crustaceans and shell-building organisms, leading to an anti-settling effect (CBP, 2015b). Photolysis and biodegradation are not important processes of removal of medetomidine from the water column, and medetomidine is regarded as a hydrolytically stable compound at all relevant pH values and temperatures (CBP, 2015b). In sediments, medetomidine is persistent, with a DT50 ranging from 50 to 120 days (CBP, 2015b).

6.3.16 N-(2,4,6-Trichlorophenyl)maleimide

N-(2,4,6-Trichlorophenyl)maleimide (TCPM) is registered for use in Japan (under the name of IT354) and China. A preliminary risk assessment conducted in Japan showed that TCPM does not pose a risk to the environment; however, the availability of toxicity data is very low, hence the reliability of predicted no effect concentrations might be low (OPRF, 2010). Information on TCPM is lacking regarding toxicity or environmental behavior, for which the available data were modeled rather than experimentally determined (Table 6.1).

6.3.17 Tralopyril

AF products containing tralopyril are used on the hulls and other immersed parts of large marine-going vessels such as commercial boats and ships, navy and other government vessels and superyachts (25 m or more in overall length), and they can also be used on immersed objects/structures to protect submerged surfaces from fouling by marine barnacles, hydroids, slimes, mussels, oysters, polychaetes and weeds (CBP, 2014c; Oliveira *et al.*, 2014).

Once tralopyril reaches aquatic systems, it is prone to rapid transformation, with short half-lives for hydrolytic and photolytic processes (Table 6.1). Tralopyril cannot be regarded as readily biodegradable, but degradation under both aerobic and anaerobic sediment–water conditions was demonstrated, with degradation being more rapid in marine than freshwater systems.

6.4 Toxicity

Following all transformation processes in the aquatic systems, the bioavailable fraction of active substances in AF biocides may be taken up by aquatic organisms, leading to toxic effects and impairing vital processes such as development/growth, reproduction, immunity and physiological processes, which may culminate in the death of the affected organism. Furthermore, some chemical substances may bioaccumulate and be transferred along the trophic web, leading to ecological effects rather than individual effects.

Many studies have been performed to investigate the toxicity and mechanisms of action of anti-foulants to representative organisms of different trophic levels. They have shown that the sensitivity of marine organisms to anti-foulants largely depends on the taxonomic group and life stage. The mode of action of the alternative biocides may present particular paths for each compound, but similar groups of anti-foulants act in general in a similar way.

Herbicides, such as Irgarol 1051 and Diuron, inhibit photosystem II electron transport, blocking important processes in photosynthesis in target and nontarget species. Studies have reported Irgarol 1051 chronic ecotoxicity ranging from 0.01 μg L^{-1} to the diatom *Skeletonema costatum* (Zhang *et al.*, 2008) to 7600 μg L^{-1} to the bacterium *Vibrio fischeri* (Mezcua *et al.*, 2002). For Diuron, reported chronic ecotoxicity ranges from 0.21 μg L^{-1} to the cyanobacterium *Synechococcus* sp. (DeVilla *et al.*, 2005) to 23,000 μg L^{-1} to *Vibrio fischeri* (Mezcua *et al.*, 2002). As expected, herbicides are more toxic to primary producers than to other organisms (Martins *et al.*, 2018).

Fungicides and broad-spectrum biocides may act by inhibiting electron transport in the mitochondrial chain, interrupting metabolic processes or being multisite stressors. Therefore, they are usually toxic to many nontarget organisms (Martins *et al.*, 2018). Many ecotoxicity tests have been performed with the fungicide chlorothalonil, for which chronic toxicity ranges from 0.57 μg L^{-1} to the diatom *Thalassiosira pseudonana* (Bao *et al.*, 2011) to 89 μg L^{-1} to the dinoflagellate *Pyrocystis lunula* (Bao *et al.*, 2008). For the fungicide dichlofluanid, reported acute toxicity ranges from 16 μg L^{-1} to the fish *Dicentrarchus labrax* (Carteau *et al.*, 2014) to >3300 μg L^{-1} to the macrophyte *Saccharina latissima* (Johansson *et al.*, 2012). For thiram, acute ecotoxicity ranges from 3.36 μg L^{-1} to the zooplankton mysid *Americamysis bahia* (US EPA, 2018) to 670 μg L^{-1} to the benthic Tubificadea *Tubifex tubifex* (US EPA, 2018).

Studies have reported acute ecotoxicity of the broad-spectrum biocide DCOIT, ranging from 0.016 μg L^{-1} to the zooplankton copepod *Acartia tonsa* (Wendt *et al.*, 2016) to 1700 μg L^{-1} to the benthic decapod *Uca pugilator* (US EPA, 2018). For TCMTB, there are a few reports on its toxicity, acutely ranging from 13.9 μg L^{-1} to the Bivalvia *Mercenaria*

mercenaria in early life stages (larvae) (Muñoz *et al.*, 2010) to 60 μg L^{-1} to the fish *Cyprinodon variegatus* (US EPA, 2006).

Pyrithione salts such as CuPT and ZnPT act by catalyzing the electroneutral exchange of H^+ and other ions with K^+ across cell membranes, resulting in the collapse of ion gradients important to cell function. This process may inhibit membrane transport of nutrients and lead organisms to starvation and eventual death (KEMI, 2014). Studies have reported ZnPT acute toxicity ranging from 0.51 μg L^{-1} to *T. pseudonana* (Bao *et al.*, 2011) to 4797.2 μg L^{-1} to the adult Bivalvia *Mytilus galloprovincialis* (Marcheselli *et al.*, 2010). CuPT acute toxicity ranged from 0.53 μg L^{-1} to embryos of the echinoid *Lytechinus variegatus* (Bao *et al.*, 2011) to 830 μg L^{-1} to the zooplankton *Artemia salina* (Koutsaftis & Aoyama, 2007).

In regard to the emerging AF co-biocides TPBP, medetomidine, capsaicin, nonivamide and tralopyril, there are few ecotoxicity data to date. TPBP's mode of action is largely unknown (Wendt *et al.*, 2016). TPBP has shown acute toxicity ranging from 0.16 μg L^{-1} to *A. tonsa* (Wendt *et al.*, 2016) to 980 μg L^{-1} to *V. fischeri* (Zhou *et al.*, 2006). Medetomidine is an octopamine receptor agonist, mimicking the action of this neurotransmitter (Lind *et al.* 2010). Studies have reported medetomidine acute toxicity ranging from 0.52 μg L^{-1} to *A. tonsa* (Wendt *et al.*, 2016) to 3800 μg L^{-1} to larvae of the ascidian *Styela clava* (Willis & Woods, 2011). For capsaicin and its derivative nonivamide, toxicity was reported from 1252 μg L^{-1} to the copepod *Tisbe battagliai* (Oliveira *et al.*, 2014) to 16,900 μg L^{-1} to the bacterium *Vibrio natriegens* (Xu *et al.*, 2005); and from 5100 μg L^{-1} to the microalga *Phaeodactylum tricornutum* (Zhou *et al.*, 2013) to 18,300 μg L^{-1} to the green alga *Chlorella vulgaris* (Liu *et al.*, 2016). They act on the nervous system through several different mechanisms, disrupt metabolism and damage membranes (Gervais *et al.* 2008). Finally, tralopyril is a broad-spectrum biocide acting by uncoupling mitochondrial oxidative phosphorylation (CBP, 2014a; International, 2014). This compound has shown a narrow range of acute toxicity among tested organisms, from 0.9 μg L^{-1} to *T. battagliai* to 3.1 μg L^{-1} to embryos of *M. galloprovincialis* (Oliveira *et al.*, 2014).

Overall, primary producers and zooplankton are more sensitive than benthic invertebrates and fish, showing that distinct biological groups respond differently according to the group of AF co-biocides (Figure 6.11). As stated in Section 6.3, it has been shown that different contaminants undergo various processes in water bodies, resulting in environmental concentrations mainly in the orders of picograms and nanograms per liter (Castro *et al.*, 2011; Dafforn *et al.*, 2011; Lee *et al.*, 2011; Yamada, 2007), concentrations that are sometimes high enough to trigger sublethal and even lethal effects in sensitive groups.

6.5 Current Use of Biocides in Marine Coatings

6.5.1 Occurrence in the Environment

After the TBT ban, many candidate tin-free co-biocides have been used widely in commercial TBT-free and copper-free AF paints, such as DCOIT, Irgarol 1051, Diuron, chlorothalonil, dichlofluanid, tolylfluanid, zineb, ziram, CuPT, ZnPT and TPBP, which

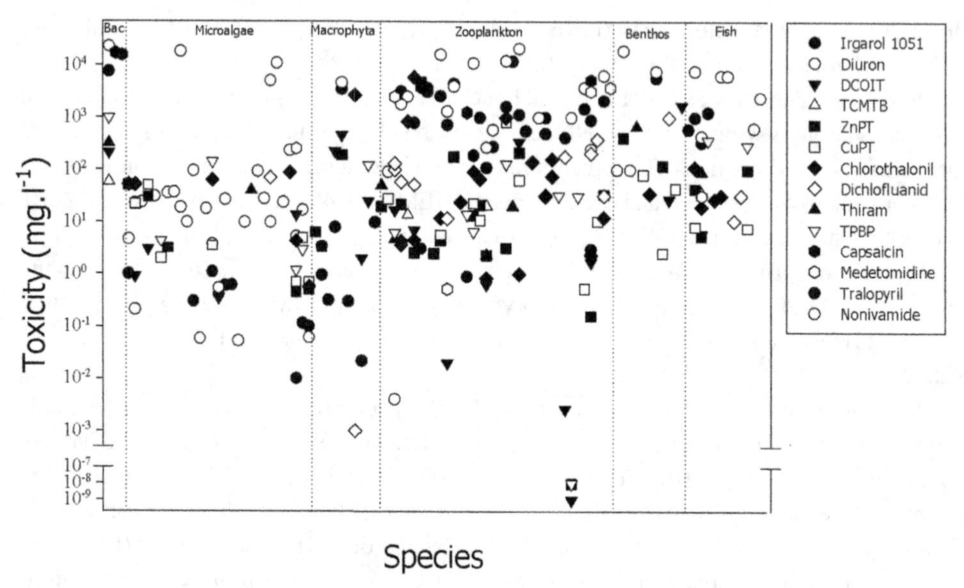

Figure 6.11 Comparative ecotoxicities of anti-foulants for marine species. Bac = bacteria. *Source:* Extracted from Martins *et al.* (2018)

have been widely used in recent years (Okamura & Mieno, 2006). Because these paints are typically applied on the hulls of ships and boats to prevent the growth of fouling organisms, they are directly released into surface waters from painted surfaces during their life service (Thomas & Langford, 2009a), contaminating the environment with their active substances. Many environmental processes will define the environmental fate of the biocides according to their physicochemical properties and environmental conditions, resulting in the presence or absence of the parental compound and/or its metabolites in abiotic and biological matrices.

The first register of contamination by co-biocides from AF paints in coastal waters dates from 1993, when Irgarol 1051 was found in concentrations up to 1700 ng L^{-1} around Cote d'Azur on the French Riviera (Readman *et al.*, 1993). This study triggered a survey of TBT-free AF active substances in other areas, and broad contamination was confirmed worldwide (Table 6.2), especially in areas of boating activities. Although the current co-biocides in use are mostly easily removed from the environment via natural processes, the constant leaching from vessels may maintain high concentrations in coastal environments (Price & Readman, 2013).

The occurrence of AF biocides in seawater samples has been mostly associated with the presence of boating activities in the sampling area. Particularly high concentrations of biocides in water and sediment have been found in marinas with high numbers of berths and low water exchange (Biggs & D'Anna, 2012; Eklund *et al.*, 2010) due to leaching of biocides from AF coatings and the sedimentation of paint particles (Daehne *et al.*, 2017). Hotspots of contamination by Irgarol 1051, whose main path of entry into the environment is via leaching of the active substance during the service life of the AF paint (CBP, 2011), are linked to areas where a high flux of embarkations is found, mainly in marinas and small fishing ports (Castro *et al.*, 2011; Yamada, 2007). Furthermore, seasonal trends in the level of Irgarol 1051 have been found in European waters according to the frequency of boating activities, highlighting the spread of the use of this anti-foulant in small boats (Bowman *et al.*, 2003). Irgarol 1051 has been phased out in many European countries and was recently disapproved as an active substance of AF biocides in the scope of the European Water Framework Directive (WFD; Legal Act EU 107/2016).

The herbicide Diuron is also well studied in regard to its occurrence worldwide (Table 6.2). However, the presence of this compound may be related to sources other than navigation activities, especially in freshwater systems, taking into account that Diuron has been widely applied in agricultural crops (Thomas, 2001). Similarly to Irgarol 1051, the high toxicity of Diuron to primary producers had led to restrictions and the banning of its use in different countries (Tables 6.3–6.5).

For most co-biocides used in AF paints, however, the detection of their traces in water and sediment samples is less frequent. This might be explained by some of the commonly used co-biocides showing high degradation rates and being rarely found in the environment at detectable levels (see Table 6.1). Their degradation products, however, are more likely to be quantified, although most of the research focuses on the parent compounds. As an example, Lee *et al.* (2015) detected DMSA in all sediment samples extracted from their survey conducted in South Korea in 2010–2011. DMSA is the main degradate of dichlofluanid, and this parent biocide was not detected in any sample. Similarly, in a survey conducted in Norway, neither zineb nor its metabolite ETU was detected in the sediment or in the dissolved aqueous phase of the water samples collected, but ETU levels in the particulate phase reached up to 15.5 ng g^{-1} (NIVA, 2012). In the marinas of New Zealand, Stewart *et al.* (2009) have found up to 88 ng g^{-1} of total dithiocarbamate in sediment samples. DCOIT has been widely applied as a co-biocide in AF paints, but its occurrence is rarely reported. In addition to its rapid degradation in seawater, the use of DCOIT is predominant in large vessels (>25 m), so its occurrence in marinas, where most of this research is focused, is unlikely (Thomas & Langford, 2009b).

Chlorothalonil is largely applied in agriculture, and only a few studies have been performed on the occurrence of chlorothalonil arising from boating activities (Thomas & Langford, 2009b). Chlorothalonil is seldom found in coastal waters and sediment samples. In Europe, low levels (<10 ng L^{-1}, limit of detection [LOD] = 5 ng L^{-1}) were found in marinas from Almeria, southern Spain, in a survey conducted during the 1990s (Mezcua *et al.*, 2002). On the other hand, studies performed in Spain and the UK (Hernando *et al.*, 2003; Martínez *et al.*, 2000; Rodríguez *et al.*, 2011a; Thomas *et al.*, 2002) did not detect chlorothalonil at levels above the LOD. In the rest of the world, chlorothalonil in coastal environments was only reported in Korea and Perth, Australia, despite being searched for in different parts of the world (Table 6.2).

In attempts to detect the presence of CuPT in coastal waters and sediment samples, this compound was found only in sediment samples from Vietnam and Otsuchi Bay, Japan, despite being searched for in other parts of Asia and Europe (Table 6.2). There were no reports of detectable levels of ZnPT in natural environments (Table 6.2), except for a

Table 6.2. *Maximum concentrations of organic AF co-biocides in coastal environments reported around the world*

Biocide	Region	Seawater (ng l^{-1})	Sediment (ng g^{-1})	References
Irgarol 1051				
Africa	Tanzania	15.44		Sheikh *et al.*, 2016
Asia	China	1620		Lam *et al.*, 2005
	Hong Kong	11.2[a]	21.26[a]	Cai *et al.*, 2006
	Indonesia		80	Harino *et al.*, 2012
	Iran	63.4	35.4	Saleh *et al.*, 2016
	Japan	296	270	Okamura *et al.*, 2000; Tsunemasa and Yamazaki, 2014
	Korea		73	Lee *et al.*, 2015
	Malaysia	2021	14	Ali *et al.*, 2013; Harino *et al.*, 2009
	Singapore	4200		Basheer *et al.*, 2002
	South Korea	14.1		Kim *et al.*, 2014
	Thailand		3.2	Harino *et al.*, 2006a
	Vietnam		4	Harino *et al.*, 2006b
Central America	Panama	5		Batista-Andrade *et al.*, 2016
	Puerto Rico	51		Carbery *et al.*, 2006
Europe	France	1700	689	Readman *et al.*, 1993; Cassi *et al.*, 2008
	Germany	440	220	Biselli *et al.*, 2000
	Italy	173		Di Landa *et al.*, 2009
	Monaco	132		Tolosa *et al.*, 1996
	Portugal	18		Gonzalez-Rey *et al.*, 2015
	Spain	2000	88	Martínez *et al.*, 2000; Martinez & Barceló 2001
	Sweden	400	9[a]	Dahl and Blanck 1996
	The Netherlands	90	<LOD	Lamoree *et al.*, 2002; Thomas *et al.*, 2000
	UK	1421	1011	Thomas *et al.*, 2001; Boxall *et al.*, 2000
	Chagos Archipelago	8		Guitart *et al.*, 2007
North America	Bermuda	590		Connelly *et al.*, 2001
	Canada	<LOD		Liu *et al.*, 1999
	Hawaii, USA	283		Knutson *et al.*, 2012
	US Virgin Islands	1300		Carbery *et al.*, 2006
	USA	2218	8.9	Hall Jr *et al.*, 1999; Sapozhnikova *et al.*, 2013
Oceania	Australia	6	1340	Government of Western Australia, 2009
	New Zealand	<LOQ		Stewart *et al.*, 2009
South America	Brazil	5700	5.8	Diniz, 2016
Diuron				
Asia	China		3.9	Xu *et al.*, 2013
	Indonesia		740	Harino *et al.*, 2012

Table 6.2. (*cont.*)

Biocide	Region	Seawater (ng l^{-1})	Sediment (ng g^{-1})	References
Irgarol 1051				
	Iran	29.1		Saleh *et al.*, 2016
	Japan	3054	897[a]	Okamura *et al.*, 2003; Kaonga *et al.*, 2016
	Malaysia	285	9.9	Ali *et al.*, 2014; Harino *et al.*, 2009
	South Korea	1360	62.3	Kim *et al.*, 2014; Kim *et al.*, 2015
	Thailand		25	Harino *et al.*, 2006a
	Vietnam		3	Harino *et al.*, 2006b
Central America	Panama	70		Batista-Andrade *et al.*, 2016
Europe	France	268		Caquet *et al.*, 2013
	Germany	131		Nödler *et al.*, 2014
	Greece	31		Nödler *et al.*, 2014
	Italy	1379	2.44	Di Landa *et al.*, 2009; Carafa *et al.*, 2007
	Portugal	15		Gonzalez-Rey *et al.*, 2015
	Spain	2000	136	Martínez *et al.*, 2000; Martinez & Barceló, 2001
	Sweden	100		Dahl & Blanck, 1996
	The Netherlands	1130	<100	Lamoree *et al.*, 2002; Thomas *et al.*, 2000
	Turkey	31		Nödler *et al.*, 2014
	UK	6742	1420	Thomas *et al.*, 2001; Thomas *et al.*, 2000
North America	USA	68	4.2	Sapozhnikova *et al.*, 2013
Oceania	Australia	2160	40	Government of Western Australia, 2009; Matthai *et al.*, 2009
	New Zealand		<LOQ	Stewart *et al.*, 2009
South America	Brazil	7800		Diniz *et al.*, 2014
DCOIT				
Asia	China	<LOD	<LOD	Liu *et al.*, 2015
	Indonesia		110	Harino *et al.*, 2012
	Japan	100	150	Tsunemasa, 2013; Harino *et al.*, 2007
	Korea		281	Lee *et al.*, 2015
	Malaysia		4.2	Harino *et al.*, 2009
	Thailand		0.09	Harino *et al.*, 2006a
	Vietnam		1.3	Harino *et al.*, 2006b
Central America	Panama	<LOD		Batista-Andrade *et al.*, 2016
Europe	Denmark	283		Thomas, 2001
	Germany	<LOQ		Daehne *et al.*, 2017
	Spain	3300	4	Martínez *et al.*, 2000; Martinez & Barceló, 2001
	UK	ND	ND	Thomas *et al.*, 2002
	Chagos Archipelago	<LOD		Guitart *et al.*, 2007

Table 6.2. (*cont.*)

Biocide	Region	Seawater (ng l^{-1})	Sediment (ng g^{-1})	References
Irgarol 1051				
OIT				
Asia	China	<LOD	<LOD	Liu *et al.*, 2015
TCMTB				
Central America	Panama	<LOD		Batista-Andrade *et al.*, 2016
Europe	France	25		Turquet *et al.*, 2010
	Spain	<10		Mezcua *et al.*, 2002
	UK	ND		Thomas, 1998
North America	USA	<LOD		Sapozhnikova *et al.*, 2007
Chlorothalonil				
Asia	Korea	29.78[a]	1065	Lee *et al.*, 2011, 2015
	Central America	<LOD		Carbery *et al.*, 2006
Europe	Spain	<10		Mezcua *et al.*, 2002
	UK	ND	ND	Thomas *et al.*, 2002
	Chagos Archipelago	<LOD		Guitart *et al.*, 2007
North America	USA	<LOD		Sapozhnikova *et al.*, 2007
Oceania	Australia	<LOD	225	Government of Western Australia, 2009
South America	Brazil	<LOQ		Diniz, 2016
Dichlofluanid				
Asia	Indonesia		<LOD	Harino *et al.*, 2012
	Japan		14	Harino *et al.*, 2007
	Korea	21.77	ND	Lee *et al.*, 2011, 2015
	Malaysia		<LOD	Harino *et al.*, 2009
	South Korea		<LOQ	Kim *et al.*, 2015
	Vietnam		13	Harino *et al.*, 2006b
Central America	Panama	<LOD		Batista-Andrade *et al.*, 2016
Europe	France		<LOD	Cassi *et al.*, 2008
	Germany	<LOQ		Daehne *et al.*, 2017
	Greece	<LOQ	ND	Schouten *et al.*, 2005
	Greece	<LOD		Hamwijk *et al.*, 2005
	Italy	<LOD		Di Landa *et al.*, 2009
	Spain	600	16.6[a]	Martínez *et al.*, 2000; Sánchez-Rodríguez *et al.*, 2011a
	UK	ND	ND	Thomas *et al.*, 2002
	Chagos Archipelago	<LOD		Guitart *et al.*, 2007
North America	USA	<LOD		Sapozhnikova *et al.*, 2007
Oceania	Australia	<LOD	ND	Government of Western Australia, 2009

Table 6.2. (*cont.*)

Biocide	Region	Seawater (ng l^{-1})	Sediment (ng g^{-1})	References
Irgarol 1051				
DMSA				
Europe	Germany	105		Daehne *et al.*, 2017
	Greece	36	ND	Schouten *et al.*, 2005
	Portugal	22		Gonzalez-Rey *et al.*, 2015
	The Netherlands	1000		CBP, 2014b
Tolylfluanid				
Europe	Germany	<LOQ		Daehne *et al.*, 2017
DMST				
Europe	Germany	110		Daehne *et al.*, 2017
	Portugal	17		Gonzalez-Rey *et al.*, 2015
	The Netherlands	<LOQ		CBP, 2014b
Zineb				
Europe	Norway	<LOD	<LOD	NIVA, 2012
ETU				
Europe	Germany	<LOQ		Daehne *et al.*, 2017
	Norway	15.5	<LOD	NIVA, 2012
Thiram				
Europe	France	<LOD		Turquet *et al.*, 2010
	Spain	<LOD		Rodríguez *et al.*, 2011b
South America	Brazil	1300		Diniz, 2016
ZnPT				
Asia	South Korea		<LOQ	Kim *et al.*, 2015
Europe	UK	33,359		Mackie *et al.*, 2004
CuPT				
Asia	Japan	<LOD	22	Harino *et al.*, 2005, 2007
	Malaysia		<LOD	Harino *et al.*, 2009
	South Korea		<LOQ	Kim *et al.*, 2015
	Vietnam		420	Harino *et al.*, 2006b
Europe	Germany	<LOQ		Daehne *et al.*, 2017
	Sweden	<LOD	<LOD	Woldegiorgis *et al.*, 2007
TPBP				
Asia	Japan	0.021		Mochida *et al.*, 2012
TCMS pyridine				
Europe	UK	<LOD		Thomas *et al.*, 2002

[a] Average value.

<LOD = below limit of detection; <LOQ = below limit of quantification; ND = not detected.

small-scale experiment conducted by Mackie *et al.* (2004), in which ZnPT was found at an average concentration of 33.4 μg L^{-1} in water sampled from a marina in the Mersey estuary in the UK.

For some co-biocides, information on environmental levels in coastal areas was not found. Usually, they are co-biocides with low frequency of use in AF paints, have only regional use restricted to a few areas or are emerging compounds as AF biocides. These include IT354, mancozeb, maneb, ziram, medetomidine and tralopyril.

6.5.2 Ecological Risks and Current Status of Regulation

The concern regarding the potential ecological impacts of AF paints is global, and different nations in the world are putting efforts into regulating the emissions of active substances into aquatic systems. Currently, a number of legislative measures and policies address the contamination of water bodies caused by hazardous substances. Within the EU, the WFD (2000/60/EC) and the Marine Strategy Framework Directive (MSFD; 2008/56/EC) establish measures against the pollution of surface waters. Many legal acts follow these directives to achieve specific goals within the scope of the law.

Biocidal Products Regulation (BPR) No. 528/2012 of the European Parliament and of the European Council was adopted to replace the Directive 98/8/CE, with the purpose of improving the functioning of the internal market through the harmonization of the rules on making available on the market and use of biocidal products, while ensuring a high level of protection of both human and animal health and the environment. Following its adoption, many regulations entered into forces to achieve the goals of BPR 528/2012.

In August 2014, Regulation No. 1062/2014 was adopted to address rules for carrying out the work program for the systematic examination of all existing active substances referred to in Article 89 of Regulation (EU) No. 528/2012. In this regard, a number of AF biocides were included in the review program as candidate substances for use in biocidal products of product-type 21 (AF biocides) (Table 6.3), and the decision regarding their inclusion or noninclusion into Annexes I or IA of BPR is taken after performing ecological risk assessments, which today assume critical importance for the commercial success of these product types (Oliveira *et al.*, 2014).

Some of the biocides listed in Table 6.3 have other uses in addition to their application as AF biocides (product-type 21). As examples, tolylfluanid is also approved in Categories 7 (film preservative) and 8 (wood preservative).

Furthermore, there are some active substances known to be used as AF biocides but not recognized within the scope of the BPR. Thiram and chromium trioxide are not applicable as product-type 21 (Tornero & Hanke, 2016), while TPBP, maneb, TCMTB and TCMS-pyridine are not identified as biocidal products by the European Chemicals Agency (ECHA). In the case of Diuron, it was regarded as a priority substance within the scope of the WFD and is not allowed in formulations placed on the market since 2008. Similarly, ziram, chlorothalonil and Folpet are forbidden in post-2008 formulations (Tornero & Hanke, 2016). The biocides bis[1-cyclohexyl-1,2-di(hydroxy-.kappa.O)diazeniumato

Table 6.3. *The EU's regulation of AF biocides (product-type 21)*

Active substance	CAS no.	Minimum purity	Decision	Legal act (EU)	Date of approval	Expiry date of approval
Dicopper oxide	1317-39-1	94.2 % w/w	Approved	1089/2016	January 1, 2018	December 31, 2025
Tralopyril	122454-29-9	975 g/kg	Approved	1091/2014	April 1, 2015	31 March 31, 2025
Cybutryne (Irgarol 1051)	28159-98-0	N/A	Not approved	107/2016	NA	NA
Zineb	12122-67-7	940 g/kg	Approved	92/2014	January 1, 2016	December 31, 2025
Medetomidine	86347-14-0	99.5% w/w	Approved	1731/2015	January 1, 2016	December 31, 2022
Tolylfluanid	731-27-1	960 g/kg	Approved	419/2015	July 1, 2016	December 31, 2025
DCOIT	64359-81-5	950 g/kg	Approved	437/2014	January 1, 2016	December 31, 2025
Copper flakes	7440-50-8	95.3% w/w	Approved	1088/2016	January 1, 2018	December 31, 2025
CuPT	14915-37-8	950 g/kg	Approved	984/2015	January 1, 2016	December 31, 2025
Copper thiocyanate	1111-67-7	99.5% w/w	Approved	1090/2016	January 1, 2018	December 31, 2025
Dichlofluanid	1085-98-9	96% w/w	Approved	796/2017	November 1, 2018	December 31, 2025
ZnPT	13463-41-7		Under review			

Sources: Official Journal of the European Union L 21/81, L 32/16, L 68/39, L 128/64, L 159/43, L 180/21, L 180/25, L 180/29, L 252/33, L 299/15; L 120/13; Tornero and Hanke (2016)

(2-)]-copper, captan, thiabendazole, zinc sulfide, fluometuron, prometryn, lignin, dodecylguanidine monohydrochloride, chlorotoluron, dimethyloctadecyl[3-(trimethoxysilyl) propyl]ammonium chloride, isoproturon, fenpropimorph, quaternary ammonium compounds (benzyl-C12–18-alkyldimethyl chlorides, benzyl-C12–14-alkyldimethyl chlorides, C12–14-alkyl[(ethylphenyl)- methyl]dimethyl chlorides), 3-benzo(b)thien-2-yl-5,6-dihydro-1,4,2-oxathiazine,4-oxide, chloromethyl n-octyl disulfide and chlorfenapyr were also not approved as AF biocides. Thus, Act 2007/565/EC concerned the noninclusion of such compounds in Annexes I, IA or IB of Directive 98/8/EC, further replaced by BPR 528/ 2012. Capsaicin is regarded as a new substance, being a proposed candidate as an active substance in AF paints (Tornero & Hanke, 2016).

Even active substances approved within the scope of the law may pose risks to the environment, especially in the surroundings of the main leaching areas. Within the scope of EU legislation, the environmental risk assessment of DCOIT identified risks in simulated commercial harbor and marine environments (using the marine antifoulant model to predict environmental concentrations (MAMPEC)). However, safe use of DCOIT could be demonstrated in the surrounding areas, both in water and sediment, drawing as an overall conclusion that it may be possible to issue authorizations of products containing DCOIT in accordance with the conditions laid down in Articles 5(1) b), c) and d) of Directive 98/8/EC (CBP, 2014a). From these results, the European Commission approved DCOIT as an active substance for use in biocidal products for product-type 21 (AF biocides), subject to certain specifications and conditions, through Regulation EU No. 437/2014 (CBP, 2014a).

Risk values in the draft risk assessment for zineb under the EU biocides directive indicate that the levels of ETU detected in suspended particulate matter may indicate a risk to the marine environment (NIVA, 2012). However, cumulative exposure of zineb and its degradation products arising from activities (application/removal phase losses plus in-service losses) associated with commercial ships does not pose a threat to the surrounding environment. In addition, no significant risk of secondary poisoning has been identified as a result of the proposed uses of zineb, and neither zineb nor its metabolites/degradates are regarded as persistent, bioaccumulative and toxic substances (CBP, 2013). Thus, the Committee on Biocidal Products approved zineb as an active substance for use in biocidal products for product-type 21 (AF biocides), subject to certain specifications and conditions, through Regulation EU no. 92/2014.

In the USA, anti-foulants are regulated along with other pesticides and require registration both federally with the Environmental Protection Agency (EPA) and with each state authority. Currently, the pesticides distributed or sold in the USA are subject to a registration review program, according to the Food Quality Protection Act (FQPA) of 1996. All pesticides must be registered based on scientific data showing that they will not cause unreasonable risks to human health or the environment when used as directed on product labeling (Table 6.4). The registration review program has being implemented pursuant to Section 3(g) of the Federal Insecticide, Fungicide, and Rodenticide Act (FIFRA), and it intends to make sure that all registered pesticides continue to meet the statutory standard of no unreasonable risks as the ability to assess risk evolves and as policies and practices change. For example, zineb was formerly registered in the USA as a general use pesticide and was rated as a pesticide of low toxicity. Following an EPA Special Review of all of the ethylene(bis)dithiocarbamate pesticides, including zineb, all registrations for zineb were voluntarily canceled by the manufacturer, followed by the total revocation of the register in 1997 (http://extoxnet.orst.edu/pips/zineb.htm). Each registered pesticide should be reviewed every 15 years to determine whether it continues to meet the FIFRA standard for re-registration (US EPA, 2013).

Some biocides commonly applied as anti-foulants elsewhere are not registered under the scope of the US Office of Pesticide Programs, such as TPBP, TCMS-pyridine, medetomidine, TCPM and dichlofluanid (https://iaspub.epa.gov/apex/pesticides/f?p=CHEMICALSEARCH:1).

In New Zealand, the document on the Decision on the Application for Reassessment of Antifouling Paints (APP201051) was adopted in 2013, in agreement with Section 63 of the Hazardous Substances and New Organisms (HSNO) Act of 1996 (New Zealand Environmental Protection Authority, 2013). APP201051 states that AF paints containing Irgarol or

Table 6.4. *US regulation of AF biocides*

Active substance	CAS no.	First register	Current regulatory action as AF	Document	Date of document
Dicopper oxide	1317-39-1	November 14, 1991	Under review	EPA-HQ-OPP-2010-0212	March 2011
Tralopyril	122454-29-9	August 5, 1993	Under review	EPA-HQ-OPP-2013-0217	June 2013
Diuron	330-54-1	May 7, 1971	NS	FR 21528/2004	April 21, 2004
OIT	26530-20-1	December 8, 2004	Under review	FR 36056/2014	June 25, 2014
Chlorothalonil	1897-45-6	December 4, 1989	Under review[a]	FR 18810/2012	March 28, 2012
Cybutryne (Irgarol 1051)	28159-98-0	March 2, 2004	Under review	EPA-HQ-OPP-2010-0003	March 30, 2010
TCMTB	21564-17-0	January 21, 1980	NS[b]	EPA-HQ-OPP-2014-0405	September 24, 2014
Zineb	12122-67-7	August 8, 1989	REV	FR 57:58384	
Maneb	12427-38-2	October 29, 1962	REV	FR 19967/2010	April 16, 2010
Mancozeb	8018-01-7	February 25, 1971	NS	EPA 738-R-04-012	September 2005
Thiram	137-26-8	February 7, 1985	NS	EPA 738-R-04-012	September 2004
Ziram	137-30-4	March 21, 2005	REG	EPA 739-R-09-001	June 2009
Tolylfluanid	731-27-1	December 5, 1972	NS[c]	PC-309200-01	September 1, 2002
DCOIT	64359-81-5	January 24, 1983	Under review	FR 57092/2014	September 24, 2014
Copper	7440-50-8	March 10, 2004	REG	EPA-HQ-OPP-2005-0558-0278	August 12, 2010
CuPT	14915-37-8	November 19, 1975	Pending registration	FR 16109/2010	March 31, 2010
Copper thiocyanate	1111-67-7	July 27, 1992	Pending registration	EPA-HQ-OPP-2005-0558-0278	August 12, 2010
Capsaicin	404-86-4	June 15, 1978	NS	EPA-HQ-OPP-2009-0121	September 15, 2010
ZnPT	13463-41-7	August 20, 2004	RED addendum[d]	EPA-HQ-OPP-2004-0147	

[a] Chlorothalonil – under review for conventional and unconventional uses, including paints, not specific for AF paints.

[b] TCMTB – use as AF biocide not specified in the document, but all labels for use as paints were removed and TCMTB can no longer be used as a paint (US EPA, 2014 – EPA-HQ-OPP-2014-0405).

[c] Imported tolylfluanid is used in some crops in the USA, but the product is not registered in the country.

[d] ZnPT – a preliminary risk assessment was made, but no further action taken.

NS = use as anti-foulant not supported; RED = reregistration eligibility decision; REG = registered; REV = revoked use of the compound in all product types.

chlorothalonil will no longer be able to be manufactured or imported, as the approvals to do so have been declined. Paints containing Diuron, OIT or ziram have a time-limited approval of 4 years and thiram of 10 years (New Zealand Environmental Protection Authority, 2013) (Table 6.5).

In Australia, AF biocides require registration with the National Registration Authority (NRA), within the Australian Pesticides and Veterinary Medicines Authority (APVMA). Under the scope of this institution, there are 46 products registered for use as AF paints in the country, most of them containing one or more of the co-biocides Diuron, CuPT, ZnPT, zineb, DCOIT, thiram and dichlofluanid (APVMA, 2017).

In Japan, three compounds were considered to have low impact on the environment based on environmental risk assessment: TCPM (IT354), butyl thiram and cuprous thiocyanate. On the other hand, TPBP, Diuron and Irgarol 1051 were stated as substances of concern. TPBP's toxicity, however, might have been overestimated due to the lack of available data and the need to use high assessment factors (OPRF, 2010). For Diuron and Irgarol, risk assessments to Japanese ecosystems have highlighted the presence of concentrations exceeding the predicted no effect concentrations in waters at ports and along the coast, so the use of paints containing these active substances should be reviewed, paying attention to the actions taken in other countries that have restricted or banned the use of these compounds as AF biocides (OPRF, 2010). As for Japan, in China and South Korea, all AF biocides used in paints must be registered on the inventory of existing chemical substances of the respective country.

For other countries, information about the current status of the regulation of AF biocides is difficult to find and is mainly restricted to reports from consulting agencies. Laws usually apply for pesticides as a whole, not considering the differences among different classes and uses of these compounds. Mainly agriculture-driven pesticides are the focus of regulation, and for AF biocides, little or nothing has evolved since the TBT ban over a decade ago.

In 2013, the Chilean Instituto de Fomento Pesquero (IFOP) published a final report on a survey of active substances of AF paints used in aquaculture in southern Chile in the Los Patos area, with a critical analysis of the effects of AF biocides on marine organisms. They found that copper oxide is the main active substance used in AF products and that some of the AF paints marketed in Chile contain organic co-biocides such as ZnPT, and possibly Irgarol 1051, Diuron, chlorothalonil and DCOIT (IFOP, 2013), but there are no current regulations regarding the authorization of such products as AF biocides. However, there are some mentions in the directive concerning the "Environmental Regulation for Aquaculture" regarding the cleaning of nets treated with AF biocides (IFOP, 2013).

Regulation of TBT-free biocides is evolving, culminating in many actions by regulatory agencies in the USA, Europe, Oceania and some Asian countries. Environmental agencies require applicants to perform various studies according to specific guidelines designed to determine environmental degradation rates under abiotic (hydrolysis and/or photolysis) and biotic (aerobic and/or anaerobic aquatic metabolism) conditions before a biocide is approved for use as an active ingredient (Liu *et al.*, 2016), and environmental risks are calculated to support the final decision for each compound.

Table 6.5. *Regulatory status of active substances in AF paints by 2011*

AF	Region/country regulation																
	Australia	New Zealand[a]	USA	California, USA	Washington, USA	Canada	Hong Kong	Japan	China	Korea	EU	Denmark	Sweden	Finland	Norway[b]	UK	The Netherlands	Germany
Capsaicin									REG									
Chlorothalonil	REG	REV	NR	NR	NR	NR	NR	REG	PA^{+++}		NS	NR	NR	NR	NS	REV	NR	
Copper	REG	REG	REG	tbPO	tbPO	REG	REG	REG			REG	REG	RU	REG	REG	REG	REG	
Copper naphthenate								REG										
Copper thiocyanate	REG	REG						REG	PA^{+++}		REG				REG			
CuPT	REG	REG	UR	NR	UR	NR	REG	REG	PA^{+++}		REG	–	RU	REG	REG	RU	REG	
DCOIT	REG	REG	REG	REG	REG	NR	REG	REG	PA^{+++}		REG	–	RU	REG	REG	REG	REG	
Dichlofluanid	REG	REG	NR	NR	NR	NR	REG	REG			REG	–	NR	REG		PA	REx	
Dicopper oxide								REG	PA^{+++}		REG				REG			
Diuron	REG	REV	NR	NR	NR	NR	REG	REG	PA^{+++}	NS	NS	RU	NR	NR	NS	REV	REx	
Irgarol 1051	NS	REV	UR	REG	UR	NR	REG	REG	PA^{+++}	NR	NS	RU	tbPO	NR		RU	REG	
Mancozeb	NR	REG	NR	NR	NR	NR	NR	NR			NS	–	NS	NR		NR	NR	
Maneb											NS					REV		
Medetomidine								REG	REG	REG	REG				REG			
OIT	NR	REV	REG	NR	REG	NR	NR	NR			NR	–	NR	NR		NR	NR	
TCMS-pyridine								REG										
TCMTB											NS							
TCPM								REG	PA^{+++}									
Thiram	REG	PA^{++}	NR	NR	NR	NR	NR	REG			NS	–	NR	NR	NS	REV	NR	

Table 6.5. (*cont.*)

	Region/country regulation																	
AF	Australia	New Zealand[a]	USA	California, USA	Washington, USA	Canada	Hong Kong	Japan	China	Korea	EU	Denmark	Sweden	Finland	Norway[b]	UK	The Netherlands	Germany
Tolylfluanid	NR	REG	NR	NR	NR	NR	NR	REG			REG	–	NR	NR	REG	NR	NR	
TPBP								REG	PA^{+++}									
Tralopyril			REG								REG				REG			
Zineb	REG	REG	NR	NR	NR	NR	REG	REG	PA^{+++}		REG	–	NR	REG	REG	PA	REG	
Ziram	NR	REV	REG	NR	REG	NR	NR	REG			NS	–	NR	NR	NS	REV	REx	
ZnPT	REG	REG	REG	REG	REG	NR	REG	REG	PA^{+++}		UR	REG	RU	REG		REG	REG	

[a] Data from 2013.

[b] Data from 2014.

REG = registered; RU = restricted use; REV = revoked; tbPO = to be phased out; NR = not registered; UR = under registration; NS = not supported/not allowed; PA = provisional approval; REx = registration expired; PA^{+} = time-limited approval until June 2017; PA^{++} = time-limited approval until June 2023; PA^{+++} = in China, approval for the use of these compounds has been applied for and there is no need to perform environmental risk assessments at the moment (CCS, 2015).

Sources: New Zealand Environmental Protection Authority (2013), *Official Journal of the European Union* L 21/81, L 32/16, L 68/39, L 128/64, L 159/43, L 180/21, L 180/25, L 180/29, L 252/33, L 299/15; Tornero and Hanke (2016); Lovdata – Norwegian Regulation FOR-2014-10-548 (https://lovdata.no/dokument/LTI/forskrift/2014-04-10-548?q=biocidforskriften)

There is a growing effort to develop effective and environmentally friendly AF biocides, and naturally extracted compounds have been tested as candidates to this end. Wang *et al.* (2014) assessed the risk of using capsaicin, a natural compound extracted from chili peppers, as an active substance for AF systems and ships, and they have concluded that it poses relatively low risk to the marine environment and might be categorized as an environmentally friendly substance.

Finding ideal AF biocides that combine effectiveness, attractive cost and low risk to the environment is a big challenge, but the first steps have been taken, and in the future, harmony might be achieved between urban/industrial development and respect for the environment, helping us to obtain the sustainability for which humankind has been searching.

References

Ali H. R., Arifin M. M., Sheikh M. A., Shazili M. A. N., Bachok Z. (2013). Occurrence and distribution of antifouling biocide Irgarol-1051 in coastal waters of Peninsular Malaysia. *Marine Pollution Bulletin* 70, 253–257.

Ali H. R., Arifin M. M., Sheikh M. A. *et al.* (2014). Contamination of Diuron in coastal waters around Malaysian Peninsular. *Marine Pollution Bulletin* 85, 287–291.

Alzieu C. (1991). Environmental problems caused by TBT in France: assessment, regulations, prospects. *Marine Environmental Research* 32, 7–17.

Alzieu C. (2000). Environmental impact of TBT: the French experience. *Science of the Total Environment* 258, 99–102.

Amey R. L., Waldron C. (2004). Efficacy and chemistry of BOROCIDETM Ptriphenylboron-pyridine, a non-metal antifouling biocide. In: *Proceedings of the International Symposium on Antifouling Paint and Marine Environment.* Tokyo: InSAfE, pp. 234–243.

Anastasiou T. I., Chatzinikolaou E., Mandalakis M., Arvanitidis C. (2016). Imposex and organotin compounds in ports of the Mediterranean and the Atlantic: is the story over? *Science of the Total Environment* 569–570, 1315–1329.

Antizar-Ladislao B. (2008). Environmental levels, toxicity and human exposure to tributyltin (TBT)-contaminated marine environment: a review. *Environment International* 34, 292–308.

APVMA (2017). Public Chemical Registration Information System Search. https://portal.apvma.gov.au/pubcris?p_auth=UdX0d7zX&p_p_id=pubcrisportlet_WAR_pubcrisportlet&p_p_lifecycle=1&p_p_state=normal&p_p_mode=view&p_p_col_id=column-1&p_p_col_pos=3&p_p_col_count=5&_pubcrisportlet_WAR_pubcrisportlet_javax.portlet.action=search

Arrhenius Å., Backhaus T., Hilvarsson A. *et al.* (2014). A novel bioassay for evaluating the efficacy of biocides to inhibit settling and early establishment of marine biofilms. *Marine Pollution Bulletin* 87, 292–299.

Artifon V., Castro Í. B., Fillmann G. (2016). Spatiotemporal appraisal of TBT contamination and imposex along a tropical bay (Todos os Santos Bay, Brazil). *Environmental Science and Pollution Research International* 23, 16047–16055.

Bao V. W. W., Koutsaftis A., Leung K. M. Y. (2008). Temperature-dependent toxicities of chlorothalonil and copper pyrithione to the marine copepod *Tigriopus japonicus* and dinoflagellate *Pyrocystis lunula. Australasian Journal of Ecotoxicology* 14, 45–54.

Bao V. W. W., Leung K. M. Y., Qiu J. W., Lam M. H. W. (2011). Acute toxicities of five commonly used antifouling booster biocides to selected subtropical and cosmopolitan marine species. *Marine Pollution Bulletin* 62, 1147–1151.

Barreiro R., Gonzalez R., Quintela M., Ruiz J. (2001). Imposex, organotin bioaccumulation and sterility of female *Nassarius reticulatus* in polluted areas of NW Spain. *Marine Ecology Progress Series* 218, 203–212.

Barroso C. M., Moreira M. (2002a). Imposex, female sterility and organotin contamination of the prosobranch *Nassarius reticulatus* from the Portuguese coast. *Marine Ecology Progress Series*, 230, 127–135

Barroso C. M., Moreira M. H. (2002b). Spatial and temporal changes of TBT pollution along the Portuguese coast: inefficacy of the EEC Directive 89/677. *Marine Pollution Bulletin* 44, 480–486.

Barroso C., Reis-Henriques M., Ferreira M., Moreira M. (2002). The effectiveness of some compounds derived from antifouling paints in promoting imposex in Nassarius reticulatus. *Journal of the Marine Biological Association of the United Kingdom*, 82(2), 249–255.

Basheer C., Tan K. S., Lee H. K. (2002). Organotin and Irgarol-1051 contamination in Singapore coastalwaters. *Marine Pollution Bulletin* 44, 697–703.

Batista R. M., Castro Í. B., Fillmann G. (2016). Imposex and butyltin contamination still evident in Chile after TBT global ban. *Science of the Total Environment*, 566–567, 446–453.

Batista-Andrade J. A., Caldas S. S., de Oliveira Arias J. L. *et al.* (2016). Antifouling booster biocides in coastal waters of Panama: first appraisal in one of the busiest shipping zones. *Marine Pollution Bulletin* 112, 415–419.

Biggs T. W., D'Anna H. (2012). Rapid increase in copper concentrations in a new marina, San Diego Bay. *Marine Pollution Bulletin* 64, 627–635.

Biselli S., Bester K., Huhnerfuss H., Fent K. (2000). Concentrations of the antifouling compound Irgarol 1051 and of organotins inwater and sediments of German North and Baltic Sea Marinas. *Marine Pollution Bulletin* 40, 233–243.

Blunden S. J., Chapman A. H. (1982). The environmental degradation of organotin compounds – a review. *Environmental Technology Letters* 3, 267–272.

Bowman J. C., Readman J. W., Zhou J. L. (2003). Seasonal variability in the concentrations of Irgarol 1051 in Brighton Marina, UK; including the impact of dredging. *Marine Pollution Bulletin* 46, 444–451.

Boxall A. B. A., Comber S. D., Conrad A. U., Howcroft J., Zaman N. (2000). Inputs, monitoring and fate modelling of antifouling biocides in UK estuaries. *Marine Pollution Bulletin* 40, 898–905.

Bryan G. W., Gibbs P. E., Hummerstone L. G., Burt G. R. (1986). The decline of the gastropod Nucella lapillus around south-west England: evidence for the effect of tributyltin from antifouling paints. *Journal of the Marine Biological Association of the United Kingdom* 66(3), 611–640.

Bureau of Land Management (2005). Diuron Ecological Risk Assessment. All U.S. Government Documents (Utah Regional Depository). Paper 105. http://digitalcommons.usu.edu/govdocs/105

Burton E. D., Phillips I. R., Hawker D. W. (2006). Tributyltin partitioning in sediments: effect of aging. *Chemosphere* 63, 73–81.

Cacciatore F., Noventa S., Antonini C. *et al.* (2018). Imposex in *Nassarius nitidus* (Jeffreys, 1867) as a possible investigative tool to monitor butyltin contamination according to the Water Framework Directive: a case study in the Venice Lagoon (Italy). *Ecotoxicology and Environmental Safety* 148, 1078–1089.

Cai Z., Fun Y., Ma W. T., Lam M., Tsui J. (2006). LC-MS analysis of antifouling agent Irgarol 1051 and its decyclopropylated degradation product in seawater from marinas in Hong Kong). *Talanta* 70(1), 91–96.

Caquet T., Roucaute M., Mazzella N. *et al.* (2013). Risk assessment of herbicides and booster biocides along estuarine continuums in the Bay of Vilaine area (Brittany, France). *Environmental Science and Pollution Research* 20, 651–666.

CAR (2014). *Work Programme for Review of Active Substances in Biocidal Products Pursuant to Council Directive 98/8/EC – Copper Pyrithione (PT 21). Document IIIA 5.* Sundbyberg: KEMI, Swedish Chemical Agency.

Carafa R., Wollgast J., Canuti E. *et al.* (2007). Seasonal variations of selected herbicides and related metabolites in water, sediment, seaweed and clams in the Sacca di Goro coastal lagoon (Northern Adriatic). *Chemosphere* 69, 1625–1637.

Carbery C., Owen R., Frickers T., Otero E., Readman R. (2006). Contamination of Caribbean coastal waters by the antifouling herbicide Irgarol 1051. *Marine Pollution Bulletin* 52, 635–644.

Carteau D., Vallée-Réhel K., Linossier I. *et al.* (2014). Development of environmentally friendly antifouling paints usingbiodegradable polymer and lower toxic substances. *Progress in Organic Coatings* 77, 485–493.

Cassi R., Tolosa I., deMora S. (2008). A survey of antifoulants in sediments from ports and marinas along the French Mediterranean coast. *Marine Pollution Bulletin* 56, 1943–1948.

Castro I. B., Westphal E., Fillmann G. (2011). Tintas anti-incrustantes de terceira geração: novos biocidas no ambiente aquático [Third-generation antifouling paints: new biocides in the aquatic environment]. *Quimica Nova* 34, 1021–1031.

CBP (2006). Assessment Report: Dichlofluanid, Product-type 8. Regulation (EU) No. 528/2012 Concerning the Making Available on the Market and Use of Biocidal Products, Inclusion of Active Substances in Annex I to Directive 98/8/EC. www.echa.europa.eu/documents/10162/17158507/consolidated_bpr_en.pdf

CBP (2009). Assessment Report: Tolylfluanid Product-type 8. Directive 98/8/EC Concerning the Placing of Biocidal Products on the Market, Inclusion of Active Substances in Annex I. https://circabc.europa.eu/sd/a/5fd06ea4-c46e-48b8-ab80-a5bb7106a807/Tolylfluanid%20assessment%20report_March_25_09.pdf

CBP (2011). Assessment Report: Cybutryne (PT21). Directive 98/8/EC Concerning the Placing of Biocidal Products on the Market, Inclusion of Active Substances in Annex I. http://dissemination.echa.europa.eu/Biocides/ActiveSubstances/1281-21/1281-21_Assessment_Report.pdf

CBP (2013). Assessment Report: Zineb, Product-type 21. Regulation (EU) No. 528/2012 Concerning the Making Available on the Market and Use of Biocidal Products, Evaluation of Active Substances. http://dissemination.echa.europa.eu/Biocides/ActiveSubstances/1409-21/1409-21_Assessment_Report.pdf

CBP (2014a). Assessment Report: 4,5-Dichloro-2-octyl-2H-isothiazol-3-one (DCOIT), Product-type 21. Regulation (EU) No. 528/2012 Concerning the Making Available on the Market and Use of Biocidal Products, Evaluation of Active Substances. https://circabc.europa.eu/sd/a/5d2b12c8-7690-4636-a962-ab277f4b183d/DCOIT%20-%20PT%2021%20(assessment%20report%20as%20finalised%20on%2013.03.2014).pdf

CBP (2014b). Assessment Report: Tolylfluanid Product-type 21. Regulation (EU) No. 528/2012 Concerning the Making Available on the Market and Use of Biocidal Products, Evaluation of Active Substances. https://echa.europa.eu/documents/10162/6dc15617-6986-8a61-3a9a-ef782a0d66f1

CBP (2014c). Assessment Report: Tralopyril. Regulation (EU) No. 528/2012 Concerning the Making Available on the Market and Use of Biocidal Products, Evaluation of Active Substances. https://echa.europa.eu/documents/10162/edf62568-dafb-73a2-51b7-b3118505dab3

CBP (2015a). Assessment Report: Copper Pyrithione, Product-type 21. Regulation (EU) No. 528/2012 Concerning the Making Available on the Market and Use of Biocidal Products, Evaluation of Active Substances. http://dissemination.echa.europa.eu/Biocides/ActiveSubstances/1275-21/1275-21_Assessment_Report.pdf

CBP (2015b). Assessment Report: Medetomidine, Product-type 21. Regulation (EU) No. 528/2012 Concerning the Making Available on the Market and Use of Biocidal Products, Evaluation of Active Substances. http://dissemination.echa.europa.eu/Biocides/ActiveSubstances/1327-21/1327-21_Assessment_Report.pdf

CCS (2015). Guideline A-01 (201510): A01 Paints. China. www.ccs.org.cn/ccswzen/font/fontAction!article.do?articleId=ff808081511f06e301518fc0c5b001d9

Champ M. A. (2000). A review of organotin regulatory strategies, pending actions, related costs and benefits. *Science of the Total Environment* 258, 21–71.

Clark E. A., Sterritt R. M., Lester J. N. (1988). The fate of tributyltin in the aquatic environment. *Environmental Science & Technology* 22, 600–604.

Commendatore M. G., Franco M. A., Gomes Costa P. *et al.* (2015). Butyltins, polyaromatic hydrocarbons, organochlorine pesticides, and polychlorinated biphenyls in sediments and bivalve mollusks in a mid-latitude environment from the Patagonian coastal zone. *Environmental Toxicology and Chemistry* 34, 2750–2763.

Connelly D. P., Readman J. W., Knap A. H., Davies J. (2001). Contamination of the coastal waters of Bermuda by organotins and the triazine herbicide Irgarol 1051. *Marine Pollution Bulletin* 42, 409–414.

Daehne D., Fürle C., Thomsen A., Watermann B., Feibicke M. (2017). Antifouling Biocides in German marinas – exposure assessment and calculation of national consumption and emission. *Integrated Environmental Assessment and Management* 13, 892–905.

Dafforn K. A., Lewis J. A., Johnston E. (2011). Antifouling strategies: history and regulation, ecological impacts and mitigation. *Marine Pollution Bulletin* 62, 453–465.

Dahl B., Blanck H. (1996). Toxic effects of the antifouling agent Irgarol 1051 on periphyton communities in coastal water microcosms. *Marine Pollution Bulletin* 32, 342–350.

de Mora S. J., Pelletier E. (1997). Environmental tributyltin research: past, present, future. *Environmental Technology* 18, 1169–1177.

Devilla R. A., Brown M. T., Donkin M., Tarran G. A., Aiken J., Readman J. W. (2005). Impact of antifouling booster biocides on single microalgal species and on a natural marine phytoplankton community. *Marine Ecology Progress Series* 286, 1–12.

Di Landa G., Parrella L., Avagliano S. *et al.* (2009). Assessment of the potential ecological risks posed by antifouling booster biocides to the marine ecosystem of the Gulf of Napoli (Italy). *Water, Air and Soil Pollution* 200, 305–321.

Diniz L. G. R. (2016). Análise cromatográfica de biocidas anti-incrustantes em amostras de água de mar e tecido de moluscos. PhD thesis, São Carlos, Brazil.

Diniz L. G. R., Jesus M. S., Dominguez L. A. E. *et al.* (2014). First appraisal of water contamination by antifouling booster biocide of 3rd generation at Itaqui Harbor (São Luiz – Maranhão – Brazil). *Journal of the Brazilian Chemical Society* 25, 380–388.

EC (2003). Regulation (EC) No. 782/2003 of the European Parliament and of the Council of 14 April 2003 on the prohibition of organotin compounds on ships. OJ L 115, 1,11. http://eur-lex.europa.eu/legal-content/EN/TXT/?uri=CELEX:32003R0782

ECHA (2014). *CLH Report for Medetomidine: Proposal for Harmonised Classification and Labelling*. York: Chemicals Regulation Directorate.

Eguchi S., Harino H., Yamamoto Y. (2009). Assessment of antifouling biocides contaminations in Maizuru Bay, Japan. *Archives of Environmental Contamination and Toxicology* 58, 684–693.

Eklund B., Elfström M., Gallego I., Bengtsson, B. E., Breitholtz, M. (2010). Biological and chemical characterization of harbour sediments from the Stockholm area. *Journal of Soils and Sediments* 10, 127–141.

Erdelez A., Furdek Turk M., Štambuk A., Župan I., Peharda M. (2017). Ecological quality status of the Adriatic coastal waters evaluated by the organotin pollution biomonitoring. *Marine Pollution Bulletin* 123, 313–323.

Fent K. (2006). Worldwide occurrence of organotins from antifouling paints and effects in the aquatic environment. *Antifouling Paint Biocides* 5, 71–100.

Fernández-Alba A. R., Hernando M. D., Piedra L., Chisti Y. (2002). Toxicity evaluation of single and mixed antifouling biocides measured with acute toxicity bioassays. *Analytica Chimica Acta* 456, 303–312.

Ferreira M., Blanco L., Garrido A., Vieites J. M., Cabado A. G. (2013). *In vitro* approaches to evaluate toxicity induced by organotin compounds tributyltin (TBT), dibutyltin (DBT), and monobutyltin (MBT) in neuroblastoma cells. *Journal of Agricultural and Food Chemistry* 61, 4195–4203.

Ferrer I., Ballesteros B., Pilar Marco M., Barceló D. (1997). Pilot survey for determination of the antifouling agent Irgarol 1051 in enclosed seawater samples by a direct enzyme-linked immunosorbent assay and solid-phase extraction followed by liquid chromatography-diode array detection. *Environmental Science and Technology* 31, 3530–3535.

Gadd G. M. (2000). Microbial interactions with tributyltin compounds: detoxification, accumulation, and environmental fate. *Science of the Total Environment* 258, 119–127.

Gervais J. A., Luukinen B., Buhl K., Stone D. (2008). Capsaicin Technical Fact Sheet. National Pesticide Information Center, Oregon State University. http://npic.orst.edu/factsheets/archive/Capsaicintech.html#mode

Gibbs P. E., Bryan G. W., Pascoe P. L., Burt G. R. (1987). The use of the dog-whelk, *Nucella lapillus*, as an indicator of tributyltin (TBT) contamination. *Journal of the Marine Biological Association of the United Kingdom* 67, 507–523.

Gibson C., Wilson S. (2003). Imposex still evident in eastern Australia 10 years after tributyltin restrictions. *Marine Environmental Research* 55, 101–112.

Gonzalez-Rey M., Tapie N., Le Menach, K. *et al.* (2015). Occurrence of pharmaceutical compounds and pesticides in aquatic systems. *Marine Pollution Bulletin* 96, 384–400.

Government of Western Australia (2009). *Antifouling Biocides in Perth Coastal Waters: A Snapshot at Select Areas of Vessel Activity. Water Science Technical Series*. Perth: Government of Western Australia.

Guardiola F. A., Cuesta A., Meseguer J., Esteban A. M. (2012). Risks of using antifouling biocides in aquaculture. *International Journal of Molecular Science* 13, 1541–1560.

Guitart C., Sheppard A., Frickers T., Price A. R. G., Readman J. W. (2007). Negligible risks to corals from antifouling booster biocides and triazine herbicides in coastal waters of the Chagos Archipelago. *Marine Pollution Bulletin* 54, 226–246.

Hall Jr. L. W., Giddings J. M., Solomon K. R., Balcomb R. (1999). An ecological risk assessment for the use of Irgarol 1051 as an algaecide for antifouling paints. *Critical Reviews of Toxicology* 29, 367–437.

Hamwijk C., Schouten A., Foekema E. M. *et al.* (2005). Monitoring of the booster biocide dichlofluanid in water and marine sediment of Greek marinas. *Chemosphere* 60, 1316–1324.

Harino H., Langston W. J. (2009). Degradation of alternative biocides in the aquatic environment. In: T. Arai H. Harino M. Ohji, W. J. Langston, eds., *Ecotoxicology of Antifouling Biocides*. Tokyo: Springer, pp. 397–412

Harino H., Kitano M., Mori Y. *et al.* (2005). Degradation of antifouling booster biocides in water. *Journal of the Marine Biological Association of the United Kingdom* 85, 33–38.

Harino H., S. Midorikawa T., Arai M. *et al.* (2006a). Concentrations of booster biocides in sediment and clams from Vietnam. *Journal of the Marine Biological Association of the United Kingdom* 86, 1163–1170.

Harino H., Ohji M. Wattayakorn G. *et al.* (2006b). Occurrence of antifouling biocides in sediment and green mussels from Thailand. *Archives of Environmental Contamination and Toxicology* 51, 400–407.

Harino H, Yamamoto Y, Eguchi S. *et al.* (2007). Concentrations of antifouling biocides in sediment and mussel samples collected from Otsuchi Bay, Japan. *Archives of Environmental Contamination and Toxicology* 52, 179–188.

Harino H., Arai T. Ohji M. Ismail A. B., Miyazaki N. (2009). Contamination profiles of antifouling biocides in selected coastal regions of Malaysia. *Archives of Environmental Contamination and Toxicology* 56, 468–478.

Harino H., Eguchi S., Arai T. *et al.* (2010). Antifouling biocides contamination in sediment of coastal waters from Japan. *Coastal Marine Science* 34, 230–235.

Harino H., Arifin Z., Rumengan I. F. M. *et al.* (2012). Distribution of antifouling biocides and perfluoroalkyl compounds in sediments from selected locations in the Indonesian coastal waters. *Archives of Environmental Contamination and Toxicology* 63, 13–21.

Hernando M. D., Ejerhoon M., Fernandez-Alba A. R., Chisti Y. (2003). Combined toxicity effects of MTBE and pesticides measured with *Vibrio fischeri* and *Daphnia magna* bioassays. *Water Research* 37, 4091–4098.

Hoch M. (2001). Organotin compounds in the environment – an overview. *Applied Geochemistry* 16, 719–743.

Horiguch, T., Shiraishi H., Shimizu M., Morita M. (1997). Imposex in sea snails, caused by organotin (tributyltin and triphenyltin) pollution in Japan: a survey. *Applied Organometallic Chemistry* 11, 451–455.

IFOP (2013). *Informe Final: Determinación y evaluación de los componentes presentes en las pinturas anti-incrustantes utilizadas em la acuicultura, sus efectos y la acumulación em sedimentos marinos de la X región de los Lagos*. Valparaíso: Instituto de Fomento Pesquero.

IMO (2002). Anti-fouling systems. www.imo.org/includes/blastDataOnly.asp/data_id%3D7986/FOULING2003.pdf

IMO (2012). *International Convention On the Control of Harmful Anti-Fouling Systems on Ships, 2001. Treaty Series n. 13*. London: The Stationery Office.

International (2014). Ficha de Informação de Segurança de Produto Químico (FISP) – Micron Navigator Preto [Material Safety Datasheet (MSDS) – Micron Navigator Black]. www.yachtpaint.com/bra/diy/produtos/anti-incrustantes/micron-navigator.aspx

Johansson, P., Eriksson, K. M., Axelsson, L., Blanck, H. (2012). Effects of seven antifouling compounds on photosynthesis and inorganic carbon use in sugar kelp Saccharina latissimi (Linnaeus). *Archives of Environmental Contamination and Toxicology* 63, 365–377.

Kaonga C. C., Takeda K., Sakugawa H. (2016). Concentration and degradation of alternative biocides and an insecticide in surface waters and their major sinks in a semi-enclosed sea, Japan. *Chemosphere*, 145, 256–264.

KEMI (2014). *Work Programme for Review of Active Substances in Biocidal Products Pursuant to Council Directive 98/8/EC: Copper Pyrithione (PT 21). Final Competent Authority Report (CAR).* Sundbyberg: KEMI, Swedish Chemical Agency.

Kim N. S., Shim W. J., Yim H. U. *et al.* (2014). Assessment of TBT and organic booster biocide contamination in seawater from coastal areas of South Korea. *Marine Pollution Bulletin* 78, 201–208.

Kim N. S., Hong S. H., An J. G., Shin K. H., Shim W. J. (2015). Distribution of butyltins and alternative antifouling biocides in sediments from shipping and shipbuilding areas in South Korea. *Marine Pollution Bulletin* 95, 484–490.

Kim N. S., Hong S. H., Shin K. H., Shim W. J. (2017). Imposex in *Reishia clavigera* as an indicator to assess recovery of TBT pollution after a total ban in South Korea. *Archives of Environmental Contamination and Toxicology* 73, 301–309.

Knutson S., Downs C. A., Richmond R. H. (2012). Concentrations of Irgarol in selected marinas of Oahu, Hawaii and effects on settlement of coral larval. *Ecotoxicology* 21(1), 1–8.

Konstantinou I. K., Albanis T. A. (2004). Worldwide occurrence and effects of antifouling paint booster biocides in the aquatic environment: a review. *Environment International* 30, 235-248.

Koutsaftis A., Aoyama I. (2007). Toxicity of four antifouling biocides and their mixtures on the brine shrimp *Artemia salina. Science of the Total Environment* 387, 166–174.

Kurouani-Harani H. (2003). Microbial and photolytic degradation of benzothiazoles in water and wastewater. Doctorate thesis. Fakultät III der Technischen Universität Berlin, Berlin, Germany.

Lam K. H., Cai Z., Wai H. Y. *et al.* (2005). Identification of a new Irgarol-1051 related s-triazine species in coastal waters. *Environmental Pollution* 136, 221–230.

Lam N. H., Jeong H.-H., Kang S.-D. *et al.* (2017). Organotins and new antifouling biocides in water and sediments from three Korean Special Management Sea Areas following ten years of tributyltin regulation: contamination profiles and risk assessment. *Marine Pollution Bulletin* 121, 302–312.

Lamoree M. H., Swart C. P., Van Der Horst A., Van Hattum B. (2002). Determination of diuron and the antifouling paint biocide Irgarol 1051 in Dutch marinas and coastal waters. *Journal of Chromatography A* 970, 183–190.

Langston W. J., Pope N. D. (1995). Determinants of TBT adsorption and desorption in estuarine sediments. *Marine Pollution Bulletin* 31, 32–43.

Langston W. J., Pope N. D., Davey M. *et al.* (2015). Recovery from TBT pollution in English Channel environments: a problem solved? *Marine Pollution Bulletin* 95, 551–564.

Laranjeiro F., Sánchez-Marín P., Barros A. *et al.* (2016). Triphenyltin induces imposex in *Nucella lapillus* through an aphallic route. *Aquatic Toxicology* 175, 127–131.

Laranjeiro F., Sánchez-Marín P., Oliveira I. B., Galante-Oliveira S., Barroso C. M. (2018). Fifteen years of imposex and tributyltin pollution monitoring along the Portuguese coast. *Environmental Pollution* 232, 411–421.

Lee M., Kim U., Lee I., Choi M., Oh J. (2015). Assessment of organotin and tin-free antifouling paints contamination in the Korean coastal area. *Marine Pollution Bulletin* 99, 157–165.

Lee S., Ching J., Won H., Lee D., Lee Y. (2011). Analysis of antifouling agents after regulation of tributyltin compounds in Korea. *Journal of Hazardous Materials* 185, 1318–1325.

Lind U., Rosenblad M. A., Frank L. H. *et al.* (2010). Octopamine receptors from the barnacle *Balanus improvisus* are activated by the alpha(2)-drenoceptor agonist medetomidine. *Molecular Pharmacology* 78, 237–248.

Liu D., Maguire R. J., Lau Y. L. *et al.* (1997). Transformation of the new antifouling compound Irgarol 1051 by *Phanerochaete chrysosporium. Water Research* 31, 2363–2369.

Liu D., Pacepavicius G. J., Maguire R. J. *et al.* (1999). Mercuric chloride-catalyzed hydrolysis of the new antifouling compound Irgarol 1051. *Water Research* 33, 155–163.

Liu S., Zhou J., Ma X. *et al.* (2016). Ecotoxicity and preliminary risk assessment of nonivamide as a promising marine antifoulant. *Journal of Chemistry* 2016, 1–4.

Liu W., Zhao J., Liu Y. *et al.* (2015). Biocides in the Yangtze River of China: spatiotemporal distribution, mass load and risk assessment. *Environmental Pollution* 200, 53–63.

Lopes-Dos-Santos R. M. A., Galante-Oliveira S., Lopes E., Almeida C., Barroso C. M. (2014). Assessment of imposex and butyltin concentrations in Gemophos viverratus (Kiener, 1834), from São Vicente, Republic of Cabo Verde (Africa). *Environmental Science and Pollution Research* 21, 10671–10677.

Mackie D. S., van den Berg C. M. G., Readman J. W. (2004). Determination of pyrithione in natural waters by cathodic stripping voltammetry. *Analytica Chimica Acta* 511, 47–53.

Marcheselli M., Rustichelli C., Mauri M. (2010). Novel anti-fouling agent zinc pyrithione: determination, acute toxicity and bioaccumulation in marine mussels (*Mytilus galloprovincialis*). *Environmental Toxicology Chemistry* 29, 2583–2592.

Martínez K., Barceló D. (2001). Determination of antifouling pesticides and their degradation products in marine sediments by means of ultrasonic extraction and HPLC–APCI–MS. *Fresenius Journal of Analytical Chemistry* 370, 940–945.

Martínez K., Ferrer I., Barceló D. (2000). Part-per-trillion level determination of antifouling pesticides and their byproducts in seawater samples by off-line solid-phase extraction followed by high-performance liquid chromatography–atmospheric pressure chemical ionization mass spectrometry. *Journal of Chromatography A* 879, 27–37.

Martins S. E., Lillicrap A., Fillmann G., Thomas K. (2018). Ecotoxicity of organic and organo-metallic antifouling co-biocides and implications for environmental hazard and risk assessments in aquatic ecosystems. *Biofouling* 34, 34–52.

Matthai C., Guise K., Coad P., McCready S., Taylor S. (2009). Environmental status of sediments in the lower Hawkesbury–Nepean River, New South Wales. *Australian Journal of Earth Science* 56, 225–243.

Matthiessen P., Waldock R., Thain J. E., Waite M. E., Scropehowe S. (1995). Changes in periwinkle (*Littorina littorea*) populations following the ban on TBT-based antifoulings on small boats in the United Kingdom. *Ecotoxicology and Environmental Safety* 30, 180–194.

Mattos Y., Stotz W. B., Romero M. S. *et al.* (2017). Butyltin contamination in Northern Chilean coast: is there a potential risk for consumers? *Science of the Total Environment* 595, 209–217.

Mezcua M., Hernando M. D., Piedra L., Agüera A., Fernandez-Alba A. R. (2002). Chromatography-Mass Spectrometry and Toxicity Evaluation of Selected Contaminants in Seawater. *Cromatographia* 56, 199–206.

Minchin D., Bauer B., Oehlmann J., Schulte-Oehlmann U., Duggan C. B. (1997). Biological indicators used to map organotin contamination from a fishing port, Killybegs, Ireland. *Marine Pollution Bulletin* 34, 235–243.

Mochida K., Ito K., Harino H., Kakuno A., Fujii K. (2006). Acute toxicity of pyrithione antifouling biocides and joint toxicity with copper to red sea bream (*Pagrus major*) and toy shrimp (*Heptacarpus futilirostris*). Environmental Toxicology Chemistry 25, 3058–3064.

Mochida K., Onduka T., Amano H. *et al.* (2012). Use of species sensitivity distributions to predict no-effect concentrations of an antifouling biocide, pyridine triphenylborane, for marine organisms. *Marine Pollution Bulletin* 64, 2807–2814.

Mohr S., Berghahn R., Mailahn W. *et al.* (2009). Toxic and accumulative potential of the antifouling biocide and TBT successor Irgarol on freshwater macrophytes: a pond mesocosm study. *Environmental Science and Technology* 43, 6838–6843.

Mukherjee A., Mohan Rao K. V., Ramesh U. S. (2009). Predicted concentrations of biocides from antifouling paints in Visakhapatnam Harbour. *Journal of Environmental Management* 90, S51–S59.

Muñoz I., Bueno M. J. M., Agüera A., Fernández-Alba A. R. (2010). Environmental and human health risk assessment of organic micro-pollutants occurring in a Spanish marine fish farm. *Environmental Pollution* 158, 1809–1816.

Nakajima K., Yasuda T. (1990). High-performance liquid chromatographic determination of zinc pyrithione in antidandruff preparations based on copper chelate formation. *Journal of Chromatography* 502, 379–384.

New Zealand Environmental Protection Authority (2012). *Antifouling Paints Reassessment – Preliminary Risk Assessment. Environmental Protection Authority, Te Mana Rauht Taiao*. Wellington: New Zealand Government.

New Zealand Environmental Protection Authority (2013). *Decision on the Application for Reassessment of Antifouling Paints (APP201051). Application for the Reassessment of a Group of Hazardous Substances, under Section 63 of the Hazardous Substances and New Organisms Act 1996*. Wellington: New Zealand Government.

Nicolaus E. E. M., Barry J. (2015). Imposex in the dogwhelk (*Nucella lapillus*): 22-year monitoring around England and Wales. *Environmental Monitoring and Assessment* 187, 736.

NIVA (2012). *Screening of Selected Alkylphenolic Compounds, Biocides, Rodenticides and Current Use Pesticides. Statlig Program for Forurensningsovervåking, Rapportnr. 1116/2012*. Oslo: Klima og Forurensnings Direktoratet.

Nödler K., Voutsa D., Licha T. (2014). Polar organic micropollutants in the coastal environment of different marine systems. *Marine Pollution Bulletin* 85, 50–59.

Oehlmann J., Markert B., Stroben E. *et al.* (1996). Tributyltin biomonitoring using prosobranchs as sentinel organisms. *Fresenius Journal of Analytical Chemistry* 354, 540–545.

Okamura H. (2002). Photodegradation of the antifouling compounds Irgarol 1051 and Diuron released from a commercial antifouling paint. *Chemosphere* 48, 43–50.

Okamura H., Mieno H. (2006). Antifouling paint biocides. In: I. Konstantinou, ed., *The Handbook of Environmental Chemistry*. Heidelberg: Springer, pp. 201–212.

Okamura H., Sugiyama Y. (2004). Photosensitized degradation of Irgarol 1051 in water. *Chemosphere* 57, 739–743.

Okamura H., Aoyama I., Liu D. *et al.* (2000). Fate and ecotoxicity of the new antifouling compound Irgarol in the aquatic environment. *Water Research* 34, 3523–3530.

Okamura H., Watanabe T., Aoyama I., Hasobe M. (2002). Toxicity evaluation of new antifouling compounds using suspension-cultured fish cells. *Chemosphere* 46, 945–951.

Okamura H., Aoyama I., Ono Y., Nishida T. (2003). Antifouling herbicides in the coastal waters of western Japan. *Marine Pollution Bulletin* 47, 59–67.

Okoro H. K., Fatoki O. S., Adekola F. A., Ximba B. J., Snyman R. G. (2016). Spatio-temporal variation of organotin compounds in seawater and sediments from Cape Town harbour, South Africa using gas chromatography with flame photometric detector (GC-FPD). *Arabian Journal of Chemistry* 9, 95–104.

Oliveira I. B., Richardson C. A., Sousa A. C. *et al.* (2009). Spatial and temporal evolution of imposex in dogwhelk *Nucella lapillus* (L.) populations from north Wales, UK. *Journal of Environmental Monitoring* 11, 1462–1468.

Oliveira I. B., Beiras R., Thomas K. V., Suter M. J. F., Barroso C. M. (2014). Acute toxicity of tralopyril, capsaicin and triphenylborane pyridine to marine invertebrates. *Ecotoxicology* 23, 1336–1344.

Omae, I. (2003). General aspects of tin-free antifouling paints. *Chemical Reviews* 103, 3431–3448.

OPRF (2010). *Final Report on the Comprehensive Management against Biofouling to Minimize Risks on Marine Environment.* Tokyo: Nippon Foundation.

Paz-Villarraga C. A., Castro Í. B., Miloslavich P., Fillmann G. (2015). Venezuelan Caribbean Sea under the threat of TBT. *Chemosphere* 119, 704–710.

Petracco M., Camargo R. M., Berenguel T. A. *et al.* (2015). Evaluation of the use of *Olivella minuta* (Gastropoda, Olividae) and *Hastula cinerea* (Gastropoda, Terebridae) as TBT sentinels for sandy coastal habitats. *Environmental Monitoring and Assessment* 187, 440.

Price A. R. G., Readman J. W. (2013). Booster biocide antifoulants: is history repeating itself? Late Lessons From Early Warnings: Science, Precaution, Innovation. Part B: Emerging Lessons From Ecosystems. European Environment Agency (EEA). www.eea.europa.eu/publications/late-lessons-2/late-lessons-chapters/late-lessons-ii-chapter-12/view

Readman J. W., Kwong L. L. W., Grondin D. *et al.* (1993). Coastal water contamination from a triazine herbicide used in antifouling paints. *Environmental Science & Technology* 27, 1940–1942.

Reemtsma T., Fiehn O., Kalnowski G., Jekel M. (1995). Microbial transformations and biological effects of fungicide-derived benzothiazoles determined in industrial wastewater. *Environmental Science & Technology* 29, 478–485.

Rodríguez Á. S., Ferrera Z. S., Rodríguez J. J. S. (2011a). An evaluation of antifouling booster biocides in Gran Canaria coastal waters using SPE-LC MS/MS. *International Journal of Environmental Analytical Chemistry* 91(12), 1166–1177.

Rodríguez Á. S., Ferrera Z. S., Rodríguez, J. J. S. (2011b). A preliminary assessment of levels of antifouling booster biocides in harbours and marinas of the island of Gran Canaria, using SPE-HPLC. *Environmental Chemistry Letters* 9, 203–208.

Ruiz J. M., Quintela M., Barreiro R. (1998). Ubiquitous imposex and organotin bioaccumulation in gastropods *Nucella lapillus* from Galicia (NW Spain): a possible effect of nearshore shipping. *Marine Ecology Progress Series* 164, 237–244.

Ruiz J. M., Barreiro R., Couceiro L., Quintela M. (2008). Decreased TBT pollution and changing bioaccumulation pattern in gastropods imply butyltin desorption from sediments. *Chemosphere* 73, 1253–1257.

Ruiz J. M., Albaina N., Carro B., Barreiro R. (2015). A combined whelk watch suggests repeated TBT desorption pulses. *Science of the Total Environment* 502, 167–171.

Saleh A., Molaei S., Fumani N., Abedi E. (2016). Antifouling paint booster biocides (Irgarol 1051 and Diuron) in marinas and ports of Bushehr, Persian Gulf. *Marine Pollution Bulletin* 105, 367–372.

Sánchez-Rodriguez A., Sosa-Ferrera Z., del Pino A. S., Santana-Rodriguez J. J. (2011). Probabilistic risk assessment of common booster biocides in surface waters of the harbours of Gran Canaria (Spain). *Marine Pollution Bulletin* 62, 985–991.

Santillo D., Johnston P., Langston W. J. (2001). Tributyltin (TBT) antifoulants: a tale of ships, snails and imposex, In: P. Harremoës, D. Gee, M. MacGarvin *et al.*, eds., *Late Lessons from Early Warnings: The Precautionary Principle 1896–2000*. Copenhagen: European Environmental Agency, pp. 135–148.

Sapozhnikova Y., Wirth E., Schiff K., Brown J., Fulton M. (2007). Antifouling pesticides in the coastal waters of Southern California. *Marine Pollution Bulletin* 54, 1962–1989.

Sapozhnikova Y., Wirth E., Schiff K., Fulton M. (2013). Antifouling biocides in water and sediments from California marinas. *Marine Pollution Bulletin* 69, 189–194.

Schouten A., Mol H., Hamwijk C. *et al.* (2005). Critical aspects in the determination of the antifouling compound dichlofluanid and its metabolite DMSA (N,N-dimethyl-N'-phenylsulfamide) in seawater and marine sediments. *Chromatographia* 62, 511–517.

Schulte-Oehlmann U., Tillmann M., Markert B. *et al.* (2000). Effects of endocrine disruptors on prosobranch snails (Mollusca: Gastropoda) in the laboratory. Part II: triphenyltin as a xeno-androgen. *Ecotoxicology* 9, 399–412.

Seymour M. D., Bailey D. L. (1981). Thin-layer chromatography of pyrithiones. *Journal of Chromatography* 206, 301–310.

Sheikh M. A., Juma F. S., Staehr P. *et al.* (2016). Occurrence and distribution of antifouling biocide Irgarol-1051 in coral reef ecosystems, Zanzibar. *Marine Pollution Bulletin* 109, 586–590.

Smith B. S. (1971). Sexuality in the American mud snail, *Nassarius obsoletus* Say. *Proceedings of the Malacological Society of London* 39, 377–378.

Sousa A., Laranjeiro F., Takahashi S., Tanabe S., Barroso C. M. (2009). Imposex and organotin prevalence in a European post-legislative scenario: temporal trends from 2003 to 2008. *Chemosphere* 77, 566–573.

Sousa A. A., Pastorinho M. R., Takahashi S., Tanabe S. (2014). History on organotin compounds, from snails to humans. *Environmental Chemistry* Letters 12, 117–137.

Stewart C., de Mora S. J. (1990). A review of the degradation of tri(*n*-butyl)tin in the marine environment. *Environmental Techology* 11, 565–570.

Stewart M., Aherns M., Olsen G. (2009). *Analysis of Chemicals of Emerging Environmental Concern in Auckland's Aquatic Sediments. Prepared by NIWA for Auckland Regional Council. Auckland Regional Council Technical Report 2009/021*. Auckland: Auckland Research Council.

Stroben E., Oehlmann J., Fioroni P. (1992). The morphological expression of imposex in *Hinia reticulata* (Gastropoda: Buccinidae): a potential indicator of tributultin pollution. *Marine Biology* 113, 625–636.

Takahashi K. (2009). Release rate of biocides from antifouling paints. In: T. Arai, H. Harino, M. Ohji, W. J. Langston, eds., *Ecotoxicology of Antifouling Biocides*. Tokyo: Springer, pp. 3–22.

Thomas K. V. (1998). Determination of selected antifouling booster biocides by high-performance liquid chromatography–atmospheric pressure chemical ionisation mass spectrometry. *Journal of Chromatography A* 825(1), 29–35.

Thomas K. V. (2001). The environmental fate and behaviour of antifouling paint booster biocides: a review. *Biofouling* 17, 73–86.

Thomas K. V. (2009). The use of broad-spectrum organic biocides in marine antifouling paints. In: C. Hellio, D Yebra, eds., *Advances in Marine Antifouling Coatings and Technologies*. Cambridge: Woodhead Publishing Limited, pp. 522–553.

Thomas K. V., Brooks S. (2010). The environmental fate and effects of antifouling paint biocides. *Biofouling* 26(1), 73–88.

Thomas K. V., Langford K. (2009a). Monitoring of alternative biocides: Europe and USA. In: T. Arai, H. Harino, M. Ohji, W. J. Langston, eds., *Ecotoxicology of Antifouling Biocides*. Tokyo: Springer, pp. 331–344.

Thomas K. V., Langford K. (2009b). The analysis of antifouling paint biocides in water, sediment and biota. In: T. Arai, H. Harino, M. Ohji, W. J. Langston, eds., *Ecotoxicology of Antifouling Biocides*. Tokyo: Springer, pp. 311–327.

Thomas K. V., Blake S., Waldock M. (2000). Antifouling paint booster biocide contamination in UK marine sediments. *Marine Pollution Bulletin* 40, 739–745.

Thomas K. V., Fileman T. W., Readman J. W., Waldock M. J. (2001). Antifouling paint booster biocides in the UK coastal environment and potential risks of biological effects. *Marine Pollution Bulletin* 42, 677–688.

Thomas K. V., McHugh M., Waldock M. (2002). Antifouling paint booster biocides in UK coastal waters: inputs, occurrence and environmental fate. *Science of the Total Environment* 293, 117–127.

Thomas K. V., McHugh M., Hilton M., Waldock M. (2003). Increased persistence of antifouling paint biocides when associated with paint particles. *Environmental Pollution* 123, 153–161.

Titley-O'Neal C. P., Munkittrick K. R., Macdonald B. A. (2011). The effects of organotin on female gastropods. *Journal of Environmental Monitoring* 13, 2360–2388.

Tolosa I., Readman J. W., Blaevoet A. *et al.* (1996). Contamination of Mediterranean (CSte d'Azur) coastal waters by organotins and Irgarol 1051 used in antifouling paints. *Marine Pollution Bulletin* 32, 335–341.

Tornero V., Hanke G. (2016). Chemical contaminants entering the marine environment from sea-based sources: a review with a focus on European seas. *Marine Pollution Bulletin* 112, 17–38.

Tsunemasa N. (2013). Impact of antifouling biocide on marine environment. In: E. Özhan, ed., *Proceedings of the Global Congress on ICM: Lessons Learned to Address New Challenges. MEDCOAST 2013 Joint Conference*, Marmaris, pp. 1065–1075.

Tsunemasa N., Yamazaki H. (2014). Concentration of antifouling biocides and metals in sediment core samples in the northern part of Hiroshima Bay. *International Journal of Molecular Science* 15, 9991–10004.

Turner A., Glegg G. (2014). TBT-based antifouling paints remain on sale. *Marine Pollution Bulletin* 88, 398–400.

Turquet J., Quiniou F., Delesmon R., Durand G. (2010). ERICOR Evaluation du risque « pesticides » pour les récifs coralliens de La Réunion [ERICOR Risk Assessment of pesticides to coral reefs de La Reúnion]. Republique Française. www.programmepesticides.fr/content/download/4531/41722/version/2/file/Turquet_synthese.pdf

US EPA (2000–2018). [ECOTOX] ECOTOXicology Knowlegdebase. https://cfpub.epa.gov/ecotox

US EPA (2003). *Reregistration Eligibility Decision (RE) for Ziram*. Washington, DC: Environmental Protection Agency.

US EPA (2005a). *Reregistration Eligibility Decision (RE) for Mancozeb*. Washington, DC: Environmental Protection Agency.

US EPA (2005b). *Reregistration Eligibility Decision (RE) for Maneb*. Washington, DC: Environmental Protection Agency.

US EPA (2006). *Reregistration Eligibility Decision for 2-(Thiocyanomethylthio)- benzothiazole (TCMTB)*. Washington, DC: Environmental Protection Agency.

US EPA (2007). *Reregistration Eligibility Decision for 2-Octyl-3 (2H)-isothiazolone (OIT)*. Washington, DC: Environmental Protection Agency.

Van Scoy A., Tjeerdema R. (2014). Environmental fate and toxicology of chlorothalonil. *Reviews of Environmental Contamination and Toxicology* 232, 89–105.

van Wezel A. P., van Vlaardingen P. (2004). Environmental risk limits for antifouling substances. *Aquatic Toxicology* 66, 427–444.

Walker W., Cripe C., Pritchard P., Bourquin A. (1988). Biological and abiotic degradation of xenobiotic compounds in *in vitro* estuarine water and sediment/water systems. *Chemosphere* 17, 2255–2270.

Wang J., Shi T., Yang X., Han W., Zhou Y. (2014). Environmental risk assessment on capsaicin used as active substance for antifouling system on ships. *Chemosphere* 104, 85–90.

Wells, F. E., Keesing, J. K., Brearley, A. (2017). Recovery of marine *Conus* (Mollusca: Caenogastropoda) from imposex at Rottnest Island, Western Australia, over a quarter of a century. *Marine Pollution Bulletin* 123, 182–187.

Wendt I., Arrhenius Å., Backhaus T. *et al.* (2013). Effects of five antifouling biocides on settlement and growth of zoospores from the marine macroalga *Ulva lactuta* L. *Bulletin of Environmental Contamination and Toxicology* 91, 426–432.

Wendt I., Backhaus T., Blanck H., Arrhenius Å. (2016). The toxicity of the three antifouling biocides DCOIT, TPBP and medetomidine to the marine pelagic copepod Acartia tonsa. *Ecotoxicology* 25, 871–879.

WFD (2012). *Proposed EQS for Water Framework Directive Annex VIII Substances: Chlorothalonil (For Consultation)*. London: Water Framework Directive – United Kingdom Technical Advisory Group (WFD-UKTAG).

WHO (1990). Environmental health criteria for tributyltin compounds. Environmental Health Criteria. World Health Organization. www.inchem.org/documents/ehc/ehc/ehc116.htm

Willis K. J., Woods C. M. C. (2011). Managing invasive *Styela clava* populations: inhibiting larval recruitment with medetomidine. *Aquatics Invasions* 6, 511–514.

Woldegiorgis A., Remberger M., Kaj L. *et al.* (2007). Results from the Swedish Screening Programme 2006 Subreport 3: Zinc Pyrithione and Irgarol 1051. www.ivl.se/webdav/files/Rapporter/B1764.pdf

Xu H., Lu A., Yu H., Sun J., Shen M. (2013). Distribution of Diuron in coastal seawater and sediments from West Sea area of Zhoushan Island. *Open Journal of Marine Science* 3, 140–147.

Xu Q., Barrios C. A., Cutright T., Newby B. Z. (2005). Evaluation of toxicity of capsaicin and zosteric acid and their potential application as antifoulants. *Environmental Toxicology* 20, 467–474.

Xu S. (2000). *Environmental Fate of Ethylenethiourea*. Sacramento, CA: California Department of Pesticide Regulation.

Yamada H. (2007). Behaviour, occurrence, and aquatic toxicity of new antifouling biocides and preliminary assessment of risk to aquatic ecosystems. *Bulletin of Fisheries Research Agencie* 21, 31–35.

Zhang A. Q., Leung K. M. Y., Kwok K. W. H., Bao V. W. W., Lam M. H. W. (2008). Toxicities of antifouling biocide Irgarol 1051 and its major degraded product to marine primary producers. *Marine Pollution Bulletin* 57, 575–586.

Zhou J., Yang C., Wang J. *et al.* (2013). Toxic effects of environment-friendly antifouling nonivamide on *Phaeodactylum tricornutum*. *Environmental Toxicology Chemistry* 32, 802–809.

Zhou X. J., Okamura H., Nagata S. (2006). Remarkable synergistic effects in antifouling chemicals against *Vibrio fischeri* by bioluminescent assay. *Journal of Health Science* 52, 243–251.

Zhou X. J., Okamura H., Nagata S. (2007). Abiotic degradation of triphenylborane pyridine (TPBP) antifouling agent in water. *Chemosphere* 67, 1904–1910.

7
Invasive Species

JOHN A. LEWIS

7.1 Aquatic Invasive Species

Through the eras since life first appeared on planet Earth, biological species have been spread around the globe, both within and between landmasses and water bodies. The current biogeographical mosaic of communities and ecosystems has evolved in response to tectonic movement, climatic fluctuations and consequent sea-level rises and falls, as well as biological adaptions to the physical and chemical environment. The environmental requisites or tolerances, competitive strengths and weaknesses, mechanisms of dispersal and reproductive strategies dictate the ability of a species to survive, propagate and expand populations, with the success, speed and distances of spread governed by not only individual species traits, but also the structure of, and ecological processes within, surrounding ecosystems. Through the evolution of the biosphere, the interactions of physical and chemical conditions – and between biological species – has led to constraints on the distributions of the myriad plants and animals, with many now defined as native, or indigenous, to specific regions. Species abundances and distributional boundaries are, however, dynamic over short and long periods of time.

A species may be spread to new locations by natural dispersal or, in the Anthropocene, through association with or as a consequence of human activity. Globally, the term 'Anthropocene' reflects that the most invasive species, in terms of both population growth and spread and ecological impact, has been *Homo sapiens*. Humans have caused vast ecological changes through direct consummation and alteration of ecosystems for habitation, sustenance (e.g., farming, resource extraction) and, in more recent times, recreation. The global redistribution of microbial, plant and animal species has been facilitated by and associated with human spread, both intentionally (crop species, ornamental plants, honeybees, domestic animals, fish, game, etc.) and unintentionally (pathogens, weeds, wasps, rodents, etc.). Anthropogenic disturbance or modification of the environment, by disrupting ecosystems or modifying habitats, enables colonization by alien species, and human-related vectors create continuously active pathways to maintain alien propagule pressure.

7.2 Terminology

A confusing array of terms has been used to categorize species introduced beyond their range of native occurrence. These include terms such as non-indigenous, exotic, alien, adventive, introduced, invasive, harmful and pest, and combinations such as invasive alien species, invasive marine species (IMS), invasive aquatic species and non-indigenous marine species (NIMS). To differentiate between species that do or do not cause harm in the new environment, whether environmental, economic, social or to human health, the following definitions of invasive species have been proposed (Russell & Blackburn, 2017; Sammarco *et al.*, 2010):

An invasive alien species is:

- 'a species that is established outside of its natural past or present distribution, whose introduction and/or spread threaten biological diversity' (IUCN, 2017);
- 'an alien species whose introduction does or is likely to cause economic or environmental harm or harm to human health' (US Executive Order 13112, 1999);
- 'a species that is non-native to the ecosystem under consideration and whose introduction causes or is likely to cause economic or environmental harm or harm to human health' (Invasive Species Advisory Committee, 2005).

Common to each of these definitions of 'invasive species' is the criterion that they cause harm within the invaded environment.

For non-native species that do not meet the above criteria as 'invasives':

- 'An alien species is a species introduced by humans – either intentionally or accidentally – outside of its natural past or present distribution' (IUCN, 2017);
- An 'alien species' is 'any species, including its seeds, eggs, spores, or other biological material capable of propagating that species, that is not native to that ecosystem' (Invasive Species Advisory Committee, 2005).

Invasive species are therefore the subset of alien species that cause harmful impacts. While seemingly adding clarity, these definitions do not assure this, as there can be differences in defining or determining what constitutes 'harm' or a 'negative impact' (Russell & Blackburn, 2017; Tassin *et al.*, 2017). Interpretations of species effects can be subjective or, even when based on experimental or quantitative field studies, uncertain in relevance beyond the small scale of the observations.

7.3 Historical Movement

Movement of species to new regions and the colonization and establishment of viable populations in new regions are natural processes. Biogeographical and ecological studies on marine invasions agree that:

- There is continuous movement of species among areas and communities.
- Such movements almost always involve migrations from locations that are relatively species rich to those that are relatively species poor.
- Species that colonize new communities are generally accommodated by the native species that occupy the appropriate habitats (Briggs, 2012).

Invasions almost entirely occur from areas of high diversity to low diversity, although even the richest communities can be invasible, and the three-step process from invasion, to accommodation, to speciation is considered to have contributed to historical increases in global marine species diversity (Briggs, 2010, 2013). Competitiveness is highest in species-rich communities, so species from these communities can have traits that favour their survival in less rich communities (Briggs, 2013; Vermeij, 2005). However, native species almost always manage to retain part of their original habitat. The biodiversity of most marine communities is dependent on continuous invasions from sources with greater richness (Briggs, 2012).

However, anthropogenic vectors have enabled faster spread of species, particularly across evolutionary biogeographical barriers, such as over oceans, mountain ranges and climatic zones. Sailing the seas is far from an exception, dating back to when mariners first embarked on sea voyages, with vessels able to carry aquatic species across and between oceans and seas. Global commercial shipping trade dates back to Chinese shipping to the Pacific and Indian oceans in the tenth century, and European shipping traffic to African, American and Asian shores in the fifteenth century (Rius *et al.*, 2015). Shipping is, however, only one of many vectors responsible for the spread of aquatic species.

7.4 Aquatic Invasive Species Impact

The introduction of non-indigenous species (NIS) has been commonly quoted as one of the greatest environmental and economic threats and, along with habitat destruction, the leading cause of extinctions and resultant biodiversity decreases worldwide (Pimentel *et al.*, 2000; Vitousek *et al.*, 1997). Similarly, alien marine species are quoted as one of the top five threats to marine ecosystem function and biodiversity, and of increasing threat to maritime industries (Carlton, 2001; Hewitt & Campbell, 2007).

For invasive species, Article 8h of the Convention on Biological Diversity obliges contracting parties to (as far as possible and appropriate) prevent the introduction of, control or eradicate those alien species which threaten ecosystems, habitats or species (www.cbd.int/convention/text/default.shtml). This obligation has prompted parties to the convention to devise policies aimed at identifying biological invasion pathways, carrying our risk assessments, listing NIS and invasive species and implementing pre-border prevention and post-border management of invasive species (Moreira *et al.*, 2014).

As a contribution to the Global Invasive Species Programme, the Invasive Species Specialist Group (ISSG) of the International Union for Conservation of Nature (IUCN) published a list of what were considered 100 of the world's worst invasive alien species (Lowe *et al.*, 2000). The aquatic species on this list, excluding those that are freshwater for the whole of their life, are the algae caulerpa seaweed (Aquarium weed; *Caulerpa taxifolia*), wakame seaweed (Asian kelp; *Undaria pinnatifida*) and the aquatic invertebrates Chinese mitten crab (*Eriocheir sinensis*), comb jelly (*Mnemiopsis leidyi*), fish hook flea (*Cercopagis pengoi*), European green crab (*Carcinus maenas*), Asian basket clam (*Potamocorbula amurensis*), Mediterranean mussel (*Mytilus galloprovincialis*) and Northern Pacific seastar (*Asterias amurensis*) (Table 7.1).

Table 7.1. *Geographical spread and suspected vectors for aquatic species on the ISSG top 100 list*

Species	Native range	Introduced range	Primary vector(s)	Secondary vector
Caulerpa taxifolia	North-eastern Australia	Mediterranean Sea (1984); California (2000); South Australia (2002); New South Wales (2004)	Aquarium trade	Vessel entanglement
Undaria pinnatifida	North-western Pacific	Mediterranean Sea (1970); New Zealand (1987); Australia (1988); Argentina (1996); California (2002)	Aquaculture (accidental); vessel biofouling; ballast water	Vessel biofouling
Eriocheir sinensis	North-western Pacific	Europe (1912)	Live import – aquaria; human consumption	Vessel biofouling, ballast water
Mnemiopsis leidyi	Western Atlantic	Black Sea (1980s); Caspian Sea (1999); north-western Europe (2006)	Ballast water	
Cercopagis pengoi	Eastern Europe	Baltic Sea (1992); Great Lakes (1998)	Ballast water	Fishing gear
Carcinus maenas	Europe	North America (1817); South America (1857); Central America (1866); Australia (late 1800s); South Africa (1983); Japan (1984)	Dry ballast	Ballast water; aquaculture (accidental); vessel fouling; fisheries (accidental); live import – aquaria; human consumption
Potamocorbula amurensis	North-western Pacific	California (1986)	Ballast water	
Mytilus galloprovincialis	Europe	Japan (1926); Hong Kong (1983); South Africa (1970s); Mexico	Vessel fouling; aquaculture (intentional)	
Asterias amurensis	North-western Pacific	Australia (1986)	Ballast water	Vessel fouling
Dreissena polymorpha	Central Europe	Russia (1801); UK (1820); north-western Europe (1826); Great Lakes (1988); Ireland (1993)	Accidental transport; ballast water	Vessel fouling (internal rivers and waterways)

Information on the ISSG species (Table 7.1) shows that the movement of some of these species predated modern shipping (e.g., *E. sinensis*, *C. maenas*, *Dreissena polymorpha*), but that an apparent pulse of invasions happened in the latter decades of the twentieth century. The movement of ballast water is linked to these latter movements.

Two of these species that are considered to have been translocated by shipping with significant consequent impacts are the North American comb jelly (*M. leidyi*) and the zebra mussel (*D. polymorpha*). The comb jelly is believed to have travelled in ships' ballast water from the eastern seaboard of the Americas to the Black, Azov and Caspian seas. By depleting zooplankton stocks and altering food web and ecosystem function, the species is considered to have contributed significantly to the collapse of Azov Sea, Black Sea and Caspian Sea fisheries in the 1990s and 2000s, with massive economic and social impact, although eutrophication, pollution and overfishing had severely reduced zooplanktivorous fish populations prior to the ctenophore outbreak and may have contributed to this impact due to the shortage of predators and competitors (Fuentes *et al.*, 2010; Purcell *et al.*, 2001).

The zebra mussel (*D. polymorpha*) has been transported from the Black Sea to western and northern Europe, including Ireland and the Baltic Sea, and the eastern half of North America. Travelling in larval form in ballast water, on release it has rapid reproductive growth and no natural predators in North America. The mussel multiplies and can foul all available hard surfaces in mass numbers. The species can alter habitats, ecosystems and the food web, and cause severe fouling problems on infrastructure and vessels. There have been high economic costs involved in unblocking water intake pipes, sluices and irrigation ditches.

7.5 Vectors for Introduction and Spread

7.5.1 Introduction Pathways

Unless an introduction is intentional, such as for the purposes of aquaculture, rarely can the introduction vector be identified with any certainty, as generally for marine species, a NIS is not detected until after colonization. The vector may be inferred; for example, aquaculture if non-indigenous macroalgae are first observed near oyster farms, or shipping if the species or similar species have been found in ballast water samples or biofouling on a hull. On rare occasions there can be reasonable certainty, such as the *Perna viridis* introduction to Cairns in tropical north-east Australia on an impounded vessel (Stafford *et al.*, 2007), the movement of *Didemnum vexillum* on a barge from the north of the North Island of New Zealand to the South Island (Coutts & Forrest, 2007) or the observation of *U. pinnatifida* on Japanese fishing boats in Wellington Harbour, New Zealand (Hay & Luckens, 1987).

However, most often the confidence in assigning a vector is an estimate of likelihood. For NIS in the Baltic Sea, in only 14 per cent of cases is the introduction pathway known with a high level of confidence, in 21 per cent is the pathway considered highly likely, 52 per cent possible and 13 per cent unknown (Ojaveer *et al.*, 2017).

7.5.2 Non-Shipping Vectors

7.5.2.1 Natural Dispersal

Most aquatic species utilize the watery medium within which they live to facilitate the maintenance and spread of their populations. Holoplanktonic species spend their entire lives drifting with the tides and currents, and many benthic species release reproductive propagules for spread in the water or have planktonic life stages to facilitate spread before the organism settles onto the sea floor or other substrate to complete its life cycle.

The extent of the natural dispersal of a species depends on the longevity of the planktonic or free-floating dispersal stage of the species, currents and tides and the physical and chemical suitability of the transited and recipient environments for species survival and reproduction. Natural dispersal mechanisms enable the localized or regional spread of all species, whether within the home range of a native species or from the point of colonization of an alien species. More distant dispersal can also be facilitated by rafting on natural debris, such as volcanic pumice and wood (Barnes, 2002) and kelp (Lewis *et al.*, 2005).

7.5.3 Aquaculture and Stocking

Some of the most significant IMS have spread following their introduction to a region for aquaculture. Of particular note is the Pacific oyster, *Magellana gigas* (syn. *Crassostrea gigas*), which was introduced from the north-west Pacific to North America, Europe and Australia to establish commercial oyster farming industries. In Australia, the oyster has spread to the wild, as it has in Ireland (Kochmann *et al.*, 2012), and it is now established in embayments around the south-eastern Australian coast and Tasmania. Similarly, the seaweeds *Gracilaria salicornia* and *Kappaphycus alvarezii*, which were introduced to Hawaii for mariculture, have since invaded reef environments (Rodgers & Cox, 1999), and the blue mussel *M. galloprovincialis*, which was introduced to the south coast of South Africa for aquaculture purposes in 1988, has since spread along more than 2000 km of the South African coastline (Rius *et al.*, 2011; Robinson *et al.*, 2005). Hybrid zones of invasive (*Mytilus edulis*, *M. galloprovincialis*) and native (*Mytilus trossulus*, *Mytilus californianus*) mussels have formed in areas of intense farming on the Pacific coast of Canada, alien mussels are expanding in some open coastal areas (Crego-Prieto *et al.*, 2015) and populations of *U. pinnatifida* in harbours of Europe are derived from cultivation area escapees (Voisin *et al.*, 2005).

Many species have also been unintentionally introduced through association with oyster transport, either as hitch-hikers on the oyster shells or in packing material. In Europe, the algae *Codium fragile* subsp. *fragile*, *Sargassum muticum*, *Grateloupia turuturu* and *U. pinnatifida* and the molluscs *Crepidula fornicata*, *Arcuatula senhousia* and *Rapana venosa* are considered to have been introduced with oysters. *M. gigas* is thought to have first been introduced to Europe from Taiwan in the sixteenth century, with more recent large-scale importations from British Columbia and Japan during the 1970s, following an industry collapse due to disease (Mineur *et al.*, 2007a). A review of marine species introductions to Europe from the north-west Pacific suggested that at least 48 of 68 species were attributable to oyster stock importation and that the discovery rate from 1966 to 2012 suggested continuous vector activity over this period (Mineur *et al.*, 2014). Macroalgae, in particular, were associated with oyster translocations, as only a few tolerant groups are linked to vessel fouling or ballast water (Flagella *et al.*, 2007; Mineur *et al.*, 2007a). Pathogens and diseases, some of which have and can greatly impact oyster production, are also associated with oyster imports and movement (Mineur *et al.*, 2014). European oyster farming continues to involve transfers of livestock between farms, and experimental studies to simulate such transfers identified 57 algal taxa on shells that had initially no visible macroalgal presence (Mineur *et al.*, 2007a). The spread of *C. fragile* subsp. *fragile* and *G. turuturu* in south-east Australia also seems to have originated from an oyster-farming region on the east coast of Tasmania (Provan *et al.*, 2005; Saunders & Withall, 2006). In the 1960s and 1970s there was a deliberate release of Ponto-Caspian sturgeons, amphipods, mysids and Pacific salmonids into the Baltic Sea (Ojaveer *et al.*, 2017). The intention of this stocking was to elevate productivity to increase commercial fish stocks, but none of the fish developed self-sustaining populations.

7.5.4 Canals

The opening of both the Suez and Panama canals provided a conduit for species to move between previously biogeographically separated seas: for the Suez between the Red and Mediterranean seas and for the Panama between the Pacific and Atlantic oceans. Many marine species are known to have migrated into the Mediterranean after the Suez Canal was opened in 1869, and to date more than 300 alien species are considered to have made this passage (Boudouresque, 1999; Rius *et al.*, 2015). Many have established thriving populations along the Levantine coast from Libya to Greece (Galil *et al.*, 2015), and some marine communities in the eastern Mediterranean have become dominated by Red Sea species (Nawrot *et al.*, 2015).

This movement is known as the Lessepsian migration or Erythrean invasion, and this migration is largely unidirectional (Boudouresque, 1999). In 2009, 138 Erythrean molluscs were known in the Mediterranean, but not one Mediterranean species was known to be established in the Red Sea (Rilov & Galil, 2009). Molluscs have become the most species-rich animal phylum of NIMS in the Mediterranean Sea (Nawrot *et al.*, 2015). A total of 84 Indo-Pacific fish species have invaded the eastern Mediterranean since the Suez Canal opened, and 55 of these have established permanent populations in the Mediterranean (Edelist *et al.*, 2013).

Far fewer species are known to have crossed between the Pacific and Atlantic oceans through the Panama Canal (Boudouresque, 1999). The canal does have a lock system that segments the canal, and also freshwater lakes along its course, which would provide a barrier to the natural movement of species. However, there are species that have moved from the Atlantic to the Pacific, such as the tubeworm *Hydroides sanctaecrucis* (Bastida-Zevala & Ten Hove, 2003; Lewis *et al.*, 2006), and from the Pacific to the Atlantic, such as the cup corals *Tubastraea coccinea*, *Tubastraea tagusensis* and *Tubastraea micranthus* (Mantelatto *et al.*, 2011; Sammarco *et al.*, 2015).

7.5.5 Aquarium Trade

A third of the world's worst aquatic invasive species have been noted to be aquarium or ornamental species (Padilla & Williams, 2004). From the multimillion dollar global aquarium trade, escape of alien species into the environment can be from the discarding of unwanted species arriving in 'live rock', drainage or breakage of tanks or release of organisms because they have grown too large, are no longer wanted or the novelty of maintaining an aquarium has gone (Calado & Chapman, 2006).

The spread through the Mediterranean Sea of the 'aquarium weed' *C. taxifolia*, a tropical Australian native green alga, is believed to have originated with the escape of the species from the Monaco Aquarium (Meinesz, 1999). This invasive species has also established in coastal embayments outside of its native range in Australia, in South Australia and along the New South Wales coast (Glasby, 2013).

A second notable aquarium escapee is the Indo-Pacific lionfish (*Pterois volitans* and *Pterois miles*), which has established and spread through reef environments along the south-east coast of America, the Caribbean and in parts of the Gulf of Mexico (Morris & Whitfield, 2009). There is speculation that people have dumped unwanted lionfish from home aquaria into the Atlantic Ocean for more than 20 years.

7.5.6 Human Transport

Separate to introductions for aquaculture, there are suggestions that aquatic species have been intentionally and illegally introduced to new locations due to their attraction as food delicacies. The Vikings are considered to have transported the American soft shell clam (*Mya arenaria*) to Europe, and Europeans the edible European periwinkle (*Littorina littorea*) to North America in the early 1800s (Carlton, 1999). The introduction of the Chinese mitten crab to Europe is another example, although it is possible that this species was alternatively transported in ship ballast water (Herborg *et al.*, 2005, 2007).

7.5.7 Flotsam

In addition to the dispersal of species on natural debris, flotsam has been ranked with ship hulls as the highest of marine anthropogenic objects in terms of size and extent of biofouling, with ship hulls estimated to account for 24 per cent of the fouled surface area and 85.5 per cent of the total biomass, and flotsam 70 per cent of the area but only 5.6 per cent of the biomass (Railkin, 2004). The flotsam is predominantly drifting plastics, with massive amounts now entering the ocean, either from land and washed into the ocean or as fishing or other vessel debris lost or dumped offshore. Animals found on plastic debris include bryozoans, barnacles, polychaete worms, hydroids and molluscs (Barnes, 2002; Barnes & Fraser, 2003; Barnes & Milner, 2005). However, particularly in high latitude regions, the predominantly oceanic source of the debris results in the transport of mostly cosmopolitan pelagic species, rather than coastal benthic species that are likely to colonize new shorelines (Lewis *et al.*, 2005). The passive dispersal of plastic debris by wind and currents is also likely to replicate natural dispersal pathways, such as on kelp, pumice or wood.

The 2011 Great East Japan Earthquake and resultant tsunami led to what could be considered an extreme example of NIS transport on flotsam. Debris, mostly non-biodegradable objects, drifted across the Pacific Ocean to the shores of North America and Hawai'i, and this is considered the longest transoceanic survival and dispersal of coastal species by rafting (Carlton *et al.*, 2017). A total of 289 living invertebrate and fish species and 49 macroalgal species have been identified on the debris (Carlton *et al.*, 2017; Hanyuda *et al.*, 2018).

7.5.8 Shipping Vectors

7.5.8.1 Dry Ballast

Ballast is used in ships to provide stability and to compensate for weight changes in cargo loads, or from fuel and water consumption. In the days of sail, beach sand, pebbles, cobbles and rocks collected from seashores were commonly used as ballast, and often landed in the destination port (Carlton, 1985). Dry ballast has been implicated in the transfer of numerous marine species (Campbell & Hewitt, 1999). The introduction of the European green or shore crab (*C. maenas*) to southern Australia from Europe in the late 1800s (Fulton & Grant, 1900) could have been associated with dry ballast, although association with heavy marine growth on the hulls of timber sailing ships was also proposed (Fulton & Grant, 1902). Semi-dry ballast is also considered the most likely vector for the introduction of the littoral amphipod *Corophium volutator* to North America from Europe in the late 1800s (Einfeldt & Addison, 2015).

7.5.9 Wet Ballast

With the introduction of steel-hulled ships, ballast tanks for holding seawater ballast became the norm, and the regular use of seawater as ballast is considered to have begun in the late 1870s or early 1980s (Carlton, 1985). Ballast water is taken aboard as ships unload and discharged as ships load. During take-up, planktonic, nektonic and free-floating organisms are entrained within the water and captured within the ballast tanks, where they may survive for the duration of the voyage.

Although the first suspicions of ballast water facilitating the translocation of marine organisms go back to 1908 (Ostenfeld, 1908), the first significant review of the history of biogeographical and ecological interest in ballast water as a dispersal mechanism, the hypothesis of ballast water transport and the evidence for such transport are attributable to Carlton in 1985. Carlton concluded that, for an increasing number of taxa in a wide range of organism groups, ballast water transport in modern times provided a hypothesis for the patterns of distribution in estuarine and shelf waters around the globe. Sampling of ballast water on ocean-going ships subsequently confirmed the presence of a diverse assemblage of organism after transoceanic voyages (Carlton & Geller, 1993; Gollasch, 2007). Unlike most other NIS vectors, ballast water entrains entire ecological communities, including viruses, bacteria, phytoplankton, zooplankton, fish and macrophytes (Minton *et al.*, 2005; Ruiz *et al.*, 2000). Ports in regions that export bulk commodities may be particularly susceptible to species incursions because tankers and bulk carriers deliver the greatest volumes of ballast water and this water is discharged at the export port prior to loading (Ruiz *et al.*, 2015).

Although Carlton's 1985 review raised awareness of the potential ballast water issue, infamous invasions in the late 1980s and early 1990s brought the issue to the fore (Davidson & Simkanin, 2012). As previously in Section 7.4, the introductions of the ctenophore *M. leidyi* from North America to Europe, firstly to the Black, Caspian and Mediterranean seas beginning in the late 1980s and then a second introduction to the North and Baltic seas starting in 2006, have been attributed to translocation in ships' ballast water (Bayha *et al.*, 2015). This holoplanktonic species has high potential reproductive, growth and feeding rates, and its invasion potential is enhanced by the ability to self-fertilize (Bayha *et al.*, 2015; Purcell *et al.*, 2001). Blooms of the species reduce plankton populations, including icthyoplankton, and particularly fish eggs (Purcell *et al.*, 2001). After the invasion of *Mnemiopsis* in the Black Sea, there was a precipitous decline in the numbers of mesozooplankton and, although the abundance and diversity of ichthyoplankton were already reduced before the first *Mnemiopsis* outbreak, further reductions in anchovy eggs and larvae were attributed to the ctenophore (Purcell *et al.*, 2001). The population size of *M. leidyi* in the Black Sea has decreased significantly during the last decade (Finenko *et al.*, 2018).

Ballast water and ballast tank sediments were also vectors for the movement of microorganisms. including viruses, bacteria and microalgae (Drake *et al.*, 2001; Hallegraeff & Gollasch, 2008; Hewitt *et al.*, 2009; Ruiz *et al.*, 2000). Microorganisms occur in the ballast water, tank sediments, residual water and biofilms on the interior tank surfaces (Drake *et al.*, 2007). Organisms introduced in this way can cause health impacts on marine life and humans, particularly as harmful algal blooms (HABs) and paralytic shellfish poisoning (PSP) (Hallegraeff, 1992, 1993). Human consumption of shellfish contaminated with PSP can lead to hospitalization and death. Outbreaks of PSP were originally restricted to the temperate waters of Europe, North America and Japan, but started to occur in the Southern Hemisphere in the 1970s (Mohammad-Noor *et al.*, 2018). The arrival of the toxic dinoflagellate *Gymnodinium catenatum* in Tasmanian waters and subsequent incidence of PSP causing aquaculture farm closures are linked to ballast water (Hallegraeff 1998; McMinn *et al.*, 1997). PSP was unknown from the Australian region until the 1980s, when outbreaks happened in the ports of Hobart, Melbourne and Adelaide (Hallegraeff, 1993). The human cholera bacterium *Vibrio cholerae* was found in the ballast of vessels arriving in the Gulf of Mexico from South America where there was a cholera epidemic (McCarthy & Khambaty, 1994; McCarthy *et al.*, 1992). Species detected in ballast water on ships arriving in Australian and British ports include the potentially toxic diatom *Pseudo-nitzschia*, the PSP dinoflagellates *Alexandrium catenella*, *Alexandrium minutum* and *Alexandrium tamarense* and the potentially ichthyotoxic dinoflagellate *Pfiesteria piscicida* (Hallegraeff & Gollasch, 2008), and in ships arriving in China, surveys found 17 red tide-causing algae (5 noxious) and 22 species of pathogenic bacteria (Wu *et al.*, 2017). Similarly, of 78 phytoplankton species identified in ballast water on ships arriving in Algeria, 11 species deemed harmful or toxic were found (Cheniti *et al.*, 2018).

Ballast water is considered the most likely vector for the introduction of the Northern Pacific seastar (*A. amurensis*) from the Northern Pacific to south-eastern Australia (Talman *et al.*, 1999). Traits conducive to uptake and transport of *Asterias* larvae in ballast water are the long larval life and abundance of these pelagic larvae in the upper 5 m of the water column. The species was first detected in the Derwent River estuary in south-east Tasmania in the early 1980s, and the population had grown to an estimated 28 million by the mid-1990s. *Asterias* was first detected in Port Phillip Bay on the southern coast of the Australian mainland in 1995 and, up until early 1998, only four adults were collected (Talman *et al.*, 1999). The population first spawned in Port Phillip Bay in 1997 and then, by 2000, the population grew to an estimated 165 million (Parry *et al.*, 2004).

Within its native range in Japan, *Asterias* has caused significant damage to commercial shellfisheries, and similar impacts were anticipated in Australia (Talman *et al.*, 1999). However, despite the size of the populations, evidence of severe impacts is yet to be documented, although it does prey on commercial species including scallops, mussels and clams (Lockhart & Ritz, 2001; Ross *et al.*, 2002). In Tasmania, scallop losses were seen in scallop spat bags (Talman *et al.*, 1999), but this was subsequently managed by ensuring bags were suspended above the sea floor. In Port Phillip Bay, declines in the abundance of several fish species have been linked to competition for food with *Asterias* (Parry & Hirst, 2016).

The abundance and diversity of zooplankton in ballast tanks decrease over time, particularly over the first days, but invariably some do survive (Desai *et al.*, 2018; Gollasch *et al.*, 2000a). In addition to some zooplankton surviving for several months, abundance can also increase through reproduction within the tank (Gollasch *et al.*, 2000b). The planktonic stages of photosynthetic diatoms and dinoflagellates have limited survival times in dark ballast tanks, but form resistant resting spores that settle and survive in the tank sediments (Hallegraeff & Bolch, 1991, 1992). Sampling of ballast tank sediments confirmed the presence of cysts, with the number of cysts in the ballast tank of one ship that had filled during a toxic dinoflagellate bloom estimated to be over 300 million (Hallegraeff, 1993; Hallegraeff & Bolch, 1991). Rough weather during a voyage can re-suspend particulate matter, and this has been associated with a subsequent increase in bacteria and then phytoplankton in the ballast water (Desai *et al.*, 2018).

Zooplankton concentrations in unmanaged ballast water have been found to have a skewed distribution, with one study finding most samples containing <3000 organisms/m^3 but 1 per cent of ship samples containing $>50,000$ organisms/m^3 (Minton *et al.*, 2005). The first ballast water management measures introduced required open ocean ballast water exchange (BWE), which results in approximately 90 per cent volumetric exchange of the water mass and 70–95 per cent removal of zooplankton (Bailey *et al.*, 2011). Osmotic shock adds to the effectiveness for transfers between freshwater bodies, leading to a further reduction of 40–88 per cent of taxa (Santagata *et al.*, 2008). At high zooplankton concentrations in marine or estuarine-sourced water, BWE can still leave a high residual concentration of zooplankton, with the inocula from a single ship proportional to the volume of water discharge. Total invasion risk further depends on the additive effect of the size and frequency of discharges into a single port system (Minton *et al.*, 2005).

Although BWE reduces the risks associated with discharge, not all source port organisms are removed, particularly if they are present in tank sediments (Hallegraeff & Gollasch, 2008; McCollin *et al.*, 2008). Although BWE reduces organism concentrations and per-ship inocula size, precise prediction of invasion risk remains uncertain because the risks associated with specific levels of propagule supply are not known (Minton *et al.*, 2005). Variability and complexity in both shipping and species composition within ballast, together with a lack of knowledge on the environmental tolerances and ecological requirements of ballast organisms, result in uncertainty regarding the dose–response relationships that would facilitate colonization and invasions (Minton *et al.*, 2005). As a consequence, ballast water treatment systems (BWTSs), which remove or kill organisms during ballast uptake, are now recommended for more effective ballast water management.

7.5.9.1 Biofouling

Any surface immersed in the sea becomes a potential settlement site for marine organisms, with the hulls of vessels being no exception. Since mariners first set sail, their boat hulls have been a welcome home for many species that like a floating, peripatetic lifestyle on substrates initially free of established communities and epibenthic predators. As observed by Hadfield (1999), the marine world of 10,000 years ago was not characterized by ships, barges, docks, floats and piling. Most of the invertebrate species typical of the fouling community are never found elsewhere. Most exist only on substrates where tidal exposure does not occur, and in the pre-maritime human environment, this habitat must have been restricted to natural floating materials, mainly the drift logs that were most abundant in bays and estuaries. The character of the hull habitat, with few natural analogues, has since globally fostered an association of organisms with a composition and characteristics unique to that community. The stresses of the peripatetic lifestyle inhibit the regular patterns of ecological succession, and complex communities often only develop in protected niche areas.

The effect of biofouling to the ship operator is an increase in hull friction, degradation of performance and added fuel and maintenance costs (Demirel *et al.*, 2017; Schultz, 2007; Schultz *et al.*, 2011). A further consequence is the increased emission to the atmosphere of environmentally harmful gases, notably sulfur and nitrogen oxides and carbon dioxide, from the added fuel. Anti-fouling coatings have been the primary defence against biofouling since the mid-nineteenth century, but despite significant advances in anti-fouling technology over the last 50 years, they still have limitations in effective life, cost and efficacy for both hulls and within hull niches. The vast array and taxonomic diversity of potential biofouling species, combined with environmental concerns and consequent limitations on the use of broad-spectrum biocides, ensure there is no simple solution.

Many translocations of marine species predated biological surveys of marine regions, so the magnitude of the phenomenon was not widely realized. At the advent of marine taxonomy in the eighteenth and nineteenth centuries, many species labelled as native are now considered likely to have arrived in historical times (Rius *et al.*, 2015). The distribution of some such species was thought to be cosmopolitan, particularly for harbour-dwelling organisms. One example is a species of the red algal genus *Grateloupia.*

Grateloupia filicina, first described from the Adriatic Sea, was subsequently recorded as occurring in Europe (England, Wales, France, the Mediterranean Sea), Africa (Ghana, South Africa, Kenya, Mauritius), Asia (India, Sri Lanka, China, Vietnam, Indonesia, Hong Kong, Japan, the Philippines), the Pacific (Australia, Hawai'i), and the Americas (the USA, Mexico, the Antilles, Trinidad, Jamaica). In 1906, *G. filicina* var. *luxurians* was described from Sydney Harbour (Gepp & Gepp, 1906), and the entity was found to be common in Port Phillip Bay in south-eastern Australia when this flora was surveyed in the 1960s (King *et al.*, 1971; Womersley, 1966), then later documented as a fouling species in the Port of Melbourne (Russ & Wake, 1975). The variety was recorded in the British Isles in 1968 and considered to be an immigrant from Australia (Farnham & Irvine, 1968). As with many groups of organisms, molecular and genetic techniques are now enabling the resolution of the true identity and phylogenetic relationships of many taxa. One such study revealed that *G. filicina* var. *luxurians* corresponded to the Japanese species *Grateloupia subpectinata* (Verlaque *et al.*, 2005). The migration story for this taxon thus changes from one of an Australian species being translocated to England, to one of a Japanese species being translocated to Australia and Europe (Guiry & Guiry, 2018). Molecular techniques have also found broader cryptic diversity within entities previously assigned to *G. filicina*, with species such as *Grateloupia asiatica*, *Grateloupia catenata*, *Grateloupia capensis* and *Grateloupia minima* split out, and the distribution of genuine *G. filicina* now limited to the Mediterranean basin (De Clerck *et al.*, 2005; Verlaque *et al.*, 2005).

Similar scenarios occur with species in the common biofouling serpulid tubeworm genus *Hydroides*, with the globally distributed *Hydroides elegans* first described from Sydney, Australia, but possibly being an introduction given earlier records in the Caribbean and Europe (Sun *et al.*, 2015; Zibrowius, 1971, 1973), and *Hydroides dianthus* described from the east coast of the USA, but with genetic diversity suggesting a possible Mediterranean native range (Sun *et al.*, 2017). *H. elegans* is now distributed through the Atlantic, Pacific and Indian oceans and the Mediterranean Sea, most likely due to dispersal on vessel hulls. *Hydroides* larvae have lifespans of less than 4 weeks, which limits spread by natural dispersal (Pettengill *et al.*, 2007), and the metamorphosis trigger of biofilms (Walters *et al.*, 1997) and the hormetic effect of copper (Lewis *et al.*, 1988, 2006) may promote settlement on ship hulls.

The occurrence of cryptic diversity in invasive species has also been detected through genetic analyses in the Ponto-Caspian hydroid *Cordylophora*, which has spread to western Europe, the east and west coasts of the USA and into the Great Lakes (Folino-Rorem *et al.*, 2009). The analyses revealed four separate lineages. Hybridization and introgression of taxa can also confound our understanding of marine species introductions. Molecular studies on the blue mussel *Mytilus* in southern Australia found that, while most displayed the southern Australian haplotype (*Mytilus planulatus*), Northern Hemisphere genotypes (*M. galloprovincialis*) were also detected (Ab Rahim *et al.*, 2016). This shows that

M. galloprovincialis has been widely introduced into Australia and both Southern and Northern Hemisphere lineages into Japan.

Shipping is cited as being the most important of all human-mediated transport pathways for NIS within the marine environment (Carlton & Geller, 1993; Hewitt *et al.*, 2009; Piola & McDonald, 2012; Ruiz *et al.*, 2015). After considerable attention was given to ballast water, attention has focused on the importance of biofouling on shipping. According to Otani (2006), biofouling on shipping is today the most important vector worldwide for marine NIS introductions. A survey of ballast water, sediment and hull samples on ships visiting German ports recorded non-native species in 38 per cent of the ballast water samples, 57 per cent of sediment samples and 96 per cent of hull samples (Gollasch, 2002). Hull fouling has also been suggested as a vector for not only organisms that attach directly to a hull surface, but also mobile species that can live within biofouling aggregations. As an example, two alien crab species were collected from within bryozoan colonies on the hulls of foreign tuna vessels in a fishing harbour in Spain (Cuesta *et al.*, 2016), a live individual of the Asian brush-clawed shore crab (*Hemigrapsus takanoi*) was found in empty barnacle shells in biofouling on a car carrier in Germany after a voyage from Japan and Korea (Gollasch, 1998) and an Indian Ocean crab (*Metopograpsus messor*) was found living in a recess on the hull of an Australian Navy ship on its return from the northern Indian Ocean (J. Lewis, pers. obs.).

Shipping has been considered an important vector for the translocation of macroalgae (e.g., Boudouresque & Verlaque, 2002; Wallentinus, 2002), but the evidence is usually indirect, based on a link between new introductions and harbours (Mineur *et al.*, 2007b). Surveys of the macroalgal flora on commercial cargo ships, however, found that the majority of fouling algae were of species with a known cosmopolitan distribution (e.g., species of brown Ectocarpales and green Cladophorales and Ulvales), and that macroalgal fouling communities on hulls had a high turnover and were short lived (Mineur *et al.*, 2007b).

More generally, the statement of importance of ship biofouling in human-mediated introductions may be true with respect to NIMS, but not so with respect to harmful IMS. Hull fouling is characterized by species that have the opportunistic traits and environmental tolerances to survive on these artificial habitats, and members of this community have, through the maritime era, been distributed into disturbed and artificial environments worldwide. These species contribute the majority of alien marine species found in any country, in their harbours and marinas and on their vessels and maritime infrastructure. Typical biofouling species include algae, barnacles, hydroids, bryozoans, polychaetes, ascidians and bivalve molluscs (Davidson *et al.*, 2009; Frey *et al.*, 2014; Inglis *et al.*, 2010). Gollasch (2002) observed that many of the non-native species found on the hulls of ships in Germany were distributed, presumably by shipping, over a wide geographical range, and many were either native or previously introduced to the north-east Atlantic Ocean. On an international ship, every fouling species is a NIS as the ship is an artificially created, newly formed, mobile island.

A comprehensive listing of marine species recorded as foulers on artificial surfaces by the Woods Hole Oceanographic Institution in 1952 listed 374 species from ships with the following most common groups (in order of species richness): hydroids, bryozoans, barnacles, bivalve molluscs and tunicates (Hutchins, 1952). In comparison, over 800 species were reported from navigational buoys, with the most common groups being hydroids, red algae, green algae, tunicates, brown algae, amphipods and decapods. The latter represent older, more structurally complex benthic communities not exposed to the rigours of shipboard life. A survey of intertidal communities in Port Phillip Bay in Australia identified 279 species, with the most common groups being red algae, brown algae, gastropod molluscs, green algae and decapods (King *et al.*, 1971), more closely aligning with buoy than ship fouling communities.

A survey of the biofouling composition on more than 500 international vessels arriving in New Zealand found that biofouling organisms present on 72 per cent (Inglis *et al.*, 2010). A total of 187 species were identified, of which 128 were not native to New Zealand, and 94 of these were not known to be established in New Zealand. The highest numbers of species on merchant and passenger ships were barnacles (51 per cent), and on yachts were bryozoans (23 per cent) and barnacles (23 per cent). The high number of species on ships deemed non-indigenous but not established in New Zealand, given the strength and continuity of the ship-fouling vector over more than a century, is suggestive of the limited association between ship biofouling and the biosecurity risk of IMS, despite this being a common assumption (e.g., Davidson *et al.*, 2016). Studies in Canada similarly found a mismatch between species composition in ports and on visiting ships (Sylvester *et al.*, 2011). Of 78 species identified in samples from ship hulls, about 90 per cent were not recorded in the sampling ports. Although it has been proposed that this indicates a substantial invasion risk, the authors also acknowledged that, despite centuries of carriage, the ship-fouling species were possibly poor colonizers incapable of establishment in new regions (Sylvester *et al.*, 2011).

A specific example is the likelihood of invasion into Australian waters of the Asian green mussel, *P. viridis*. This species has been treated by Australian authorities as a major risk due to the species invading several regions elsewhere in the world and temperate matching indicating that large parts of the Australian coastline would be suitable for the species to establish (Heersink *et al.*, 2014). However, despite high propagule pressure over more than 50 years, only a single incursion is known, and the species did not establish. By analysing various information sources, Heersink *et al.* (2014) demonstrated that the likelihood of *P. viridis* was low and declining, due either due to hurdles faced during the initial colonization phase or to a factor causing the environment to be unsuitable to the species.

Gollasch (2002) propounded that all non-native species are potentially harmful unless it is shown that the risks involved are low or that the introduction of the species is beneficial, and that this uncertainty emphasized the need for a precautionary approach in order to minimize unintentional transport of species. Of species introduced to the North Sea, most are considered to have been introduced by ships (47 species) or by aquaculture (36 species), with close to two-thirds of the ship-associated introductions attributed to fouling.

However, most fouling-vectored alien marine species that colonize new regions do not cause significant environmental, economic, human health or social impacts, except as a contribution to the overall biofouling impact on vessel performance, mariculture infrastructure, coastal industrial plants, etc. In Europe, of 19 alien marine species considered invasive in DAISIE (2009), only two are considered to have been introduced to Europe by vessel biofouling: the bay barnacle *Amphibalanus improvisus* in 1844 and the club sea squirt *Styela clava* in the 1950s (Lewis, 2016). Other introductions were associated with aquaculture (nine species), ballast water (three species), the aquarium trade (two species), through the Suez Canal (two species) and intentionally for human consumption (one species).

In Australia, from an assessment through business cases of the impacts of more than 300 documented NIMS, the development of species management plans was recommended for only six: the Northern Pacific seastar (*A. amurensis*), European green crab (*C. maenas*), Asian bag mussel (*A. senhousia*), Mediterranean fan worm (*Sabella spallanzanii*), Asian kelp (*U. pinnatifida*) and the European basket shell clam (*Corbula gibba*). Of these, three are possible or likely introductions via fouling, *A. senhousia*, *S. spallanzanii* and *U. pinnatifida*. Harmful impacts from these species in the >20 years since their arrival are not well documented.

Transocean IMS introductions appear to be extremely rare events considering the number of ships trading on international routes. Two main factors influencing the abundance of biofouling on a vessel hull are the time since dry-docking, on a scale of months to years, and port residence time, on a scale of hours to days, although other factors including ship age, hull complexity, voyage history and speed are likely to also have influences (Davidson *et al.*, 2009). Today's container ships travel at relatively high speeds with short port stays (<1 day), which is likely to reduce the risk of species transfers on a per-ship basis (Davidson *et al.*, 2009). However, the operational profile of bulk carriers, which mostly operate on charter, may spend extended periods queuing to load (Otani *et al.*, 2007) or laid up in the absence of work (Coutts & Taylor, 2004; Floerl & Coutts, 2009), which can increase the time window for colonization.

Determining the risk of a particular vessel introducing an invasive species has been related to both the vessel origin and pathway (Hewitt *et al.*, 2011) and the abundance of biofouling present on the hull (Floerl *et al.*, 2005a). The reliability of both depends on a common underlying assumption: that the vessel or biofouling will harbour a potential IMS. If biofouling is present, then it is almost a surety that NIS will be part of this community. However, as discussed earlier in this section, the likelihood of the presence of a potential IMS is extremely low for vessels on international voyages. Coastal and recreational vessels, particularly low-activity vessels and/or vessels with long residence times, on short voyages within or between coastal bioregions do pose a risk of spreading nuisance biofouling NIS between regional ports and embayments (Minchin *et al.*, 2006).

The total wetted surface area (WSA) of active commercial ships in the world fleet was estimated, as this was considered an 'important metric' that would represent the potential for transport of 'aquatic nuisance species' (Moser *et al.*, 2016). The result was 325×10^6 m^2, with bulk carriers and tankers accounting for the majority of the WSA, and that this greater than previously estimated WSA was consistent with biofouling being a major vector and 'contributing strongly, if not driving, the observed rapid increase in marine invasions in

North America and elsewhere around the globe' (Moser *et al.*, 2016, p. 276). However, in discussing this result, the authors do note that biofouling on active vessels is generally present on only a small portion of the WSA, due largely to the use of anti-fouling coatings, and that growth in niche areas may represent a greater risk for species transfers than more uniform hull surfaces. A study of container ships found that, on almost all vessels surveyed, less than 1 per cent of the hull was colonized by macroorganisms, although taxonomic richness was high (Davidson *et al.*, 2009). The fouling present was not evenly distributed, with a wider range of organism in niche areas than on exposed hulls.

WSA was also calculated in terms of annual flux to marine bioregions of the USA from international vessels and vessels on coastal voyages that transited US bioregions (Miller *et al.*, 2018). WSA was considered a useful proxy for direct biological measurement by representing the potential for ships to transfer biofouling organisms and how this may be partitioned by vessel type, source or recipient regions and time, although the extent of biofouling on WSA is highly variable and influenced by the operational history and biofouling management measures on each vessel. The calculated result was an annual mean flux from overseas of 333 km^2/year and between US bioregions of 177 km^2/year.

Biofouling abundance and species richness are not evenly distributed over the submerged surface of a vessel hull. Although the outer, immersed hull of a vessel presents the greatest potential surface for colonization, anti-fouling paints usually prevent the recruitment and survival of biofouling macroalgae and invertebrate species on surfaces coated with an active anti-fouling system. Because anti-fouling coatings limit macrofouling settlement and growth over most of the WSA, WSA is likely to be a better proxy measure for microbial organism transfers, as biofilms develop on almost all wetted surfaces (Miller *et al.*, 2018). However, as for macro-fouling communities on ships, the biofilm composition on hulls is structured by the type of surface and biocides present on that surface (Callow, 1986; Molino & Wetherbee, 2008; Molino *et al.*, 2009).

However, biofouling can accumulate on coated surfaces if the anti-fouling biocide is depleted, biocide release is obstructed by insoluble surface precipitates or deposits or the paint is mechanically damaged or otherwise physically unsound. The highest levels of biofouling accumulation and richness are commonly found in 'niche' areas on and around the hull, such as: (1) on unpainted surfaces such as cathodic protection anodes, propellers and propeller shafts, rudder stocks, box coolers and docking block positions; (2) where anti-fouling paint is lost or damaged in areas of high turbulence and cavitation, such as propeller stocks, rudders, intake grates, bilge keels and around bow thrusters; and (3) in protected areas of low flow in hull recesses, such as sea chests, thruster tunnels, moon pools and spud cans, and around recessed complex equipment, such as box coolers, thrusters and emergency propulsion units (ASA, 2006; Coutts & Taylor, 2004; Davidson *et al.*, 2016; Lewis, 2002b; J. Lewis, pers. obs.). The latter are prone to biofouling because effective anti-fouling paint application is difficult or water flow is insufficient to maintain effective biofouling release from self-polishing anti-fouling or foul-release coatings (Otani *et al.*, 2007). Vessel sea chests have been found to be particularly vulnerable to biofouling (Coutts & Dodgshun, 2007; Frey *et al.*, 2014; Minchin, 2002). Of the species collected on international vessels arriving in New Zealand, more than 80 per cent were recorded from niche areas, including around bow thrusters, gratings, steering and propulsion gear and docking block marks (Inglis *et al.*, 2010).

The total surface area of niche areas in the global commercial ship fleet has been estimated from shipping data archives and standard ship designs for nine ship types (Moser *et al.*, 2017). The niches considered were bilge keels, dock block marks, propellers, rudders, sea chests and gratings and thrusters, thruster tunnels and tunnel gratings. The total surface area of these niches amounted to approximately 10 per cent of the total hull WSA of the global fleet, with thruster tunnels, dock block marks and bilge keels contributing over three-quarters of the calculated area (Moser *et al.*, 2017). As acknowledged by these authors, surface area alone does not necessarily correlate with risk, which requires a deeper understanding of the biota colonizing these various niches and the factors that influence biofouling settlement, survival and growth, such as the type and efficacy of anti-fouling coatings, ship speeds, port residence times and environmental conditions in ports of origin and destination and during the voyage.

7.6 The Invasion Process on Shipping

For marine species to be successfully translocated from a donor region and to establish and cause impact in a recipient region, they must pass through and overcome a sequence of events (Floerl, 2002; Lewis & Coutts, 2009; Schimanski *et al.*, 2017). The species must: (1) be entrained in ballast water during uptake or colonize and establish on a vessel's hull in a donor region (recruitment); (2) survive the voyage within or on a vessel from the donor to the recipient region (translocation); (3) be discharged, reproduce or be dislodged in the recipient region (transfer); (4) colonize available habitat in the recipient region (colonization); (5) be able to complete its life cycle in the recipient region (establishment); and (6) spread and cause harm to the environment, economy, human health or society in the recipient region (invasion). At each step of the sequence, there is attrition or selective filters (factors) that affect the total number of organisms and species that transition successfully (survive) to the next stage.

The biofouling process on vessels is generally similar to that on other immersed natural and artificial surfaces: initial surface conditioning and microbial biofilm formation, followed by primary surface colonization by sessile macroalgae and invertebrates, then secondary colonization of the biofouling-formed macrohabitat by epibionts, mobile invertebrates and sometimes small fish (Lewis, 1998; Richmond & Seed, 1991; Wahl, 1989, 1997). On a vessel hull, significant environmental influences that differ from other substrata are the presence of anti-fouling treatments and the movement and activity of the vessel. Anti-fouling treatments can prevent the recruitment of most microbial, algal and invertebrate species, but favour a few with natural or evolved resistance to anti-fouling biocides (e.g., Callow, 1986; Evans, 1981; Lewis, 2002a; Reed & Moffat, 1983; Russell & Morris, 1973).

However, a range of factors are capable of influencing the success of biofouling translocation, including port residency time and transit speed, duration and route (Coutts,

1999; Floerl, 2005; Johnston *et al.*, 2017; Lewis, 2002a; Takata *et al.*, 2006). Greater fouling biomass, diversity and survival occur on vessels that spend extended periods stationary in ports and harbours, move slowly, travel short distances and/or ply similar latitudes than high-activity, faster vessels that undertake long voyages through different environmental climes (e.g., trans-equatorial) (Coutts, 1999; Coutts & Taylor, 2004; Coutts *et al.*, 2007; Floerl *et al.*, 2005a; Lewis, 2002a; Rainer, 1995; Visscher, 1928; Wood & Allen, 1958). This is a consequence of the effects of hydrodynamics engendered by the vector movement, environmental fluctuations across water bodies and unfavourable feeding conditions or food supply on organism survival.

Many common biofouling species have life forms that enable their survival on vessels. Barnacles, for example, have a hard shell that protects soft body parts, a conic shape that is hydrodynamic resistant and a strong adhesive and adhesion mechanism to hold on to surfaces. Encrusting bryozoans can survive by lying within the boundary layer at the hull surface, protected from turbulent flow that could damage or dislodge the organism. Macroalgal biofouling species commonly have a fast-growing filamentous or sheet-like thallus facilitating rapid maturation, a crustose habit, a heteromorphic life history with a microscopic filamentous or encrusting phase and/or the ability to perennate vegetatively from basal crusts, stolons or thallus fragments, all traits that would facilitate their initial attachment to a vessel hull, survival of transoceanic transport and successful colonization of a foreign port (Lewis, 1999). Other species are less well adapted to high-flow conditions, but can survive because of strong attachment mechanisms or by hunkering in recesses, niches or within fouling aggregations that provide protection from high-flow conditions.

For a vessel-biofouling species, the likelihood of spread to a new location beyond its native range is expected to diminish as the route crosses discrete bioregions of differing environmental characteristics, with increasing distance and as an interaction of both factors. For most species, the likelihood of organism survival on a vessel that crosses a sea or ocean latitudinally, such as from Australia to New Zealand or from western Europe to the east coast of North America, is considered greater than on a vessel tracking longitudinally, such as from Europe to southern Africa or from the Far East to southern Australia. This is attributable to the greater seawater temperature gradient moving between north and south than between east and west.

Some of the most significant marine species invasions are of species that have successfully crossed the equatorial barrier from Northern Hemisphere temperate waters to Southern Hemisphere temperate waters or the reverse. Examples include the spread of the European green crab (*C. maenas*) and Mediterranean fan worm (*S. spallanzanii*) from Europe to Australia, and the Northern Pacific seastar (*A. amurensis*) and Asian kelp (*U. pinnatifida*) from north-west Pacific coasts to southern Australia. The American and Asian/African continents functioned similarly as barriers between the Pacific and Atlantic oceans.

Perhaps surprisingly, there are examples of IMS establishing in countries far from the native range but not at nearer locations, despite similar transport vectors operating. One example is the Asian brush-clawed shore crab (*H. takanoi*), for which molecular studies showed that European populations were a genetic admixture derived from populations in Japan and the Yellow Sea, but there was no evidence of gene flow between the two Asian regions (Makino *et al.*, 2018). The New Zealand green shell mussel (*Perna canaliculus*) has also never established in Australia and only several small incursions have been detected. This is despite high vessel traffic between the two countries and detections on vessels in Australian waters. Similarly, the lack of establishment of the Asian green mussel (*P. viridis*) in Australia appears counter to the numerous detections of this mussels on vessels arriving in Australia (Heersink *et al.*, 2014). *P. viridis* has, however, established in the south-eastern USA (Firth *et al.*, 2011).

To colonize a recipient location, the biofouling species must release reproductive material in that location. This can be through sporulation or spawning of reproductively mature individuals, dislodgement and fragmentation or, for mobile species, movement off the vector. Many fouling species brood their larvae, retaining fertilized eggs to develop in internal brood structures (Keough & Ross, 1999). Development often takes weeks, so animals do not need to depend on encountering suitable mates when they arrive in a new port. The release of a brood, or brooded larvae, which are thought to have higher survival than larvae that have a long planktonic development, could provide significant initial inocula to a new port.

Establishment success improves with increasing frequency and density of translocations (Kolar & Lodge, 2001; Roughgarden, 1986; Simberloff, 1989), so the greater the quantity, quality and frequency of NIMS translocated to inshore regions with similar environments, the greater the likelihood of establishment. The incursion frequency is termed 'propagule pressure', which is the combination of the number of individuals introduced and the frequency of immigration events (Drake & Lodge, 2006). Relating this to shipping, propagule pressure is a function of the number of reproductive organisms or propagules on one ship and the number of ships visiting the location. Propagule pressure can also relate to organism fitness during and subsequent to the voyage, as this can influence the spawning success in the recipient environment (Schimanski *et al.*, 2017). The success of establishment can also be affected by Allee effects, which represent the diminished population growth rate at low population densities (Drake & Lodge, 2006).

Environmental conditions generally influence the triggering of sporulation, or spawning, and the environment must be suitable to the species to enable larval survival and settlement. Temperature change can induce spawning activity in mussels (Buchanan & Babcock, 1997; Utting & Spencer, 1997), and the water temperature change as a vessel enters a port can trigger spawning of reproductively mature fouling species (Apte *et al.*, 2000; Minchin & Gollasch, 2003). The invasion of fouling organisms can also be facilitated by the artificial structures in ports, such as pilings and pontoons, by creating habitats that are unsuitable to native species (Glasby & Creese, 2007; Glasby *et al.*, 2007; Minchin *et al.*, 2006). Such human-produced habitats can be colonized by communities that are quite different from those of natural rocky reefs, and these habitats are often poorly utilized by native species (Glasby & Connell, 1999; Glasby *et al.*, 2007). Transient disturbance events and pollution in the recipient environment can also facilitate the establishment, persistence and spread of NIS (Clarke & Johnston, 2005; Dafforn *et al.*, 2007; Glasby & Creese, 2007; Hutchings *et al.*, 2002; Johnston, 2007; Johnston & Keough, 2002; Johnston *et al.*, 2017).

Invasibility, or the susceptibility of communities to colonization by alien species, originates in the biotic resistance hypothesis of Elton (1958) that proposed a negative relationship between native species diversity and invasion success (Briggs, 2013; Johnston *et al.*, 2017). However, invasion success can also depend on the availability of open space, and factors other than species diversity can be influential, including disturbance, competitive release, resource availability and propagule pressure (Richardson & Pyšek, 2006; Stachowicz & Byrnes 2006). In estuaries, anthropogenic disturbance and resource exploitation, together with increased propagule pressure from opportunistic species associated with shipping, result in greater numbers of aliens than on open coasts (Briggs, 2012). Examples include species considered alien constituting more than 15 per cent of biota inside Elkhorn Sound, California compared to less than 4 per cent on surrounding rocky coasts (Wasson *et al.*, 2005), fewer than 10 species on the outer coasts outside San Francisco Bay compared to more than 240 species inside the bay (Ruiz *et al.*, 1997) and the paucity of NIS in the outer coast habitats of central California (Zabin *et al.*, 2018).

There is also evidence that biotic resistance can limit the abundance of holoplanktonic species. In a tropical offshore port in Brazil, the density of a non-indigenous copepod was negatively correlated with the species richness and evenness of the native community, with higher numbers inside the port compared to outside and lower copepod diversity inside the port (Soares *et al.*, 2018). Some degree of biotic resistance was inferred in the outer, more diverse community.

The biological traits common to many marine NIS include high dispersal ability, high reproductive rate and ecological generalisation (Cardeccia *et al.*, 2018). However, these authors found no clear relationship between biological traits and their pathways of introduction, in particular for vessel pathways, indicating that all pathways can transport NIS with different biological trait profiles and ecological functions. For the majority of common ship-fouling species, common traits include broad environmental tolerance, high reproductive output, rapid growth rate and relatively small size, facilitating opportunistic colonization of available space. Organism morphology also contributes to survivorship on ship hulls, with colonial, encrusting, hard and/or flexible forms being more resilient, particularly with increasing speed (Coutts *et al.*, 2010). This has resulted in the global distribution of many of these species on artificial structures and vessels in harbours and marinas and in other disturbed or modified environments worldwide. Most are poor competitors for space and so are unable to invade established communities or persist on surfaces colonized by larger, slower-growing and longer-lived species, particularly on natural substrates (Lezzi *et al.*, 2018). The establishment of larger, longer-living marine species that spread widely in the new environment after crossing hemispheres or oceans on the hull of a vessel appears to be a rare event.

7.7 Distribution of Invasive Marine Species by Ships

The rate of introduction of NIS has been assessed as increasing over recent decades, an increase attributed to the increase in global shipping (Cohen & Carlton, 1998; Thresher *et al.*, 1999). Introduction rates have been estimated to be an average of one new species every 14 weeks in San Francisco Bay and one new species every 1.3 years in Port Phillip Bay. In surveys on NIS numbers in Osaka Bay, Japan, the number of introduced species remained essentially constant over the period from 2007 to 2013, despite the presence of major shipping ports within the Bay (Otani & Willan, 2017). The greatest numbers of species were from near the shipping ports in the inner bay. The two factors correlating with NIS abundance were salinity and native species richness. Salinity was rejected as a limiting factor due to the wide tolerance range of the NIS, leaving native species richness as the limiting factor. This is consistent with the ecosystem resistance hypothesis that strong biotic resistance by native species reduces the success of the establishment of non-native species, but contrasts with terrestrial systems where rich native diversity correlates positively with exotic diversity at a broad geographical scale (Wasson *et al.*, 2005). The limiting factor may be, however, the relative saturation of community maturity rather than absolute native species abundance, with interaction strengths between species being less and resources being more available in undersaturated communities (Wasson *et al.*, 2005).

Ship history, including the duration of voyages and ports visited, is important in determining the biological community in ballast water and on hulls, but accurately predicting and identifying the small proportion of ships that pose the highest risk remain challenging due to the multiple interacting factors of ship history, environmental seasonality and stochasticity and species biology and ecology (Bailey, 2015). Miller *et al.* (2018) traced the history of ships arriving in the USA back to ports prior to the last port of call, which showed that, as ports of call accumulate, a much greater range of biotic regions are visited, but a greater numbers of inter-ocean transits occur, which can physiologically stress hull-fouling organisms, such as when traversing the low-salinity waters of the Panama Canal or the hypersaline waters of the Suez Canal. Ships moving from place to place are considered by these authors to provide continual opportunities for the colonization and release of biofouling species, and to also experience the continual development and replacement of species assemblages that integrate across the places they visit. Tracing voyages back over maritime history explains the origin and common diversity and composition of ship-biofouling communities globally. Unlike ballast water, which samples whole ecological communities in the source location, the biofouling species contributing to the accumulated hull-fouling community in a particular port are not necessarily species native to that bioregion, but a subset of the cosmopolitan biofouling community established on artificial structures in the disturbed environment of that port. Native species from mature communities on natural, undisturbed surfaces rarely colonize ship hulls within a port.

Successful colonization of a species in a foreign location is the consequence of a process selecting traits for survival, adaptability and environmental tolerance, and establishment and spread can be enhanced by propagule pressure, phenotypic plasticity and preadaptation and the effects of selection and genetic admixture (Rius *et al.*, 2015). Once established, life history traits of high levels of fecundity or clonal reproduction can combine with environmental tolerance to facilitate further local and regional spread in the open coastal environments without transoceanic or climate barriers. Significant site-specific adaptation may also occur (Sherman *et al.*, 2016), further enhancing invasiveness.

Following initial introduction, NIS often spread along coasts and, given that most established populations are within bays and estuaries, human-aided dispersal facilitates this secondary spread (Ruiz *et al.*, 2015; Zabin *et al.*, 2014). Commercial ships, recreational craft and fishing vessels are all potential vectors for intra-coastal NIS spread. Small commercial and recreational vessels are also more likely to have biofouling accumulate on their hulls due to factors including lower speeds, longer residence times and less efficacious anti-fouling systems (Minchin *et al.*, 2006). On international vessels arriving in New Zealand, the greatest densities of species per vessel were on yachts and fishing vessels (Inglis *et al.*, 2010).

Voyages by small vessels are often infrequent with long periods spent in boat harbours, marinas or at moorings, and it has been found that invertebrate assemblages on heavily fouled vessels reflect the composition of biotic assemblages within the marina in which they were moored (Floerl & Inglis, 2005). Boat harbours and marinas are also characterized by extensive infrastructure, such as piers, pilings and pontoons, and poor flushing rates that favour the establishment of NIS (Glasby *et al.*, 2007; Johnston *et al.*, 2017; Rivero *et al.*, 2013). Marinas, and also aquaculture sites, are characterized by artificial structures and surfaces held at a constant depth, which can suit the growth of NIS such as the Asian kelp *U. pinnatifida* (Epstein & Smale, 2018). Although many NIS remain within these modified environments, high population abundance and reproduction can lead to not only more likely colonization of vessel hulls in or near the infested habitat, but also 'spillover' into nearby natural rocky reef and other communities (Epstein & Smale, 2018; James & Shears, 2016).

Standards of biofouling management on smaller vessels can also be lower on small craft due to the application of less effective (cheaper) anti-fouling products, poor application and irregular maintenance. Yacht anti-fouling paints have advertised service lives of between 9 and 18 months, compared to anti-fouling systems for merchant vessels that require service intervals of between 36 and 60 months, and some recreational vessels have been observed to foul within 3 months (Floerl & Inglis, 2005). Colonization of hulls can also increase when not regularly slipped for anti-fouling renewal and when copper release rates drop below critical levels, or on copper alloy propellers through a hermetic stimulation of settlement (Johnston & Keough, 2000, 2002; Lewis *et al.*, 1988; Pyefinch & Downing, 1949; Stebbing, 1981). Simultaneous spawning events can also be initiated by in-water hull cleaning and recruitment promoted on surfaces that have not been thoroughly cleaned of biofouling residues (Floerl *et al.*, 2005b).

Unlike international merchant ship voyage profiles, which usually involve long voyages and short port stays, domestic or regional vessel voyages are generally of shorter durations with varying lengths of port stays. Slow-moving vessels with long residency times have been considered as high risk for NIS transfer, but experimental studies have shown that vessels with short port stays and frequent voyages are capable of facilitating IMS transfer, and possibly more successfully than vessels with long voyages and port stays (Schimanski *et al.*, 2017). For the bryozoan *Bugula neritina*, recruitment to a vessel is possible in 1 day, but slightly older 1-week-old recruits showed greater ability to survive voyages than older or younger recruits (Schimanski *et al.*, 2016, 2017). Survival was not influenced by differing speeds of 6 and 18 knots or durations of 2 or 8 days.

Ballast water carried by intra-coastal shipping is seen as a significant vector for the secondary transfer of NIS due largely to the shorter voyages and greater organism survivorship (Adebayo *et al.*, 2014; Desai *et al.*, 2018; Simkanin *et al.*, 2009). Although ballast discharges by vessels operating within the exclusive economic zone of a single country are generally exempt from regulation, the exclusive economic zone can contain distinct ecoregions with diverse biological communities (Adebayo *et al.*, 2014).

Invasive macroinvertebrates, including the Europeans green crab (*C. maenas*), mud crab (*Rhithropanopeus harrisii*), common periwinkle (*L. littorea*), soft shell clam (*M. arenaria*) and blue mussel *(M. galloprovincialis*), were all found in ballast tanks, after deballasting, of ships arriving in east coast Canadian ports, but only in ships completing shorter voyages (average 4.5 days) (Briski *et al.*, 2012). Mid-ocean exchange did not affect macroinvertebrate occurrence. Significantly, a live gravid female *R. harrisii* was found, a species that has the potential to release up to 4000 eggs per clutch. Decapods were considered to pose a higher invasive risk than the bivalve molluscs because the simultaneous presence of both sexes within a tank is not a prerequisite to the release of free-swimming larvae. Domestic ballast water transported from St Lawrence River to the Great Lakes has been determined to pose a risk for survival and future invasion of the Great Lakes, particularly for euryhaline species (Adebayo *et al.*, 2014).

For species transported in ballast water, the requisite long larval lives or short reproductive cycles can enable natural dispersal away from the colonization site, while, for biofouling species, the abundance of artificial substrates around shipping ports and boat harbours can allow founder populations to establish. Coastal development results in the replacement of natural ecosystems developed on sand or soft sediment with human-made hard surfaces that are suited to immigrant biofouling species, which results in native biodiversity being threatened by both habitat destruction and the introduction of NIS (Ardura & Planes, 2017; Bulleri & Airoldi, 2005; Bulleri & Chapman, 2010; Glasby *et al.*, 2007). Habitat degradation is therefore seen as a cause rather than a result of biological invasions through habitat modification and loss of native biodiversity that facilitates invasions (Ardura & Planes, 2017; MacDougall & Turkington 2005; Stachowicz *et al.*, 1999, 2002).

The role of commercial harbours as sink and source habitats for NIS and the role of recreational boating in secondary spread was also investigated in the western Mediterranean Sea to test the hypothesis that assemblages in recreational marinas would be subsets of those in harbours (Ferrario *et al.*, 2017). This hypothesis was not consistently supported by the data, with the number of NIS in some marinas similar to those in harbours and the composition differing between marinas and harbours. Several new records of species were restricted to marinas, adding to the conclusion regarding the significance of recreational boating in facilitating NIS spread and that marinas can be beachheads for new species incursions. In California, there is a strong degree of connectivity between San Francisco Bay and nearby coastal bays in terms of small boat traffic (Zabin *et al.*, 2014). Additionally, hull surveys of boats moving between bays found macrofouling on more than 80 per cent of boats. In British Columbia, recreational boating was determined to be the most probable pathway for NIS spread through the region and the determinant of the current distribution (Murray *et al.*, 2014).

The spread of *U. pinnatifida* in Port Phillip Bay in southern Australia highlights the role of small boats in facilitating spread. *Undaria* was first found in 1996 in the Corio Arm of Port Phillip Bay, not within a harbour, but near a shipping lane. Over the subsequent 10 years, the alga spread eastwards along the northern bay coastline to Melbourne and inner Melbourne bayside coasts, presumably by natural dispersal. First detections on the Port of Melbourne were in 1999, and then not within the main international shipping harbour. No populations were detected elsewhere in or outside of the Bay until 2007, when fronds were found on small boat wharves in the western reach and on the southern coast of the bay. Satellite populations have since established at four other locations, all small boat marinas or berthing facilities. In California, the spread of *U. pinnatifida* along the coast is also attributed to small boat traffic (Zabin *et al.*, 2014). After its discovery at sites near Los Angeles in 2000, it was subsequently found in Monterey Harbor in 2001, Baja California in 2003 and San Francisco Bay in 2009, illustrating the connectivity of bays within and outside of California.

In the USA, the total WSA of ships travelling coastal routes and crossing bioregional boundaries was estimated to be 177 km^2 (Miller *et al.*, 2018). Such voyages can potentially transfer native species to new bioregions or secondarily spread introduced NIS. In the Mediterranean, where non-indigenous ascidians are dominant nuisance organisms, such ascidians were detected on 51 per cent of recreational, commercial and military craft surveyed in shipyards in Israel (Gewing & Shenkar, 2017). Most of the species were well established in the Levant Basin.

7.8 Regulations, Requirements and Guidance

7.8.1 Ballast Water

7.8.1.1 International

In the late 1980s, Australia, together with Canada, raised concerns on ballast water at the International Maritime Organization (IMO) Marine Environment Protection Committee (MEPC), and in 1991, the MEPC adopted the non-mandatory International Guidelines for preventing the introduction of unwanted aquatic organisms and pathogens from ships' ballast water and sediment discharges. Continued review of the issue consequently led the IMO to adopt the legally binding International Convention for the Control and Management of Ships' Ballast Water and Sediments (the 'BWM Convention') on 13 February 2004 (IMO, 2009).

The BWM Convention included a two-stage introduction of management measures; firstly, up until at latest 2016, ships would be required to meet a BWE standard (Regulation D-1), then, after this, to meet a ballast water performance standard (Regulation D-2) (David & Gollasch, 2008). Regulation D-1 required ships on international voyages to exchange ballast water taken up in ports with open ocean water with an efficiency of at least 95 per cent volumetric exchange. Regulation D-2 requires ships to have a BWTS installed that will ensure that ballast water discharged contains:

- Less than 10 viable organisms per cubic metre greater than or equal to 50 μm in minimum dimension;
- Less than 10 viable organisms per millilitre less than or equal to 10 μm in minimum dimension; and
- Indicator microbe concentrations as a human health standard of:
  - Toxicogenic *V. cholerae* less than 1 colony-forming unit (cfu) per 100 mL;
  - *Escherichia coli* less than 250 cfu per 100 mL; and
  - Intestinal enterococci less than 100 cfu per 100 mL.

As with all IMO conventions, entry into force of the BWM Convention would be 12 months after a required minimum number of IMO Member States ratified the Convention and, for this Convention, the requirement was not less than 30 Member States, the combined merchant fleets of which constituted not less than 35 per cent of the gross tonnage of the world's merchant shipping. Acceptance of the Convention by Member States was slow, and the requirement was only met in September 2016, so the date for entry into force of the Convention was 8 September 2017. However, it should be noted that ballast water discharge from international ships arriving in Australian ports has been managed since 2001 under the Australian Ballast Water Management Requirements that require deep ocean BWE.

Meeting the D-2 standard for BWTSs is challenging due to factors including the range of organisms to be rendered non-viable, the volume of water to be treated, the water chemistry and quality (which may encompass salinity ranges from freshwater to fully marine and high water turbidity) and that the standard is a discharge standard, meaning that organisms are not able to regrow in the tanks during the voyage. The IMO requires that a commercially developed BWTS is approved by a flag administration in accord with IMO guidelines that require land-based testing, ship-board trials and, for systems using active substances, approval of the environmental impact of discharged ballast water by the MEPC. At the end of 2016, 68 systems had received type approval certification in line with IMO requirements, 30 of which made use of active substances.

With entry into force of the BWM Convention, ships not already fitted with a BWTS will be required to do so by 2021. Estimates of the number of ships needing retrofit BWTS installations over the next 5 years vary from 35,000 to 120,000 ships. One calculation based on the lower number and an estimate that there are 250 shipyards worldwide capable of doing the work results in each yard having to install a system every 16 days. Installation of a BWTS can cost up to US$5 million per ship.

The BWM Convention entered in to force on 8 September 2017, when the requirement of at least 30 countries, representing a minimum of 35 per cent of the gross tonnage of the world's merchant, was met. With Qatar's accession in February 2018, 68 countries, representing more than 75 per cent of the world's merchant fleet tonnage, are now signatories to the Convention.

7.8.1.2 Regional Requirements: North America

Canada first drew up guidelines to prevent species transfers to the Great Lakes and St Lawrence Seaway in 1989, primarily in response to the successful invasion by the Eurasian ruffe (*Gymnocephalus cernua*), a freshwater fish native to temperate regions of Europe and northern Asia (Bailey, 2015). The guidelines recommended that all vessels should exchange their ballast water in the mid-ocean, where ocean depths are greater than 2000 m. The Canadian action was followed a year later by the USA with the enactment of the US Nonindigenous Aquatic Nuisance Prevention and Control Act (NANCPA) (Albert *et al.*, 2013; Davidson & Simkanin, 2012).

Canada is now a signatory to the IMO BWM Convention, but the USA is not. Instead, the US Coast Guard (USCG) and US Environmental Protection Agency (US EPA) developed national ballast water regulations through the NANCPA and the Clean Water Act (CWA). These entered into force in June 2012 and required all ships calling at US ports and intending to discharge ballast water to have either carried out BWE or to have an operational BWTS compliant with the IMO D-2 treatment standard (Cohen *et al.*, 2017). The USCG regulations require ballast water treatment from the first dry-docking after 1 January 2016 and at delivery for new builds. However, the legislation requires BWTSs to be type approved by the USCG independently of the IMO approval process, and the USCG type-approval process is more stringent that than that of the IMO. The first US type approval of a BWTS was not issued until December 2016 and, at the end of 2017, only five systems had been approved. As an interim measure, the USCG will allow a foreign type-approved system to be used for up to 5 years after the date at which the vessel is required to comply, after which the vessel owner must ensure the system is USCG approved.

The State of California adopted new legislation on ballast water management in July 2017. Under the California Marine Invasive Species Act, the requirements apply to vessels over 3000 gross tonnage that are capable of carrying ballast water that arrive in Californian waters from a location outside the Pacific Coast Region (PCR) or from a location within the PCR. Compliance with the requirements is by:

- Retention of all ballast on board;
- Exchange of ballast water either mid-ocean (more than 200 nautical miles from land and at least 2000 m deep) for vessels arriving from outside the PCR or in PCR near-coastal waters (more than 200 nautical miles from land and at least 200 m deep) for vessels arriving from elsewhere in the PCR;
- Discharge of ballast at the same location it was taken up;
- Use of an alternative environmentally sound and approved method of treatment; or
- Discharge to an approved reception facility.

7.8.1.3 Regional Requirements: Australia

Australia first addressed concerns on the introduction of harmful organisms in the ballast water of ships by, in 1990, introducing a range of voluntary guidelines for ships entering Australian ports from overseas (Hallegraeff & Bolch, 1991). These led to, in 2001, the introduction of mandatory ballast water management requirements for all international shipping, which required that vessels not discharge high-risk ballast water in Australian ports or waters (AFFA, 2001). The requirements were enforced under the Quarantine Act

1908 and administered by the then Australian Quarantine Inspection Service (AQIS). Options for the treatment of high-risk ballast water were:

- No discharge of ballast water to occur in Australian ports or water;
- Tank-to-tank transfer to prevent the discharge of high-risk ballast water in Australian waters; or
- Undertake a full BWE at sea prior to arrival in Australian waters using either or a combination of the sequential, flow-through or dilution method. The method was required to achieve at least 95 per cent volumetric exchange.

The Biosecurity Act 2015 (Australian Government, 2015), which superseded the Quarantine Act 1908 in 2016, included ballast water provision that implemented and aligned with the BWM Convention. Under the new Act, vessels had the option of managing their ballast water by either BWE or by use of a type-approved BWTS. However, in line with the implementation schedule of the BWM Convention, BWE would be phased out in favour of a method compliant with the D-2 discharge standard (DAWR, 2017). The new Act extended ballast water management requirements to vessels sailing between Australian ports, as well as international vessels arriving in Australian seas.

Within Australia, the State of Victoria enacted the Waste Management Policy (Ships' Ballast Water) in 2006 'to minimise the introduction of marine pests into Victorian State waters from high-risk domestic ballast water discharge' (EPA Victoria, 2008). Ships' masters were required to assess the risk status of any domestic ballast water through use of an online risk assessment tool. The risk assessment tool performed a biological risk assessment to predict if harmful aquatic organisms and pathogens could be in the ballast water in each tank. Management options for high-risk ballast water were retention on board/tank-to-tank transfer, BWE outside of Victorian waters (at least 12 nautical miles off the Australian coast) or treatment by an approved method.

7.8.1.4 Regional Requirements: New Zealand

New Zealand mandated that vessels conduct BWE in 1998 (Albert *et al.*, 2013). The minimum requirement for vessel ballast water loaded in countries other than New Zealand and intended for discharge in New Zealand water are detailed in an import health standard (MPI, 2016). The Import Health Standard (IHS) provides three options for ballast water management:

- Mid-ocean exchange en route to New Zealand;
- The ballast water is freshwater; or
- Use of an approved BWTS.

The discharge into New Zealand waters of sediment from the cleaning of ballast tanks is also prohibited.

7.8.2 Biofouling

7.8.2.1 IMO Biofouling Guidelines

On 15 July 2011, the Marine Environment Protection Committee of the IMO adopted Resolution MEPC.207(62) – Guidelines for the control and management of ships' biofouling to minimize the transfer of invasive aquatic species (IMO, 2012a). These guidelines were developed to address concerns that biofouling was a significant vector for the transfer of invasive aquatic species, and that 'ships entering the waters of States may result in the establishment of invasive aquatic species that may pose threats to human, animal and plant life, economic and cultural activities and the aquatic environment'.

The objectives of the Guidelines were to provide practical guidance on measures to minimize the risk of translocating invasive aquatic species in biofouling through the implementation of biofouling management practices, including the use of anti-fouling systems, to keep ships' submerged surfaces and internal seawater cooling systems as free of biofouling as possible. They are intended to provide a globally consistent approach to the management of biofouling. The Guidelines propose that every ship should have a biofouling management plan and a Biofouling Record Book. The plan would detail procedures for effective biofouling management and consider measures including the selection and application of suitable anti-fouling systems, identification of hull locations susceptible to biofouling and schedules of planned inspections, repairs, maintenance and renewal of anti-fouling systems. The Biofouling Record Book would contain details of the biofouling management measures and systems undertaken, including details of the anti-fouling systems, dates and locations of dry-dockings or slippings, in-water inspections, inspections and maintenance of internal seawater systems and the work completed.

Sections of the Guidelines provide guidance on anti-fouling installation and maintenance, choosing the anti-fouling system, installing, reinstalling or repairing the anti-fouling system, in-water inspection, in-water cleaning and maintenance and design and construction (IMO, 2012b).

The Guidelines are just that – guidelines – and their uptake on international shipping is encouraged, but not mandatory. There are also no current proposals for the IMO to develop a convention or other mechanism to mandate biofouling management measures. However, as detailed in the following sections, some governments have made such moves.

In addition to the Guidelines, the IMO Marine Environment Protection Committee also circulated guidance for minimizing the transfer of invasive aquatic species as biofouling for recreational craft (IMO, 2012b). This provides guidance on the selection of anti-fouling coating systems, minimization of biofouling in niche areas and in- and out-of-water cleaning.

7.8.2.2 Regional Requirements: New Zealand

The New Zealand Craft Risk Management Standard (CRMS) on Biofouling on Vessels Arriving to New Zealand entered into force in May 2018 (MPI, 2014). The intended objective of the regulation is to reduce the introduction of marine pests and diseases into New Zealand as – or harboured within – biofouling, and the CRMS specifies the requirements for the management of biofouling risk associated with vessels entering New Zealand territorial waters.

The requirement of the CRMS is that every type of vessel (including floating platforms) must arrive in New Zealand with a 'clean hull', with a 'clean hull' considered to be a hull with no biofouling of live organisms present beyond defined thresholds. These thresholds are separately defined for 'short-stay' and 'long-stay' vessels. 'Short-stay vessels' are vessels intending to remain in New Zealand for 20 days or fewer and to only visit the 15 ports designated as 'Places of First Arrival'; 'long-stay vessels' are vessels intending to remain in New Zealand for 21 days or longer or to visit areas other than the ports designated as 'Places of First Arrival'.

For 'long-stay vessels', the threshold for allowable biofouling on all hull surfaces is a slime layer and/or goose barnacles, with 'slime layer' defined as 'a layer of microscopic organisms, such as bacteria and diatoms, and the slimy substances they produce'. For 'short-stay vessels', there is an additional allowance for:

- At the wind and water line, green algae growth of unrestricted cover and no more than 50 mm in frond, filament or beard length; brown and red algal growth of no more than 4 mm in length; and incidental (maximum of 1 per cent) coverage of a single species, or what appears to be the same species, of one organism type of either tubeworms, bryozoans or barnacles, occurring as isolated individuals or small clusters;
- On hull areas, algal growth no more than 4 mm in length in continuous strips and/or patches of no more than 50 mm in width; scattered (maximum of 5 per cent) coverage of a single species, or what appears to be the same species, of one organism type of either tubeworms, bryozoans or barnacles, occurring as widely spaced individuals and/or infrequent, patchy clusters that have no algal overgrowth; and
- In niche areas, algal growth no more than 4 mm in length in continuous strips and/or patches of no more than 50 mm in width; incidental (maximum of 1 per cent) coverage of a single species, or what appears to be the same species, of one organism type of either tubeworms, bryozoans or barnacles, occurring as isolated individuals or small clusters that have no algal overgrowth; and incidental (maximum of 1 per cent) coverage of a single species, or what appears to be the same species, of a second organism type of either tubeworms, bryozoans or barnacles, occurring as isolated individuals or small clusters that have no algal overgrowth.

Acceptable measures for meeting the CRMS are stated as:

(1) Cleaning before visit to New Zealand (or immediately on arrival in a facility or by a system, approved by the Ministry for Primary Industries (MPI)). All biofouling must be removed from all parts of the hull and this must be carried out less than 30 days before arrival to New Zealand or within 24 hours after time of arrival;
(2) Continual maintenance using best practice, including: application of appropriate anti-foul coatings; operation of marine growth prevention systems on sea chests; and in-water inspections with biofouling removal as required. Following the IMO Biofouling Guidelines is recognized as an example of best practice; or
(3) Application of approved treatments. Treatments are approved and listed under the Approved Biosecurity Treatments MPI-STD-ABTRT.

As an alternative to the acceptable measures above, a vessel operator may submit, for MPI approval, a Craft Risk Management Plan that includes steps that will be taken to reduce risk to the equivalent degree as meeting the requirements of this standard.

7.8.2.3 Regional Requirements: USA

The State of California has enacted Biofouling Management Regulations to Minimize the Transport on Nonindigenous Species from Vessels Arriving at California Ports (CSLC, 2017). The regulations came into effect on 1 October 2017 and require, after a vessel's first regularly scheduled out-of-water maintenance after 1 January 2018 or upon delivery on or after this date, development and maintenance of a Biofouling Management Plan, development and maintenance of a Biofouling Record Book, mandatory biofouling management of the vessel's wetted surfaces and mandatory biofouling management for vessels that undergo an extended residency period, such as remaining in one location for 45 days or more.

The USCG, in addition to requiring ballast water discharge standards consistent with the IMO BWM Convention, requires the regular cleaning of ballast tanks to remove sediments, rinsing of anchors and chains when the anchor is retrieved and regular removal of biofouling form the hull, piping and tanks.

7.8.2.4 Regional Requirements: Australia

The Australian Government is proposing to introduce requirements for the management of biofouling on international ships entering Australian waters, but the form of these has yet to be finalized. National Biofouling Management Guidelines, which predated the IMO Guidelines, were developed for different maritime sectors, including commercial vessels, non-trading vessels and petroleum production and exploration vessels (Australian Government, 2009a, 2009b, 2009c).

The State of Western Australia imposed Ministerial Conditions on oil and gas development projects within the state that required vessels arriving from overseas or other Australian states to be free of marine pests. In part, this required vessels to be inspected by a Western Australia-approved biofouling inspector in the last overseas port of call before mobilizing to Western Australia to ensure freedom from marine pests. A list of more than 70 alien species considered of concern and that could be spread by biofouling or ballast water underpinned this requirement (Department of Fisheries, 2014).

7.9 Shipping and Invasive Marine Species

That the introduction of alien species or biological invasion is one of the major causes of biodiversity loss is one of the most frequent claims in the scientific literature. This loss is attributed to the negative effects of species interactions, native species extinction, changes in community structure, habitat degradation and ecosystem functioning (e.g., IUCN, 2017; Riul *et al.*, 2013; Sax *et al.*, 2007; Shiganova, 1998; Silva *et al.*, 2011; Simberloff *et al.*, 1997; Stachowicz *et al.*, 2002; Vilà *et al.*, 2011; Vitousek *et al.*, 1997). The intensity of ecological change caused by invasive alien species is also claimed to increase over time, reducing the natural capacity of ecosystems to recover to their pre-invasion states (Silva *et al.*, 2011). The invasion of non-native species has also been cited as the second largest ongoing global ecological disaster after climate change (Precht *et al.*, 2014).

Although there are examples of alien species causing loss of biodiversity, this is not a widespread phenomenon, and most are in ecosystems that have been biogeographically isolated and in locations that are extremely space limited, such as on oceanic islands or in freshwater lakes and streams (Briggs, 2013; Zalba & Ziller, 2007). One source comments that 80 per cent of known extinctions are on islands, with invasive alien species considered the cause (CBD, 2016). No examples of loss of biodiversity in marine ecosystems are known and, despite occurring relatively frequently, competition rarely limits immigration or leads to local extinction (Bruno *et al.*, 2003). One reason proposed for this is the interconnectivity within the marine environment and the absence of barriers to the movement of nektonic and planktonic larvae and other reproductive propagules (Coutinho *et al.*, 2013). More broadly, the widely cited claim that alien species constitute the second greatest threat to biodiversity after habitat destruction has been scrutinized and determined to have no scientific basis (Chew, 2015; Davis & Chew, 2017).

The economic costs of invasive species are claimed to be high and, for example, in the USA, the economic impact of invasive species has been estimated at about US$120 billion per year (Pimentel *et al.*, 2000, 2005). Besides the intrinsic loss of ecological services and values that invasive species may cause, the cost to remove exotic species that have become naturalized can be prohibitive (Precht *et al.*, 2014). However, 94 per cent of the US$120 billion is derived from six broad groups of aliens: crop weeds, US$27 billion; rats, US$19 billion; cats, US$17 billion; crop pests, US$14.4 billion; crop plant pathogens, US$21.5 billion; and livestock diseases, US$14 billion (Pimentel *et al.*, 2005). The only aquatic species to be considered are the freshwater zebra mussel (*D. polymorpha*, US$1 billion), Asian clam (*Corbicula fluminea*, US$1 billion), marine shipworms (*Teredo navalis*, US$205 million) and European green crab (*C. maenas*, US$44 million).

Shipping is a major vector for the movement of aquatic species along coasts and across seas and oceans, and this has been occurring since humans took to boats. Spread of species as biofouling on vessel hulls has been a continuous process through maritime history, with the added vector of seawater ballast over the last century. The strength of the translocation would have varied over time, influenced by the types of vessel, voyage patterns, levels of anti-fouling protection, ship speeds and ship numbers. In the days of sail, wooden ships with slow speeds and limited anti-fouling protection would have been highly susceptible to fouling colonization and growth, and these would account for the cosmopolitan distribution of many fouling macroalgae, barnacles, serpulid tubeworms, bryozoans, etc. This era led to the formation of a global biofouling community of opportunistic species capable of colonizing vessel hulls and other maritime structures in and around harbours and ports. With passing time, the increasing speeds of vessels and the effective lives of anti-fouling coating systems increasing from 12 months, to 24 months, to 36 months, to 60 months, the colonization and development of biofouling communities were slowed, along with the selection for more copper-tolerant fouling species among algae, tubeworms and bryozoans. The counterbalance was the ever-increasing shipping trade, meaning more ships with low-level biofouling compared to the earlier, fewer ships with high levels of biofouling.

On surfaces with no fouling defence, the initial, relatively small and fast-growing opportunists are progressively displaced by larger, slower-growing species, and the physical structures these form provide a habitat for mobile species and epibionts that increase the complexity and diversity of the biofouling community. This process progresses more rapidly on port and harbour structures such as pilings, floating buoys and pontoons. On vessel hulls, the development is slowed by anti-fouling coating systems, and until the coating is spent, biofouling remains predominantly of microbial slime or, as biocide release rates diminish, a small number of scattered, low-profile, copper-tolerant macroorganisms (Floerl *et al.*, 2004).

On modern ships, the greatest diversity and local abundance of biofouling is not usually on the outer hull, but in nooks and crannies protected from turbulent water flow and/or where anti-fouling coatings are difficult to apply or do not function effectively. In these niches, and on areas of damaged or ineffective anti-fouling, the composition of biofouling is restricted by the tolerance of settled organisms to the suitability of the hull surface, the time window for settlement and voyage pressures, such as hydrodynamics and environmental fluctuation.

The greatest levels of biofouling commonly occur on non-trading and small vessels. The operational profiles of these vessels can combine low speeds, long residence times, less regular maintenance and/or lower-performance anti-fouling coatings. A primary selection factor for anti-fouling choice on these vessels is commonly paint cost, with the economic benefits of more expensive high-performance coatings not evident on vessels with short docking cycles and little impact of biofouling on operational costs such as fuel consumption and speed. This may increase the risk of IMS spread in two ways: firstly, by biofouling abundance on small boat hulls providing both brood populations and spreading species regionally; and secondly, the greater risk of international translocation in the mobilization/demobilization of global construction vessels, such as dredgers, or mobile infrastructure, such as drilling rigs.

Unlike biofouling, the ballast water vector is a relatively recent vector for the translocation of marine species. The uptake of ballast water differs in entraining a sample of all plankton and a sub-sample of nekton (those unable to actively evade the suction) in the water body from which the water is drawn. The one selection process for species translocation is survival in the ballast tank throughout the voyage. The absence of light is a major inhibitor for photosynthetic algae, but this is countered by the ability of some microalgal species to form resting cysts with long survival potential. An additional difference from the external biofouling community is that ballast water organisms have the potential to reproduce during transit and prolong species survival.

As has now been commonly cited, a census of NIMS in any harbour or developed coastal embayment will show that most of the species are likely to have been introduced via biofouling on ships. However, the majority of these species are relatively benign, with impacts largely that of biofouling (i.e., colonization of vessels and immersed coastal and aquaculture infrastructure). Biofouling per se, and not exclusively NIS, represents the problem in this regard, and the solution is best practice biofouling management. Minimizing biofouling on an installation not only benefits that installation, but also protects nearby installations by reducing propagule pressure.

Marine NIS can have negative and positive effects on local biota (Thomsen *et al.*, 2015). Commonly, the presence of alien species increases biodiversity by either colonizing disturbed or modified environments that are unfavourable for natives (Briggs, 2007, 2012, 2013; Gurevitch & Padilla, 2004; Thomsen *et al.*, 2014) or enhancing ecosystem services by creating habitats for epifaunal and infaunal organisms (Bruno *et al.*, 2003; Buschbaum *et al.*, 2006; Escapa *et al.*, 2004; Polte & Buschbaum, 2008; Tassin *et al.*, 2017; Viejo, 1999).

NIS can be drivers or passengers of ecological change (Thomsen *et al.*, 2015), and there is often not a clear understanding of this. Loss of native marine biodiversity is more typically a consequence of urbanization of the coastal zone (Connell & Glasby, 1999; Glasby & Connell, 1999; Stuart-Smith *et al.*, 2015), and establishment of alien species a secondary impact of coastal modification and disturbance (Briggs, 2012; Glasby, 1999; Vaselli *et al.*, 2008). Urban development is considered one of the major threats to Brazilian coral reefs (Leão *et al.*, 2016).

IMS associated with ballast water are documented to cause significant economic and human health impacts. Introduced microalgae can lead to HABs that cause fish kills and PSP with potentially fatal human consequences, and cholera can also be spread in ballast water. Some of the most significant marine invasions of macroinvertebrates, notably the comb jelly *M. leidyi* and the North Pacific sea star *A. amurensis*, are also linked to ballast. The need to manage these risks resulted in regional, then global requirements for ballast water management. Deep ocean BWE, though not perfect, provided an important first step in reducing ballast water risk with the implementation of the BWM Convention and the requirement for ships to have BWTSs providing a much greater level of protection against harmful microalgae, bacteria and viruses.

With respect to ship biofouling and the associated spread of marine species, there is a long history of movements that are readily observed in ports, harbours and marinas throughout the world. The impacts of the majority of these species appear to be low, with many remaining within disturbed or artificially constructed environments (Dafforn *et al.*, 2012). However, although not impacting on natural ecosystems, their contribution as opportunistic biofouling organisms can increase regional biofouling severity, which can impact on infrastructure and aquaculture and increase maintenance requirements for vessels.

Although the levels of risk and impact associated with biofouling on international ships appear generally low, this does not negate the potential environmental benefits of best practice biofouling management as laid out in the IMO Guidelines for the Control and Management of Ships' Biofouling. Reduced levels of biofouling on ship hulls will improve efficiency and reduce fuel consumption, with consequent reductions in the release of harmful air emissions. Reduction of biofouling growth in internal seawater systems will additionally improve the operational performance of equipment through unimpeded water flow and cooling capacity. Biofouling management would also reduce the risk, even if small, of the transport of a harmful marine species.

In regard to biofouling NIS, the greatest economic impacts are their contributions to fouling of maritime infrastructure and vessels. For natural ecosystems, high-value assets such as marine protected areas or conservation zones warrant protection from NIS incursions (Zabin *et al.*, 2018). Ports, harbours and marinas, and the resident vessels therein, are hotspots for NIS. As observed by Zabin *et al.* (2014), while management efforts to reduce biotic transfers by commercial ships can decrease new invasions to commercial hubs, this will not stop the continuing secondary spread and human-aided expansion. Management of biofouling on structures within facilities and the application of best practice biofouling management to vessels within or in proximity to these waterbodies and their regional movements can significantly reduce both the impact of biofouling on the vessels themselves and the risks of NIS escape to natural ecosystems by natural dispersal or regional dispersal on small craft. Management of recreational and regional fishing boats is, however, considered to pose a much greater challenge than commercial vessels, as the owners are more numerous and widely distributed, hull maintenance and operational patterns are less predictable and they are generally not regulated by a single authority (Zabin *et al.*, 2014).

As proposed by Johnston *et al.* (2017), high-priority management actions that can mitigate the invasive impacts of marine NIS include: reducing contamination levels in the marine environment; designing artificial structures to reduce NIS colonization; managing regional transport pathways; and protecting native ecological communities to maintain biotic resistance.

7.10 Conclusions

The anthropogenic movement of IMS and the perceived potential economic and environmental harm of these have precipitated global actions to minimize further introductions and spread of NIS. Translocations and biogeographical spread have been associated with aquaculture, the opening of inter-ocean canals, the aquarium industry and vessels. International shipping has been targeted as a major vector for the transportation of marine species in either ballast water or as hull fouling, and the need for management of these has been promoted through the IMO Ballast Water Management Convention and the Guidelines for the Control and Management of Ships' Biofouling.

The risks posed by ballast and biofouling differ. Ballast water uptake is essentially a random sample of the biota within the water column at point of uptake and includes not only larval and holoplanktonic invertebrates, but also microalgae and bacteria, of which some can directly or indirectly harm marine and human life. In contrast, ship biofouling is an accumulation of species – both from the history of maritime transport and the in-service interval of an individual ship – that are opportunistic occupiers of immersed surfaces and tolerant of the often biocidal and hydrodynamic stresses of peripatetic life on a ship hull. Over time, unmanaged biofouling can approach the structural complexity and greater richness of native benthic communities, including the harbouring of mobile species that can be invasive. The IMO BWM Convention and Biofouling Guidelines are therefore appropriate measures for managing invasive species risks. The latter also serve to contribute to the reduction of harmful gas emissions through improved hull and fuel efficiency.

The introduction of the majority of NIMS found in ports around the world is attributable to ship biofouling, but these predominantly establish on only artificial substrates or in anthropologically disturbed environments. They can therefore be considered a consequence, not a cause, of environmental disturbance. The impacts of biofouling NIS are predominantly that of fouling: fouling vessels, port and marina infrastructure, aquaculture nets, lines and poles and seawater pipework. The problem is one of biofouling per se, and not NIS. The question therefore arises as to whether the expenditure of resources on all-encompassing regulatory systems directed at international trading ships is justified, or whether a more focused approach that targets higher-risk vessels and vectors for the protection of high-value marine and conservation assets is a more judicious approach. Biofouling management on coastal infrastructure and higher standards of biofouling management on recreational and other coastal vessels may then become a priority.

References

Ab Rahim ES, Nguyen TTT, Ingram B *et al.* (2016). Species composition and hybridisation of mussel species (Bivalvia: Mytilidae) in Australia. *Marine and Freshwater Research* **67**, 1955–1963.

Adebayo AA, Zhan A, Bailey SA, MacIsaac HJ (2014). Domestic ships as a potential pathway of nonindigenous species from the Saint Lawrence River to the Great Lakes. *Biological Invasions* **16**, 793–801.

AFFA (2011). *Australian Ballast Water Management Requirements, Version 5.3.* Canberra: Department of Agriculture, Fisheries and Forestry.

Albert RJ, Lishman JM, Saxena JR (2013). Ballast water regulations and the move toward concentration-based numeric discharge limits. *Ecological Applications* **23**, 289–300.

Apte S, Holland BS, Godwin LS, Gardner PA (2000) Jumping ship: a stepping stone event mediating transfer of a non-indigenous species via a potentially unsuitable environment. *Biological Invasions* **2**, 75–79.

Ardura A, Planes S (2017). Rapid assessment of non-indigenous species in the era of the eDNA barcoding: a Mediterranean case study. *Estuarine, Coastal and Shelf Science* **188**, 81–87.

ASA (2006). *Assessment of Introduced Marine Pest Risks Associated with Niche Areas in Commercial Shipping: Final Report. Report to Invasive Marine Species Program, Department of Agriculture, Fisheries and Forestry, Canberra.* Melbourne: Australian Shipowners Association.

Australian Government (2009a). *National Biofouling Management Guidelines for Commercial Vessels.* Canberra: Commonwealth of Australia.

Australian Government (2009b). *National Biofouling Management Guidance for Non-Trading Vessels.* Canberra: Commonwealth of Australia.

Australian Government (2009c). *National Biofouling Management Guidance for the Petroleum Production and Exploration Industry.* Canberra: Commonwealth of Australia.

Australian Government (2015). Biosecurity Act 2015. No. 61, 2015. Authorized Version C2015A00061. Federal Register of Legislation. www.legislation.gov.au

Bailey SA (2015). An overview of thirty years of research on ballast water as a vector for aquatic invasive species to freshwater and marine environments. *Aquatic Ecosystem Health & Management* **18**, 261–268.

Bailey SA, Deneau MG, Jean L *et al.* (2011). Evaluating efficacy of an environmental policy to prevent biological invasions. *Environmental Science & Technology* **45**, 2554–2561.

Barnes DKA (2002). Invasions by marine life on plastic debris. *Nature* **416**, 808–809.

Barnes DKA, Fraser KPP (2003). Rafting by five phyla on man-made flotsam in the Southern Ocean. *Marine Ecology Progress Series* **262**, 289–291.

Barnes DKA, Milner P (2005). Drifting plastic and its consequences for sessile organism dispersal in the Atlantic Ocean. *Marine Biology* **146**, 815–825.

Bastida-Zavala JR, ten Hove HA (2003). Revision of Hydroides Gunnerus, 1768 (Polychaeta: Serpulidae) from the Eastern Pacific Ocean and Hawaii. *Beaufortia* **53**, 67–110.

Bayha KM, Chang MH, Mariani CL *et al.* (2015). Worldwide phylogeography of the invasive ctenophore *Mnemiopsis leidyi* (Ctenophora) based on nuclear and mitochondrial DNA data. *Biological Invasions* **17**, 827–850.

Boudouresque CF (1999). The Red Sea–Mediterranean link: unwanted effects of canals. In: OT Sandlund, PJ Schei, Å Viken, eds., *Invasive Speciss and Biodiversity Management*. Dordrecht: Kluwer Academic Publishers, pp. 213–228.

Boudouresque CF, Verlaque M (2002). Assessing scale and impact of ship-transported alien macrophytes in the Mediterranean Sea. In: F Briand, ed., *Alien Marine Organisms introduced by Ships in the Mediterranean and Black Seas*. Workshop Monographs No. 20. Monaco: CIESM, pp. 53–62.

Briggs JC (2007). Marine biogeography and ecology: invasions and introductions. *Journal of Biogeography* **34**, 193–198.

Briggs JC (2010). Marine biology: the role of accommodation in shaping marine biodiversity. *Marine Biology* **157**, 2117–2126.

Briggs JC (2012). Marine species invasions in estuaries and harbors. *Marine Ecology Progress Series* **449**, 297–302.

Briggs JC (2013). Invasion ecology: origin and biodiversity effects. *Environmental Skeptics and Critics* **2**(3), 73–81.

Briski E, Ghabooli S, Bailey S, MacIsaac H (2012). Invasion risk posed by macroinvertebrates transported in ships' ballast tanks. *Biological Invasions* **14**, 1843–1850.

Bruno JF, Fridley JD, Bromberg KD, Bertness MD (2003). Insights into biotic interactions from studies of species invasions. In: DF Sax, JJ Stachowicz, SD Gaines, eds., *Species Invasions: Insights into Ecology, Evolution and Biogeography*. Sunderland, MA: Sinauer Associates, Inc., pp. 13–40.

Buchanan S, Babcock R (1997) Primary and secondary settlement of the greenshell mussel *Perna canaliculus*. *Journal of Shellfish Research*, **16**, 71–76.

Bulleri F, Airoldi L (2005). Artificial marine structures facilitate the spread of a non-indigenous green alga, *Codium fragile* ssp. *tomentosoides*, in the north Adriatic Sea. *Journal of Applied Ecology* **42**, 1063–1072.

Bulleri F, Chapman MG (2010). The introduction of coastal infrastructure as a driver of change in marine environments. *Journal of Applied Ecology* **47**, 26–35.

Buschbaum C, Chapman AS, Saier B (2006). How an introduced seaweed can affect epibiota diversity in different coastal systems. *Marine Biology* **148**(4), 743–754.

Calado R, Chapman PM (2006). Aquarium species: deadly invaders. *Marine Pollution Bulletin* **52**, 599–601.

Callow ME (1986). Fouling algae from 'in-service' ships. *Botanica Marina* **29**, 351–357.

Campbell ML, Hewitt CL (1999). Vectors, shipping and trade. In: CL Hewitt, ML Campbell, RE Thresher, RB Martine, eds., *Marine Biological Invasions of Port Phillip Bay,*

Victoria. Centre for Research on Introduced Marine Pests. Technical Report No. 20. Hobart: CSIRO Marine Research, pp. 45–60.
Cardeccia A, Marchini A, Occhipinti-Ambrogi A *et al.* (2018). Assessing biological invasions in European Seas: biological traits of the most widespread non-indigenous species. *Estuarine, Coastal and Shelf Science* **201**, 17–28.
Carlton JT (1985). Transoceanic and interoceanic dispersal of coastal marine organisms: the biology of ballast water. *Oceanography and Marine Biology: An Annual Review* **23**, 313–371.
Carlton JT (1999). The scale and ecological consequences of biological invasions in the world's oceans. In, OT Sandlund, PJ Schei, Å Viken, eds., *Invasive Speciss and Biodiversity Management*. Dordrecht: Kluwer Academic Publishers, pp. 195–212.
Carlton JT (2001). *Introduced Species in U.S. Coastal Waters: Environmental Impacts and Management Priorities*. Arlington, VA: Pew Oceans Committee.
Carlton JT, Geller JB (1993). Ecological roulette: the global transport of non-indigenous species. *Science* **261**(5117), 78–82.
Carlton JT, Chapman JW, Geller JB *et al.* (2017). Tsunami-driven rafting: transoceanic species dispersal and implications for marine biogeography. *Science* **357**, 1402.
CBD (2016). Press Release: The Honolulu Challenge: an ambitious initiative to tackle the threat of invasive alien species. 29 November 2016. United Nations Environment Programme, Convention on Biological Diversity. www.cbd.int/doc/press/2016/pr-2016-11-29-HonoluluChallenge-en.pdf
Cheniti R, Rochon A, Frihi H (2018). Ship traffic and the introduction of diatoms and dinoflagellates via ballast water in the port of Annaba, Algeria. *Journal of Sea Research* **133**, 154–165.
Chew MK (2015). Ecologists, environmentalists, experts, and the invasion of the 'second greatest threat'. *International Review of Environmental History* **1**, 7–40.
Clarke GF, Johnston EL (2005) Manipulating larval supply in the field: a controlled study of marine invisibility. *Marine Ecology Progress Series*, **298**, 9–19.
Cohen AN, Carlton JT (1998). Accelerating invasion rate in a highly invaded estuary. *Science* **279**, 555.
Cohen AN, Dobbs FC, Chapman PM (2017). Revisiting the basis for US ballast water regulations. *Marine Pollution Bulletin* **118**, 348–353.
Connell SD, Glasby TM (1999). Do urban structures influence local abundance and diversity of subtidal epibiota? A case study from Sydney Harbour, Australia. *Marine Environmental Research* **47**(4), 373–387.
Coutinho R, Gonçalves JEA, Messano LVR, Ferreira CEL (2013). Avaliação Crítica das Bioinvasões por Bioincrustação. *A Ressurgência*. No. 7: 11–18. Instituto de Estudos do Mar Almirante Paulo Moreira – IEAPM. Marinha do Brasil. www.redebim.dphdm .mar.mil.br/vinculos/000009/00000910.pdf
Coutts ADM (1999). Hull fouling as a modern vector for marine biological invasions: investigation of merchant vessels visiting northern Tasmania. MSc thesis, Australian Maritime College.
Coutts ADM, Dodgshun TJ (2007). The nature and extent of organisms in vessel sea-chests: a protected mechanism for marine bioinvasions. *Marine Pollution Bulletin* **54**, 875–886.
Coutts ADM, Forrest BM (2007). Development and application of tools for incursion response: lessons learned from the management of the fouling pest *Didemnum vexillum*. *Journal of Experimental Marine Biology and Ecology* **342**, 154–162.
Coutts ADM, Taylor MD (2004). A preliminary investigation of biosecurity risks associated with biofouling on merchant vessels in New Zealand. *New Zealand Journal of Marine & Freshwater Research* **38**, 215–229.

Coutts ADM, Taylor MD, Hewitt CL (2007). Novel method for assessing the *en route* survivorship of biofouling organisms on various vessel types. *Marine Pollution Bulletin* **54**, 97–116.

Coutts ADM, Piola RF, Hewitt CL, Connell SD, Gardner JPA (2010). Effect of vessel voyage speed on survival of biofouling organisms: implications for translocation of non-indigenous marine species. *Biofouling: The Journal of Bioadhesion and Biofilm Research* **26**, 1–13.

Crego-Prieto V, Ardura A, Juanes F *et al.* (2015). Aquaculture and the spread of introduced mussel genes in British Columbia. *Biological Invasions* **17**, 2011–2026.

CSLC (2017). Title 2. Administration. Division 3. State Property Operation. Chapter 1. State Lands Commission. Article 4.8. Biofouling Management to Minimize the Transfer of Nonindigenous Species from Vessels Arriving at California Ports. California State Lands Commission, Sacramento, CA. http://slc.ca.gov/laws-regulations/current-regulations

Cuesta JA, Bruon A, Pérez-Dieste J, Trigo JE, Bañón B (2016). Role of ships' hull fouling and tropicalization process on European carcinofauna: new records in Galiciain waters (NW Spain). *Biological Invasions* **18**, 619–630.

Dafforn KA, Glasby TM, Johnston EL (2007). Differential effects of tributyltin and copper anti-foulants on recruitment of non-indigenous species. *Biofouling* **24**, 23–33.

Dafforn KA, Glasby TM, Johnston EL (2012). Comparing the invisibility of experimental 'reefs' with field observations of natural reefs and artificial structures. *PLoS ONE* 7, e38124.

DAISIE (2009). *Handbook of Alien Species in Europe*. Dordrecht: Springer.

David KAB, Keiron PPF (2003). Rafting by five phyla on man-made flotsam in the Southern Ocean. *Marine Ecology Progress Series* **262**, 289–291.

David M, Gollasch S (2008). EU shipping in the dawn of managing the ballast water issue. *Marine Pollution Bulletin* **56**, 1966–1972.

Davidson I, Simkanin C (2012). The biology of ballast water 25 years later. *Biological Invasions* **14**, 9–13.

Davidson IC, Brown CW, Sytsma MD, Ruiz GM (2009). The role of containerships as transfer mechanisms of marine biofouling species. *Biofouling* **25**, 645–655.

Davidson I, Scianni C, Hewitt C *et al.* (2016). Mini-review: assessing the drivers of ship biofouling management – aligning industry and biosecurity goals. *Biofouling* **32**, 411–428.

Davis MA, Chew MK (2017). 'The denialists are coming!' Well, not exactly: a response to Russell and Blackburn. *Trends in Ecology & Evolution* **32**, 229–230.

DAWR (2017). *Australian Ballast Water Management Requirements. Version 7*. Canberra: Australian Government Department of Agriculture and Water Resources.

De Clerck O, Gavio B, Fredericq S, Bárbara I, Coppejans E (2005). Systematics of *Grateloupia filicina* (Halymeniaceae, Rhodophyta), based on *rbcL* sequence analyses and morphological evidence, including the reinstatement of *G. minima* and the description of *G. capensis* sp. nov. *Journal of Phycology* **41**, 391–410.

Demirel YK, Uzun D, Zhang Y *et al.* (2017). Effect of barnacle fouling on ship resistance and powering. *Biofouling* **33**, 819–834.

Department of Fisheries (2014). *Western Australian Prevention List for Introduced Marine Pests*. Perth: Government of Western Australia, Department of Fisheries.

Desai DV, Narale D, Khandeparker L, Anil AC (2018). Potential ballast water transfer of organisms from the west to the east coast of India: insights through on board sampling. *Journal of Sea Research* **133**, 88–99.

Drake JM, Lodge DM (2006). Allee effects, propagule pressure and the probability of establishment: risk analysis for biological invasions. *Biological Invasions* **8**, 365–375.

Drake LA, Choi KH, Ruiz GM, Dobbs FC (2001). Global redistribution of bacterioplankton and virioplankton communities. *Biological Invasions* **3**, 193–199.

Drake LA, Doblin MA, Dobbs FC (2007). Potential microbial bioinvasions via ships' ballast water, sediment, and biofilm. *Marine Pollution Bulletin* **55**, 333–341.

Edelist D, Rilov G, Golani D, Carlton JT, Spanier E (2013). Restructuring the sea: profound shifts in the world's most invaded marine ecosystem. *Diversity and Distributions* **19**, 69–77.

Einfeldt AL, Addison JA (2015). Anthropocene invasion of an ecosystem engineer: resolving the history of Corophium volutator (Amphipoda: Corophiidae) in the North Atlantic. *Biological Journal of the Linnean Society* **115**, 288–304.

Elton C (1958). *The Ecology of Invasions by Animals and Plants*. London: Methuen.

EPA Victoria (2008) *Protocol for Environmental Management: Domestic Ballast Water Management in Victorian State Waters*. Southbank: EPA Victoria.

Epstein G, Smale DA (2018). Environmental and ecological factors influencing the spillover of the non-native kelp, *Undaria pinnatifida*, from marinas into natural rocky reef communities. *Biological Invasions* **20**, 1049–1072.

Escapa M, Isacch JP, Daleo P *et al.* (2004). The distribution and ecological effects of the introduced Pacific oyster *Crassostrea gigas* (Thunberg, 1793) in northern Patagonia. *Journal of Molluscan Research* **23**, 765–772.

Evans LV (1981) Marine algae and fouling: a review, with particular reference to ship-fouling. *Botanica Marina*, 24, 167–171.

Farnham WF, Irvine LM (1968). Occurrence of unusually large plants of *Grateloupia* in the vicinity of Portsmouth. *Nature* **219**, 744–746.

Ferrario J, Caronni S, Occhipinti-Ambrogi A, Marchini A (2017). Role of commercial harbours and recreational marinas in the spread of non-indigenous fouling species. *Biofouling* **33**, 651–660.

Finenko GA, Anninsky BE, Datsyk NA (2018). *Mnemiopsis leidyi* A. Agassiz, 1865 (Ctenophora: Lobata) in the inshore areas of the Black Sea: 25 years after its outbreak. *Russian Journal of Biological Invasions* **9**, 86–93.

Firth LB, Knights AM, Bell SS (2011). Air temperature and winter mortality: Implications for the persistence of the invasive mussel, Perna viridis in the intertidal zone of teh south-eastern United States. *Journal of Experimental Marien Biology and Ecology* **400**, 250–256.

Flagella MM, Verlaque M, Soria A, Buia MC (2007). Macroalgal survival in ballast water tanks. *Marine Pollution Bulletin* **54**, 1395–1401.

Floerl O (2002). Intracoastal spread of fouling organisms by recreational vessels. PhD thesis, James Cook University.

Floerl O (2005). Factors that influence hull fouling on ocean-going vessels. In: LS Godwin, ed., *Hull Fouling as a Mechanism for Marine Invasive Species Introductions: Proceedings of a Workshop on Current Issues and Potential Management Strategies, February 12–13, 2003*. Honolulu, HI: Bishop Museum.

Floerl O, Coutts ADM (2009). Potential ramifications of the global economic crisis on human-mediated dispersal of marine non-indigenous species. *Marine Pollution Bulletin* **58**, 1595–1598.

Floerl O, Inglis GJ (2005). Starting the invasion pathway: the interaction between source populations and human transport vectors. *Biological Invasions* **7**, 589–606.

Floerl O, Pool TK, Inglis GJ (2004). Positive interactions between non-indigenous species facilitate transport by human vectors. *Ecological Applications* **14**, 1724–1736.

Floerl O, Inglis GJ, Hayden BJ (2005a). A risk-based predictive tool to prevent accidental introductions of nonindigenous marine species. *Environmental Management* **35**, 765–778.

Floerl O, Inglis GJ, Marsh HM (2005b). Selectivity in vector management: an investigation of the effectiveness of measures used to prevent transport of non-indigenous species. *Biological Invasions* **7**, 459–475.

Folino-Rorem N, Darling J, D'Ausilio C (2009). Genetic analysis reveals multiple cryptic invasive species of the hydrozoan genus *Cordylophora*. *Biological Invasions* **11**, 1869–1882.

Frey MA, Simard N, Robichaud DD, Martin JL, Therriault TW (2014). Fouling around: vessel sea-chests as a vector for the introduction and spread of aquatic invasive species. *Management of Biological Invasions* **5**, 21–30.

Fuentes V, Angel D, Bayha K *et al.* (2010). Blooms of the invasive ctenophore, *Mnemiopsis leidyi*, span the Mediterranean Sea in 2009. *Hydrobiologia* **645**, 23–37.

Fulton SW, Grant FE (1900). Note on the occurrence of the European crab, Carcinus maenas, Leach, in Port Phillip. *Victorian Naturalist* **17**, 145–146.

Fulton SW, Grant FE (1902). Some little known Victorian decapod crustacea with description of a new species. *Proceedings of the Royal Society of Victoria* **14**, 55–64.

Galil B, Boero F, Campbell M *et al.* (2015). 'Double trouble': the expansion of the Suez Canal and marine bioinvasions in the Mediterranean Sea. *Biological Invasions* **17**, 973–976.

Gepp A, Gepp ES (1906). Some marine algae from New South Wales. *Journal of Botany, London* **44**, 249–261, Plate 481.

Gewing MT, Shenkar N (2017). Monitoring the magnitude of marine vessel infestation by non-indigenous ascidians in the Mediterranean. *Marine Pollution Bulletin* **121**, 52–59.

Glasby TM (1999). Effects of shading on subtidal epibiotic assemblages. *Journal of Experimental Marine Biology and Ecology* **234**, 275–290.

Glasby TM (2013) *Caulerpa taxifolia* in seagrass meadows: killer or opportunistic weed? *Biological Invasions* **15**, 1017–1035.

Glasby TM, Connell SD (1999). Urban structures as marine habitats. *Ambio* **28**, 595–598.

Glasby TM, Creese RG (2007). Invasive marine species management and research. In: SD Connell, BM Gillanders, eds., *Marine Ecology*. Melbourne: Oxford University Press, pp. 569–594.

Glasby T, Connell S, Holloway M, Hewitt C (2007). Nonindigenous biota on artificial structures: could habitat creation facilitate biological invasions? *Marine Biology* **151**, 887–895.

Gollasch S (1998). The asian decapod *Hemigrapsus penicillatus* (de Haan, 1835) (Grapsidae, Decapoda) introduced in European waters: status quo and future perspective. *Helgoländer Meeresuntersuchungen* **52**, 359–366.

Gollasch S (2002). The importance of ship hull fouling as a vector of species introductions into the North Sea. *Biofouling* **18**, 105–121.

Gollasch S (2007). Is ballast water a major dispersal mechanism for marine organisms? In: W Nentwig, ed., *Biological Invasions*. Berlin: Springer-Verlag, pp. 49–57.

Gollasch S, Lenz J, Dammer M, Andres HG (2000a). Survival of tropical ballast water organisms during a cruise from the Indian Ocean to the North Sea. *Journal of Plankton Research* **22**, 923–937.

Gollasch S, Rosenthal H, Botnen H *et al.* (2000b). Fluctuations of zooplankton taxa in ballast water during short-term and long-term ocean-going voyages. *International Review of Hydrobiology* **85**, 597–608.

Guiry MD, Guiry GM (2018). AlgaeBase. Worldwide electronic publication, National University of Ireland, Galway. www.algaebase.org

Gurevitch J, Padilla DK (2004). Are invasive species a major cause of extinctions? *Trends in Ecology and Evolution* **19**, 470–474.

Hadfield MG (1999). Macrofouling processes: a developmental and evolutionary perspective. In: *Book of Abstracts*. International Congress on Marine Corrosion and Fouling, Melbourne, Australia, p. 27 (unpublished).

Hallegraeff GM (1992). Harmful algal blooms in the Australian region. *Marine Pollution Bulletin* **25**, 186–190.

Hallegraeff GM (1993). A review of harmful algal blooms and their apparent global increase. *Phycologia* **32**, 79–99.

Hallegraeff GM (1998). Transport of toxic dinoflagellates via ships ballast water: bioeconomic risk assessment and efficacy of possible ballast water management strategies. *Marine Ecology Progress Series* **168**, 297–309.

Hallegraeff GM, Bolch CJ (1991). Transport of toxic dinoflagellate cysts via ships' ballast water. *Marine Pollution Bulletin* **22**, 27–30.

Hallegraeff GM, Bolch CJ (1992). Transport of diatom and dinoflagellate resting spores in ships' ballast water: implications for plankton biogeography and aquaculture. *Journal of Plankton Research* **14**, 1067–1084.

Hallegraeff GM, Gollasch S (2008). Anthropogenic introductions of microalgae. In: E Granéli, JT Turner, eds., *Ecology of Harmful Algae*. Berlin: Springer-Verlag, pp. 379–390.

Hanyuda T, Hansen GI, Kawai H (2018). Genetic identification of macroalgal species on Japanese tsunami marine debris and genetic comparisons with their wild populations. *Marine Pollution Bulletin* **132**, 74–81.

Hay CH, Luckens PA (1987). The Asian kelp *Undaria pinnatifida* (Phaeophyta: Laminariales) found in a New Zealand Harbour. *New Zealand Journal of Botany* **25**, 329–332.

Heersink A, Paini D, Caley P, Barry S (2014). *Asian Green Mussel: Estimation of Approach Rate and Probability of Invasion via Biofouling. Research Services Final Report*. Bruce: Plant Biosecurity Cooperative Research Centre.

Herborg L-M, Rushton SP, Clare AS, Bentley MG (2005). The invasion of the Chinese mitten crab (*Eriocheir sinensis*) in the United Kingdom and its comparison to continental Europe. *Biological Invasions* **7**, 959–968.

Herborg L-M, Weetman D, Van Ooosterhout C, Hänfling B (2007). Genetic population structure and contemporary dispersal patterns of a recent European invader, the Chinese mitten crab, *Eriocheir sinensis*. *Molecular Ecology* **16**, 231–242.

Hewitt CL, Campbell ML (2007). Mechanisms for the prevention of marine bioinvasions for better biosecurity. *Marine Pollution Bulletin* **55**, 395–401.

Hewitt C, Gollasch S, Minchin D (2009). The vessel as a vector – biofouling, ballast water and sediments. In: G. Rilov, J Crooks, eds., *Biological Invasions in Marine Ecosystems*. Berlin: Springer-Verlag, pp. 117–131.

Hewitt C, Campbell M, Coutts A, Rawlinson N (2011). *Vessel Fouling Risk Assessment*. Canberra: Department of Agriculture, Fisheries and Forestry.

Hutchings PA, Hilliard RW, Coles SL (2002) Species introductions and potential for marine pest invasions into tropical marine communities, with special reference to Indo-Pacific. *Pacific Science* **56**, 223–233.

Hutchins LW (1952). Species recorded from fouling. In: Woods Hole Oceanographic Institution, *Marine Fouling and Its Prevention*. Annapolis, MD: United States Naval Institute, pp. 165–207.

IMO (2009). *Ballast Water Management Convention and the Guidelines for Its Implementation*. 2009 Edition. London: International Maritime Organization.

IMO (2012a). *Guidelines for the Control and Management of Ships' Biofouling to Minimize the Transfer of Invasive Aquatic Species*. 2012 Edition. London: International Maritime Organization.

IMO (2012b). *Guidance for Minimizing the Transfer of Invasive Aquatic Species as Biofouling (Hull Fouling) for Recreational Craft. MEPC.1/Circ. 792*. London: International Maritime Organization.

Inglis GJ, Floerl O, Ayhong S *et al.* (2010). *The Biosecurity Risks Associated with Biofouling on International Vessels Arriving in New Zealand: Summary of the Patterns and Predictors of Fouling*. Wellington: Ministry of Agriculture and Forestry.

Invasive Species Advisory Committee (2005). *Progress Report on Meeting the Invasive Species Challenge: National Invasive Species Management Plan*. Washington, DC: Invasive Species Advisory Committee.

IUCN (2017). Invasive species: what is an invasive alien species? International Union for Conservation of Nature. www.iucn.org/theme/species/our-work/invasive-species

James K, Shears NT (2016). Proliferation of the invasive kelp *Undaria pinnatifida* at aquaculture sites promotes spread to coastal reefs. *Biological Invasions* **17**, 3393–3408.

Johnston EL (2007). Biological invasions and pollution. In: SD Connell, BM Gillanders, eds., *Marine Ecology*. Melbourne: Oxford University Press, p. 581.

Johnston EL, Keough MJ (2000). Field assessment of effects of timing and frequency of copper pulses on settlement of sessile marine invertebrates. *Marine Biology* **137**, 1017–1029.

Johnston EL, Keough MJ (2002) Direct and indirect effects of repeated pollution events on marine hard substrate assemblages. *Ecological Applications* **12**, 1212–1228.

Johnston EL, Dafforn KA, Clark GF, Rius M, Floerl O (2017). How anthropogenic activities affect the establishment and spread of non-indigenous species post-arrival. *Oceanography and Marine Biology: An Annual Review* **55**, 389–420.

Jousson O, Pawlowski J, Zaninetti L *et al.* (2000). Invasive alga reaches California. *Nature* **408**, 157–158.

Keough MJ, Ross J (1999) Introduced fouling species in Port Phillip Bay. In: CL Hewitt, ML Campbell, RE Thresher, RB Martin, eds., *Marine Biological Invasions of Port Phillip Bay, Victoria*. Centre for Research on Introduced Marine Pests, Technical Report No. 20. Hobart: CSIRO Marine Research, pp. 193–226.

King RJ, Black JH, Ducker SC (1971). Intertidal ecology of Port Phillip Bay with systematic list of plants and animals. *Memoirs of the National Museum of Victoria* **32**, 93–128.

Kochmann J, Carlsson J, Crowe TP, Mariani S (2012). Genetic evidence for the uncoupling of local aquaculture activities and a population of an invasive species – a case study of Pacific oysters (*Crassostrea gigas*). *Journal of Heredity* **103**, 661–671.

Kolar CS, Lodge DM (2001). Progress in invasion biology: predicting invaders. *Trends in Ecology and Evolution* **16**, 199–204.

Leão Z, Kikuchi R, Ferreira B *et al.* (2016). Brazilian coral reefs in a period of global change: a synthesis. *Brazilian Journal of Oceanography* **64**, 97–116.

Lewis JA (1998). Marine biofouling and its prevention on underwater surfaces. *Materials Forum* **22**, 41–61.

Lewis JA (1999). A review of the occurrence of exotic macroalgae in southern Australia, with emphasis on Port Phillip Bay. In: CL Hewitt, ML Campbell, RE Thresher, RB Martin, eds., *Marine Biological Invasions of Port Phillip Bay, Victoria*. Centre for Research on Introduced Marine Pests, Technical Report No. 20. Hobart: CSIRO Marine Research, pp. 61–87.

Lewis JA (2002a). *Hull Fouling as a Vector for the Translocation of Marine Organisms: Report 1 – Hull Fouling Research*. Ballast Water Research Series Report No. 14. Canberra: Department of Agriculture, Fisheries and Forestry.

Lewis JA (2002b). *Hull Fouling as a Vector for the Translocation of Marine Organisms: Report 2 – The Significance of the Prospective Ban on Tributyltin Antifouling Paints on the Introduction and Translocation of Marine Pests in Australia*. Ballast Water Research Series Report No. 15. Canberra: Department of Agriculture, Fisheries and Forestry.

Lewis JA (2016). Ship biofouling: what are the biosecurity risks? 18th International Congress on Marine Corrosion and Fouling, Toulon, France, June 2016. Unpublished poster.

Lewis JA, Coutts ADM (2009). Biofouling invasions. In: S Dürr, J Thomason, eds., *Biofouling*. Oxford: Blackwell Publishing, pp. 348–365.

Lewis JA, Smith BS, Taylor RJ, Batten JJ (1988). Fouling of RAN seawater systems and a comparison of electrochemical control methods. 8th Inter-Naval Corrosion Conference, Plymouth, UK, April 1988, Paper No. 7.

Lewis JA, Watson C, ten Hove HA (2006). Establishment of the Caribbean serpulid tubeworm *Hydroides sanctaecrucis* Krøyer [in] Mörch, 1863, in northern Australia. *Biological Invasions* **8**, 665–671.

Lewis PN, Riddle MJ, Smith SDA (2005). Assisted passage or passive drift: a comparison of alternative transport mechanisms for non-indigenous coastal species into the Southern Ocean. *Antarctic Science* **17**, 183–191.

Lezzi M, Del Pasqua M, Pierri C, Giangrande A (2018). Seasonal non-indigenous species succession in a marine macrofouling community. *Biological Invasions* **20**, 937–961.

Lockhart SJ, Ritz DA (2001). Preliminary observations of the feeding periodicity and selectivity of the introduced seastar, *Asterias amurensis* (Lutken), in Tasmania, Australia. *Papers and Proceedings of the Royal Society of Tasmania* **135**, 25–33.

Lowe S, Browne M, Boujelas S, De Poorter M (2000). *100 of the World's West Invasive Alien Species. A Selection from the Global Invasive Species Database*. Published by the Invasive Species Specialist Group (ISSG) a specialist group of the Species Survival Commission (SSC) of the World Conservation Union (IUCN). Auckland: ISSG Office.

MacDougall AS, Turkington R (2005). Are invasive species the drivers or passengers of change in degraded ecosystems? *Ecology* **86**, 42–55.

Makino W, Miura O, Kaiser F, Geffray M, Katsube T, Urabe J (2018). Evidence of multiple introductions and genetic admixture of the Asian brush-clawed crab Hemigrapsus takanoi (Decapoda: Brachyura: Varunidae) along the Northern European coast. *Biological Invasions* **20**, 825–842.

Mantelatto MC, Creed JC, Mourão GG, Migotto AE, Lindner A (2011). Range expansion of the invasive corals *Tubastraea coccinea* and *Tubastraea tagusensis* in the southwest Atlantic. *Coral Reefs* **30**, 397.

McCarthy SA, Khambaty FM (1994). International dissemination of epidemic *Vibrio cholerae* by cargo ship ballast and other nonpotable waters. *Applied and Environmental Microbiology* **60**, 2597–2601.

McCarthy SA, Mcphearson RM, Guarino AM, Gaines JL (1992). Toxigenic *Vibrio cholerae 01* and cargo ships entering Gulf of Mexico. *Lancet* **339**, 624–625.

McCollin T, Shanks AM, Dunn J (2008). Changes in abundance and diversity after ballast water exchange in regional seas. *Marine Pollution Bulletin* **56**, 834–844.

McMinn A, Hallegraeff GM, Thomson P, Jenkinson AV, Heijnis H (1997). Cyst and radionucleotide evidence for the introduction of the toxic dinoflagellate *Gymnodinium catenatum* into Tasmanian waters. *Marine Ecology Progress Series* **161**, 165–172.

Meinesz A (1999). *Killer Algae*. Chicago and London: University of Chicago Press.

Miller AW, Davidson IC, Minton MS *et al.* (2018). Evaluation of wetted surface area of commercial ships as biofouling habitat flux to the United States. *Biological Invasions* **20**, 1977–1990.

Minchin D (2002). Shipping: global changes and management of bioinvasions. In: F Briand, ed., *Alien Marine Organisms Introduced by Ships in the Mediterranean and Black Seas*. CIESM Workshop Monographs No. 20. Monaco: CIESM, pp. 99–102.

Minchin D, Gollasch S (2003). Fouling and ships' hulls: how changing circumstances and spawning events may result in the spread of exotic species. *Biofouling* **19**, 111–122.

Minchin D, Floerl O, Savini D, Occhipinti-Ambrogi A (2006). Small craft and the spread of exotic species. In: J Davenport, JL Davenport, eds., *The Ecology of Transportation: Managing Mobillity for the Environment*. Amsterdam: Springer, pp. 99–118.

Mineur F, Belsher T, Johnson MP, Maggs CA, Verlaque M (2007a). Experimental assessment of oyster transfers as a vector for macroalgal introductions. *Biological Conservation* **137**, 237–247.

Mineur F, Johnson M., Maggs CA, Stegenga H (2007b). Hull fouling on commercial ships as a vector of macroalgal introduction. *Marine Biology* **151**, 1299–1307.

Mineur F, Le Roux A, Maggs CA, Verlaque, M (2014). Positive feedback loop between introductions of non-native marine species and cultivation of oysters in Europe. *Conservation Biology* **28**, 1667–1676.

Minton MS, Verling E, Miller AW, Ruiz GM (2005). Reducing propagule supply and coastal invasions via ships: effects of emerging strategies. *Frontiers in Ecology and the Environment* **3**, 304–308.

Mohammad-Noor N, Adam A, Lim PT *et al.* (2018). First report of paralytic shellfish poisoning (PSP) caused by *Alexandrium tamiyavanichii* in Kuantan Port, Pahang, East Coast of Malaysia. *Phycological Research* **66**, 37–44.

Molino PJ, Wetherbee R (2008). The biology of biofouling diatoms and their role in the development of microbial slimes. *Biofouling* **24**, 365–379.

Molino PJ, Campbell E, Wetherbee R (2009). Development of the initial diatom microfouling layer on antifouling and fouling-release surfaces in temperate and tropical Australia. *Biofouling* **25**, 685–694.

Moreira PL, Ribeiro FV, Creed JC (2014). Control of invasive marine invertebrates: an experimental evaluation of the use of low salinity for managing pest corals (*Tubastraea* spp.). *Biofouling* **30**, 639–650.

Morris JAJ, Whitfield PE (2009). *Biology, Ecology, Control and Management of the Invasive Indo-Pacific Lionfish: An Updated Integrated Assessment*. NOAA Technical Memorandum NOS NCCOS 99. Beaufort, NC: NOAA.

Moser CS, Wier TP, Grant JF *et al.* (2016). Quantifying the total wetted surface area of the world fleet: a first step in determining the potential extent of ships' biofouling. *Biological Invasions* **18**, 265–277.

Moser CS, Wier TP, First MR *et al.* (2017). Quantifying the extent of niche areas in the global fleet of commercial ships: the potential for 'super-hot spots' of biofouling. *Biological Invasions* **19**, 1745–1759.

MPI (2014). *Craft Risk Management Standard: Biofouling on vessels arriving to New Zealand*. Wellington: Ministry for Primary Industries.

MPI (2016). *Import Health Standard: Ballast Water from All Countries. New Zealand Government Ministry for Primary Industries*, Wellington: Ministry for Primary Industries.

Murray CC, Gartner H, Gregr EJ *et al.* (2014). Spatial distribution of marine invasive species: environmental, demographic and vector drivers. *Diversity and Distributions* **20**, 824–836.

Nawrot R, Chattopadhyay D, Zuschin M (2015). What guides invasion success? Ecological correlates of arrival, establishment and spread of Red Sea bivalves in the Mediterranean Sea. *Diversity and Distributions* **21**, 1075–1086.

Ojaveer H, Olenin S. Naršcius A *et al.* (2017). Dynamics of biological invasions and pathways over time: a case study of a temperate coastal sea. *Biological Invasions* **19**, 799–813.

Ostenfeld CH (1908). On the immigration of *Biddulphia sinensis* Grev. and its occurrence in the North Sea during 1903–1907 and on its use for the study of the direction and rate of flow of the currents. *Meddelelser fra Kommissionen for Danmarks Fiskeri- og Havundersøgelser: Serie Plankton* **6**, 1–44.

Otani M (2006). Important vectors for marine organisms unintentionally introduced to Japanese waters. In: F Koike, MN Clout, M Kawamichi, M De Poorter, K Iwatsuki, eds., *Assessment and Control of Biological Invasion Risks*. Kyoto and Gland: Shoukadoh Booksellers and IUCN, pp. 92–103.

Otani M, Willan RC (2017). Osaka Bay in Japan as a model for investigating the factors controlling temporal and spatial persistence among introduced marine and brackish species in a heavily industrialized harbor. *Sessile Organisms* **34**, 28–37.

Otani M, Oumi T, Uwai S *et al.* (2007). Occurrence and diversity of barnacles on international ships visiting Osaka Bay, Japan, and the risk of their introduction. *Biofouling* **23**, 277–286.

Padilla DK, Williams SL (2004). Beyond ballast water: aquarium and ornamental trades as sources of invasive species in aquatic ecosystems. *Frontiers in Ecology and the Environment* **2**, 131–138.

Parry GD, Hirst AJ (2016). Decadal decline in demersal fish biomass coincident with a prolonged drought and the introduction of an exotic starfish. *Marine Ecology Progress Series* **544**, 37–52.

Parry GD, Heislers S, Werner G (2004). *Changes in the Distribution and Abundance of Asterias amurensis in Port Phillip Bay 1999–2003*. Queenscliff: Primary Industries Research Victoria.

Pettengill JB, Wendt DE, Schug MD, Hadfield MG (2007) Biofouling likely serves as a major mode of dispersal for the polychaete tubeworm *Hydroides elegans* as inferred form microsatellite loci. *Biofouling* **23**, 161–169.

Pimentel D, Lach L, Zuniga R, Morrison D (2000). Environmental and economic costs of nonindigenous species in the United States. *Bioscience* **50**, 53–65.

Pimentel D, Zuniga R, Morrison D (2005). Update on the environmental and economic costs associated with alien-invasive species in the United States. *Ecological Economics* **52**, 273–288.

Piola RF, McDonald JI (2012). Marine biosecurity: the importance of awareness, support and cooperation in managing a successful incursion response. *Marine Pollution Bulletin* **64**, 1766–1773.

Polte P, Buschbaum C (2008). Native pipefish *Entelurus aequoreus* are promoted by the introduced seaweed *Sargassum muticum* in the northern Wadden Sea, North Sea. *Aquatic Biology* **3**, 11–18.

Precht WF, Hickerson EL, Schmahl GP, Aronson RB (2014). The invasive coral *Tubastraea coccinea* (Lesson, 1829): implications for natural habitats in the Gulf of Mexico and the Florida Keys. *Gulf of Mexico Science* **2014**, 55–59.

Provan J, Murphy S, Maggs CA (2005). Tracking the invasive history of the green alga *Codium fragile* ssp *tomentosoides*. *Molecular Ecology* **14**, 189–194.

Purcell JE, Shiganova TA, Decker MB, Houde ED (2001). The ctenophore *Mnemiopsis* in native and exotic habitats: U.S. estuaries versus the Black Sea basin. *Hydrobiologia* **451**, 145–176.

Pyefinch KA, Downing FS (1949). Notes on the general biology of *Tubularia* Larynx Ellis & Solander. *Journal of the Marine Biological Association of the United Kingdom* **28**, 21–43.

Railkin AI (2004). *Marine Biofouling: Colonization Processes and Defenses*. TA Ganf, OG Manylov (translators). Boca Raton, FL: CRC Press.

Rainer SF (1995) *Potential for the Introduction and Translocation of Exotic Species by Hull Fouling: A Preliminary Assessment*. Centre for Research on Introduced Marine Pests, Technical Report No. 1. Hobart: CSIRO Marine Research.

Reed RH, Moffat L (1983). Copper toxicity and copper tolerance in *Enteromorpha compressa* (L.) Grev. *Journal of Experimental Marine Biology and Ecology* **69**, 85–103.

Richardson DM, Pyšek P (2006). Plant invasions: merging the concepts of species invasiveness and community invasibility. *Progress in Physical Geography* **30**, 409–431.

Richmond CA, Seed R (1991). A review of marine macrofouling communities with special reference to animal fouling. *Biofouling* **3**, 151–168.

Rilov G, Galil B (2009). Marine bioinvasions in the Mediterranean Sea – history, distribution and ecology. In: G Rilov, JA Crooks, eds., *Biological Invasions in Marine Ecosystems*. Berlin: Springer-Verlag, pp. 549–575.

Riul P, Targino CH, Júnior LAC *et al.* (2013). Invasive potential of the coral *Tubastraea coccinea* in the southwest Atlantic. *Marine Ecology Progress Series* **480**, 73–81.

Rius M, Heasman KG, McQuaid CD (2011). Long-term coexistence of non-indigenous species in aquaculture facilities. *Marine Pollution Bulletin* **62**, 2395–2403.

Rius M, Turon X, Bernardi G, Volckaert F, Viard F (2015). Marine invasion genetics: from spatio-temporal patterns to evolutionary outcomes. *Biological Invasions* **17**, 869–885.

Rivero NK, Dafforn KA, Coleman MA, Johnston EL (2013). Environmental and ecological changes associated with a marina. *Biofouling* **29**, 803–815.

Robinson TB, Griffiths CL, McQuaid CD, Rius M (2005). Marine alien species of South Africa – status and impacts. *African Journal of Marine Science* **27**, 297–306.

Rodgers SKU, Cox EF (1999). Rate of spread of introduced rhodophytes *Kappaphycus alvarez*, *Kappaphycus striatum*, and *Gracilaria salicornia* and their current distributions in Kane'ohe Bay, O'ahu, Hawai'i. *Pacific Science* **53**, 232–241.

Ross DJ, Johnson CR, Hewitt CL (2002). Impact of introduced seastars *Asterias amurensis* on survivorship of juvenile commercial bivalves *Fulvia tenuicostata*. *Marine Ecology Progress Series* **241**, 99–112.

Roughgarden J (1986). Predicting invasions and rates of spread. In: HA Mooney, JA Drake, eds., *Ecology of Biological Invasions of Northern America and Hawaii.* New York: Springer-Verlag, pp. 179–190.

Ruiz GM, Carlton JT, Grosholz ED, Hines AH (1997). Global invasions of marine and estuarine habitats by non-indigenous species: mechanisms, extent, and consequences. *American Zoologist* **37**, 621–632.

Ruiz GM, Rawlings TK, Dobbs FC *et al.* (2000). Global spread of microorganisms by ships. *Nature* **408**, 49–50.

Ruiz GM, Fofonoff PW, Steves BP, Carlton JT (2015). Invasion history and vector dynamics in coastal marine ecosystems: a North American perspective. *Aquatic Ecosystem Health & Management* **18**, 299–311.

Russ GR, Wake LV (1975). *A Manual of the Principal Australian Marine Fouling Organisms. Report 644.* Maribyrnong: Australian Defence Scientific Service.

Russell G, Morris OP (1973). Ship-fouling as an evolutionary process. In: RF Acker, B Floyd Brown, JR DePalma, WP Iverson, eds., *Proceedings of the Third International Congress on Marine Corrosion and Fouling, Gaithersburg, Maryland, October 2–6, 197.* Gaithursburg, MD: National Bureau of Standards, pp. 719–730.

Russell JC, Blackburn TM (2017). The rise of invasive species denialism. *Trends in Ecology & Evolution* **32**, 3–6.

Sammarco PW, Porter SA, Cairns SD (2010). A new coral species introduced into the Atlantic Ocean – *Tubastraea micranthus* (Ehrenberg, 1834) (Cnidaria, Anthozoa, Scleractinia): an invasive threat? *Aquatic Invasions* **5**, 131–140.

Sammarco PW, Porter SA, Genazzio M, Sinclair J (2015). Success in competition for space in two invasive coral species in the western Atlantic – *Tubastraea micranthus* and *T. coccinea*. *PLoS ONE* **10**(12), e0144581.

Santagata S, Gasiunaite ZR, Verling E *et al.* (2008). Effect of osmotic shock as a management strategy to reduce transfers of non-indigenous species among low-salinity ports by ships. *Aquatic Invasions* **3**, 61–76.

Saunders GW, Withall RD (2006). Collections of the invasive species *Grateloupia turuturu* (Halymeniales, Rhodophyta) from Tasmania, Australia. *Phycologia* **45**, 711–714.

Sax DF, Stachowicz JJ, Brown JH, Bruno JF (2007). Ecological and evolutionary insights from species invasions. *Trends in Ecology and Evolution* **22**, 465–471.

Schimanski KB, Piola RF, Goldstien SJ *et al.* (2016). Factors influencing the en route survivorship and post-voyage growth of a common ship biofouling organism, *Bugula neritina*. *Biofouling* **32**, 969–978.

Schimanski KB, Goldstien SJ, Hopkins GA, Atalah J, Floerl O (2017). Life history stage and vessel voyage profile can influence shipping-mediated propagule pressure of non-indigenous biofouling species. *Biological Invasions* **19**, 2089–2099.

Schultz MP (2007). Effects of coating roughness and biofouling on ship resistance and powering. *Biofouling* **23**, 331–341.

Schultz MP, Bendick JA, Holm ER, Hertel WM (2011). Economic impact of biofouling on a naval surface ship. *Biofouling* **27**, 87–98.

Sherman CDH, Lotterhos KE, Richardson MF *et al.* (2016). What are we missing about marine invasions? Filling in the gaps with evolutionary genomics. *Marine Biology* **163**, 198.

Shiganova TA (1998). Invasion of the Black Sea by the ctenophore *Mnemiopsis leidyi* and recent changes in pelagic community structure. *Fisheries Oceanography* **7**, 305–310.

Silva AG, Lima RP, Gomes AN, Fleury BG, Creed JC (2011). Expansion of the invasive corals *Tubastraea coccinea* and *Tubastraea tagusensis* into the Tamoios Ecological Station Marine Protected Area, Brazil. *Aquatic Invasions* **6**, S105–S110.

Simberloff DS (1989). Which insect introductions succeed and which ones fail? In: JA Drake, HA Mooney, F di Castri *et al.*, eds., *Biological Invasions: A Global Perspective*. New York: John Wiley and Sons, pp. 61–67.

Simberloff D, Schmatz DC, Brown TC (1997). *Strangers in Paradise: Impact and Management of Non-Indigenous Species in Florida*. Washington, DC: Island Press.

Simkanin C, Davidson I, Falkner M, Sytsma M, Ruiz G (2009). Intra-coastal ballast water flux and the potential for secondary spread of non-native species on the US West Coast. *Marine Pollution Bulletin* **58**, 366–374.

Soares MO, Campos CC, Santos NMO *et al.* (2018). Marine bioinvasions: differences in tropical copepod communities between inside and outside a port. *Journal of Sea Research* **134**, 42–48.

Stachowicz JJ, Byrnes JE (2006). Species diversity, invasion success, and ecosystem functioning: disentangling the influence of resource competition, facilitation, and extrinsic factors. *Marine Ecology Progress Series* **311**, 251–262.

Stachowicz JJ, Whitlatch RB, Osman RW (1999). Species diversity and invasion resistance in a marine ecosystem. *Science* **286**, 1577–1579.

Stachowicz JJ, Fried H, Osman RW, Whitlach RB (2002). Biodiversity, invasion resistance, and marine ecosystem function: reconciling pattern and process. *Ecology* **83**, 2575–2590.

Stafford H, Willan RC, Neil KM (2007). The invasive Asian Green Mussel, *Perna viridis* (Linnaeus, 1758) (Bivalvia: Mytilidae), breeds in Trinity Inlet, tropical northern Australia. *Molluscan Research* **27**, 105–109.

Stebbing ARD (1981). Hormesis – stimulation of colony growth in *Campanularia flexuosa* (hydrozoa) by copper, cadmium and other toxicants. *Aquatic Toxicology* **1**, 227–238.

Stuart-Smith RD, Edgar GJ, Stuart-Smith JF *et al.* (2015). Loss of native rocky reef biodiversity in Australian metropolitan embayments. *Marine Pollution Bulletin* **95**, 324–332.

Sun Y, Wong E, ten Hove HA *et al.* (2015). Revision of the genus *Hydroides* (Annelida: Serpulidae) from Australia. *Zootaxa* **4009**, 1–99.

Sun Y, Wong E, Keppel E, Williamson JE, Kupriyanova EK (2017). A global invader or a complex of regionally distributed species? Clarifying the status of an invasive calcareous tubeworm *Hydroides dianthus* (Verrill, 1873) (Polychaeta: Serpulidae) using DNA barcoding. *Marine Biology* **164**, 28.

Sylvester F, Kalaci O, Leung B *et al.* (2011). Hull fouling as an invasion vector: can simple models explain a complex problem? *Journal of Applied Ecology* **48**, 415–423.

Takata L, Falkner M, Gilmore S (2006). *Commercial Vessel Fouling in California: Analysis, Evaluation, and Recommendations to Reduce Nonindigenous Species Release from the Non-Ballast Water Vector*. Report to the California State Legislature. Sacramento, CA: Marine Facilities Division, California State Lands Commission.

Talman S, Bité JS, Campbell SJ *et al.* (1999). Impacts of some introduced marine species found in Port Phillip Bay. In: CL Hewitt, ML Campbell, RE Thresher, RB Martin, eds., *Marine Biological Invasions of Port Phillip Bay, Victoria*. Centre for Research on Introduced Marine Pests, Technical Report No. 20. Hobart: CSIRO Marine Research, pp. 261–274.

Tassin J, Thompson K, Carroll SP, Thomas CD (2017). Determining whether the impacts of introduced species are negative cannot be based solely on science: A response to Russell and Blackburn. *Trends in Ecology & Evolution* **32**, 230–231.

Thomsen MS, Byers JE, Schiel DR *et al.* (2014). Impacts of marine invaders on biodiversity depend on trophic position and functional similarity. *Marine Ecology Progress Series* **495**, 39–47.

Thomsen MS, Wernberg T, Schiel DR (2015). Invasions by non-indigenous species. In: TP Crowe, CLJ Frid, eds., *Marine Ecosystems: Human Impacts on Biodiversity, Functioning and Services*. Cambridge and New York: Cambridge University Press, pp. 274–331.

Thresher RE, Hewitt CL, Campbell ML (1999). Synthesis: introduced and cryptogenic species in Port Phillip Bay. In: CL Hewitt, ML Campbell, RE Thresher, RB Martin, eds., *Marine Biological Invasions of Port Phillip Bay, Victoria*. Centre for Research on Introduced Marine Pests, Technical Report No. 20. Hobart: CSIRO Marine Research, pp. 283–295.

US Executive Order 13112 (1999). Invasive species. *Federal Register* **64**, 6183–6186.

Utting SD, Spencer BD (1997). *The Hatchery Culture of Bivalve Mollusc Larvae and Juveniles*. Laboratory Leaflet No. 68. Lowestoft: Ministry of Agriculture, Fisheries and Food.

Vaselli S, Bulleri F, Benedetti-Cecchi L (2008). Hard coastal-defence structures as habitats for native and exotic rocky-bottom species. *Marine Environmental Research* **66**, 395–403.

Verlaque M, Brannock PM, Komatsu T, Villalard-Bohnsack M, Marston M (2005). The genus *Grateloupia* C. Agardh (Halymeniaceae, Rhodophyta) in the Thau Lagoon (France, Mediterranean): a case study of marine plurispecific introductions. *Phycologia* **44**, 477–496.

Vermeij MJA (2005). A novel strategy allows *Tubastraea coccinea* to escape small-scale adverse conditions and start over again. *Coral Reefs* **24**, 442.

Viejo RM (1999). Mobile epifauna inhabiting the invasive *Sargassum muticum* and two local seaweeds in northern Spain. *Aquatic Botany* **64**(2), 131–149.

Vilà M, Espinar JL, Hejda M *et al.* (2011). Ecological impacts of invasive alien plants: a meta-analysis of their effects on species, communities and ecosystems. *Ecology Letters* **14**, 702–708.

Visscher JP (1928). Nature and extent of fouling on ships' bottoms. *Bulletin of the Bureau of Fisheries* **43**, 193–252.

Vitousek PM (1990). Biological invasions and ecosystem processes: towards an integration of population biology and ecosystem studies. *Oikos* **57**, 7–13.

Vitousek PM, Dantonio CM, Loope LL, Rejmanek M, Westbrooks R (1997). Introduced species: a significant component of human-caused global change. *New Zealand Journal of Ecology* **21**, 1–16.

Voisin M, Engel CR, Viard F (2005). Differential shuffling of native genetic diversity across introduced regions in a brown alga: aquaculture vs. maritime traffic effects. *Proceedings of the National Academy of Sciences of the United States of America* **102**, 5432–5437.

Wahl M (1989). Marine epibiosis. I. Fouling and antifouling: some basic aspects. *Marine Ecology Progress Series*, **58**, 175–189.

Wahl M (1997). Living attached: aufwuchs, fouling, epibiosis. In: R Nagabhushanam, MF Thompson, eds., *Fouling Organisms of the Indian Ocean: Biology and Control Technology*. Rotterdam: A.A. Balkema, pp. 31–83.

Wallentinus I (2002). Introduced marine algae and vascular plants in European aquatic environments. In: E Leppäkoski, S Gollasch, S Olenin, eds., *Invasive Aquatic Species of Europe: Distribution, Impacts and Management*. Alphen aan den Rijn: Kluwer, pp. 27–52.

Walters LJ, Hadfield MG, Del Carmen KA (1997). The importance of larval choice and hydrodynamics in creating aggregations of *Hydroides elegans* (Polychaeta: Serpulidae). *Invertebrate Biology* **116**, 102–114.

Wasson K, Fenn K, Pearse J (2005). Habitat differences in marine invasions of central California. *Biological Invasions* **7**, 935–948.

Wilcove DS, Rothstein D, Dubow J, Phillips A, Losos E (1998). Quantifying threats to imperiled species in the United States: assessing the relative importance of habitat destruction, alien species, pollution, overexploitation, and disease. *BioScience* **48**, 607–615.

Womersley HBS (1966). Port Phillip Bay Survey 1957–1963. Algae. *Memoirs of the National Museum of Victoria* **27**, 133–156.

Wood EJF, Allen FE (1958). *Common Marine Fouling Organisms of Australian Waters.* Melbourne: Department of the Navy, Navy Office.

Wu H, Chen C, Wang Q, Lin J, Xue J (2017). The biological content of ballast water in China: a review. *Aquaculture and Fisheries* **2**, 241–246.

Zabin CJ, Ashton GV, Brown CW *et al.* (2014). Small boats provide connectivity for nonindigenous marine species between a highly invaded international port and nearby coastal harbors. *Management of Biological Invasions* **5**, 97–112.

Zabin CJ, Marraffini M, Lonhart SI *et al.* (2018). Non-native species colonization of highly diverse, wave swept outer coast habitats in Central California. *Marine Biology* **165**, 131.

Zalba S, Ziller SR (2007). Manejo adaptativo de espécies exóticas invasoras: colocando a teoria em prática. *Natureza & Conservação* **5**, 16–22.

Zibrowius H (1971). Les espèces Méditerranéennes du genre *Hydroides* (Polychaeta Serpulidae): remarques sur le prétendu polymorphisme de *Hydroides uncinata. Tethys* **2**, 691–746.

Zibrowius H (1973). Remarques sur trois espèces de Serpulidae acclimatées en Méditerranée: Hydroides dianthus, (Verrill, 1873), Hydroides dirampha Mörch, 1863, et Hydroides elegans (Haswell, 1883). *Rapports et process-verbaux des réunions Commission international pour l'exploration scientifique de la Mer Méditerranée* 21, 683–686.

8

Physical Effects of Ships on the Environment

TIMOTHY FILEMAN, STEPHEN DE MORA AND THOMAS VANCE

8.1 Introduction

The physical impacts that vessels may cause have not generally been emphasized in the past. However, they are becoming more and more apparent (Roberts, 2011), and this chapter offers a brief overview of the physical impacts that ships have on the marine environment. Numerous mechanisms are considered, together with examples of deleterious consequences. Coral reefs and seagrass communities are highlighted. The impacts of various processes, such as sediment suspension, aeration and microorganism mortality, are described.

Ships can physically affect the marine environment indirectly in a number of ways. For example, the impact of shipbreaking activities is discussed in Chapter 12. However, the management of ports and harbours together with the maintenance issues associated with dredging and the disposal of dredged material are not within the scope of that chapter.

8.2 Categories of Environmental Disturbance

Ships exert a physical effect on the marine environment through several mechanisms. Ship noise is considered in Chapter 9. The topics covered in this chapter comprise the following: ship wake and squat disturbance; propeller wash and aeration; anchoring; and grounding and sinking.

8.2.1 Ship Wake and Squat Disturbance

Ships create transverse and diverging waves as a result of their movement through water (Figure 8.1). This is known as the ship's wake or Kelvin wake pattern because it was first explained mathematically by Lord Kelvin (Thomson, 1887). These waves will interact with other vessels, the seabed, the shoreline and infrastructure.

There is a transfer of energy from the vessel to these waves. As the waves generated by the ship spread on the water surface, their energy is also transported horizontally. In general, the larger the wave, the more powerful it is and therefore the more effect it will have when it interacts with something. The highest waves generated by commercial vessels

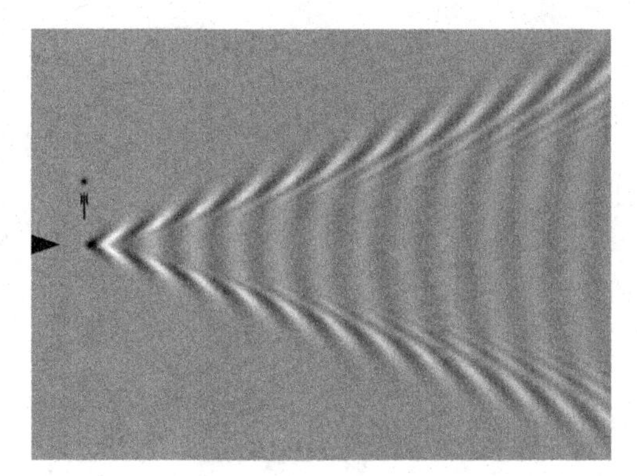

Figure 8.1 Kelvin ship's wake simulation.
Source: L3erdnik – own work, CC BY-SA 4.0: https://commons.wikimedia.org/w/index.php?curid=71068896

are estimated to be about 0.3 m (smaller than those generated by a 25-knot wind) (AMOG Consulting, 2010). As they approach the shore, waves are affected by the bathymetry, number of vessels, their distance from the shore and their speed and movements. When waves hit something, they are dissipated (i.e., the energy they contain is absorbed) or reflected (either partially or completely) back away from the thing that they hit. Wake waves differ from wind-generated waves in that they have regular periods and short durations. Wake waves build to a peak height close to a vessel and reduce in size as the vessel moves further away from them. The wave height, direction and period together with water depth will determine the effects such waves have when they interact with another vessel, the shoreline or a structure.

Studies show that while vessels travelling at very low and very high speeds cause very little bottom stirring, even at fairly shallow depths, vessels that travel at transitional speeds (near-plane) can cause significant stirring (the speed at which this occurs is a function of boat size and water depth), even in several metres of water (Beachler & Hilla, 2003; Hill & Beachler, 2002). Ships can also squat as they move through shallow waters, and may even ground in very shallow channels. This is a hydrodynamic phenomenon that is caused when a vessel moving through shallow water creates an area of lowered pressure under the ship that sucks it closer to the seabed than would otherwise be expected. The potential impacts of ship-generated waves and squatting are:

- Resuspension of sediment in shallow waters and subsequent shadowing (which affects the photosynthesis of plants and algae), smothering and/or loss of vegetation through uprooting
- Shoreline erosion
- Disturbance of wildlife (in and out of the water)
- Disturbance inside ports and marinas
- Damage to structures and moored vessels
- Reduced safety of other vessels (small craft in particular)

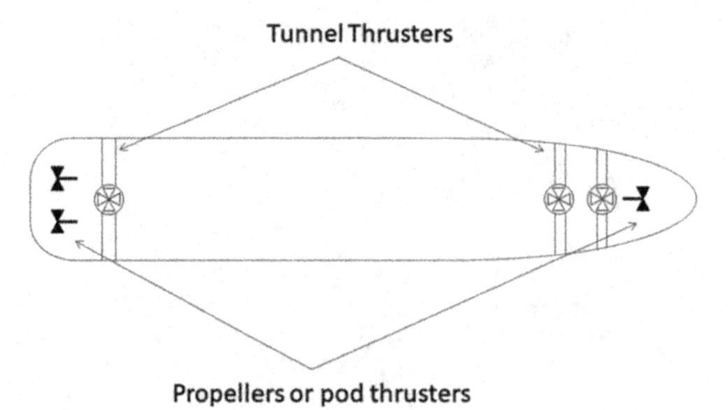

Figure 8.2 Common positions of propellers and/or thrusters on ships. The numbers of each may vary depending on the size and function of the ship.

8.2.2 Propeller Wash

When ships navigate shallow water areas their propellers will stir up sediments, water is accelerated through a spinning propeller, creating a swirling jet of water (propeller wash). This could be from any propeller-based propulsion system (e.g., normal propeller, tunnel thruster, podded/azimuth thrusters), and as shown in Figure 8.2, they could be at either end of the ship. Modern ships may have a number of propellers at bow and stern locations in the form of traditional propellers at the stern and tunnel thrusters at the bow and stern. Some modern ships may instead have pod (or azimuth) thrusters at the stern and sometimes at the bow too.

When propellers are used in shallow water, the jet of water generated can cause physical damage to animals and plants (e.g., uprooting), sediment resuspension and may also result in scouring (the removal of seabed sediment) of the bed sediments. Scour is an issue in some ports where continuous mooring and unmooring of vessels results in accumulated scouring with undermining and subsequent instability of port structures. Propeller wash can also impact plants through mechanical damage and uprooting combined with the removal of fine sediment, which often leaves large unvegetated regions (Essink, 1999; Kenworthy *et al.*, 2002).

8.2.3 Anchoring and Anchor Damage

Most ships regularly use anchors to maintain a position when stopped in shallow waters at sea or in a designated anchorage. When the anchor is dropped from the ship, the crown of the anchor (Figure 8.3) is the first to hit the seabed, and as the ship stops and drops back, pushed by the wind or tide, the flukes will dig into the seabed and 'anchor' the ship (Figure 8.4a). When the ship wants to move on, the anchor is picked up as its chain is winched onto the ship and the anchor's flukes are broken out of the seabed (Figure 8.4b).

By its very nature, anchoring will affect any seabed area. The extent and duration of the damage will depend upon the size of the anchor and the nature of the substrate. The seabed

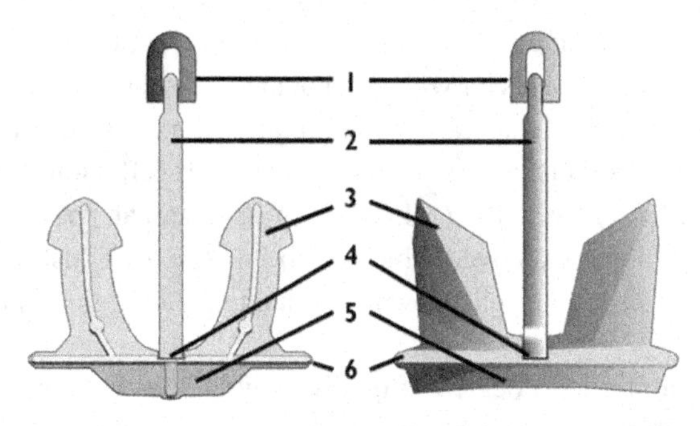

Conventional Stockless Anchor **New Stockless Anchor**

1. Anchor ring or shackle
2. Shank
3. Fluke (the point is called the Bill and the bottom section is the Arm)
4. Crown pin or Hinge Pin
5. Anchor Head or Crown
6. Ear

Figure 8.3 Diagram of the most common type of ship anchor; the stockless anchor.
Source: Attributed to Tosaka~commonswiki: https://commons.wikimedia.org/wiki/File:Stockless_anchors_Old%26New_NT.PNG

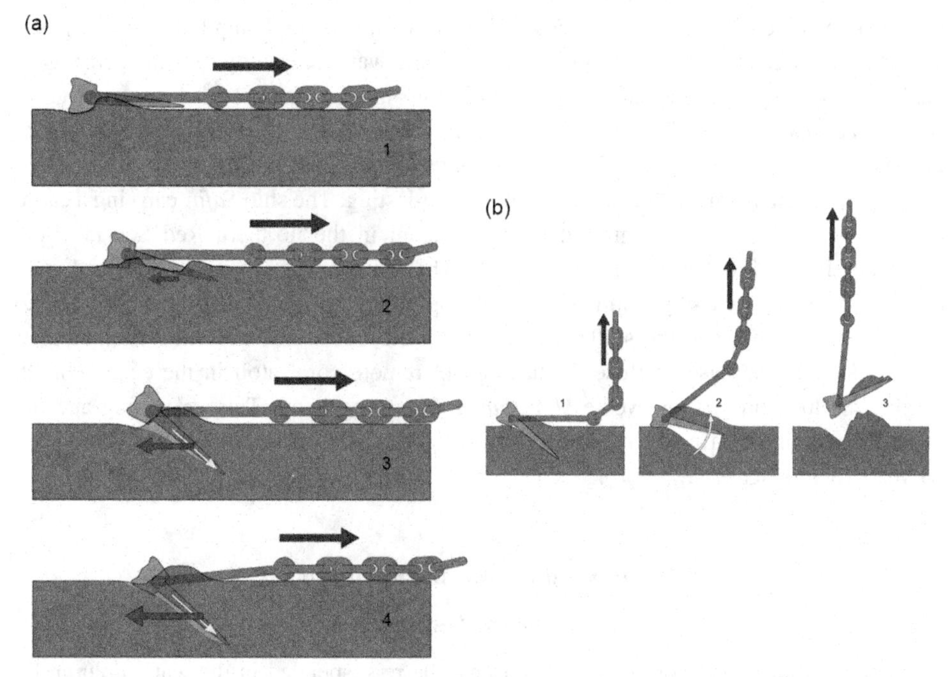

Figure 8.4 (a) The action of a stockless anchor being set (1 – anchor is dropped and the ship begins to drop back and pull on the chain; 2 – the bills on each fluke bite into the seabed and the flukes begin to dig in; 3 – the flukes dig in further and push deeper into the seabed; 4 – the anchor is fully bedded and holding). (b) An anchor being 'broken out' from the seabed (1 – the chain is winched up to the ship; 2 – the shank is lifted upwards until it hits the stop on the crown and begins to break the flukes out of the seabed; 3 – the anchor if lifted clear of the seabed).
Sources: (a) Attributed to Tosaka~commonswiki: https://en.wikipedia.org/wiki/Anchor#/media/File:Stockless_anchor_1_NT.PNG. (b) Attributed to Tosaka~commonswiki: https://en.wikipedia.org/wiki/Anchor

and its associated benthic community vary enormously in type, biodiversity and sensitivity to damage. Anchor damage has been widely recognized as a significant issue in coral reef areas, which are very sensitive habitats. For example, an extensive staghorn coral, *Acropora cervicornis*, reef was damaged by boat anchors in Fort Jefferson National Monument, Dry Tortugas, FL, USA (Davis, 1977). Ecosystem damage can manifest as a permanent scar and a decrease in live coral cover, which has been seen to persist for decades after the damage occurred in at least one well-known case due to the dragging of an anchor (Rogers & Garrison, 2001).

Seagrass beds exhibit a high degree of biodiversity and provide important nursery and juvenile habitats. Damage due to anchoring has been recognized in a marine protected area in Italy (Milazzo *et al.*, 2004), as well as in boat moorings in Perth, Australia (Walker *et al.*, 1989).

8.2.4 Ship Grounding and Sinking

Similarly to anchor damage, the environmental impact depends on where grounding or sinking happens. Most noticeable and of concern to the public is the accidental spillage of oil, considered in Chapter 3.

Coral reefs have received particular attention due to their vulnerability. The adverse effects on benthic habitats and species resulting from abrasion by ship hulls and propeller scarring are obviously restricted to relatively shallow water areas (e.g., submerged ridges, banks and bars, harbours, canals, bays and inlets, navigation channels and straits). Significant impacts can also result from spillage of cargo (Hawkins *et al.*, 1991; Schroeder *et al.*, 2008) or contamination from anti-fouling coatings (Jones, 2007; Negri *et al.*, 2002).

The impact on the marine ecosystem can be long-lasting. The ship *Safir*, carrying a cargo of crushed fluorapatite, ran aground at Ras Nasrani in the northern Red Sea in 1989, damaging the coral reef (Hawkins *et al.*, 1991). The damage was severe but localized, with benthic life being virtually eliminated within approximately 500 m of the reef due to crushing by the ship and the spillage of its cargo. Recovery was evident 7 months after the accident. In the case of Rose Atoll, a small, remote coral atoll in the central South Pacific, the long-line fishing vessel *Jin Shiang Fa* ran aground. This grounding and the spilled contaminant affected benthic communities and fish assemblages for more than 13 years (Schroeder *et al.*, 2008).

8.3 Effects on the Marine Environment

8.3.1 Sediment Resuspension

Sediment that has settled on the sea bottom can be re-suspended in the water column by ships' wake, wash or anchoring activities. Re-suspending contaminated bottom sediments leads to the release and increased bioavailability of a range of contaminants from bacteria (e.g., faecal coliforms) to chemical contaminants (e.g., metals, anti-fouling booster biocides and hydrocarbons) (Hill & Beachler, 2002; Oscarson *et al.*, 1980; Shipley *et al.*, 2011).

Generally, contaminant remobilization due to sediment disturbance is a highly complex process. The primary means by which contaminants are remobilized during sediment resuspension is through oxidation and desorption. For organic contaminants, desorption from particulates and transfer to the dissolved state are strongly influenced by the solubility of the contaminant, which is determined by its molecular structure (Latimer *et al.*, 1999). Contaminants that are remobilized into the water column become bioavailable to more organisms. To put this into context, however, biological reworking of sediments by benthic fauna plays a similar if not larger role than physical resuspension in the remobilization of sediment-bound contaminants (Pettibone *et al.*, 1996).

Sediment resuspension can also adversely affect benthic communities both through direct burial and through alterations to habitats due to changes in sediment structure (Roberts, 2012). Light extinction due to suspended sediment can be severe, although heavy sediment loads from propeller wash are short-term phenomena. This effect is also highly dependent on sediment type and depth, together with the speed and power of the vessel. However, it has been found that the relationship between sediment resuspension and depth may be exponential rather than linear (Goossens & Zwolsman, 1996; Hedman *et al.*, 2009). Thus, for a given vessel, small changes in depth may cause large differences in the sediment resuspension they cause. Repeated disturbances will maintain sufficient sediment load in the water column (turbidity) to sustain elevated light extinction if the sediment is fine enough. Increased turbidity results in low levels of transmitted light and can therefore negatively affect the functioning of light-dependent organisms such as phytoplankton, eelgrass and visual predators, such as fish and fish-eating birds (Zimmerman *et al.*, 2003). Increased turbidity may also interfere with the food intake of benthic invertebrates and copepods, and the functioning of fish gills may be impaired due to clogging (Zimmerman *et al.*, 2003). However, these effects will be localized in space and time and therefore have little impact on the total primary production of the estuary or tidal basin where a port is located. Naturally occurring turbidity elevations induced by tides, wind and storm events will have a longer-lasting and more significant effect than the periodic increased levels caused by vessel activities (Zimmerman *et al.*, 2003).

8.3.2 Microorganism Mortality

Very few studies have been conducted to quantify plankton mortality from propellers. There is also literature that examines the stresses that could affect entrained eggs and larvae in turbines (Gucinski, 1981). The shear forces, pressure changes and blade contact from propellers can damage plankton, entrained eggs and larvae in the surface layers of water (Gucinski, 1981; Klein, 2007; Paitkowski, 1983). The scale of this depends upon the propellers' rotational velocity (i.e., damage is low at low velocities). Many propellers also cavitate during their normal operation; this is particularly true of water jet propulsion systems. Cavitation damages cells in organisms, particularly bacteria and soft phytoplankton. These effects will be localized in space and time and therefore are likely to have little impact overall.

8.3.3 Habitat Damage

The physical effects of ship activities on seagrass beds and coral reefs from such things as propeller scarring, anchoring and groundings have been reported. Anchoring and grounding in particular have been shown to damage precious seagrass beds in shallow coastal waters. If an anchor is set into a seagrass bed and broken out (Figure 8.4), it uproots plants, reducing the density of shoots and leading to patchy bed cover. Anchor chains can cause significant damage to seagrass beds and reefs. Anchoring on rocky bottoms is also a threat to the species that live there (Panigada *et al.*, 2008).

8.3.4 Aeration

Aeration of water is generally defined as the process of adding oxygen. Aquatic organisms use dissolved oxygen (DO) in water when they respire (the metabolic process by which an organism obtains energy by reacting oxygen with glucose to produce water, carbon dioxide and energy). DO aids in the breakdown of decaying vegetation and other sources of nutrients that can facilitate harmful algal blooms and excess aquatic plant growth. In these respects, aeration by propellers is generally a good thing. Maintaining healthy levels of DO in water is one of the most – if not *the* most – important water quality parameter. However, it has been shown that the oxidation of certain sulfides in surface sediments increases the release and toxicity of copper to the benthic amphipods (Simpson *et al.*, 2012); in other words, DO may oxidize metal sulfides in sediments and release metals into more bioavailable forms (Shipley *et al.*, 2011). Conversely, there is also evidence of the reduced toxicity of arsenic through abiotic oxidation in aquatic environments (Oscarson *et al.*, 1980). In considering the overall impact of aeration of water by ships' propellers, one needs to bear in mind the relative scale of the issue with respect to the wider marine environment and the natural processes it undergoes (i.e., diffusion, photosynthesis, waves, tides and wind and storm events).

8.4 Conclusions

It is clear that any significant physical effects of ships on the environment will be in near-shore and/or relatively shallow areas. Many of these are localized and some take years to repair, if they can be repaired at all.

References

AMOG Consulting (2010). *BHP Billiton Olympic Dam EIS Project Report Appendix H11. Assessment of Waves and Propeller Wash Associated with Shipping: Final Report.* Notting Hill: AMOG Consulting.

Beachler, M. M. & Hilla, D. F. (2003). Stirring up trouble? Resuspension of bottom sediments by recreational watercraft. *Lake and Reservoir Management*, **19**, 15–25.

Davis, G. E. (1977). Anchor damage to a coral reef on the coast of Florida. *Biological Conservation*, **11**, 29–34.

Essink, K. (1999). Ecological effects of dumping of dredged sediments; options for management. *Journal of Coastal Conservation*, **5**, 69–80.

Goossens, H. & Zwolsman, J. J. G. (1996). An evaluation of the behaviour of pollutants during dredging activities. *International Journal on Public Works, Ports & Waterways Developments – Terra et Aqua*, **62**, 20–28.

Gucinski, H. (1981). *Sediment Suspension and Resuspension from Small-Craft Induced Turbulence*. Washington, DC: US Environmental Protection Agency.

Hawkins, J. P., Roberts, C. M. & Adamson, T. (1991). Effects of a phosphate ship grounding on a Red Sea coral reef. *Marine Pollution Bulletin*, **22**, 538–542.

Hedman, J. E., Stempa Tocca, J. & Gunnarsson, J. S. (2009). Remobilization of polychlorinated biphenyl from Baltic Sea sediment: comparing the roles of bioturbation and physical resuspension. *Environmental Toxicology and Chemistry*, **28**, 2241–2249.

Hill, D. F. & Beachler, M. M. (2002). ADV measurements of planing boat prop wash in the extreme near field. In: T. L. Wahl, C. A. Pugh, K. A. Oberg & T. B. Vermeyen, eds., *Hydraulic Measurements and Experimental Methods*. Reston, VA: American Society of Civil Engineers, pp. 1–10.

Jones, R. J. (2007). Chemical contamination of a coral reef by the grounding of a cruise ship in Bermuda. *Marine Pollution Bulletin*, **54**, 905–911.

Kenworthy W. J., Fonseca M. S., Whitfield P. E. & Hammerstrom K. K. (2002). Impacts of motorized watercraft on shallow estuarine and coastal marine environments. *Journal of Coastal Research*, **37**, 75–85.

Klein, R. (2007). The Effects of Marinas and Boating Activity upon Tidal Waterways. Community & Environmental Defence Services. http://tidewatercurrent.com/Marinas.pdf

Latimer J. S., Davis W. R. & Keith D. J. (1999). Mobilization of PAHs and PCBs from in-place contaminated marine sediments during simulated resuspension events. *Estuarine, Coastal and Shelf Science*, **49**, 577–595.

Milazzo, M., Badalamenti, F., Ceccherelli, G. & Chemello, R. (2004). Boat anchoring on beds in a marine protected area (Italy, western Mediterranean): effect of anchor types in different anchoring stages. *Journal of Experimental Marine Biology and Ecology*, **299**, 51–62.

Negri, A. P., Smith, L. D., Webster, N. S. & Heyward, A. J. (2002). Understanding ship-grounding impacts on a coral reef: potential effects of anti-foulant paint contamination on coral recruitment. *Marine Pollution Bulletin*, **44**, 111–117.

Oscarson, D. W., Huang, P. M. & Liaw, W. K. (1980). The oxidation of arsenite by aquatic sediments. *Journal of Environmental Quality*, **9**, 700–703.

Panigada, S., Pavan, G., Borg, J. A., Galil, B. S. & Vallini, C. (2008). Biodiversity impacts of ship movements, noise, grounding and anchoring. In: A. Abdullah & O. Lindend, eds., *Maritime Traffic Effects on Biodiversity in the Mediterranean Sea – Volume 1: Review of Impacts, Priority Areas and Mitigation Measures*. Gland: IUCN, p. 10.

Pettibone G. W., Irvine K. N. & Monahan K. M. (1996). Impact of a ship passage on bacteria levels and suspended sediment characteristics in the Buffalo River, New York. *Water Research*, **30**, 2517–2521.

Piatkowski, U. (1983). *Joint biological Expedition on RRS 'John Biscoe', February 1982 (II), Reports on Polar Research*. Bremerhaven: Alfred Wegener Institute for Polar and Marine Research.

Roberts, D. A. (2012). Causes and ecological effects of resuspended contaminated sediments (RCS) in marine environments. *Environment International*, **40**, 230–243.

Roberts, J. (2011). *Maritime Traffic in the Sargasso Sea: An Analysis of International Shipping Activities and Their Potential Environmental Impacts*. Hampshire: IUCN Sargasso Sea Alliance Legal Working Group by Coastal & Ocean Management.

Rogers, C. S. & Garrison, V. H. (2001). Ten years after the crime: lasting effects of damage from a cruise ship anchor on a coral reef in St. John, U.S. Virgin Islands. *Bulletin of Marine Science*, **69**, 793–803.

Schroeder, R. E., Green, A. L., DeMartini, E. E. & Kenyon, J. C. (2008). Long-term effects of a ship-grounding on coral reef fish assemblages at Rose Atoll, American Samoa. *Bulletin of Marine Science*, **82**, 345–364.

Shipley, H. J., Gao, Y., Kan, A. T. & Tomson, M. B. (2011). Mobilization of trace metals and inorganic compounds during resuspension of anoxic sediments from Trepangier Bayou, Louisiana. *Journal of Environmental Quality*, **40**, 484–491.

Simpson, S. L., Ward, D., Strom, D. & Jolley, D. F. (2012). Oxidation of acid-volatile sulfide in surface sediments increases the release and toxicity of copper to the benthic amphipod *Melita plumulosa*. *Chemosphere*, **88**, 953–961.

Thomson, W. (1887). On ship waves. *Institution of Mechanical Engineers, Proceedings*, **38**, 409–434; illustrations, pp. 641–649.

Walker, D. I., Lukatelich, R. J., Bastyan, G. & McComb, A. J. (1989). Effect of boat moorings on seagrass beds near Perth, Western Australia, *Aquatic Botany*, **36**, 69–77.

Zimmerman, J. K. H., Vondracek B. & Westra J. (2003). Agricultural land use effects on sediment loading and fish assemblages in two Minnesota (USA) watersheds. *Environmental Management*, **32**, 93–105.

9

Ship Noise

SALVATORE VIOLA AND VIRGINIA SCIACCA

9.1 Introduction

Anthropogenic underwater noise has severely increased over the last century and a significant component of noise in marine environments is due to ship traffic. Every year at sea, we observe the continuous movement of more than 60,000 medium to very large commercial vessels, such as cargo ships, bulk carriers, container vessels, tankers, cruise ships and ferries (Equasis, 2015). The incredible increase in commercial maritime trade and the related increase in vessel speed of the last 40 years have raised the amount of noise that ship traffic is spreading throughout the oceans. From the 1960s, when the first measures of noise levels were reported (Wenz, 1962), until the 1990s, underwater noise has almost doubled every 10 years (Andrew *et al.*, 2002; McDonald *et al.*, 2006a; Merchant *et al.*, 2012). While some recent studies describe slowly decreasing low-frequency ocean noise levels at different oceanic locations during the early 2000s (Andrew *et al.*, 2011; Miksis-Olds *et al.*, 2016), the typical and long-term trends for ship noise are still unknown in many regions of the world (Viola *et al.*, 2017).

What is the contribution to ambient noise by ship traffic and how does it affect marine life? Is worldwide legislation going towards an efficient management of the 'ship noise' issue? Taking our start from these two questions, in this chapter we will give a general overview of the noise produced by ships and of its effects on marine life. Lastly, we will briefly review the main international and national regulatory efforts for the mitigation of ship noise impacts.

9.2 Environmental Acoustic Noise

The environmental noise in seawater has been mapped out in detail by military and marine science researchers. At frequencies lower than 1 kHz, the noise is dominated by seismic and shipping noise. Between 1 and 50 kHz, the underwater acoustic noise is strongly dependent on the sea-state condition: the main contribution is due to wind and rain interactions with the sea surface. The power spectral density (PSD) of the acoustic noise in seawater as a function of the frequency (f) is usually approximated by the Knudsen formula (Urick, 1982):

$$PSD(f_{Hz}, SS) = 94.5 - 10 \log f^{5/3} + 30 \log (SS + 1) re\ 1\ \mu Pa^2/Hz$$

where *SS* (sea state) indicates the conditions of the sea surface on a scale from 0 (calm sea) to 9. This approximation can be applied in the frequency range between 1 kHz and a few tens of kHz. Above 50 kHz, the main contribution to the noise spectrum is due to the thermal vibration of the water molecules that is proportional to f^2.

9.3 Radiated Noise from Ships

Ship traffic can be considered as the main source of anthropogenic noise in the underwater environment. Ship noise typically affects the frequency band from 50 to 150 Hz, although it can extend to some tens of kHz (Ross, 1976). At low frequency (below a few hundred hertz), sound attenuation can be attributed almost entirely to geometrical spreading, since the absorption due to salts dissolved in seawater can be neglected (Fisher & Simmons, 1977).

The most prominent part of what is generally indicated as 'ship noise' underwater is actually produced by propeller rotation. Propellers can produce both non-cavitating and cavitating noise. In a non-cavitation condition, the propeller noise comes from the vibration of blades. This results in the production of distinct narrowband tones at very low frequencies with a broadband component at higher frequencies. The frequency of the tones depends on the shaft speed and on the number of propeller blades. The broadband component is generated by the inflow turbulence into the propeller and by edge effects (Carlton, 2007).

The propeller becomes noisier in the case of cavitation. Cavitation is the predominant source of noise from vessels (Ross, 1976). Cavitation noise occurs when propeller blades pass through the uneven wake field behind the ship. When the pressure on the suction side is below evaporation pressure, steam-filled bubbles are created, and their collapse produces a 'white noise' that can extend up to around 1 MHz (Carlton, 2007).

The speed of cavitation inception depends on inflow speed variation, propeller loading, propeller submergence and the quality of the propeller design. Blade sheet cavitation also produces tonal noise at harmonics of the blade passing rate. Typical pressure maxima are 2–10 kPa for a first harmonic, which usually decreases with the higher harmonics. Tonals at multiples of the blade rate can be very much influenced by the stern and propeller design and the application of appendages improving the wake field (Wittekind, 2014).

For fixed pitch propellers, cavitation is generally directly related to the shaft rotational speed and usually decreases with reducing ship speed (Traverso *et al.*, 2015). In the case of controllable pitch propellers, cavitation increases when the blades rotate away from the design condition. In this case, cavitation phenomena may be stronger at low speeds.

Since the propeller is typically placed a few metres under the sea surface, the propagation of propeller noise is affected by Lloyd's mirror effects. The composition of the direct waves and their reflection on the sea surface results in dipole interference, producing a low-frequency cut-off depending on the depth of the propeller (Brekhovskikh,

2003). Lloyd's mirror effects influence the amount of radiated noise in the horizontal direction as a function of the propeller depth. They are lower when the source depth is shallower. This means that the radiated noise of an individual ship changes as a function of its load. Unloaded ships radiate less noise because of the shallower propeller depth (McKenna *et al.*, 2012).

Propeller rotation is not the only source of ship noise. Sounds from noisy machinery such as engines, generators and winches can be transmitted via the hull into the sea (Zimmer, 2011). The propeller is typically driven by a low-speed two-stroke-cycle engine rigidly mounted in the ship. This kind of engine rotates at the same speed as the propeller and may contribute to the overall radiated noise. For smaller ships, medium-speed propulsion diesels are commonly used. These kinds of engine rotate at constant speed and the ship velocity is controlled by varying the pitch of the propeller's blades. This propulsion system is noisier at lower speeds, when the pitch is far from optimal performance.

Other sources of acoustic noise in water are the diesel generators installed aboard ships. The most common diesel generators at 720 rpm produce tonal sounds at 6 Hz and multiple frequencies up to the kilohertz range (Wittekind, 2014). Radiation of noise from ships is typically asymmetrical. Blockage by the hull and absorption by wake bubbles can modify the far-field acoustic output near the bow and stern aspects (Trevorrow, 2008). It was observed that in modern commercial ships the stern aspect noise levels are from 5 to 10 dB higher than the bow aspect noise levels (McKenna *et al.*, 2012). Ship noise is obtained by the composition of multiple acoustic sources (propeller, engines, generators, etc.). Each ship has its own acoustic signature and the levels of noise mainly depend on the size and on the operating speed.

Noise from commercial shipping is mainly peaked between 10 and 50 Hz. In this frequency band, noise spectral levels go from 195 dB re 1 $\mu Pa^2/Hz$ at 1 m for fast-moving supertankers to 140 dB re 1 $\mu Pa^2/Hz$ at 1 m for small fishing vessels. Noise levels from small boats are typically lower, but their spectral contribution in 1–5-kHz band is higher (Hildebrand, 2009). Broadband hydrodynamic noise radiated from fast ferries can extend up to 10 kHz, with narrow peaks in the 1–2-kHz range (Abdulla & Linden, 2008). A study performed in the Salish Sea, in Washington State and in British Columbia based on the acoustic measurement of 2809 isolated transits reports mean broadband (20–40,000 Hz) source levels for the ship population of 173 $\pm$ 7 dB re 1 μPa at 1 m. In this study, the noisiest ship classes were identified as carriers, cargo ships, tankers and bulk carriers (Veirs *et al.*, 2016).

9.4 Effects of Ship Noise on Marine Life

Exposure to ship noise can produce a wide range of harmful effects on marine mammals (Weilgart, 2007), fishes and invertebrates (Buscaino *et al.*, 2010; Celi *et al.*, 2014; Slabbekoorn *et al.*, 2010), ranging from behavioural modifications to physiological and auditory effects.

The monitoring of the impact of ship noise on marine species requires a highly interdisciplinary approach, with the application of tools and methodologies typical of different disciplines (e.g., bioacoustics, ecology, ethology, physical and biological oceanography, etc.). Furthermore, the monitoring methodology to be applied strongly depends on the characteristics of the species of interest and on expected effects, but it also depends on the characteristics of the noise source.

As an example, in the case of offshore noisy activities, such as drilling and seismic monitoring campaigns, surveys can be conducted before, during and after the exposition. In contrast, understanding of the effects of low-frequency diffuse ship noise remains particularly challenging due to the numerous difficulties in determining the degree of cumulative exposure and due to the many uncertainties in the physiology, behaviour, distribution and hearing sensitivity of the different species.

In recent decades, responses to noise have been mainly studied in marine mammals, particularly in cetacean species. Only in recent years has a growing concern regarding ship noise impact led to an increased number of studies on the effects that noise could have on fish. More than 50 families of fish use sound (typically in the frequency range between 100 Hz and 3 kHz) to perceive information from the environment and to communicate (Obrist *et al.*, 2010; Panigada *et al.*, 2008).

9.5 Auditory and Physiological Effects

9.5.1 Acoustic Masking

Acoustic masking is probably the most diffuse consequence of the introduction of ship noise in the marine environment. The masking of biologically important sounds by ship noise affects animals' acoustic communication underwater, but it also compromises predator–prey interactions, perceptions and interpretations of other environmental or anthropic sounds (Clark *et al.*, 2009). In particular, 'communication masking' can be defined as a loss of 'communication space' (i.e., the volume of space within which acoustic communication among conspecifics is expected to occur; Clark *et al.*, 2009) as a result of the presence of an 'external' interfering noise. This means that a given receiver has a reduced ability to detect important signals in the same frequency range as the noise (Clark *et al.*, 2009; Southall *et al.*, 2007).

During the last century, because of the global increase in shipping, rises in background noise levels seem to have caused a severe reduction of the acoustic communication space of several species. However, the degree of interference from acoustic masking varies with the spatial, spectral and temporal relationships between signal and noise (Erbe *et al.*, 2016; Southall *et al.*, 2007). Hence, it is still not possible to quantify globally the ecological consequences of acoustic masking, due to the many remaining uncertainties regarding the species-specific biological and ecological costs of the reduction in acoustic communication capabilities.

In marine mammals, masking is considered the most common auditory effect of diffuse ship noise (Southall *et al.*, 2007). In these highly sonic species, acoustic masking potentially interferes with many vital functions, such as echolocation, predator–prey interactions and long-range acoustic communication needed to facilitate social interactions and mating (Clark *et al.*, 2009; Croll *et al.*, 2002; Edds-Walton, 1997). The acoustic masking associated with the increase in shipping noise is particularly relevant to baleen whales such as the fin whale (*Balaenoptera physalus*) and the blue whale (*Balaenoptera musculus*) that use low-frequency sounds (10–200 Hz) to communicate over long distances (Clark *et al.*, 2009; Edds-Walton, 1997; McDonald *et al.* 2006b; Širović *et al.*, 2013). A relevant reduction of the acoustic communication space has been demonstrated in the highly endangered North Atlantic right whale (Clark *et al.*, 2009; Tennessen & Parks, 2016).

Nevertheless, the high-frequency components of ship noise may also mask the echolocation clicks of odontocete species, such as the Cuvier's beaked whale, *Ziphius cavirostris* (Aguilar Soto *et al.*, 2006) and the killer whale, *Orcinus orca* (Veirs *et al.*, 2016). High-frequency components of ship noise typically increase with vessel speed due to the increase in broadband cavitation noise (Aguilar Soto *et al.*, 2006; Southall, 2005; Wright *et al.*, 2007).

A growing body of literature demonstrates that ship noise can also mask fish communication and their use of the acoustic 'soundscape' to learn about the environment (Slabbekoorn *et al.*, 2010; Wahlberg & Westerberg, 2005). As an example, Vasconcelos *et al.* (2007) demonstrated that ship noise affects the intra-species acoustic communication of the Lusitanian toadfish (*Halobatrachus didactylus*) by masking its mating and agonistic vocalizations. Moreover, Codarin *et al.* (2009) observed under controlled conditions that noise emanating from a cabin cruiser can significantly reduce the auditory sensitivity and increase the thresholds for detection of conspecific sounds in two fish species: the brown meagre (*Sciaena umbra*) and the Mediterranean damselfish (*Chromis chromis*).

Other important auditory effects directly involve the auditory system, leading to what is called a 'noise-induced threshold shift' (TS). A TS occurs when animals are exposed to a sound that is intense enough to cause a shift in the hearing threshold and hence a reduction in auditory sensitivity. The magnitude of the TS depends on several factors, such as the duration of the exposure, the frequency, the received level and the energy distribution. The extent of the TS usually decreases with time after the exposure. If the TS goes rapidly back to zero, it can be defined a temporary TS (TTS). If, after a long period, the TS value does not return to zero, the animal will suffer of a so-called permanent TS (PTS). A PTS is a physical injury that implies irreversible damage to the auditory system (e.g., damage to the sensory hair cells of the inner ear or changes in the chemical composition of the inner ear fluids). To date, there is still a lack of evidence regarding the occurrence and causes of physical injury to the auditory system after exposure to ship nose, especially in marine mammals. Nevertheless, both temporary and permanent hearing losses may occur as an irreversible consequence of long-term exposure to ship noise, as has been demonstrated in fish (Panigada *et al.*, 2008; Southall *et al.*, 2007).

The main obstacle in assessing the auditory effects of ship noise on many marine species remains the lack of information concerning: (1) species-specific hearing sensitivity; (2) how animals derive information from received sounds in a complex sound field (frequency discrimination, resolution and detection; see Southall *et al.*, 2007); and (3) the history of exposure. The hearing sensitivity of many marine organisms is still poorly understood, with the exception of those studied under controlled conditions. Behavioural studies and electrophysiological measurements can be performed and anatomical models constructed to study the hearing sensitivity of different species or functional hearing groups (Southall *et al.*, 2007). Audiograms are the conventionally displayed graphs of hearing ability, represented as frequency versus sensitivity measured as sound pressure or intensity (Ketten, 2002). While in laboratory conditions behavioural thresholds are excellent ways of measuring hearing, they also have considerable constraints (Nachtigall *et al.*, 2007). Moreover, each species hears differently a sound received at the same frequency. Thus, analytical methodologies such as frequency-weighting functions are often used to study the responses from different species and to compensate for varying sensitivity (Southall *et al.*, 2007).

9.5.2 Physiological Effects

Physiological effects related to the exposure to noise are described for several species of cetaceans, but also for fish and invertebrates (Celi *et al.*, 2016; Nedelec *et al.*, 2014; Weilgart, 2007; Wright *et al.*, 2007). A main cause of physiological alterations in marine species is the *stress* induced by the exposure to ship noise. The chronic stress caused by noise induces different physiological effects and, above all, the inhibition of the immune system, with the potential impairment of the health of the affected specimens (Weilgart, 2007; Wright *et al.*, 2007). To acquire information on the auditory and physiological status of a given specimen in relation to ship noise, many cues can be measured, such as heart rate, auditory response sensitivity, TTS and PTS (Kull & McGarrity, 2002), but also the noise-induced production or increase in production of stress hormones or stress-related hormone metabolites.

An increase in the levels of stress hormones (norepinephrine, epinephrine and dopamine) in relation to the increase of noise levels has been observed in species of cetaceans in captivity (*Delphinapterus leucas* and *Tursiops truncatus*) after exposure to impulsive sounds produced by a water gun (Romano *et al.*, 2004). Rolland *et al.* (2012) described a reduction in ship noise levels that was associated with a decrease in the baseline levels of stress-related faecal hormone metabolites (glucocorticoids) in endangered North Atlantic right whales (*Eubalaena glacialis*). The registered decrease of about 6 dB in low-frequency noise levels (<150 Hz), measured in a Canadian port, was due to a massive reduction in ship traffic for a few days after 9/11 (Rolland *et al.*, 2012).

In fish, the stress derived from long-term exposure to noise can alter their regular physiological equilibrium, influencing their behaviour and altering their normal energy consumption processes. As for other types of effects, the degree of influence by noise-induced stress depends significantly on the degree and duration of the exposure to the source and on the health of the specimen (Celi *et al.*, 2016).

A brief review of auditory and physiological effects in response to boating and shipping noise is reported for different marine species in Table 9.1.

Table 9.1. *Auditory and physiological effects of boating and shipping noise observed in different marine organisms*

Family	Species	Effects	Reference
Balaenidae	*Eubalaena glacialis*	Decreased baseline levels of stress-related faecal hormone metabolites (glucocorticoids; noise levels decrease)	Rolland *et al.*, 2012
Monodontidae Delphinidae	*Delphinapterus leucas* *Tursiops truncatus*	Increased levels of stress hormones (norepinephrine, epinephrine and dopamine)	Romano *et al.*, 2004
Portunidae	*Carcinus maenas*	Increased metabolism	Wale *et al.*, 2013
Moronidae Sparidae	*Dicentrarchus labrax* *Sparus aurata*	Increased metabolism and induced motility	Buscaino *et al.*, 2010
Palinuridae	*Palynurus elephas*	Decreased immunity	Celi *et al.*, 2014
Phocidae	*Mirounga angustirostris*	Increased background noise constrains acoustic communication	Southall *et al.*, 2003
Gobiidae Pomacentridae Sciaenidae	*Gobius cruentatus* *Chromis chromis* *Sciaena umbra*	Reduced auditory sensitivity and hearing thresholds shifted	Codarin *et al.*, 2009
Batrachoididae	*Halobatrachus didactylus*	Acoustic communication constrained and hearing thresholds shifted	Vasconcelos *et al.*, 2007

Source: Extended from Peng *et al.* (2015)

9.5.3 Behavioural Responses

When animals are exposed to either natural or anthropogenic sounds, they experience many physical and psychological effects, depending on several variables (Southall *et al.*, 2007). These variables may change in relation to operational and environmental factors, influencing the interaction between an animal and the sound source (e.g., source level, distance of the animal from the source, duration of the exposure, sound propagation paths, etc.). Other variables depend on the physiological and psychological characteristics of the animal, age, sex and history of exposure, but may also depend on the season and on the behaviour at the time of exposure.

Several types of behavioural responses have been observed in marine mammals and they can be summarized in three categories:

- *Displacement* and *avoidance* reactions;
- Changes into the surface and diving behaviour;
- Changes in the acoustic behaviour.

The removal from the noise source (*avoidance*) and the dispersion of individuals from a group in a given area (*displacement*) often cause a decrease in foraging activities, the temporary or permanent abandonment of important feeding or breeding areas and also shifts in migration paths (Castellote *et al.*, 2012). Displacement and avoidance reactions have been documented in a number of marine mammal species exposed to noise (Weilgart, 2007).

Short-term changes in surface (e.g., swim speed, respiration rate, orientation, etc.) and diving behaviour in relation to the presence of small boats and ship traffic are also well documented for dolphins (e.g., *T. truncatus*) (Nowacek *et al.*, 2001) and for killer whales, who reduce swimming predictability in presence of approaching 'whale-watching' boats (Williams *et al.*, 2002). Similar responses to the approach of vessels have been also observed in the humpback whale (*Megaptera novaeangliae*) (Frankel *et al.*, 2000). Modifications in the acoustic behaviour in response to ship noise are also described in several cetacean species. In particular, due to the overlap in frequency of ship noise and baleen whale signals, substantial changes in the acoustic behaviour of these species have been observed in response to ship noise. Decreases in call emission rates and changes into the frequencies of emission, with implications for intra-species communication and possibly abnormal energetic consumption, are reported for different populations of fin and blue whales (Castellote *et al.*, 2012; Watkins, 1986; Weilgart, 2007). Changes have been also observed in the vocal behaviour of North Atlantic right whales (*E. glacialis*) and South Atlantic right whales (*Eubalaena australis*) (Parks *et al.*, 2007; Tennessen & Parks, 2016). In particular, using acoustic propagation modelling, Tennessen and Parks (2016) showed that North Atlantic right whales might improve their communication range in the presence of low-frequency ship noise by increasing the amplitude and frequency of the contact 'upcall' emitted. Melcón *et al.* (2012) described an acoustic response in blue whales in reaction to the presence of ship noise. They observed that the presence of nearby ships triggered the emission of more intense 'D-calls' by the blue whales. This could mean that the animals raise the emission level of their vocalizations with increased background noise, suggesting what is known as the 'Lombard effect' (Lane *et al.*, 1971).

Changes in the vocal behaviour of dolphins (Delphinidae spp.) have been observed in response to contextual increases in background noise levels (Azzolin *et al.*, 2013; La Manna *et al.*, 2013; Papale *et al.*, 2015). In particular, Papale *et al.* (2015) showed that dolphins adjusted the species-specific frequency parameters of their whistles by varying their vocalizing frequencies to compensate for increased levels of background noise in the same frequency bands.

Concerning behavioural responses to noise in fish, few studies indicate the negative correlation between the presence of noise and fish presence (Slabbekoorn *et al.*, 2010). Atlantic herring (*Clupea harengus*), Atlantic cod (*Gadus morhua*) and bluefin tuna (*Thunnus thynnus*) have been observed changing their behaviour and swimming fast towards the surface or towards the sea bottom in response to ship noise (Handegard *et al.*, 2003; Sarà *et al.*, 2007; Vabø *et al.*, 2002). In particular, Sarà *et al.* (2007) reported that tuna observed in a fixed tuna trap near shipping routes exhibited uncoordinated behaviour and swam fast individually in the presence of approaching ferries. Avoidance reactions have also been observed in net-penned Pacific herring (*Clupea pallasii*) in response to the sound of large vessels approaching at constant speed and smaller vessels approaching at accelerating speed (Schwarz & Greer, 1984). Besides startle responses, attention distraction and increases in errors associated with food handling and discrimination between food and non-food items have been observed in several species of fish (Peng *et al.*, 2015). In other marine species, such as the Caribbean hermit crab (*Coenobita clypeatus*) and the shore crab (*Carcinus maenas*), noise disturbance increased the specimens' vulnerability to predation (Chan *et al.*, 2010; Peng *et al.*, 2015; Wale *et al.*, 2013).

A brief review of behavioural effects in response to boating and shipping noise is reported for different marine species in Table 9.2.

As already discussed in this section, behavioural responses to sound are highly variable and context specific (Southall *et al.*, 2007). In addition to the studies conducted in captivity, behavioural studies in wild cetacean species can be performed by means of visual or passive acoustic techniques or by a combination of these two methods.

Visual observations are relevant to our understanding of the possible short-term responses to noise, whereas several obstacles still prevent the performance of efficient long-term monitoring of wild, free-ranging species. Visual (e.g., boat, aerial or shore-based) methods for cetacean observations imply the sighting and prolonged observation of the animals while they are swimming close to the sea surface. Cetaceans spend only a small amount of their life at the water surface, with some species being more elusive than others (e.g., Cuvier's beaked whale; Tyack *et al.*, 2006). As a result, these species are particularly difficult to intercept and to observe; furthermore, visual methods are limited to daylight hours and good weather conditions.

For this reason, and thanks to the rapid technological development of the last two decades, passive acoustic monitoring (PAM) represents an increasingly distributed and a well-established tool in cetacean monitoring and conservation (Notarbartolo di Sciara & Gordon, 1997; Pavan *et al.*, 1997; Zimmer, 2011). Acoustic sensors and acquisition systems for PAM monitoring are set according to the characteristics of the signal of interest and the site of installation. As to the underwater deployment, hydrophones can be either mobile or stationary and used in both single- and multi-sensor configurations (Figure 9.1). In mobile configurations, surveys are usually performed during navigation. Towed arrays of hydrophones, autonomous vehicles equipped with acoustic sensors or recording devices attached on the backs of the animals (e.g., digital acoustic recording tags) (Johnson & Tyack, 2003) are preferably used during at-sea operations.

In a stationary configuration, the sensors are installed on autonomous or radio-linked buoys at different operational depths or on cabled fixed stations that directly send the acquired signals to on-shore storage systems. Moreover, the acoustic sensors may also be installed in a configuration that allows the acoustic tracking of the sound sources. Fixed configurations also allow for the round-the-clock and long-term acquisition of acoustic data, with several advantages in the study of marine mammals. Studies can be performed on population composition and abundance (Caruso *et al.*, 2015) and on the seasonal presence

Table 9.2. *Brief summary of the main behavioural effects described for individuals of different marine species*

Family	Species	Effects	Reference
Delphinidae	*Tursiops truncatus*	Changes in surface and diving behaviour; changes in vocal behaviour	Azzolin *et al.*, 2013 La Manna *et al.*, 2013 Nowacek *et al.*, 2001 Papale *et al.*, 2015
	Orcinus orca	Reduced swimming predictability	Williams *et al.*, 2002
Balaenopteridae	*Megaptera novengliae*	Reduced swimming predictability	Frankel *et al.*, 2000
	Balaenoptera physalus *Balaenoptera musculus*	Changes in vocal behaviour (calling rate and frequencies)	Castellote *et al.*, 2012 Melcón *et al.*, 2012 Weilgart *et al.*, 2007
Balaenidae	*Eubalaena glacialis* *Eubalaena australis*	Modified calling behaviour; increased amplitude and frequency of contact calls	Parks *et al.*, 2007 Tennessen and Parks, 2016
Clupeidae	*Clupea pallasii*	Avoidance responses	Schwarz and Greer, 1984
Cichlidae	*Neolamprologus pulcher*	Reduced defence capabilities; increased aggression	Bruintjes and Radford, 2013
Paguroidea (superfamily)	*Coenobita clypeatus*	Reduced defence capabilities	Chan *et al.*, 2010
Portunidae	*Carcinus maenas*	Reduced defence capabilities	Wale *et al.*, 2013
Gobiidae	*Gobius cruentatus*	Decreased time in nest caring; increased time in shelter	Picciulin *et al.*, 2010
Pomacentridae	*Chromis chromis*		

Source: Extended from Peng *et al.* (2015)

and distribution of the observed animals (Sciacca *et al.*, 2015), with fewer limitations compared to short-term dynamic observations. PAM applications with cabled observatories also allow for the monitoring of the animals' movements. For distant sound sources, the definition of the direction of arrival can be defined by recording with acoustic sensors placed in particular geometrical configurations (Zimmer, 2011). By having an acoustic array with at least four synchronized, omni-directional hydrophones displaced in a tetrahedral shape, it is possible to estimate the time difference of arrival (TDOA) of the sound wave to each couple

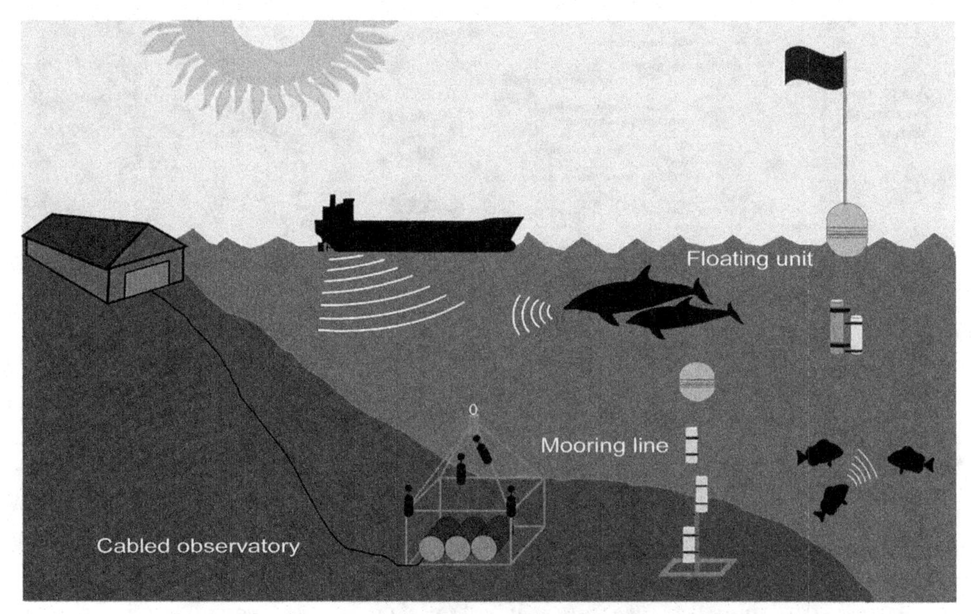

Figure 9.1 Schematic illustration of three types of PAM recording devices. Moorings and buoys are two examples of autonomous recording systems. Cabled observatories can be equipped with a fixed array of hydrophones continuously sending data to shore.

of hydrophones. According to the distance between each hydrophone and the frequency of emission, the TDOAs of sounds from different species can be studied, and the continuous movements of individuals could potentially be tracked. Swimming speed and diving depth can be monitored, together with intra-species relations and noise source–animal interactions.

Much information on the effects of ship noise may also come from an increased knowledge of the characteristics of the typical sounds emitted by different species. In particular, important outcomes may derive from the reliable measurement of the spectral features and source levels of the target signals and from the evaluation of the received levels (i.e., the levels of noise exposure of the studied specimens) (Moore *et al.*, 2012).

Besides visual and PAM methods, recent technological advances have brought marine mammal scientists to use new systems for cetacean behavioural studies. In particular, unmanned aerial systems, satellite-linked telemetry and multi-sensor high-resolution acoustic recording tags are increasingly used to reveal the responses of cetaceans to anthropogenic activities such as ship noise (for a review, see Nowacek *et al.*, 2016).

In recent years, PAM methods have been increasingly applied to the monitoring of fishes and invertebrates and to the study of the soundscape (Buscaino *et al.*, 2011, 2016; Rountree *et al.*, 2006). Soundscape ecology is an emerging discipline that focuses on the entire acoustic community – considered as an ensemble of *biophonies*, *geophonies* and *anthropophonies* (Pijanowski *et al.*, 2011) – to obtain ecological information such as biodiversity measurements and interactions between anthropogenic and biological sounds. Initially applied to terrestrial habitats, soundscape ecology is being progressively applied to the

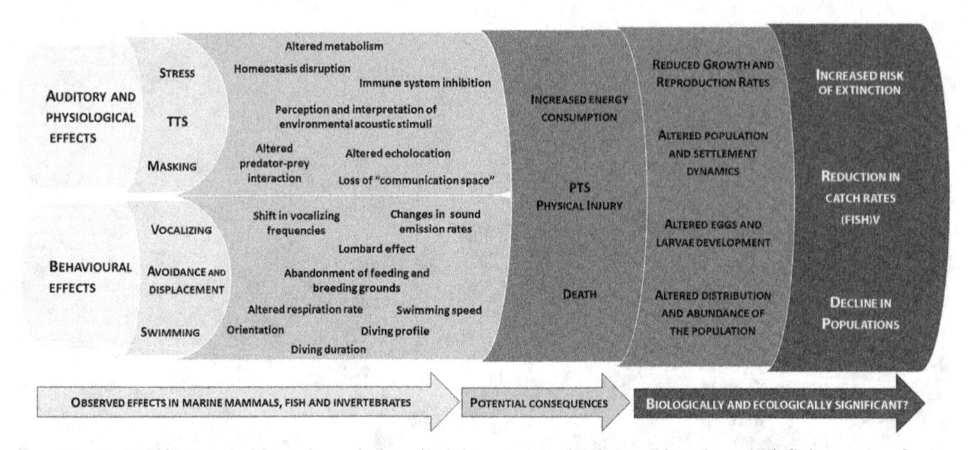

Figure 9.2 'Matryoshka scheme' showing the expected (observed and potential) impacts of ship noise on marine species.

marine environment, in combination with PAM methods, to assess the health status of ecosystems (Buscaino *et al.*, 2016). The study of the soundscape for ecological monitoring purposes aims to characterize the overall sound field in terms of spectral levels and temporal trends. The dominant acoustic sources are identified and the main temporal (diurnal, inter-seasonal, tidal) patterns and spatial variations of the sound field are investigated. Eventual relationships among the different acoustic elements are also studied to define the characteristics of the soundscape of the studied area (Guan *et al.*, 2015).

9.5.4 Potential Impact on Population Dynamics and Fish Catch Rates

The overall response to noise has direct energetic costs for individual vital rates (e.g., survival or reproductive success) and, ultimately, for population dynamics (Merchant *et al.*, 2014), with potential long-term effects on foraging, navigation and breeding activities, as observed in marine mammals (Wright *et al.*, 2008). The cumulative energetic costs of behavioural responses to noise may also endanger the survival of eggs and the reproductive growth rates of fish species and can have negative effects on fish larvae, with implications for the dynamics of settlement and population, as demonstrated by Holles *et al.* (2013) in coral reef habitats (Celi *et al.*, 2016).

Reduced successful development of embryos and increased mortality of recently hatched larvae have also been observed in the invertebrate sea hare (*Stylocheilus striatus*) after exposure to boat noise playback compared to ambient noise playback (Nedelec *et al.*, 2014).

Reductions in catch rates may be another indirect effect of ship noise on the abundance of populations of marine fishes. However, to date, direct evidence of this effect is mainly reported after exposure to seismic shooting noise rather than to ship noise (for reviews, see Peng *et al.*, 2015; Weilgart, 2013). A review scheme summarizing the impacts of noise on the marine environment and resources is given in Figure 9.2. In this scheme, the risks of population declines, species extinctions and economically relevant reductions in fishing rates represent the largest potential impacts of ship noise on marine organisms.

9.6 Sound Propagation Models

The growing interest in the spatial and temporal distribution of ship noise around the world has prompted the development of several models to predict the acoustic energy transferred from ship traffic to the marine environment. Many of these predictive models are based on Automatic Identification System (AIS) data. Indeed, it has been demonstrated that most of the noise emission in the 0.1–10 kHz frequency band is produced by vessels operating with AISs (Merchant *et al.*, 2012). Each ship equipped with an AIS communicates, through a very-high-frequency antenna, its Maritime Mobile Service Identity (MSSI) code, length, position, navigation course and speed. This information can be used to estimate through semi-empirical relationships the power spectral levels at the source.

AIS data do not provide information about depth, which would be useful for evaluating the spectrum distortion due to Lloyd's mirror effects. However, this information can be derived if we know the ship category and length (Erbe *et al.*, 2012). Predictive models should describe not only noise levels at the source, but also noise propagation into the marine environment at different ranges. They must consider both the geometrical spreading and the absorption due to molecular relaxation, whose effects on sound propagation are more evident at higher frequencies (Ainslie *et al.*, 1998).

In a simplistic approach, the propagation loss (*PL*) at low frequencies can be approximated by the spreading law:

$$PL = N \log_{10} R$$

where R is the distance in metres from the source and N is a coefficient that goes from 20 for spherical spreading to 10 for cylindrical spreading.

Models that are more accurate include bathymetry, seabed composition and sound speed profile information. Variations in the speed of sound along the water column may cause the occurrence of shadow areas or of acoustic ducts that trap the acoustic waves. In a duct, sound can travel further with less spreading and without interaction with the seafloor and sea surface (Wang, 2014).

Etter *et al.* (2012) grouped loss propagation models into five big families based on ray tracing theory, normal modes, multipath expansion, fast field or parabolic equations. Each model is optimized for specific working conditions. Ray tracing models, which are the most suited models for predicting noise levels at short range and high frequency in deep waters, are not applicable to simulating low-frequency propagation in shallow water, where modelling approaches based on normal modes or parabolic equations are preferable (Farcas, 2016).

Models are also classified in range-independent or range-dependent models. Range-independent models assume that the properties of the propagation medium vary only as a function of depth (Hernandez, 2009). These models are used when the environment is considered to have cylindrical symmetry. Range-dependent models are suitable for predicting noise propagation over very large distances, where the environmental conditions may change. These kinds of models are typically more accurate, but are computationally more expensive than range-independent models. In Table 9.3, the applicability of the five categories of models to range-independent and range-dependent environments is reported

Table 9.3. *Possible application of five model categories to range-independent and range-dependent environments. Noise components below 500 Hz are considered as low frequency and components above 500 Hz are consider as high frequency. Dark grey colour refers to the good applicability of the model family in terms of predictability and computational costs. Light grey colour refers to the partial applicability of the model family. Black colour means that the family model is not suitable for that condition*

	Shallow water, low frequency		Shallow water, high frequency		Deep water, low frequency		Deep water, high frequency	
	RI	RD	RI	RD	RI	RD	RI	RD
Ray theory								
Normal modes								
Multipath expansion								
Fast field								
Parabolic equations								

RD = range-independent environment; RI = range-independent environment.
Source: Adapted from Etter (2012)

according to Etter (2012). Most of the propagation models are two-dimensional (depth, distance). This means that they are applied to a transect and do not include horizontal refraction, reflection or diffraction. In many cases, three-dimensional propagation can be obtained by combining a number of two-dimensional transects. In the presence of obstacles such as small islands, an accurate prediction needs the application of a fully three-dimensional model (Wang, 2014).

However, the predictability of a model must be assessed through comparison with measurement data, yet few studies evaluate the performance of a model or its limits via comparison with experimental data (Chion *et al.*, 2017; Erbe, 2012).

The development of increasingly accurate models will support decision-makers in the management of future maritime transports and in the implementation of the necessary mitigation measures. The creation of noise maps based on predictive models will reduce the time required to establish noise trends underwater. This will also limit the number of measurement stations, thus cutting the costs of monitoring (Dekeling *et al.*, 2013).

9.7 Regulation

The ever-increasing injection of anthropogenic noise into the marine environment and the growing awareness about the potential effects of noise pollution led to the development of international and regional operational guidelines to mitigate and minimize the impact of noise pollution on marine life.

In 2014, the International Maritime Organization (IMO) ratified non-mandatory guidelines for designers, shipbuilders and ship operators to reduce noise from commercial shipping (IMO, 2014). According to these guidelines, the largest opportunities for a reduction of underwater noise come from the initial design of ships. In order to reduce cavitation, the diameter, blade number, pitch, skew and sections of the propeller should be designed or selected in order to ensure water flow that is as uniform as possible into the propellers. Moreover, the ship hull should be designed such that the wake field is as homogeneous as possible. This could be achieved by reducing the cavitation of the propeller that operates in the wake field generated by the hull. The selection of the inboard machinery must consider vibrational control measures, and their contribution to radiated noise must be evaluated. Operational and maintenance considerations are also included within the guidelines. In particular, the IMO considers propeller cleaning and maintenance of the underwater hull surface. The document identifies the proper selection of the ship speed as an effective measure to reduce radiated noise by cavitation. Moreover, it contemplates the possibility of re-planning routes to reduce ship noise impacts on specific marine areas.

The IMO guidelines recommend to shipyards, shipowners and ship surveyors the measurement of ship noise through objective standard procedures, as defined by the International Organization for Standardization (ISO). In particular, the use of the standard ISO 17208-1 is recommended. This standard can be applied to any underway surface vessel, under prescribed operating conditions, but it is limited to ships transiting at speeds no greater than 50 km/hour. The standard also provides requirements for an ocean test site in deep water.

Despite suggesting potentially effective measures to reduce ship noise, the IMO guidelines are applied voluntarily, and they are only addressed to commercial shipping. Their application is not foreseen for naval and war ships. In addition, the IMO guidelines do not deal with the introduction of other types of noise (i.e., the noise produced by sonar and seismic activities) (IMO, 2014).

In Europe, the Marine Strategy Framework Directive 2008/56/EC (MSFD) established a framework for community actions in the field of marine environmental policy. The Directive defines underwater acoustic noise as 'the intentional or accidental introduction of acoustic energy in the water column from impulsive and diffuse sources'. Moreover, it expressly recognizes acoustic noise as a form of pollution. According to the MSFD, Member States shall take the necessary measures to achieve or maintain a good environmental status in the marine environment by the year 2020 at the latest.

A good environmental status at the level of the marine region or sub-region is determined by means of 11 quality descriptors. In particular, the 11th descriptor says that 'the introduction of energy, including underwater noise, should be at levels that do not adversely affect the marine environment'. The European Commission, through Decision 2010/477/EU, selected for Member States two criteria that are useful for assessing the fulfilment (implementation) of Descriptor 11, further extending these with the specifications expressed by the TSG Noise Group (Dekeling *et al.*, 2013):

- Criterion 11.1 describes the requirements to measure the spatial and temporal distribution of loud, low- and mid-frequency impulsive sounds.

- Criterion 11.2 describes the requirements to measure continuous low-frequency sound. Trends in the ambient noise level within the one-third of an octave bands of 63 and 125 Hz (centre frequency) (re 1 μPa root mean square; average noise level in these octave bands over a year) should be measured by observation stations and/or with the use of appropriate models.

In particular, this last criterion lays the foundations for a broader understanding of the distribution and trend levels of diffuse ship noise. An increasing number of research studies underline the need to improve the MSFD noise monitoring criteria in order to gain a more inclusive view of different scenarios. In this view, a key factor would be to assess whether higher- or lower-frequency bands, compared to those required by Criterion 11.2 of the MSFD, might be appropriate indicators for noise exposure from shipping. As an example, Merchant *et al.* (2014) compared mean noise levels in the one-third of an octave frequency bands centred on 63, 125, 250 and 500 Hz with daily broadband sound exposure levels in the 0.05–1 kHz range. For the area monitored in this study, all four bands were highly correlated with noise exposure levels. However, this relationship was strongest at 125 Hz and lower in the 63-Hz band. The authors indicate that this lower correlation could be caused either by the noise related to tidal flows or by the low-frequency propagation effects typical of shallow waters. These effects may limit the efficacy of the 63-Hz band as an indicator of anthropogenic noise exposure in other shallow water and coastal sites.

The selection of the most appropriate indicators related to the criteria for assessment of progress towards good environmental status is supported by the MSFD, as underlined in the Annex of the Commission Decision of 1 September 2010. Moreover, it is preliminary to the development of specific tools that can support an ecosystem-based approach to the management of human activities. 'Such tools include spatial protection measures and measures in the list in Annex VI to Directive 2008/56/EC, notably spatial and temporal distribution controls, such as maritime spatial planning' (European Commission, 2010).

Nevertheless, the MSFD must be implemented at the level of marine regions or sub-regions by relevant Member States. As an example, in Italy, the establishment of a register of noise emissions from ships has been proposed by researchers (Maccarrone *et al.*, 2015). This registry should be filled with both the data provided by manufacturers or shipowners and the experimental data taken at sea.

Similar data could be strategically useful for the development of propagation noise models and mitigation strategies, such as quieting technologies and route modifications. This is also the intention of integrating the MSFD Directive and Commission Decision with the aims and recommendations of the IMO (2014).

In 2016, the US National Oceanic and Atmospheric Administration (NOAA) released the Ocean Noise Strategy Roadmap. This document is intended to provide a cross-line office roadmap of the agency over the next 10 years to address the effects of acoustic noise on protected species (marine mammals, fish, invertebrates and sea turtles) and acoustic habitats. In the Ocean Noise Strategy Roadmap, the species-level impacts of ocean noise and associated management actions are reviewed, and two place-based case studies that highlight the Roadmap's science and management recommendations are presented.

At the local scale, municipal governments and ports have the opportunity to apply measures to reduce marine acoustic disturbances. These measures can also be based on voluntary mechanisms. In the St Lawrence Estuary (Canada), voluntary mitigation measures, including a speed reduction area and a no-go area, have been applied to reduce the risks of lethal collisions with four species of baleen whales. Simulations of beluga whales' soundscapes demonstrate that these protection measures could also reduce the levels of noise received by beluga whales. A statistically significant 1.6 per cent decrease in the total amount of noise received by beluga whales in their critical habitat has been estimated (Chion *et al.*, 2017). Another interesting voluntary mechanism is the mechanism proposed by the Green Marine Environmental Program. The Green Marine Environmental Program offers a detailed framework for maritime companies to first establish and then reduce their environmental footprint. Each participant (shipowners, ports, seaway corporations, terminals and shipyards) is invited to make a self-evaluation that is independently verified and certified every 2 years. Each evaluation is made public. The programme foresees for shipowners the reduction of noise through ship operations such as regular hull cleaning and propeller blade maintenance or incorporating applicable vessel-quieting technologies during refits and new vessel construction. Performance indicators for ports consider the management of the underwater noise sources during ongoing activities, development/construction and port maintenance (Green Marine Environmental Program, 2017). Another effective measure to protect the marine environment from anthropogenic noise is the establishment of marine protected areas. Marine protected area should be large enough to safeguard essential habitat and migration corridors and to accommodate highly mobile species (Weilgart, 2006).

In the last 30 years, a number of international and regional conventions have recognized the significance of marine noise pollution for marine mammals. The Agreement on the Conservation of Small Cetaceans in the Baltic, North East Atlantic, Irish and North Seas (ASCOBANS) and the Agreement on the Conservation of Cetaceans of the Black Sea, Mediterranean Sea and Contiguous Atlantic Area (ACCOBAMS) consider marine anthropogenic noise to be an issue to mitigate. In 2007, ACCOBAMS produced two resolutions: the 'Guidelines to address the impact of anthropogenic noise on marine mammals in the ACCOBAMS area' and Resolution 4, which requested parties and range states to develop mitigation, conduct research and 'develop and implement procedures to assess the effectiveness of any guidelines or management measures introduced' (Simmonds *et al.*, 2014).

In the USA, the NOAA established Acoustic Guidelines in 1995 defining what constitutes a 'take' of marine mammals under the US Marine Mammal Protection Act (MMPA) and Endangered Species Act (Simmonds *et al.*, 2014).

In order to provide managers with a concrete initial guide to address the implications of the MMPA, Southall *et al.* (2007) offered initial scientific guidance based on the available literature that has been used worldwide by regulators and industries. This guidance provides criteria to avoid injurious exposure effects in marine mammals, such as TTS and PTS. However, given the high context dependency already discussed in Section 9.5.3, threshold criteria were not suggested for behavioural responses (Simmonds *et al.*, 2014). An update of the criteria proposed by Southall *et al.* (2007) has been proposed more recently by Tougaard *et al.* (2015). In 2016, the NOAA produced the Technical Guidance for Assessing the Effects of Anthropogenic Sound on Marine Mammal Hearing: Underwater Acoustic Thresholds for Onset of Permanent and Temporary Threshold Shifts. However, this guidance defines acoustic thresholds to avoid hearing shifts (TTS and PTS) for acute, incidental exposure to underwater anthropogenic sound sources. Hence, it is only partially related to the exposure of animals to diffuse noise and cumulative exposure caused by ship traffic (National Marine Fisheries Service, 2016).

9.8 Conclusions

In order to gain a proper understanding of and to prevent the impacts of ship noise on the marine environment, the monitoring and detailed investigation of the chronic and cumulative effects on the most sensitive and key marine species need to be improved. The risks related to ship noise at the population and species levels could be defined by developing increasingly accurate prediction models. This will be possible only through the validation of models with experimental acoustic data. Long-term studies of ambient noise and ship noise trends in marine habitats are increasingly required, considering that historical series (over several years) of ambient noise data are limited to few oceanic regions and coastal areas (Merchant *et al.*, 2014).

Future scenarios of ship noise could be predicted through models, and these may be useful for the low-impact implementation of new routes and plans for shipping traffic, such as for the future European Motorways of the Sea (Neenan, 2016). Moreover, by comparing modelled noise distributions and trends with animal density data, it will be possible to identify noise hotspots for the most sensitive species and in particular for marine mammals (Erbe, 2014).

As discussed, a further increase in ship noise levels potentially has long-term negative effects at all levels of the food web. As a natural consequence, this may also affect the management of aquatic resources, with the most serious socio-economic consequences observed on fisheries.

References

Abdulla, A., Linden, O. (2008). *Maritime Traffic Effects on Biodiversity in the Mediterranean Sea: Review of Impacts, Priority Areas and Mitigation Measures*. Malaga: IUCN Centre for Mediterranean Cooperation.

Aguilar Soto, N., Johnson, M., Madsen, P. T. *et al.* (2006). Does intense ship noise disrupt foraging in deep diving Cuvier's beaked whales (*Ziphius cavirostris*)? *Marine Mammal Science*, **22**(3), 690–699.

Ainslie, M. A., McColm, J. G. (1998). A simplified formula for viscous and chemical absorption in sea water. *Journal of the Acoustical Society of America*, **103**(3), 1671–1672.

Andrew, R. K., Howe, B. M., Mercer, J. A., Dzieciuch, M. A. (2002). Ocean ambient sound: comparing the 1960s with the 1990s for a receiver off the California coast. *Acoustics Research Letters Online*, **3**(2), 65–70.

Andrew, R. K., Howe, B. M., Mercer, J. A. (2011). Long-time trends in ship traffic noise for four sites off the North American west coast. *Journal of the Acoustical Society of America*, **129**(2), 642–651.

Au, W. W., Hastings, M. C. (2008). *Principles of Marine Bioacoustics*. Berlin: Springer.

Azzolin, M., Papale, E., Lammers, M. O., Gannier, A., Giacoma, C. (2013) Geographic variation of whistles of striped dolphin (*Stenella coeruleoalba*) within the Mediterranean Sea. *Journal of the Acoustical Society of America*, **134**, 694–705.

Brekhovskikh, L. M., Lysanov, Yu. P. (2003). *Fundamentals of Ocean Acoustics*, 3rd Edition. New York: Springer New York.

Bruintjes, R., Radford, A. N. (2013). Context-dependent impacts of anthropogenic noise on individual and social behaviour in a cooperatively breeding fish. *Animal Behaviour*, **85**, 1343–1349.

Buscaino, G., Filiciotto, F., Buffa, G. *et al.* (2010) Impact of an acoustic stimulus on the motility and blood parametres of European sea bass (*Dicentrarchus labrax* L.) and gilthead sea bream (*Sparus aurata* L.). *Marine Environmental Research*, **69**, 136–142.

Buscaino, G., Filiciotto, F., Gristina, M. *et al.* (2011). Acoustic behaviour of the European spiny lobster *Palinurus elephas*. *Marine Ecology Progress Series*, **441**, 177–184.

Buscaino, G., Ceraulo, M., Pieretti, N. *et al.* (2016). Temporal patterns in the soundscape of the shallow waters of a Mediterranean marine protected area, *Scientific Reports*, **6**, 34230.

Carlton, J. (2007). *Marine Propellers and Propulsion*, 2nd Edition. Amsterdam: Elsevier.

Caruso, F., Sciacca, V., Bellia, G. *et al.* (2015). Size distribution of sperm whales acoustically identified during long term deep-sea monitoring in the Ionian sea. *PLoS ONE*, **10**(12), e0144503.

Castellote, M., Clark, C. W., Lammers, M. O. (2012). Acoustic and behavioural changes by fin whales (*Balaenoptera physalus*) in response to shipping and airgun noise. *Biological Conservation*, **147**, 115–122.

Celi, M., Filiciotto, F., Parrinello, D. *et al.* (2013). Physiological and agonistic behavioural response of *Procambarus clarkii* to an acoustic stimulus. *Journal of Expimental Biology*, **216**, 709–718.

Celi, M., Filiciotto, F., Vazzana, M. *et al.* (2014). Shipping noise affecting immune responses of European spiny lobster *Palinurus elephas* (Fabricius, 1787). *Canadian Journal of Zoology*, **93**, 113–121.

Celi, M., Filiciotto, F., Quinci E. M. *et al.* (2016). Vessel noise pollution as a human threat to fish: assessment of the stress response in gilthead sea bream (*Sparus aurata*, Linnaeus 1758). *Fish Physiology and Biochemistry*, **42**, 631–641.

Chan, A. A., Giraldo-Perez, P., Smith, S., Blumstein, D. T. (2010). Anthropogenic noise affects risk assessment and attention: the distracted prey hypothesis. *Biology Letters*, **6**, 458–461.

Chion, C., Lagrois, D., Dupras, J. *et al.* (2017). Underwater acoustic impacts of shipping management measures: results from a social–ecological model of boat and whale movements in the St. Lawrence River Estuary (Canada). *Ecological Modelling*, **354**, 72–84.

Clark, C. W., Ellison, W. T., Southall, B. L. *et al.* (2009). Acoustic masking in marine ecosystems: intuitions, analysis, and implication. *Marine Ecology Progress Series*, **395**, 201–222.

Codarin, A., Wysocki, L. E., Ladich, F., Picciulin, M. (2009). Effects of ambient and boat noise on hearing and communication in three fish species living in a marine protected area (Miramare, Italy). *Marine Pollution Bulletin*, **58**(12), 1880–1887.

Croll, D. A., Clark, C. W., Acevedo, A. *et al.* (2002). Bioacoustics: only male fin whales sing loud songs. *Nature*, **417**, 809–809.

Dekeling, R. P. A., Tasker, M. L., Van der Graaf, A. J. *et al.* (2013). *Monitoring Guidance for Underwater Noise in European Seas. Part I: Executive Summary. Part II: Monitoring Guidance Specifications. Part III: Background Information and Annexes. Joint Research Centre Scientific and Policy Reports.* Luxembourg: Publications Office of the European Union.

Edds-Walton, P. L. (1997). Acoustic communication signals of mysticete whales. *Bioacoustics*, **8**(1–2), 47–60.

Engås, A., Misund, O. A., Soldal, A. V., Horvei, B., Solstad, A. (1995). Reactions of penned herring and cod to playback of original, frequency-filtered and time-smoothed vessel sound. *Fisheries Research*, **22**, 243–254

Equasis (2015). *The World Merchant Fleet in 2014.* Lisbon: European Maritime Safety Agency.

Erbe, C., MacGillivray, A., Williams, R. (2012). Mapping cumulative noise from shipping to inform marine spatial planning. *Journal of the Acoustical Society of America*, **132**(5), EL423.

Erbe, C., Williams, R., Sandilands, D., Ashe, E. (2014). Identifying modeled ship noise hotspots for marine mammals of Canada's Pacific region. *PLoS ONE*, **9**(3), e89820.

Erbe, C., Reichmuth, C., Cunningham, K., Lucke, K., Dooling, R. (2016). Communication masking in marine mammals: a review and research strategy. *Marine Pollution Bulletin*, **103**(1), 15–38.

Etter, P. C. (2012). Advanced applications for underwater acoustic modeling. *Advances in Acoustics and Vibration*, **2012**, 214839.

European Commission (2010). Commission decision of 1 September 2010 on criteria and methodological standards on good environmental status of marine waters (2010/477/EU).

Farcas, A., Thompson, P. M., Merchant, N. D. (2016). Underwater noise modelling for environmental impact assessment. *Environmental Impact Assessment Review*, **57**, 114–122.

Fisher, F. H., Simmons, V. P. (1977). Sound absorption in seawater. *Journal of the Acoustical Society of America*, **62**, 558–564.

Frankel, A. S., Clark, C. W. (2000). Behavioral responses of humpback whales (*Megaptera novaeangliae*) to full-scale ATOC signals. *Journal of the Acoustical Society of America*, **108**(4), 1930–1937.

Guan, S., Suite, S., Spring, S. *et al.* (2015). Dynamics of soundscape in a shallow water marine environment: a study of the habitat of the Indo-Pacific humpback dolphin. *Journal of the Acoustical Society of America*, **137**, 2939–2949.

Green Marine Environmental Program (2017). Performance Indicators for Ship Owners and Performance Indicators for Ports & St. Lawrence Seaway Corporations. www.green-marine.org/wp-content/uploads/2017/01/2017-Summary_PortsSeaway_FINAL.pdf

Handegard, N. O., Michalsen, K., Tjøstheim, D. (2003). Avoidance behavior in cod, *Gadus morhua*, to a bottom-trawling vessel. *Aquatic Living Resources*, **16**, 265–270.

Hernández, J. V. (2009). The Significance of Passive Acoustic Array-Configurations on Sperm whale Range Estimation when using the Hyperbolic Algorithm. PhD thesis, Heriot-Watt University.

Hildebrand, J. A. (2009). Anthropogenic and natural sources of ambient noise in the ocean. *Marine Ecology Progress Series*, **395**, 5–20.

Holles, S., Simpson, S. D., Radford, A. N., Berten, L., Lecchini, D. (2013). Boat noise disrupts orientation behaviour in a coral reef fish. *Marine Ecology Progress Series*, **485**, 295–300.

IMO (2014). *Guidelines for the Reduction of Underwater Noise from Commercial Shipping to Address Adverse Impacts on Marine Life*. MEPC.1 Circ. 833. London: International Maritime Organization.

Johnson, M. P., Tyack, P. L. (2003). A digital acoustic recording tag for measuring the response of wild marine mammals to sound. *IEEE Journal of Oceanic Engineering*, **28**, 3–12.

Ketten, D. R. (2002) Marine mammal auditory systems: a summary of audiometric and anatomical data and implications for underwater acoustic impacts. *Polarforschung*, **72**, 79–92.

Kull, R. C., McGarrity, C. (2002) Noise effects on animals: 1998–2002 review. *Proceedings of the 8th International Congress on Noise as a Public Health Problem*, **6**, 291–298.

La Manna, G., Manghi, M., Pavan, G., Lo Mascolo, F., Sarà, D. G. (2013). Behavioural strategy of common bottlenose dolphins (*Tursiops truncatus*) in response to different kinds of boats in the waters of Lampedusa Island (Italy). *Aquatic Conservation Marine Freshwater Ecosystems*, **23**, 745–757.

Lane, H., Tranel, B. (1971). The Lombard sign and the role of hearing in speech. *Journal of Speech Hearing Research*, **14**, 677–709.

Maccarrone, V., Filiciotto, V., de Vincenzi, G., Mazzola, S., Buscaino, G. (2015). An Italian proposal on the monitoring of underwater noise: relationship between the EU Marine Strategy Framework Directive (MSFD) and marine spatial planning directive (MSP). *Ocean & Coastal Management*, **118**, 215–224.

Marine Mammal Commission (2007). Marine mammals and noise. A sound approach to research and management. A report to Congress from the Marine Mammal Commission. In: *Congress from the Marine Mammal Commission*. Bethesda, MD: Marine Mammal Commission, p. 370.

McDonald, M. A., Hildebrand, J. A., Wiggins, S. M. (2006a). Increases in deep ocean ambient noise in the Northeast Pacific west of San Nicolas Island, California. *Journal of the Acoustical Society of America*, **120**(2), 711– 718.

McDonald, M. A., Mesnick, S. L., Hildebrand, J. A. (2006b). Biogeographic characterization of blue whale song worldwide: using song to identify populations. *Journal of Cetacean Research and Management*, **8**, 55–65.

McKenna, M. F., Ross, D., Wiggins, S. M., Hildebrand, J. A. (2012). Underwater radiated noise from modern commercial ships. *Journal of the Acoustical Society of America*, **131**(1), 92.

Melcón, M. L., Cummins, A. J., Kerosky, S. M. *et al.* (2012). Blue whales respond to anthropogenic noise. *PLoS ONE*, **7**, e32681.

Merchant, N. D., Witt, M. J., Blondel, P., Godley, B. J., Smith, G. H. (2012). Assessing sound exposure from shipping in coastal waters using a single hydrophone and Automatic Identification System (AIS) data. *Marine Pollution Bulletin*, **64**(7), 1320–1329.

Merchant, N. D., Pirotta, E., Barton, T. R., Thompson, P. M. (2014). Monitoring ship noise to assess the impact of coastal developments on marine mammals. *Marine Pollution Bulletin*, **78**, 85–95.

Miksis-Olds, J. L., Nichols, S. M. (2016). Is low frequency ocean sound increasing globally? *Journal of the Acoustical Society of America*, **139**(1), 501–511.

Moore, S. E., Reeves, R. R., Southall, B. L. *et al.* (2012). A new framework for assessing the effects of anthropogenic sound on marine mammals in a rapidly changing Arctic. *BioScience*, **62**, 289–295.

Morley, E. L., Jones, G., Radford, A. N. (2014). The importance of invertebrates when considering the impacts of anthropogenic noise. *Proceedings of the Royal Society B: Biological Sciences*, **281**(1776), 20132683.

Nachtigall, P. E., Mooney, A., Taylor, K. A., Yuen, M. M. L. (2007). Hearing and auditory evoked potential methods applied to odontocete cetaceans. *Aquatic Mammals*, **33**, 6–13.

National Marine Fisheries Service (2016). *Technical Guidance for Assessing the Effects of Anthropogenic Sound on Marine Mammal Hearing: Underwater Acoustic Thresholds for Onset of Permanent and Temporary Threshold Shifts*. Silver Spring, MD: US Department of Commerce, NOAA.

Nedelec, S. L., Radford, A. N., Simpson, S. D. *et al.* (2014). Anthropogenic noise playback impairs embryonic development and increases mortality in a marine invertebrate. *Scientific Reports*, **4**, 5891.

Neenan, S. T. V., White, P. R., Leighton, T. G., Shaw, P. J. (2016). Modeling vessel noise emissions through the accumulation and propagation of Automatic Identification System data. *Proceedings of the Meetings on Acoustics*, **27**, 70017.

Notarbartolo di Sciara, G., Gordon, J. (1997). Bioacoustics: a tool for the conservation of cetaceans in the Mediterranean Sea. *Marine & Freshwater Behaviour & Physiology*, **30**, 125–146.

Nowacek, S. M., Wells, R. S., Solow, A. R. (2001). Short-term effects of boat traffic on bottlenose dolphins, *Tursiops truncatus*, in Sarasota Bay, Florida. *Marine Mammal Science*, **17**(4), 673–688.

Nowacek, D. P., Christiansen, F., Bejder, L., Goldbogen, J. A., Friedlaender, A. S. (2016). Studying cetacean behaviour: new technological approaches and conservation applications. *Animal Behaviour*, **120**, 235–244.

Obrist, M. K., Pavan, G., Sueur, J. *et al.* (2010). Bioacoustics approaches in biodiversity inventories. In: J. Eymann, J. Degreef, C. Häuser, J. C. Monje, Y. Samyn, D. VandenSpiegel, eds., *ABC Taxa. Manual on Field Recording Techniques and Protocols for All Taxa Biodiversity Inventories*. Brussels: The Belgian Development Cooperation, pp. 68–99.

Panigada, S., Pavan, G., Borg, J. A., Galil, B. S., Vallini, C. (2008). Shipping noise, a challenge for the survival and welfare of marine life? In: A. Abdulla, O. Linden, eds., *Maritime Traffic Effects on Biodiversity in the Mediterranean Sea: Review of Impacts, Priority Area and Mitigation Measures*. Malaga: IUCN Centre for Mediterranean Cooperation, pp. 10–21.

Papale, E., Gamba, M., Perez-Gil, M., Martin, V. M., Giacoma, C. (2015). Dolphins adjust species-specific frequency parametres to compensate for increasing background noise. *PLoS ONE*, **10**(4), e0121711.

Parks, S. E., Clark, C. W., Tyack, P. L. (2007). Short- and long-term changes in right whale calling behavior: the potential effects of noise on acoustic communication. *Journal of the Acoustical Society of America*, **122**, 3725–3731.

Pavan, G., Borsani, J. F. (1997). Bioacoustic research on cetaceans in the Mediterranean Sea. *Marine & Freshwater Behaviour & Physiology*, **30**, 99–123.

Pavan, G., La Manna, G., Zardin, F. *et al.*; and the NEMO Collaboration (2008). *Short Term and Long Term Bioacoustic Monitoring of the Marine Environment. Results from NEMO ONDE Experiment and Way Ahead. Computational Bioacoustics for Assessing Biodiversity. Proceedings of the International Expert Meeting on*

IT-based Detection of Bioacoustical Patterns. Bonn: Federal Agency for Nature Conservation.

Parsons, E., Dolman, S. (2003). The use of sound by cetaceans. In: *Oceans of Noise*, Chippenham: Whale and Dolphin Conservation Society, pp. 44–52.

Peng, C., Zhao, X., Liu, G. (2015). Noise in the sea and its impacts on marine organisms. *International Journal of Environmental Research and Public Health*, **12**, 12304–12323.

Picciulin, M., Sebastianutto, L., Codarin, A. *et al.* (2010). In situ behavioural responses to boat noise exposure of *Gobius cruentatus* (Gmelin, 1789; fam. Gobiidae) and Chromis chromis (Linnaeus, 1758; fam. Pomacentridae) living in a Marine Protected Area. *Journal of Experimental Marine Biology and Ecology*, **386**, 125–132.

Pijanowski, B. C., Villanueva-Rivera, L. J., Dumyahn, S. L. *et al.* (2011). Soundscape ecology: the science of sound in the landscape. *Bioscience*, **61**, 203–216.

Rolland, R. M., Parks, S. E., Hunt, K. E. *et al.* (2012). Evidence that ship noise increases stress in right whales. *Proceedings. Biological Sciences*, **279**, 2363–2368.

Romano, T. A., Keogh, M. J., Kelly, C. *et al.* (2004). Anthropogenic sound and marine mammal health: measures of the nervous and immune systems before and after intense sound exposure. *Canadian Journal of Fisheries and Aquatic Sciences*, **61**, 1124–1134.

Ross, D. (1976). *Mechanics of Underwater Noise*. Westport, CT: Peninsula Publishing.

Rountree, R. A., Gilmore, R. G., Goudey, C. A. *et al.* (2006). Listening to fish: applications of passive acoustics to fisheries science. *Fisheries*, **31**(9), 433–446.

Sarà, G., Dean, J. M., D'Amato, D. *et al.* (2007). Effect of boat noise on the behaviour of bluefin tuna *Thunnus thynnus* in the Mediterranean Sea. *Marine Ecology Progress Series*, **331**, 243–253.

Schwarz, A. L., Greer, G. L. (1984). Responses of Pacific herring, *Clupea harengus pallasi*, to some underwater sounds. *Canadian Journal of Fisheries and Aquatic Science*, **41**, 1183–1192.

Sciacca, V., Caruso, F., Beranzoli, L. *et al.* (2015). Annual acoustic presence of fin whale (*Balaenoptera physalus*) offshore eastern Sicily, central Mediterranean Sea. *PLoS ONE*, **10**, e0141838.

Simmonds, M. P., Dolman, S. J., Jasny, M. *et al.* (2014). Marine noise pollution – increasing recognition but need for more practical action. *Journal of Ocean Technology*, **9**(1), 71–90.

Širović, A., Williams, L. N., Kerosky, S. M., Wiggins, S. M., Hildebrand, J. A. (2013). Temporal separation of two fin whale call types across the eastern North Pacific. *Marine Biology*, **160**(1), 47–57.

Slabbekoorn, H., Bouton, N., Van Opzeeland, I. *et al.* (2010). A noisy spring: the impact of globally rising underwater sound levels on fish. *Trends in Ecology & Evolution*, **25**(7), 419–427.

Southall, B. L. (2005). Shipping Noise and Marine Mammals: A Forum for Science, Management, and Technology. www.beamreach.org/wiki/images/4/47/2004NoiseReport.pdf

Southall, B. L., Schusterman, R. J., Kastak, D. (2003). Acoustic communication ranges for northern elephant seals (*Mirounga angustirostris*). *Aquatic Mammals*, **29**, 202–213.

Southall, B. L., Bowles, A. E., Ellison, W. T. *et al.* (2007). Marine mammal noise exposure criteria: initial scientific recommendations. *Aquatic Mammals*, **33**, 411.

Tennessen, J. B., Parks, S. E. (2016). Acoustic propagation modeling indicates vocal compensation in noise improves communication range for North Atlantic right whales. *Endangered Species Research*, **30**, 225–237.

Tougaard, J., Wright, A. J., Madsen, P. T. (2015). Cetacean noise criteria revisited in the light of proposed exposure limits for harbour porpoises. *Marine Pollution Bulletin*, **90**, 196–208.

Traverso, F., Gaggero, T., Rizzuto, E., Trucco, A. (2015). Spectral analysis of the underwater acoustic noise radiated by ships with controllable pitch propellers. In: *OCEANS 2015 – Genova*. Piscataway, NJ: IEEE, pp. 1–6.

Trevorrow, M. V., Vasiliev, B., Vagle, S. (2008). Directionality and manoeuvring effects on a surface ship underwater acoustic signature. *Journal of the Acoustical Society of America*, **124**, 767–778.

Tyack, P. L., Johnson, M., Soto, N. A., Sturlese, A., Madsen, P. T. (2006). Extreme diving of beaked whales. *Journal of Expimental Biology*, **209**, 4238–4253.

Urick, R. J. (1982). *Sound Propagation in the Sea*. Westport, CT: Peninsula Publishing.

Vabø, R., Olsen, K., Huse, I. (2002). The effect of vessel avoidance of wintering Norwegian spring spawning herring. *Fisheries Research*, **58**, 59–77.

Vasconcelos, R. O., Amorim, M. C. P., Ladich, F. (2007). Effects of ship noise on the detectability of communication signals in the Lusitanian toadfish. *Journal of Experimental Biology*, **210**, 2104–2112.

Veirs, S., Veirs, V., Wood, J. D. (2016). Ship noise extends to frequencies used for echolocation by endangered killer whales. *PeerJ*, **4**, e1657.

Viola, S., Grammauta, R., Sciacca, V. *et al.* (2017). Continuous monitoring of noise levels in the Gulf of Catania (Ionian Sea). Study of correlation with ship traffic. *Marine Pollution Bulletin*, **121**, 97–103.

Wahlberg, M., Westerberg, H. (2005). Hearing in fish and their reactions to sounds from offshore windfarms. *Marine Ecology Progress Series*, **288**, 295–309.

Wale, M. A., Simpson, S. D., Radford, A. N. (2013). Noise negatively affects foraging and antipredator behaviour in shore crabs. *Animal Behaviour*, **86**, 111–118.

Wang, L., Heaney, K., Pangeric, T. *et al.* (2014). *Review of Underwater Acoustic Propagation Models*. NPL Report AC 12. Teddington: National Physical Laboratory.

Watkins, W. A. (1986). Whale reactions to human activities in Cape Cod waters. *Marine Mammal Science*, **2**(4), 251–262.

Weilgart, L. S. (2006). Managing Noise through Marine Protected Areas around Global Hot Spots. http://whitelab.biology.dal.ca/lw/publications/8.%20Weilgart%202006.%20Managing%20noise%20PAs..pdf

Weilgart, L. S. (2007). A brief review of known effects of noise on marine mammals. *International Journal of Comparative Psychology*, **20**, 159–168.

Weilgart, L. (2013). A review of the impacts of seismic airgun surveys on marine life. Submitted to the CBD Expert Workshop on Underwater Noise and Its Impacts on Marine and Coastal Biodiversity, 25–27 February 2014, London, UK. www.cbd.int/doc/?meeting=MCBEM-2014-01

Wenz, G. M. (1962). Acoustic ambient noise in the ocean: spectra and sources. *Journal of the Acoustical Society of America*, **34**(12), 1936–1956.

Williams, R., Trites, A. W., Bain, D. E. (2002). Behavioural responses of killer whales (*Orcinus orca*) to whale-watching boats: opportunistic observations and experimental approaches. *Journal of Zoology*, **256**(2), 255–270.

Wittekind, D. K. (2014). A simple model for the underwater noise source level of ships. *Journal of Ship Production and Design*, **30**(1), 1–8.

Wright, A. J. (2008). *International Workshop on Shipping Noise and Marine Mammals, Hamburg, Germany, 21–24 April 2008*. Darmstadt: Okeanos – Foundation for the Sea.

Wright, A. J., Soto, N. A., Baldwin, A. L. *et al.* (2007). Do marine mammals experience stress related to anthropogenic noise? *International Journal of Comparative Psychology*, **20**(2), 274–316.

Zimmer, W. (2011). *Passive Acoustic Monitoring of Cetaceans*. Cambridge: Cambridge University Press.

10

Vessel Strikes and North Atlantic Right Whales

CAROLINE H. FOX AND CHRISTOPHER T. TAGGART

10.1 Introduction

Vessel strikes – collisions between ships and whales – occur throughout the world's oceans (Figure 10.1) (Laist *et al.*, 2001; Van Waerebeek *et al.*, 2007). First reported in the late nineteenth century as ships reached speeds greater than 13–15 knots, lethal vessel strikes remained relatively infrequent until the mid-twentieth century (Laist *et al.*, 2001). Since then, as the number, speed and size of vessels increased, reported vessel strikes have similarly grown (Figure 10.2) (Laist *et al.*, 2001; Vanderlaan *et al.*, 2009). Vessel strikes involving large whales have emerged as a global conservation concern, largely due to evidence that vessel traffic is increasing (Tournadre, 2014), vessel strikes are increasing (Laist *et al.*, 2001; Vanderlaan *et al.*, 2009), vessel strikes are hampering the recovery of certain endangered whale species (e.g., North Atlantic right whales; Knowlton & Kraus, 2001) and, where enacted, the success of mitigation efforts to reduce the threat of ship strikes may be mixed (e.g., Lagueux *et al.*, 2011; Silber & Bettridge, 2012; Silber *et al.*, 2012a, 2012b; Vanderlaan *et al.*, 2008; van der Hoop *et al.*, 2013).

All baleen and toothed whales are vulnerable to vessel strikes, with frequently reported victims of vessel strikes including fin (*Balaenoptera physalus*), humpback (*Megaptera novaeangliae*), North Atlantic right (*Eubalaena glacialis*), southern right (*Eubalaena australis*), sperm (*Physeter catodon*) and gray whales (*Eschrichtius robustus*) (Laist *et al.*, 2001; Van Waerebeek *et al.*, 2007). However, baleen whales experience elevated risk of vessel strike. Unlike toothed whales, such as sperm and killer whales, which use sonar to echolocate, baleen whales do not echolocate (i.e., baleen whales are not acoustically looking) and often demonstrate little to no avoidance behaviours to approaching vessels (McKenna *et al.*, 2015; Nowacek *et al.*, 2004).

Although many species of whales are struck by vessels across the world's oceans, in this chapter we focus on vessel strikes involving North Atlantic right whales (hereafter referred to as 'right whales') for several reasons (Figure 10.3). First, on a per capita basis using population size estimates and constraining reports to recent events (1960–2002), right whales are two orders of magnitude more commonly killed relative to other large whales (Figure 10.4) (Vanderlaan & Taggart, 2007), with vessel strikes representing the leading cause of death for right whales (1970–2009; van der Hoop *et al.*, 2013). Second, as one of

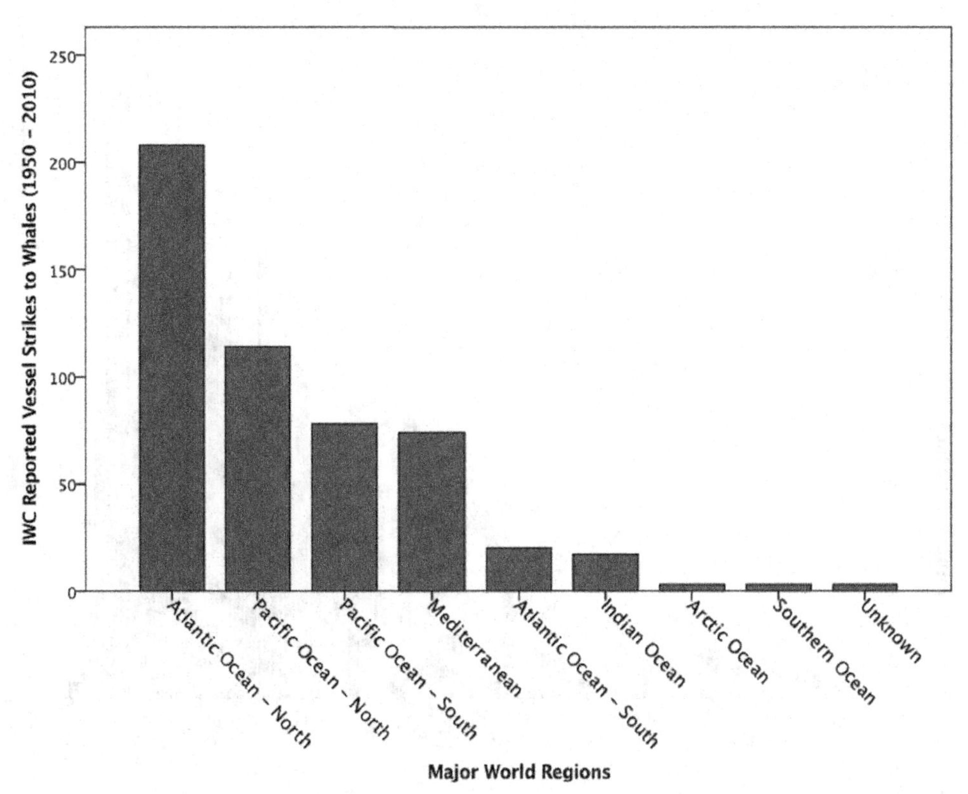

Figure 10.1 Number of International Whaling Commission (IWC) reported vessel strikes to whales (1950–2010) by ocean region.

the most intensively studied baleen whale species, with decades of scientific research, management and outreach undertaken with regards to vessel strike mitigation, right whales represent an important case study with regards to vessel strikes and strike mitigation efforts. Third, with 17 documented right whale deaths in Canadian and US waters in 2017 and a declining population totalling ~451 individuals (Corkeron, 2017), right whales are in crisis; limited time remains to effectively reduce anthropogenic threats, including lethal vessel strikes, and to stave off potential extinction.

Following an overview of the fleet and an introduction to the biology and ecology of right whales, this chapter focuses on the ability (or lack thereof) of right whales to avoid vessels, vessel avoidance of whales, consequences to whales and vessels following a strike, mitigation measures to reduce the threat of vessel strikes, fleet compliance and considerations for the future. Where appropriate, and to provide important context and information, we also draw upon more general knowledge of vessels and large whales. Lastly, for the purposes of this chapter, vessel strikes are defined as collisions between any part of a vessel, often the bow or propeller, and a living whale. Instances of whales breaching on vessels or acts that are interpreted as accidentally or intentionally bumping, ramming or

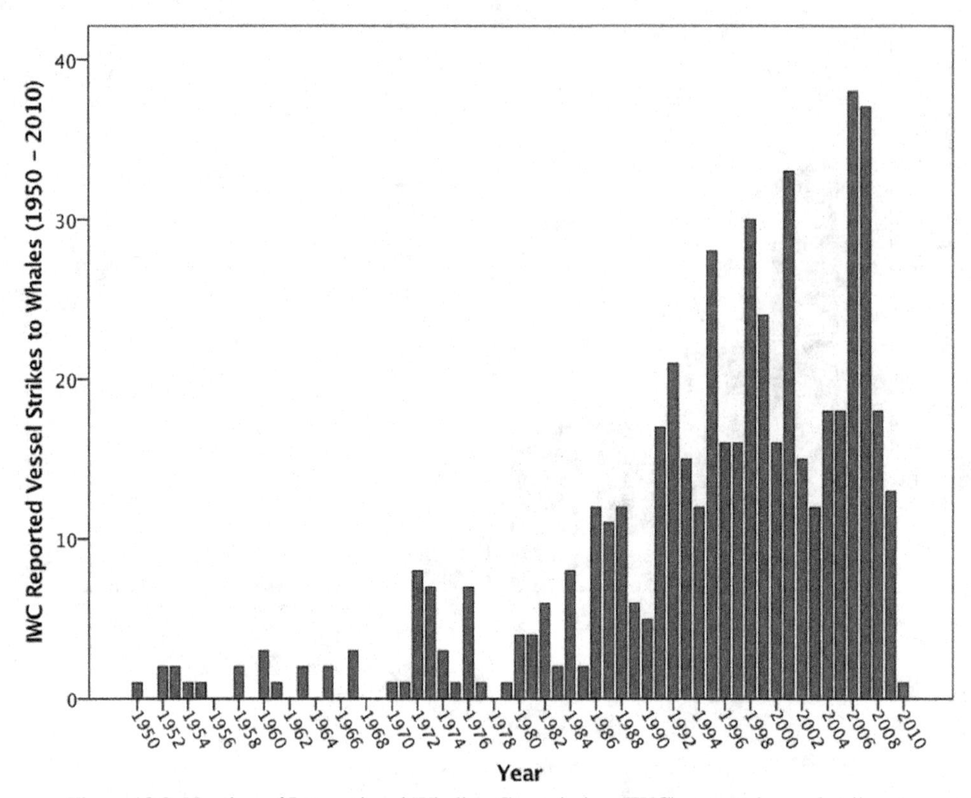

Figure 10.2 Number of International Whaling Commission (IWC) reported vessel strikes to whales (1950–2010).

Figure 10.3 A North Atlantic right whale surfaces with a commercial vessel in the distance. Image attribution: New England Aquarium.

otherwise making contact with moving, anchored or drifting vessels have been reported but represent a small component of known vessel strikes (e.g., Neilson *et al.*, 2012; Van Waerebeek *et al.*, 2007; Zappes *et al.*, 2013). Although entanglement in fishing gear or vessel components (e.g., anchor chains or lines) may result in contact between whales and vessels, such occurrences are categorized here as a form of by-catch.

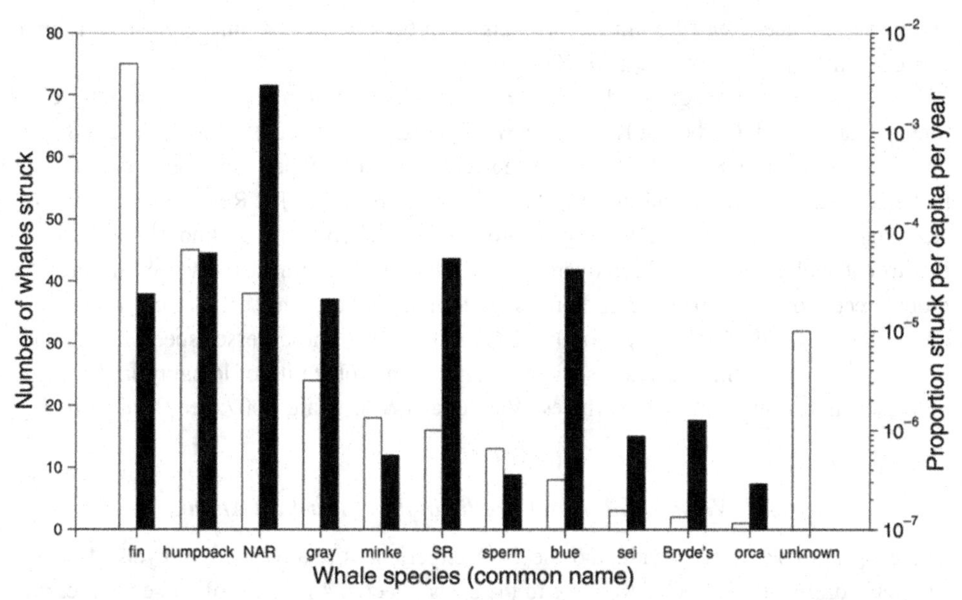

Figure 10.4 Frequency histograms of worldwide documented (Jensen & Silber, 2003; Laist *et al.*, 2001) numbers of large whales, including North Atlantic right (NAR) and southern right (SR) whales reported struck by vessels for the period 1960–2002 (open bars), and the same information presented as an adjusted per capita rate (solid bars; $\log_{10}$ scale) using population size estimates for each species where the proportion struck per capita per year = (number of species-specific whales struck/contemporary species-specific population size)/ 43 years. Where a range in population size was provided, we use the midpoint of the range.
Source: Reproduced from Vanderlaan and Taggart (2007)

10.2 The Fleet

10.2.1 The Global Fleet

The global fleet has grown considerably, and currently an estimated 90 per cent of global trade travels by sea (IMO, 2012). Using information from Lloyds Register of Shipping, the number of vessels over 100 gross tonnes (GT) more than tripled since World War II (1947–2009; Frisk, 2012). Furthermore, according to altimeter (nadir-looking radar) information, vessel traffic in the world's oceans quadrupled from 1992 to 2012, with virtually all ocean regions experiencing an increase, the exception being near the coast of Somalia (Tournadre, 2014). In 2010, there were more than 100,000 vessels greater than 100 GT in the global fleet (Frisk, 2012; IMO, 2012). In 2015, there were more than 87,000 merchant vessels (excluding military, recreational and research vessels) of four class sizes; small (37 per cent, <500 GT), medium (44 per cent, 500–25,000 GT), large (13 per cent, 25,000–60,000 GT) and very large (6 per cent, >60,000 GT; EMSA, 2015). Of these four categories of vessels, 19 per cent were general cargo vessels, 18 per cent were tankers and 13 per cent bulk carriers, most of which are medium to very large vessels (EMSA, 2015). In 2015, cargo-carrying vessels (bulk carriers, container ships, cargo ships, passenger ships and tankers) made over 1.6 million port calls, with 53.6 billion ton-miles (1 ton of freight carried 1 mile) achieved (UNCTAD, 2016).

These increases in the global fleet have been matched by an increase in vessel size (e.g., tonnage) and speed (Corbett & Koehler, 2003; Frisk, 2012). However, during the 2008 economic recession (2008–2011), vessel speed reductions of 15 per cent were reported for tankers, container ships and most bulk carriers (PWC, 2011). Referred to as 'slow steaming', vessels reduce speeds from approximately 20–25 to 16–19 knots as a mechanism to cut fuel costs and reduce overcapacity; although the practice emerged during the global recession, it remains common as part of dynamic commercial fleet operations (Boersma *et al.*, 2015; Bonney, 2010; UNCTAD, 2016). These vessel speed reductions, however, are still within ranges associated with a high probability of lethality in the event of vessel strikes involving large whales (Vanderlaan & Taggart, 2007; see Section 10.4).

10.2.2 Vessels and Vessel Traffic in the North-East Atlantic

The eastern seaboard of Canada and the USA experiences relatively high levels of vessel traffic (Endresen *et al.*, 2003). Relative to the global fleet, 6.4 per cent of vessels representing 7.5 per cent of global gross tonnage navigate the eastern North Atlantic, from coastal Florida to the Arctic (EMSA, 2015). Overall, vessel traffic in the North Atlantic Ocean has increased, with certain areas within the range of right whales estimated to have increased by over 200 per cent (1992–2012; Tournadre, 2014).

Vessels in the eastern North Atlantic include commercial, recreational, research and government vessels. Commercial vessel traffic is diverse, including cargo-carrying vessels, fishing vessels, tugs and piloting vessels, vessels associated with oil and gas extraction and other vessels. Most commercial shipping involves port-to-port movements over regional and global scales (Bouwman, 2016), and major sea routes, including great circle routes (i.e., the shortest distance between points over the Earth's surface), are located along the eastern North American seaboard. These habitual traffic patterns are distinct from traffic separation schemes (TSSs) and traffic lanes; near major ports, such as New York and New Jersey, and on entrance to the Great Lakes/St Lawrence Seaway via the Gulf of St Lawrence, vessel traffic service zones regulate vessel traffic, including traffic lanes and TSSs. Combined, regulated and habitual traffic patterns result in elevated vessel densities in many coastal areas.

10.3 North Atlantic Right Whales

10.3.1 Population Biology, Status and Trends

Following centuries of whaling that severely impacted the population of right whales (Aguilar, 1986; Reeves *et al.*, 2007), international efforts to protect the species were initiated in the 1930s, when the species was legally protected from whaling. Despite decades of international, national and provincial/state conservation and management efforts, right whales remain one of the most endangered marine mammals on the planet.

Listed as Endangered under Canada's Species at Risk Act (SARA), the US Endangered Species Act (ESA) and the International Union for Conservation of Nature (IUCN), the species faces possible extinction.

The population of right whales is currently in decline. Following population increases of approximately 2.8 per cent on an annual basis from 270 animals in 1990 to 483 in 2010, the population declined to an estimated 458 in 2015 (Pace *et al.*, 2017) and to 451 in 2016 (Corkeron, 2017). An unprecedented 17 right whale deaths were documented in Canada (12) and the USA (5) in 2017, which represents a minimum estimate of total mortality. Sixteen of these 2017 mortalities were later designated unusual mortality events (UMEs) under the US Marine Mammal Protection Act (NOAA, 2017). Of the 2017 right whale deaths, 12 occurred in the Gulf of St Lawrence (Daoust *et al.*, 2017). Primary diagnoses for six autopsied animals in Canada indicate that three died of acute internal haemorrhage compatible with blunt trauma (two suspected, one probable), one died of a skull fracture due to blunt trauma (probable), one died of chronic entanglement with fishing gear (confirmed) and one cause of death was undetermined due to advanced decomposition, although blunt trauma was suspected (Daoust *et al.*, 2017). Although preliminary, the deaths due to blunt trauma may be attributable to vessel strikes. Notwithstanding births, the population in 2017 is anticipated to be lower than 2016 estimates.

In addition to overall population declines occurring since 2010, sex-related survival differences have resulted in divergent abundance trends between males and females (Pace *et al.*, 2017). Females can live to at least 70 years, but most are dying between 20 and 30 years of age (Corkeron, 2017). These reduced adult female survival rates have resulted in a model-estimated male:female sex ratio of 1.46 in 2015 (Pace *et al.*, 2017). Approximately 100 reproductively aged females, which represent the reproductive capital of the population, are estimated to persist in 2017 (Corkeron, 2017). These females also represent roughly 25 per cent of the total population.

Contributing to the ongoing population decline is the variable and low average annual per capita calving rate (estimated 4.4 per cent), which is lower than that of the congeneric southern right whale (*E. australis*; Pace *et al.*, 2017). Three periods of poor calving have been noted since 1990 (1993–1995, 1998–2000 and 2012–2015; Pace *et al.*, 2017). Linked to a number of often interrelated factors, including poor whale health (Rolland *et al.*, 2016), anthropogenic stressors (see Section 10.3.3) and environmental conditions (e.g., climate-driven changes in prey abundance; Meyer-Gutbrod *et al.*, 2015), the direct mechanism(s) underlying reduced calving rates are not well understood (Pace *et al.*, 2017).

10.3.2 Diet, Habitats and Migration Patterns

Right whales move along the eastern coasts of Canada and the USA, from Florida to the Gulf of St Lawrence and Newfoundland, with additional habitats including the mid-North Atlantic near Greenland and Iceland. In any given year, the locations of approximately 50 per cent of the population are unknown. Recent research has demonstrated that right whale habitat use changes over time and that animals are distributed across their entire range year-round (Davis *et al.*, 2017). Although our knowledge of right whale movement patterns is still evolving, areas of known seasonal use include four habitats in US and Canadian coastal waters. Winter calving ground critical habitats are located in the southern portion of their range, in the waters adjacent to Florida and Georgia (south-eastern US calving area). Northern feeding ground critical habitats are located in US and Canadian coastal waters, some of which also represent nursery grounds: in the USA, these include the Gulf of Maine and adjacent waters (north-eastern US foraging area), encompassing the Great South Channel and Cape Cod Bay, and in Canada, these include the Grand Manan Basin in the Bay of Fundy and Roseway Basin on the Scotian Shelf (Figure 10.5). In recent years, shifts away from known, reliably used northern habitats have been documented. Although previous survey effort was limited, right whales have been occasionally sighted in the Gulf of St Lawrence for decades (Daoust *et al.*, 2017). However, since 2015, dedicated surveys and passive acoustic monitoring (PAM) have documented larger numbers of right whales in the Gulf of St Lawrence in summer (Hamilton & Davis, 2017), with over 100 unique individuals documented in the region during 2017 (Daoust *et al.*, 2017). Furthermore, there is also evidence that whales are spending more time in the mid-Atlantic (Davis *et al.*, 2017).

These recent northwards shifts in feeding habitat use by right whales are likely linked in part to changes in their food supply. Right whales are highly mobile, specialized predators primarily targeting the later juvenile stages of the lipid-rich copepod *Calanus finmarchicus* (Baumgartner *et al.*, 2007; Mayo *et al.*, 2001), a species that is vulnerable to the effects of climate change (Grieve *et al.*, 2017). The north-east Atlantic, including traditional right whale feeding grounds in the Gulf of Maine, the Bay of Fundy and the Scotian Shelf, is experiencing rapid and unprecedented warming, with consequences to the distribution and abundance of *C. finmarchicus* and other zooplankton (Mills *et al.*, 2013). With *C. finmarchicus* predicted to have significantly reduced densities in the Gulf of Maine near the end of this century (Grieve *et al.*, 2017), or potentially to be virtually absent south of the Gulf of St Lawrence by 2050 (Reygondeau & Beaugrand, 2011), the quality of right whale traditional feeding grounds is changing, with implications for right whales and the suite of conservation and management measures enacted within and outside of management areas, including critical habitats.

10.3.3 Anthropogenic Threats

Vessel strikes are one of several anthropogenic threats faced by right whales. Of 122 documented deaths from 1970 to 2009, vessel strike was the leading known cause of death for right whales (n = 38; van der Hoop *et al.*, 2013). Using the right whale photo-catalogue (n = 357 animals), 6 per cent of right whales have scars from vessel strikes (Hamilton *et al.*, 1998) and females appear to be struck more often than males (Moore *et al.*, 2007; NOAA, 2008). Factors contributing to the vulnerability of right whales to vessel strikes include: their occurrence in regions that experience heavy vessel traffic (Ward-Geiger *et al.*, 2005); large body size; buoyant bodies (Nowacek, 2001); slow swimming speeds (Hain *et al.*,

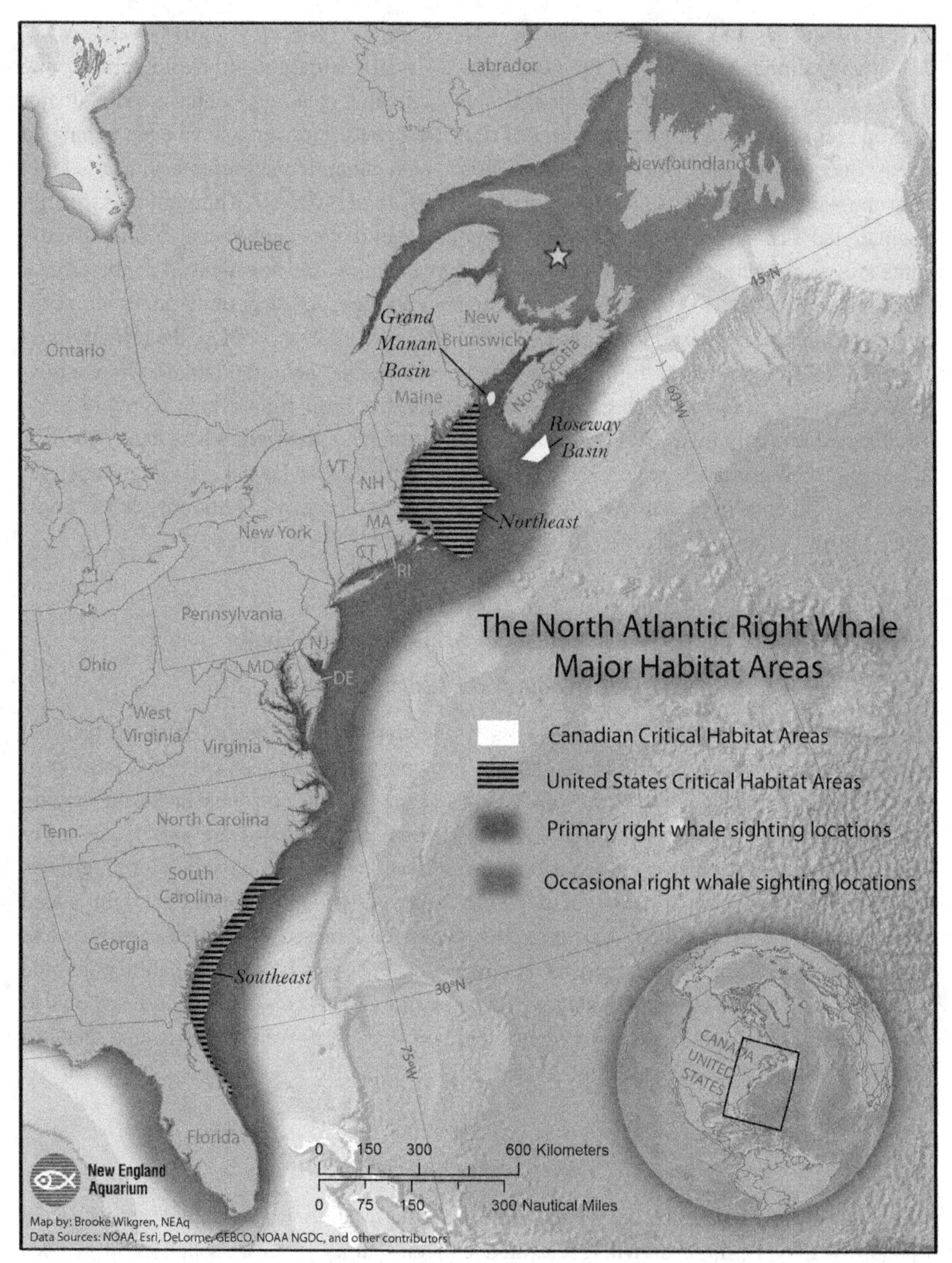

Figure 10.5 North Atlantic right whale range, migration movements and critical habitats in Canada and the USA. The Gulf of St Lawrence represents additional primary right whale sighting locations in recent years, including 2017. Map attribution: Brooke Wikgren, New England Aquarium (figure modified for greyscale reproduction).

2013); a tendency to spend time near the water's surface (Kot *et al.*, 2014; Parks *et al.*, 2011); and, similar to a 'deer in the headlights', virtually no ability to avoid approaching vessels (Nowacek *et al.*, 2004). Entanglement in fishing gear represents another major anthropogenic threat; a large proportion of right whales (83 per cent) have been entangled at least once (1980–2009; Knowlton *et al.*, 2012), with entanglement representing a leading anthropogenic source of mortality (van der Hoop *et al.*, 2013). Although less readily disentangled from natural processes, anthropogenic climate change is also undoubtedly influencing right whales and their habitats; as a key example, the north-east Atlantic Ocean, including right whale feeding grounds in Canada and the USA, has undergone unprecedented warming, with consequences to *C. finmarchicus* (Mills *et al.*, 2013), the primary prey of right whales. Combined, these anthropogenic threats may act cumulatively or synergistically, in addition to interacting with other factors that limit right whale recovery (e.g., variable and low calving rates; Pace *et al.*, 2017). Ultimately, the right whale extinction risk is thought to be primarily governed by vessel traffic (vessel strike risk), fishing (entanglement risk) and environmental conditions such as prey availability (Melbourne & Hastings, 2008; Meyer-Gutbrod & Greene, 2017).

10.4 Vessels and North Atlantic Right Whales

10.4.1 Whale–Vessel Intersections

The spatiotemporal overlap between vessel traffic, typically concentrated along habitual shipping routes and regulated shipping lanes near port entrances, and the distribution of right whales along the eastern coasts of Canada and the USA is a primary factor underlying right whale vessel strikes. Although locations where right whales are struck are infrequently known, ship-killed right whale carcasses have been found close to shipping lanes and other areas of high-density vessel traffic, both within and adjacent to right whale critical habitats and migratory corridors (Knowlton & Kraus, 2001). Vessel strikes can occur anywhere where vessels and whales intersect, but vessel strike probability may be highest in areas where elevated densities of whales and vessels co-occur. Vessel strikes also occur in areas with high densities of whales and few vessels, as well as the reverse scenario of high densities of vessels and few whales.

10.4.2 Vessel Strikes and Strike Risk

Right whales are vulnerable to vessel strikes throughout their range, but vessel strike probability may be highest where elevated concentrations of vessels and right whales intersect; the number of known collisions involving right whales appears to be related to the overlap between important right whale habitats where right whales occur in greater numbers and shipping routes (NOAA, 2008). However, vessel strikes are not always lethal, and lethal strike risk estimation is a function of vessel strike probability and the probability of strike lethality (Vanderlaan & Taggart, 2007). Lethal vessel strikes involving large whales were first reported in the late 1880s, were relatively infrequent until the mid-1900s and have increased since the 1950s (Laist *et al.*, 2001). All sizes and classes of vessels strike whales, but most lethal strikes are caused by large vessels (i.e., >80 m in length) and by vessels travelling at higher speeds (Laist *et al.*, 2001; Vanderlaan & Taggart, 2007).

The momentum of any object, including vessels and whales, is a product of its mass and velocity. As a consequence, when vessels and whales collide, the resultant impact force is a function of their respective momentums. At above 500 tons, which includes commercial cargo-carrying vessels and other large ships, vessel speed and not vessel size was found to be the dominant factor affecting the impact forces experienced by a whale (Wang *et al.*, 2007). Vessel speed has been demonstrated to influence whale injury severities, with increasing whale lethalities associated with greater vessel speeds (Laist *et al.*, 2001; Vanderlaan & Taggart, 2007); because impact forces received by whales increase with vessel speed (Silber *et al.*, 2010; Wang *et al.*, 2007), the probability of strike lethality also increases with vessel speed (Silber *et al.*, 2010; Vanderlaan & Taggart, 2007).

Using published records of vessel strikes involving large whales, Vanderlaan and Taggart (2007) examined the relationship between vessel speed and probability of lethal injuries using four injury classes (killed, severe, minor and none apparent), as per Laist *et al.* (2001). On a cumulative percentage basis, the killed and severe injury classes increase with vessel speed, as do the minor and none apparent injury classes, although at a lower level (Vanderlaan & Taggart, 2007). At vessel speeds less than 12 knots, the probability of lethality is estimated at less than 0.5; at speeds greater than 12 knots, the probability of lethality asymptotically approaches 1.0 (Figure 10.6) (Vanderlaan & Taggart, 2007). This positive relationship between vessel speed and the probability of lethal injury to whales was confirmed by Conn and Silber (2013), although it was less pronounced than was reported by Vanderlaan and Taggart (2007). Using models of a container ship and a right whale in flow tanks, Silber *et al.* (2010) similarly found that greater vessel speed was associated with an increased magnitude of whale accelerations, but they also described a propeller suction effect (hydrodynamic draw), which drew the model whale towards the ship's hull. Potentially causing large whales (and presumably other species) to be pulled towards vessels, this hydrodynamic draw effect may increase the probability of propeller strikes and strike impacts to whales (Silber *et al.*, 2010).

The probability of vessel encounters with whales also varies with vessel speed. Using a model to approximate average encounter probabilities, Vanderlaan and Taggart (2007) demonstrated that the encounter probability increases gradually as speed decreases from 24 knots, before increasing rapidly at low vessel speeds (Figure 10.7). Although the mechanism is not well understood and further investigation is needed, Gende *et al.* (2011) found that the average distance between ships and humpback whales decreased with increasing vessel speeds, which could influence the probability of vessel strikes. Furthermore, a recent analysis determined that vessel strike rate increases as a function of vessel speed (Conn & Silber, 2013), although further examination of this issue and potential bias of the underlying data are warranted.

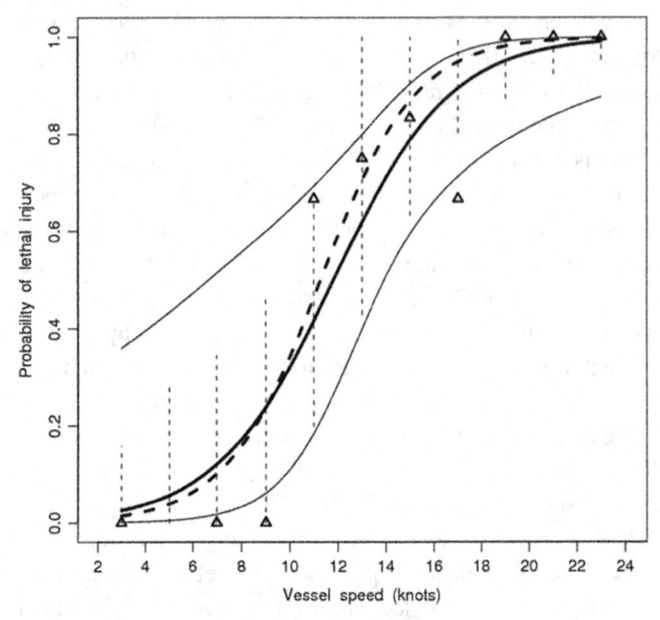

Figure 10.6 Probability of a lethal injury (whales killed or severely injured) resulting from a vessel strike to a large whale in relation to vessel speed based on the simple logistic regression (solid heavy line) and 95 per cent confidence interval (solid thin lines) and the logistic fitted to the bootstrapped predicted probability distributions (heavy dashed line) and 95 per cent confidence intervals for each distribution (vertical dashed lines) where each datum (triangles) is the proportion of whales killed or severely injured when struck by a vessel navigating within a given 2-knot speed class. No data are available in the 4–6-knot speed class.
Source: Reproduced from Vanderlaan and Taggart (2007)

10.4.3 Consequences to Vessels

Vessel strikes, for the purposes of this chapter, are defined as impacts between any part of a vessel, often the bow or propeller, and any body part of a living whale. Consequences of whale–vessel collisions to the vessel, its crew and passengers, if present, range from none or negligible to severe, including human fatalities and vessel loss. Detailed documentation of consequences to vessels and whales is challenging to compile; mariners may be unaware of the collision itself, unaware that it should be reported or unwilling to report due to fear of negative consequences, such as damage to reputation and reprisals (Neilson *et al.*, 2012; Zappes *et al.*, 2013). As a consequence, collision accounts are biased but nonetheless provide useful information and insight (Laist *et al.*, 2001).

Several studies of vessel strikes detail consequences to vessels and people on board. The first, an analysis of 108 Alaskan vessel strikes of whales, reported that 34 per cent of events resulted in no harm to the vessel or passengers aboard, 33 per cent resulted in human and/or vessel damage and 32 per cent had unknown outcomes (Nielson *et al.*, 2012). Of the collisions that resulted in negative effects to people, these included relatively minor (e.g., knocking people over), unspecified injuries and people being thrown into the water

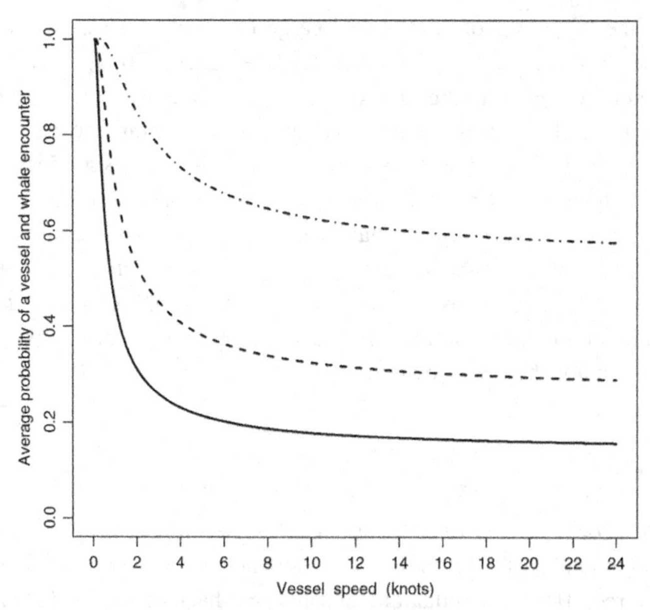

Figure 10.7 The probability of a vessel and whale encounter as a function of vessel speed within a 1-km^2 domain estimated using a random walk model in two dimensions of a 16.5-m whale swimming at 1.5 m/s in the presence of an example vessel (125-m length and 20-m beam). The lines represent the domain with one whale and one vessel (solid line), two vessels (dashed line) and five vessels (dashed–dotted line).

Source: Reproduced from Vanderlaan and Taggart (2007)

(Neilson *et al.*, 2012). In that same analysis, three small recreational vessels sank, although two of those cases involved anchored or drifting vessels that were reportedly rammed by humpback whales (Neilson *et al.*, 2012). Second, a study of 111 reported cetacean–sailing vessel collisions, including 5 whales either breaching on or 'attacking' sailing vessels (5 per cent), found damage to vessels occurred in 57 per cent of events (Ritter, 2012). Of events with reported damage, this ranged from minor (46 per cent) to major (43 per cent), with seven strikes (11 per cent) resulting in vessel loss (Ritter, 2012). Third, of 292 reported vessel strikes, 13 reported vessel damage that ranged from minor to severe, including a cracked hull and other structural damage (Jensen & Silber, 2003). Fourth, an examination of five collisions between high-speed hydrofoil vessels in Japan and objects that were likely cetaceans occurred; two of those events resulted in minor vessel damage, with one event injuring 13 passengers and severely damaging the vessel (Honma *et al.*, 1997). Collisions involving human fatalities are rare, but they do occur (e.g., a high-speed ferry collision with a sperm whale in the Canary Islands killed one passenger; de Stephanis & Urquiola, 2006). Lastly, detailed descriptions of collisions involving right whales are uncommon, but two non-lethal vessel strikes that injured both whales resulted in no reported damage to the two research vessels involved (Wiley *et al.*, 2016).

In addition to vessel damage and injury to people arising from a collision with a whale, harm may also occur to the vessel owner's and/or operator's reputation (Neilson *et al.*,

2012). Due to the large size of commercial cargo-carrying vessels, vessel operators are likely rarely aware of collisions with whales (Conn & Silber, 2013). Yet of reported ship strikes, commercial vessels arriving in port with large whales – particularly long, slender-bodied species such as fin whales – draped over the bow are common (Laist *et al.*, 2001). In recent decades, bow-draped whales found in port often result in significant and typically negative media attention. Furthermore, for documented violators of vessel speed restrictions enacted to reduce vessel strike lethality, consequences may include fines, warnings and/or education, but also increased attention arising from federal agency press releases (e.g., Silber *et al.*, 2014), industry trade publications (e.g., Silber *et al.*, 2014) and associated media reports that may identify the vessel, the mariners involved, the relevant industry (if any) and/or the vessel owner.

10.4.4 Consequences to Whales

For large whales, consequences of being struck by vessels range from no discernible effect to serious injuries and death. Injuries include crushed skulls, fractured bones, propeller lacerations, severed arteries, amputations and internal haemorrhaging (Campbell-Malone *et al.*, 2008; Costidis *et al.*, 2013; Laist *et al.*, 2001). In extreme cases, animals may be severed in two (Campbell-Malone *et al.*, 2008).

In a study of 40 right whales necropsied between 1970 and 2006, of which vessel strikes were identified as the ultimate cause of death for 21 animals (52.5 per cent), injuries were divided into two non-distinct categories: sharp trauma, often caused by contact with the propeller or rudder; and blunt trauma, often resulting from contact with the vessel hull (Campbell-Malone *et al.*, 2008). Sharp traumas may range from non-lethal nicks and cuts to severe, immediately fatal wounds (Figure 10.8a–c) (Campbell-Malone *et al.*, 2008). Sharp trauma includes propeller wounds to soft tissues and bones that often present as obvious external injuries, including amputations and single or multiple, shallow to deep and approximately parallel incising or chopping wounds delivered by a sharp, rotating propeller (Campbell-Malone *et al.*, 2008; Costidis *et al.*, 2013). Chronic sequelae, meaning lasting complications arising from the initial injury, may also result from sharp traumas associated with vessel strikes (Campbell-Malone *et al.*, 2008). In one known example, a female right whale received three deep propeller lacerations from a vessel strike as a calf and died 14 years later when the wounds reopened and became infected during her first pregnancy (Campbell-Malone *et al.*, 2008; Costidis *et al.*, 2013).

In contrast to sharp trauma, blunt trauma arising from a vessel strike often leaves little or no discernible evidence due to the dark colouration of right whale skin and the thickness of their external tissues, although swelling and bruising are externally apparent for some struck right whales (Campbell-Malone *et al.*, 2008). Four categories of blunt trauma, as defined by DiMaio and DiMaio (2001), include abrasions, contusions (bruises and haemorrhages), lacerations and bone fractures (Figure 10.9a–c). Although varying in their presentation, common characteristics of vessel strike blunt force trauma involve well-defined areas of subcutaneous oedema and haemorrhage, torn or otherwise disrupted

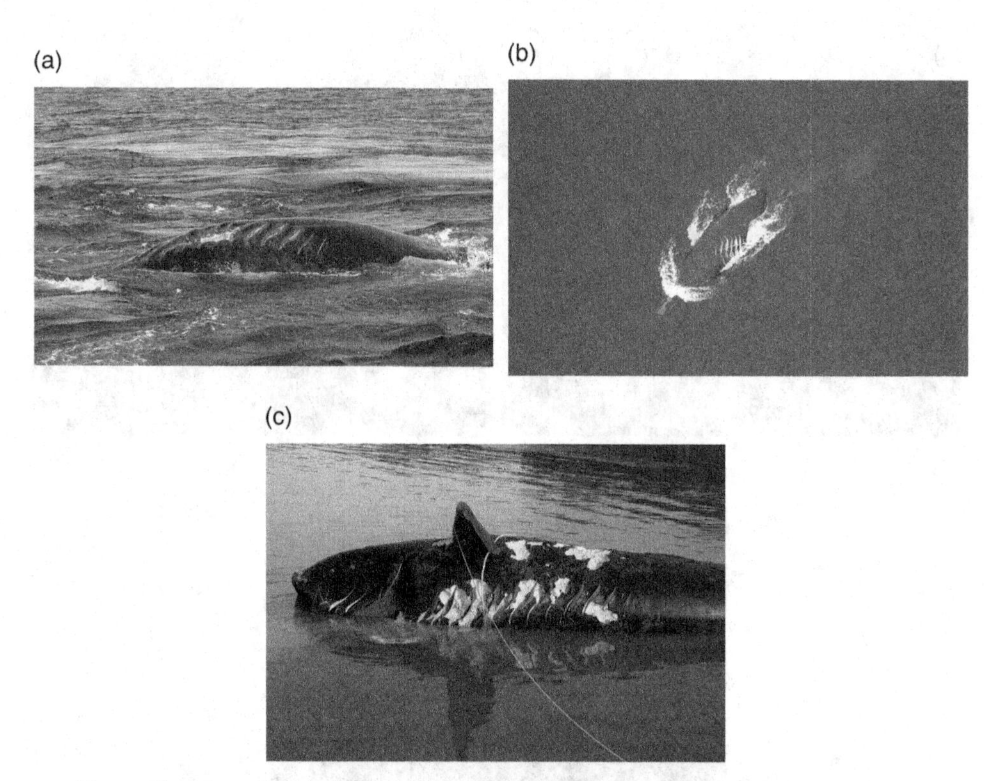

Figure 10.8 Examples of North Atlantic right whale sharp traumas resulting from vessel strikes: (a) photographed in 2007 and seen again as recently as 2016, right whale 3503 survived her interaction with a vessel (image attribution: New England Aquarium); (b) photographed alive, with serious injuries resulting from propeller wounds, right whale 3522 was not seen again following this sighting (image attribution: New England Aquarium); (c) right whale 3508 died as a result of lethal sharp trauma injuries resulting from a propeller, with cause of death determined by necropsy being rapid exsanguination and pneumothorax from a catastrophic interaction with a vessel (NOAA permit # 932-1905/ MA-009526, photo attribution: Katie Jackson, Florida).

muscle and/or disrupted internal organs (Campbell-Malone *et al.*, 2008; McLellan *et al.*, 2013). Due to the frequent lack of obvious external injuries, complete forensic necropsies remain a fundamental diagnostic tool for identifying probable cause of death (Campbell-Malone *et al.*, 2008).

Animal welfare issues associated with vessel strikes are apparent. While suffering may be intense but short-lived for animals killed immediately following a vessel strike, likely pronounced suffering of large whales with severe injuries have been reported (e.g., a probably vessel-struck humpback whale with a 'grossly inflated tongue and deformed head' survived for several days or more; Neilson *et al.*, 2012). After being struck by a small vessel travelling at high speed, a right whale was reported thrashing and bleeding at the surface with propeller cuts that mangled its fluke; 6 months later, the whale was observed to be emaciated and in such poor body condition that it was predicted to die

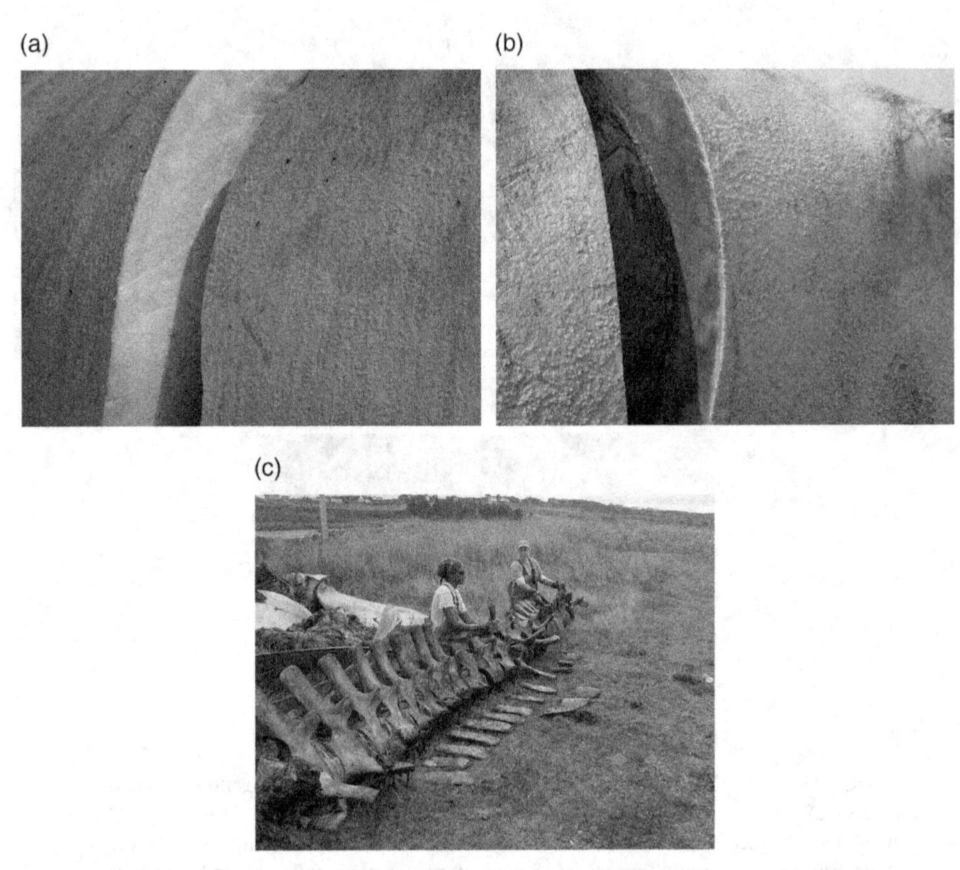

Figure 10.9 North Atlantic right whale blunt traumas resulting from a vessel strike and comparison with normal blubber: (a) diagnostic incision into right flank blubber showing normal blubber coat; (b) diagnostic incision into left flank blubber showing bruising of the blubber coat (skin missing due to decomposition); (c) reassembled axial skeleton showing multiple fractures of the transverse processes in the left lumbar region, with the vertebra held in place. Photographs attributed to Michael Moore, Woods Hole Oceanographic Institution.

within weeks (Knowlton & Brown, 2007). Evidence of healing processes (i.e., woven bone) was found in a vessel-killed right whale with a fractured skull, broken jaw, internal haemorrhaging and oedema, and these healing processes were likely attributable to the animal having survived days or weeks following the strike or, alternatively, it having survived an earlier strike that fractured its jaw (Campbell-Malone *et al.*, 2008). Lastly, in one described non-lethal vessel strike involving a right whale, the animal was reported to be bleeding from lacerations along its fluke and was observed 'flailing its flukes' at the surface; hours later, the animal was observed to be behaving normally, and the wound later healed with part of the fluke missing (Wiley *et al.*, 2016). For the whales that ultimately survive a vessel strike, many will have temporary or permanent injuries and/or associated complications that may interfere with locomotion, feeding, breeding, pregnancy and conspecific interactions, with potential consequences for individual and population-level fitness.

10.4.5 Whales Avoiding Vessels

Right whales demonstrate virtually no ability to avoid approaching vessels. Exposed to approaching vessels and the sounds of approaching vessels, right whales demonstrated no behavioural change (Nowacek *et al.*, 2004). Furthermore, when exposed to an alert signal, right whales responded by swimming strongly to the surface and spending additional time near the surface, responses that would be likely to increase vessel strike risk (Nowacek *et al.*, 2004). In two described right whale vessel collisions, no reaction to avoid the vessels was reported, although this lack of avoidance response should not be confused with a failure to detect or react to vessels (Wiley *et al.*, 2016). During tagging of right whales, observations of animals in close proximity (~10 m) to an approaching vessel indicated that some animals detected the vessel and reacted with behavioural responses that might not reduce the risk of being struck, such as moving away from the vessel with an open, water-filled mouth, which would reduce escape speed and manoeuvrability (Wiley *et al.*, 2016). Importantly, the vessel had to be very close (~10 m) for the animal to elicit such a response.

10.4.6 Vessels Avoiding Whales

With evidence that right whales demonstrate virtually no ability to avoid vessels, vessel avoidance of right whales is therefore a major component of strike risk reduction. In a review of vessel strikes, many reports indicate that large whales struck by vessels are not detected beforehand or are only detected immediately prior to a strike (Laist *et al.*, 2001). Furthermore, for larger vessels such as commercial cargo-carrying vessels, operators are likely rarely aware of vessel strikes occurring at all (Conn & Silber, 2013). Even with dedicated observers, experienced crews, small, manoeuvrable vessels and good to moderate sighting conditions, vessel strikes involving right whales have occurred, with the whale either not detected until struck or detected immediately prior to the collision, with insufficient time for observers to alert the vessel operator (Wiley *et al.*, 2016).

In order to successfully avoid a collision with a whale, the whale must first be detected, and second the vessel operator must successfully manoeuvre to avoid the whale. Whale detection is dependent on multiple factors, including weather conditions, whale behaviour (e.g., diving versus time spent at surface; Hain, 1997), the surface profile of the whale (with no dorsal fin, right whales have a low surface profile), light levels, observer bias and characteristics of the observation platform (e.g., observer height and view obstructions). For right whales, detection by trained observers can be low. As an example, an estimated 33 and 55 per cent of mother and calf pairs present and at the surface, respectively, were sighted when within 1.5 nautical miles of a survey aeroplane (Hain *et al.*, 1999). Even when sighted, the success of the operator in manoeuvring the vessel depends on several factors, including time available, knowledge of and willingness to undertake avoidance measures, anticipation of whale movements, considerations for navigation and safety, mariner skill and vessel type, speed and manoeuvrability.

Given the frequent inability of vessels to detect and avoid right whales in their path coupled with evidence that right whales have virtually no ability to avoid vessels, efforts to improve vessel avoidance of right whales and subsequently reduce vessel strike risk are key mitigation strategies in Canada and the USA. Measures that focus on vessel avoidance of right whales are numerous and varied, and include both voluntary and mandatory vessel requirements, in addition to efforts aimed at educating and raising mariner awareness about right whales and measures mariners can take to avoid collisions with whales. Vessel avoidance of right whales as a mitigation strategy is discussed in Section 10.5.2.

10.5 Mitigation Strategies and Policies

10.5.1 Mitigation Overview

Over the past two decades, three primary, non-exclusive conservation measures have been enacted to protect right whales from vessel strikes (Mullen *et al.*, 2013). Both Canada and the USA have implemented vessel avoidance and vessel speed restriction measures, which represent a practical approach to reduce the risk that vessels pose to right whales. Vessel avoidance measures, where vessel traffic has been rerouted around important right whale habitats, are aimed at reducing vessel strike probability. Vessel speed restriction measures, in contrast, are intended to reduce the probability of lethality in the event of a vessel strike. A second approach includes educating and raising mariner awareness about right whales, their migratory movements and important habitats and measures mariners can take to avoid collisions with right whales (Knowlton & Brown, 2007). A third approach focuses on alerting mariners to the near-real-time locations of right whales, with consequences that may include vessel manoeuvring to avoid whale locations and/or reducing vessel speed.

10.5.2 Vessels Avoiding Whales

Vessel avoidance measures, where vessels are rerouted around areas of importance to right whales, are intended to reduce vessel strike risk by reducing the probability of vessel–whale co-occurrences. These efforts include the rerouting of regulated and habitual traffic patterns. In Canada, the Bay of Fundy TSS was amended in 2003 to reroute vessels around important right whale habitats in an effort to reduce the probability of right whales being struck by vessels. The amended TSS was 1.5 km (0.8 nautical miles) longer than the original and increased lane transit time by 1.4 per cent, based on a 12-knot vessel speed (Vanderlaan *et al.*, 2008). This relatively minor rerouting was estimated to reduce lethal vessel strike risk by 90 per cent in the area where the TSS overlapped with the Grand Manan Basin Conservation Area (now designated as a Critical Habitat) and by 62 per cent throughout the study area (Vanderlaan *et al.*, 2008).

Similar rerouting efforts have been achieved in the USA. In 2007, the Boston TSS was shifted and narrowed to reduce spatial overlap between vessel traffic and important right whale habitats in an effort to reduce vessel strikes (IMO, 2006); this rerouting was estimated to reduce vessel strike risk to right whales by 46 per cent (Merrick, 2005). The Boston TSS was further amended in 2009 to reduce right whale–vessel overlap (IMO, 2008). There are also recommended traffic routes that are intended to shift traffic away from right whale habitats in the USA. As an example, voluntary rerouting combined with reduced vessel speeds in the southern calving grounds (coastal Georgia and Florida) was estimated to reduce right whale mortalities arising from vessel strikes by ~72 per cent, with an estimated vessel rerouting compliance of 96 per cent at the end of the study (Lagueux *et al.*, 2011).

In addition to permanent TSS amendments and voluntary traffic pattern rerouting, Areas to Be Avoided (ATBA) are also applied. Implemented in 2008, an IMO-adopted seasonal (1 June–31 December) ATBA on Canada's Scotian Shelf reroutes traffic around Roseway Basin, which was recently designated as a Critical Habitat (Canada Gazette, 2016). Rerouting around the ATBA does not add substantial distances to overall vessel routes; as an example, for vessels navigating between Halifax, Nova Scotia, and the New York TSS, compliance with the Roseway Basin ATBA adds 3.7 km (2 nautical miles) to a 419-km (226-nautical mile) route (Vanderlaan *et al.*, 2008). Using simulations with constant whale abundance, voluntary vessel compliance with the Roseway Basin ATBA resulted in a 69 per cent reduction in the per capita lethal strike rate for right whales (van der Hoop *et al.*, 2012). In recent years, however, voluntary compliance with the Roseway Basin ATBA has declined (Taggart *et al.*, 2017). An IMO-endorsed ATBA was also established and has been demonstrated to be effective in the USA; the seasonal (1 April–31 July) ATBA in the Great South Channel was estimated to result in a 63 per cent reduction in relative right whale vessel strike risk (Merrick & Cole, 2007).

When vessel strike risk reductions that rely on vessel avoidance are achieved, benefits to right whales are clear: fewer numbers of animals are struck. Although not verified, vessel avoidance may also reduce underwater vessel noise, which has been demonstrated to negatively impact right whales (Rolland *et al.*, 2012). Consequences to the fleet may also be minimal. As examples, for the Bay of Fundy TSS and Roseway Basin ATBA, rerouting caused minimally increased transit times and/or distances travelled; both measures were associated with reductions in the lethal vessel strike risk (Vanderlaan *et al.*, 2008). However, significant vessel rerouting could substantially increase transit times and subsequent costs. When vessels cannot avoid whales through avoidance measures, mitigating the threat of vessels to right whales by slowing vessel speeds is an attractive and often effective alternative measure (Wiley *et al.*, 2011).

10.5.3 Vessel Speed Reduction

In contrast to vessel avoidance measures, vessel speed restrictions are not intended to influence vessel strike risk. Speed restrictions are intended to reduce vessel strike lethality in the event of a vessel strike (NOAA, 2008), with whales struck by vessels travelling at lower speeds more likely to survive the event (Vanderlaan & Taggart, 2007). See Section

10.4.4 for a description of sharp and blunt trauma injuries to struck right whales. Implemented as voluntary and mandatory measures to protect right whales throughout the eastern USA, and much more recently in Canada, some but not all speed restriction measures have been demonstrated to reduce the probability of vessel strike lethality.

In the USA, seasonal management areas (SMAs) impose mandatory vessel speed restrictions in 10 areas where elevated right whale occurrences and vessel activities occur at various times of the year: for vessels greater than 19.8 m (65 ft), vessel speed is restricted to no more than 10 knots (NOAA, 2008). Implemented in 2008 with a 5-year sunset clause that has since been removed (NOAA, 2008, 2013), speed restrictions within SMAs were estimated to have reduced the right whale vessel strike mortality risk within SMAs by 80–90 per cent (Conn & Silber, 2013). Furthermore, Laist *et al.* (2014) found that following the 2008 speed restriction rule, average annual discovery rates of vessel-struck right whale carcasses located within or adjacent to active SMAs declined significantly from 0.72 to 0 carcasses per year. However, Silber *et al.* (2015) cautioned that the reduction in the total number of vessels attributed to the 2008–2009 economic downturn may also have had a role in reducing right whale vessel strikes. Initially, SMA vessel compliance was low but later improved, although foreign vessel compliance was consistently low (Silber & Bettridge, 2012).

In contrast to SMAs, dynamic management areas (DMAs) are short-term (15-day) management areas for right whales found outside active SMAs (NOAA, 2008). Established for aggregations of three or more right whales, the size of the DMA is commensurate with the number of whales observed (NOAA, 2008). Allowing for small, short-term management measures that are directly linked to the presence of right whales, DMAs come with a minimized burden to industry in comparison to larger-area and longer-term SMAs (NOAA, 2008). Announced via various maritime communication media, vessels are requested to voluntarily reduce speed to less than 10 knots or avoid the DMA (NOAA, 2008). Compliance with voluntary DMAs has been very poor; in an analysis by Silber *et al.* (2012a), mean vessel speeds exceeded 10 knots within DMAs, vessel speeds did not differ substantially from speeds outside DMAs and few vessel transits appeared to involve attempts to manoeuvre around DMAs. Silber *et al.* (2012a) ultimately concluded that DMAs had only ‘modest’ consequences on vessel strike risk to right whales.

Canada recently implemented voluntary and mandatory speed restrictions to protect right whales in response to the 2017 right whale deaths in the Gulf of St Lawrence. Due to significant numbers of right whales in this region being concurrent with several right whale deaths, some of which were attributed to blunt trauma (Daoust *et al.*, 2017) consistent with vessel strikes, speed restrictions were enacted. In July 2017, a temporary voluntary slowdown zone, where vessels >19.8 m (65 ft) were requested to reduce to speeds of ≤10 knots, was implemented for the western Gulf of St Lawrence (DFO, 2017a). Given additional right whale deaths in the Gulf of St Lawrence, vessel compliance with the slowdown zone was made mandatory in August 2017 (DFO, 2017b). Several vessels, including a coast guard vessel, were fined for not adhering to mandatory vessel speeds, with violating vessels publicly identified (e.g., Transport Canada, 2017). Vessel compliance with the voluntary vessel slowdown was low (42 per cent), but increased to 98 per cent with mandatory vessel slow down (Taggart *et al.*, 2017).

For vessels and maritime industries reliant on vessels, speed restrictions raise economic and safety concerns. Large vessels travelling at ≤10 knots become less controllable, and manoeuvres relating to right whale sightings can increase the probability of vessel-grounding accidents (Convertino & Valverde, 2018), which raises safety and navigation issues for vessels in confined and/or shallow areas. Furthermore, while some course corrections as part of vessel avoidance measures enacted to protect right whales have been shown to result in minimal effect on vessel transit times (Vanderlaan *et al.*, 2008), speed reductions result in increased and more costly transit times. Following the 2017 temporary mandatory vessel slowdown in the Gulf of St Lawrence, estimates of millions of dollars of lost revenue due to cancelled cruise ship port visits and other economic impacts were reported in news media. These costs were likely exacerbated due to a relative lack of advance notice of the speed restrictions; however, situations involving unanticipated and unacceptably high risks to highly mobile, endangered right whales warrant rapid management responses. In the USA, the maximum direct and indirect estimates of the economic effects of right whale speed restrictions were between US$52 million and US$79 million, assuming 100 per cent compliance (Silber & Bettridge, 2012).

10.5.4 Mariner Education and Awareness

For over two decades, efforts aimed at reducing the threat of vessels to right whales have included mariner education about right whales and precautionary measures to avoid vessel strikes. Mariner education and awareness efforts also extend to communications directed at affected maritime industries and communities regarding right whale protection measures in Canada (e.g., the Roseway Basin ATBA and Gulf of St Lawrence mandatory slowdown zone) and the USA (e.g., SMAs and DMAs) in order to increase compliance. Outreach and education efforts have been numerous, varied and often coordinated across Canada and the USA. Furthermore, there have been additional education and awareness efforts from international bodies such as the IMO and the International Whaling Commission (IWC).

In the USA, one of the first formal actions to reduce right whale vessel strikes was the IMO-endorsed Mandatory Ship Reporting (MSR) system, the goal of which is to provide information about right whales and their vulnerability to vessel strikes as vessels enter important right whale habitats (Silber *et al.*, 2015). Implemented in 1999, commercial vessels greater than 300 GT are required to report vessel information (e.g., name, course, speed and destination) to shore-based stations as they enter two regions on the US east coast (off the coast of Massachusetts and off the coasts of Florida and Georgia); reporting vessels then receive information regarding right whale sightings, vessel strike avoidance measures and additional vessel strike mitigation regulations (Silber *et al.*, 2015). Vessel compliance with the MSR system was initially below 70 per cent, but later ranged between 70 and 80 per cent (Silber *et al.*, 2015). Since 2006, non-reporting vessels are no longer fined, with education and non-monetary penalties instead being applied in order to encourage compliance (Silber *et al.*, 2015).

Available in more detail in Knowlton and Brown (2007), Silber and Bettridge (2012) and Mullen *et al.* (2013), outreach and education activities regarding right whales, their movements and habitats, measures to avoid vessel strikes and/or the locations of whales in near-real-time in the USA are extensive and include: notifications and updates in the US Coast Pilot, international navigational publications, radio broadcast Notices to Mariners, NAVTEX (a shipboard communication system), email and interactive online maps (e.g., www.nefsc.noaa.gov); announcements on National Oceanic and Atmospheric Administration (NOAA) weather radio and websites; information and announcements on NOAA and other government agency websites; National Marine Fisheries Service (NMFS) advisory letters to vessel owners; NOAA-issued Notices of Violation and Assessment (NOVAs); MSR system messages (described above); delineation of Critical Habitats and right whale management areas (e.g., SMAs) on navigational charts; email reminders and summaries of vessel operations to maritime industries via distribution lists; compliance guides distributed by maritime industries and agencies; interactive training CDs; the Whale Alert app (described in Section 10.5.4); Right Whale Protection Program notebook distribution; Merchant Marine training curricula modules incorporated in marine academies and available for free online; right whale workshops; multi-stakeholder meetings and discussions; industry trade journal articles; press releases; oral presentations to communities and maritime industries; numerous media interview and stories; and flyers, brochures and stickers.

Similar but less extensive than in the USA, efforts to educate, raise awareness and otherwise communicate about right whales, their habitats and movements, the measures in place to reduce the risk of vessel strikes, and/or the limited near-real-time information regarding whale locations in Canada include: information on right whales published in Sailing Directions for the Canadian Maritimes; Shipping Notices and Annual Notice to Mariners (NOTMAR) regarding right whales and their habitats and temporary and mandatory slowdown zones; the development of the Mariner's Guide to Whales in the Northeast Atlantic; Critical Habitat and right whale management area delineation on navigational charts; updates, notifications and information on DFO, Transport Canada and other government agency websites; the Whale Alert app (described in Section 10.5.4); near-real-time sightings of right whales available on multiple online websites and via Vessel Traffic Control; the Let's Talk Whales outreach programme; statements and press releases; right whale workshops; multi-stakeholder meetings and discussions; direct communication with vessel operators; oral presentations to communities and maritime industries; numerous media interviews and stories; and flyers, brochures and stickers. For additional detail, see Knowlton and Brown (2007), Duff *et al.* (2013) and DFO (2017c).

The effectiveness of individual education and awareness activities is largely impossible to disentangle and, furthermore, is inherently linked with other measures taken to reduce the risks vessels pose to right whales. As a key example, high fleet compliance with respect to demonstrably successful right whale avoidance or vessel speed restrictions is reliant on adequate communication, education and awareness, in addition to penalties and/or non-monetary notices of violation. While education and awareness of right whales and the risks they face from vessels are important components of vessel strike reduction strategies, education and awareness efforts must be coincident with meaningful management and conservation measures to reduce the risks posed by vessels to right whales.

10.5.5 Technological Advances: Current and Future

Near real-time information regarding whale locations detected using visual (i.e., aerial- or vessel-based) monitoring or PAM are directly communicated to vessels using various maritime communication systems; upon receiving an alert, vessels may (or may not) comply and either alter course to avoid the whale(s) or slow down and reduce the risk of a lethal strike. Visual detections, whether aerial or vessel based, are constrained by weather, daylight hours and financial costs for dedicated survey time. In contrast, PAM, a method for detecting sound-producing animals such as right whales, is a cost-effective approach that can be undertaken in any weather conditions and for any time period required. Currently in use on fixed platforms (e.g., bottom mounted) and more recently developed for mobile platforms (i.e., ocean gliders; see Baumgartner & Mussoline, 2011; Baumgartner *et al.*, 2013; Klinck *et al.*, 2012), PAM represents a major technological advancement for efforts to mitigate vessel strike risk to right whales. As a key example, a network of fixed-PAM buoys was installed in the Boston TSS in 2008, with coverage for an 89-km (55-mile) portion of the shipping lanes on entrance to Boston. When on-board software detects a right whale call, an analyst verifies the detection, and an alert will subsequently be sent to vessels in the area; time from detection to posting the alert may be as short as 20 minutes (www.listenforwhales.org).

In conjunction with acoustic monitoring advancements, there have been substantial advances in communications technologies. Launched in 2012, the Whale Alert app (www.whalealert.org) is a free mobile-based app that provides near-real-time information regarding whale locations derived from visual sightings and acoustic detections in conjunction with information regarding the current vessel location, whale management areas (e.g., SMAs and ATBAs) and speed restrictions. Members of the maritime community can now also report sightings of live whales, whales in distress and dead whales, with the option for direct contact with management professionals. Initially developed for right whales on the US Atlantic coast with the shipping industry as the target audience, the Whale Alert app has expanded to include multiple baleen species and is available in US and Canadian Atlantic and Pacific coast waters.

In addition to the recent expansion of the Whale Alert app to Canadian waters, a near-real-time alert system reliant on Automated Identification System (AIS) messaging to vessels is also under development by Dr Chris Taggart (Dalhousie University) and partners. Previously, a survey of Canadian Shipping Federation members indicated that most would prefer to receive whale alerts via NAVTEX (messaging every few hours) or AIS (near-real-time messaging; Reimer *et al.*, 2016). Given the receptivity of the fleet to AIS messaging, in addition to a lack of cellular service in offshore waters, an AIS-based whale alert system could contribute additional power to right whale vessel strike mitigation efforts. Furthermore, it is anticipated that this AIS-based alert will integrate PAM-mounted ocean gliders (mobile autonomous underwater vehicles) as a mobile detection system.

If the previous few years are an indication of the future, right whale movements and aggregations may become increasingly unpredictable, scenarios where large numbers of right whales occur in areas without vessel strike mitigation measures may become more frequent and the threat of vessel strikes may worsen. In these scenarios, near-real-time monitoring of right whales using newer technologies such as ocean gliders could improve vessel strike mitigation effectiveness, as ocean glider deployments are more adaptable than fixed PAM and could achieve monitoring over greater spatial scales (Reimer *et al.*, 2016). When linked to an AIS- and/or mobile app-based alert system and received by a compliant fleet, near-real-time monitoring derived in part from mobile platforms could provide an effective means of reducing vessel strikes.

10.5.6 Overall Effectiveness and Optimization of Mitigation Measures

The success of reducing vessel strike risk or lethal vessel strike risk to right whales via mandatory and voluntary vessel avoidance and speed reduction measures has been mixed. While fleet compliance with some voluntary measures, including ATBAs, TSSs and recommended routes in the south-eastern USA, has been relatively high (e.g., Lagueux *et al.*, 2011; Silber *et al.*, 2012b; Vanderlaan *et al.*, 2008), longer-term monitoring is important as compliance may wane (e.g., Roseway Basin ATBA; Taggart *et al.*, 2017). For other voluntary measures, such as DMAs and the voluntary slowdown zone in the Gulf of St Lawrence, compliance has been low (Silber *et al.*, 2012a; Taggart *et al.*, 2017). Compliance with mandatory measures is also variable. For example, fleet compliance with the mandatory slowdown zone in the Gulf of St Lawrence was high (Taggart *et al.*, 2017). In contrast, foreign vessel compliance with SMAs was consistently low (Silber & Bettridge, 2012).

Despite concerted mitigation measures that have often been effective at reducing vessel strike risk or vessel strike lethality risk for right whales, right whales are still being killed by vessels (NMFS, 2017; van der Hoop *et al.*, 2013). No significant change in vessel strike mortalities involving large whales was detected before and after 2003, the year in which several mitigation measures were implemented in the USA (van der Hoop *et al.*, 2013). However, prior to the 2017 UME, decreases in the number (Laist *et al.*, 2014) and probability (Conn & Silber, 2013) of lethal vessel strikes involving right whales in the USA were attributed to the 2008 vessel speed restrictions, with potential additional reductions via vessel avoidance measures in the USA (e.g., the modified Boston TSS and recommended routes in the south-eastern calving grounds; Laist *et al.*, 2014). The reduction in vessel transits due to economic constraints may also have been a contributing factor (Silber *et al.*, 2015). In Canada, reduced lethal vessel strike risk has been demonstrated for two areas important to right whales (van der Hoop *et al.*, 2012; Vanderlaan *et al.*, 2008). However, and although investigations are still ongoing at the time of writing, the number of known lethal vessel strikes of right whales in Canadian waters in 2017 will be high and without known precedent.

In order to reduce the risk that vessels pose to right whales, the primary goals of ATBAs, amended TSSs, recommended routes and speed restrictions should be to maximize the reduction in right whale vessel strike risk balanced by minimized disruption to vessel operations, while still maintaining safe navigation (Vanderlaan *et al.*, 2008). As the

2017 UME demonstrated, right whale movements outside of areas with enacted mitigation measures can result in significant mortalities. In addition to requiring a compliant and aware fleet, dynamic, adaptable and rapidly implemented mitigation measures will become increasingly important if right whale movements continue to become less predictable in future.

10.6 The Future

10.6.1 Options

Technological advances in maritime communications, ocean monitoring platforms (e.g., ocean gliders) and PAM offer substantial opportunities to contribute to right whale vessel strike mitigation efforts and to complement existing vessel avoidance and speed reduction measures. Given the urgency with regards to reducing the threats posed by vessels to right whales, demonstrably effective approaches should be expanded, with additive measures using new technologies, such as the Whale Alert app and near-real-time monitoring using mobile and fixed-PAM platforms, allowing for increasing spatial coverage, the evaluation of new approaches and/or additive measures in areas of particular concern. As an example, the recent expansion of the Whale Alert app beyond US waters and to species other than right whales represents a potentially powerful and complementary mitigation tool. Although still under development, AIS-based whale alerts are planned for Canadian waters, with multiple monitoring platforms, including ocean gliders, offering previously unattainable spatial coverage, which may prove crucial if right whales continue to demonstrate reduced affinity for traditional northern feeding grounds. Ultimately, the success of these measures will be largely dependent on an aware and compliant fleet. As such, optimized mitigation efforts should maximize vessel strike risk reduction and minimize costs and disturbance to the fleet in order to support high vessel compliance.

10.6.2 Fleet Compliance

Will the fleet comply will future changes in policy to protect right whales from vessel strikes? If the past is a predictor of the future, fleet compliance with future mitigation measures will be mixed; such responses may not be sufficient to reduce right whale vessel strikes, with potentially profound consequences for the persistence of the species. In addition to compliance with existing measures (e.g., voluntary and mandatory vessel avoidance and speed reduction measures), the development and implementation of new, technologically innovative and effective approaches are reliant on a receptive, participatory and compliant fleet. An increase in fleet participation, collaboration and ultimately compliance with mitigation measures is needed; the question is, how can this be achieved?

In addition to being grounded in science, species conservation policies must be economically feasible and socially acceptable (Stankey & Shindler, 2006), with acceptance by the fleet being of particular relevance to right whale vessel strike mitigation policies. As 'inherently social phenomena', conservation policies are meant to induce changes in human behaviour (Mascia *et al.*, 2003), with vessel operators modifying their behaviour by either slowing down or by adopting a course intended to avoid whales. Factors that shape acceptance are complex and include cognitive (knowledge, information), affective (emotions, aesthetics, beliefs) and contextual (interests, values at risk, uncertainty) components; these interactions are considered alongside the policy's scientific soundness and economic feasibility (Stankey & Shindler, 2006).

Vessel strike mitigation approaches are largely reliant on top-down management; all too often, top-down endangered species conservation and other biodiversity initiatives are incapable of engaging different sectors of society, including key stakeholders (Aburto-Oropeza *et al.*, 2018). Perpetuating top-down management strategies for right whale vessel strike reduction may result in continued mixed compliance. As the primary target of conservation efforts aimed at reducing vessel strike risk, the success of vessel strike mitigation is largely dependent on fleet acceptance and subsequent compliance with policy and practices in areas where right whales are present. As such, the diverse needs, preferences, values, constraints and other acceptance components of the fleet must be considered as central to the development and implementation of mitigation measures. In particular, knowledge or awareness of endangered species and their conservation challenges has been linked to support for conservation measures (Aipanjiguly *et al.*, 2003; Sawchuk *et al.*, 2015), a willingness to pay for conservation (Ressurreição *et al.*, 2012) and other pro-environmental behaviours (Sawchuk *et al.*, 2015). As a consequence, mariners who are informed, engaged as active partners and aware of endangered right whales and the threats they face from vessel strikes may be more willing to comply with vessel strike mitigation measures and to participate in the development and application of new, technologically innovative mitigation approaches.

10.7 Conclusion

Coupled with ongoing population declines (Corkeron, 2017), the known deaths of 17 right whales in 2017, which represented ~4 per cent of the population, signals an urgent need to reduce anthropogenic mortalities in order to avert potential extinction. With only ~100 reproductively aged females remaining, right whale conservation efforts will need to result in meaningful consequences for the population within years, not decades. Because vessel strikes represent a leading source of right whale mortality (van der Hoop *et al.*, 2013), with overall extinction risk thought to be governed largely by vessel traffic strike risk, entanglement risk and environmental conditions (Meyer-Gutbrod & Greene, 2017), efforts to mitigate the threat of vessels to right whales will partially determine whether right whales continue to persist or decline towards eventual extinction. As one of the most intensively studied marine mammals on the planet, and largely occurring in the coastal waters of two developed nations with established endangered species legislation, whether efforts to reduce vessel strikes are ultimately successful at reducing right whale fatalities will also prove to be an important test for addressing whale–vessel strike issues worldwide.

While solutions to reduce the risk of lethal vessel strikes appear straightforward – vessels should avoid whales or slow down – these measures are complicated by numerous factors, including mixed compliance by the fleet and unknown or unpredictable right whale movements outside of areas with mitigation measures in place. Although often successful at reducing lethal right whale vessel strikes, mitigation efforts have not been sufficient; right whales are still being killed by vessels. Even when combined, SMAs and DMAs leave 25 per cent of right whale sightings in US waters without additional protections (Convertino & Valverde, 2018), and in any given year, the locations of approximately 50 per cent of the right whale population are unknown. With recent lowered fidelity to northern feeding grounds with enacted mitigation measures, a large proportion of right whales are unprotected from vessel strikes; if right whales increasingly stray from traditional habitats in future, the number of right whales killed by vessels could increase.

More than two decades of scientific research, technological innovation and conservation intervention to reduce the risk of vessel strikes to right whales have already occurred, with often effective vessel strike mitigation measures enacted in key right whale habitats. In addition to existing measures, such as DMAs, ATBAs and amended TSSs, technological innovations such as the Whale Alert app and glider-mounted near-real-time monitoring offer promise for urgently required dynamic and expanded protective measures. Both existing and emergent measures are united by their reliance on an aware, compliant and engaged fleet; as such, the fleet must be central to the development and implementation of vessel strike mitigation measures. With only years available, conservation actions to mitigate vessel strikes, in addition to coupled efforts to reduce other right whale anthropogenic mortalities, will play major roles in determining whether North Atlantic right whales persist beyond this century.

References

Aburto-Oropeza, O., López-Sagástegui, C., Moreno-Báez, M. *et al.* 2018. Endangered species, ecosystem integrity, and human livelihoods. *Conservation Letters*. 11: e12358.

Aguilar, A. 1986. A review of old Basque whaling and its effect on the right whales (*Eubalaena glacialis*) of the North Atlantic. *Report of the International Whaling Commission*. 10: 191–199.

Aipanjiguly, S., Jacobson, S. K., Flamm, R. 2003 Conserving manatees: knowledge, attitudes, and intentions of boaters in Tampa Bay, Florida. *Conservation Biology*. 17: 1098–1105.

Baumgartner, M. F., Mussoline, S. E. 2011. A generalized baleen whale call detection and classification system. *Journal of the Acoustical Society of America*. 129(5): 2889–2902.

Baumgartner, M. F., Mayo, C. M., Kenney, R. D. 2007. Enormous carnivores, microscopic food, and a restaurant that's hard to find. In: S. D. Kraus, R. M. Rolland, eds., *The Urban Whale*. Cambridge, MA: Harvard University Press, pp. 138–171.

Baumgartner, M. F., Fratantoni, D. M., Hurst, T. P. *et al.* 2013. Real-time reporting of baleen whale passive acoustic detections from ocean gliders. *Journal of the Acoustical Society of America*. 134: 1814–1823.

Boersma, K. F., Vinken, G. C., Tournadre, J. 2015. Ships going slow in reducing their NO_x emissions: changes in 2005–2012 ship exhaust inferred from satellite measurements over Europe. *Environmental Research Letters*. 10(7): 074007.

Bonney, J. 2010. Built for speed. *The Journal of Commerce Online*. www.joc.com/maritime/built-speed

Bouwman, K. 2016. An analysis of Scotia–Fundy vessel users and what this means for the North Atlantic right whales. Master's thesis, Dalhousie University.

Campbell-Malone, R., Barco, S. G., Daoust, P. Y. *et al.* 2008. Gross and histologic evidence of sharp and blunt trauma in North Atlantic right whales (*Eubalaena glacialis*) killed by vessels. *Journal of Zoo and Wildlife Medicine*. 39(1): 37–55.

Canada Gazette. 2016. Canada Gazette Part 1. 150(20): 1424–1530.

Conn, P. B., Silber, G. K. 2013. Vessel speed restrictions reduce risk of collision-related mortality for North Atlantic right whales. *Ecosphere*. 4(4): 43.

Convertino, M., Valverde, L. 2018. Probabilistic analysis of the impact of vessel speed restrictions on navigational safety: accounting for the right whale rule. *Journal of Navigation*. 71(1): 65–82.

Corbett, J. J., Koehler, H. W. 2003. Updated emissions from ocean shipping. *Journal of Geophysical Research: Atmospheres*. 108(D20): 10.1029/2003JD003751.

Corkeron, P. 2017. North Atlantic right whales: species decline and life expectancy. North Atlantic Right Whale Consortium Meeting, Halifax, NS, 22 October 2017 (oral presentation).

Costidis, A. M., Berman, M., Cole, T. *et al.* 2013. Sharp trauma induced by vessel collisions with pinnipeds and cetaceans. *Diseases of Aquatic Organisms*. 103: 251–256.

Daoust, P.-Y., Couture, E. L., Wimmer, T., Bourque, L. 2017. *Incident Report: North Atlantic right whale mortality event in the Gulf of St. Lawrence, 2017*. Saskatoon: Canadian Wildlife Health Cooperative, Marine Animal Response Society, and Fisheries and Oceans Canada.

Davis, G. E., Baumgartner, M. F., Bonnell, J.M. *et al.* 2017. Long-term passive acoustic recordings track the changing distribution of North Atlantic right whales (*Eubalaena glacialis*) from 2004 to 2014. *Scientific Reports*. 7(1): 13460.

de Stephanis, R., Urquiola, E. 2006. *Collisions between Ships and Cetaceans in Spain*. SC/58/BC5. Cambridge: International Whaling Commission Scientific Committee.

DFO. 2017a. Written notices to shipping – maritimes, M1409/17, issued on 12 July 2017. https://nis.ccg-gcc.gc.ca

DFO. 2017b. Written notices to shipping – maritimes, M1970/17, issued on 22 September 2017. https://nis.ccg-gcc.gc.ca

DFO. 2017c. North Atlantic right whale, a science based review of recovery actions for three at-risk whale populations. Fisheries and Oceans, Let's Talk Whales. https://waves-vagues.dfo-mpo.gc.ca/Library/40679652.pdf

DiMaio, V. J., DiMaio, D. 2001. *Forensic Pathology: Practical Aspects of Criminal and Dorensic Investigation*, 2nd edition. Boca Raton, FL: CRC Press.

Duff, J., Dean, H., Gazit, T., Taggart, C. T., Cavanagh, J. H. 2013. On the right way to right whale protections in the Gulf of Maine – case study. *Journal of International Wildlife Law & Policy*. 16: 229–265.

EMSA. 2015. *The World Merchant Fleet in 2015, Statistics from Equasis*. Lisbon: European Maritime Safety Agency.

Endresen, Ø., Sørgård, E., Sundet, J. K. *et al.* 2003. Emission from international sea transportation and environmental impact. *Journal of Geophysical Research: Atmospheres*. 108(D17): ACH 14-1-22.

Frisk, G. V. 2012. Noiseonomics: the relationship between ambient noise levels in the sea and global economic trends. *Scientific Reports*. 2: 437.

Gende, S. M., Hendrix, A. N., Harris, K. R. *et al.* 2011. A Bayesian approach for understanding the role of ship speed in whale–ship encounters. *Ecological Applications*. 21(6): 2232–2240.

Grieve, B. D., Hare, J. A., Saba, V. S. 2017. Projecting the effects of climate change on *Calanus finmarchicus* distribution within the US Northeast Continental Shelf. *Scientific Reports*. 7(1): 6264.

Hain, J. 1997. Detectability of right whales in Cape Cod Bay relative to NE EWS monitoring flights. In: A. Knowlton, S. Kraus, D. Meck, M. Mooney-Sues, eds., *Shipping/Right Whale Workshop Report*. Boston, MA: The New England Aquarium, pp. 113–118.

Hain, J. H., Ellis, S. L., Kenney, R. D., Slay, C. K. 1999. Sightability of right whales in coastal waters of the southeastern United States with implications for the aerial monitoring program. In: G. W. Garner, S. C. Amstrup, J. L. Laake, B. F. J. Manly, L. L. McDonald, D. G. Robertson, eds., *Marine Mammal Survey and Assessment Methods*. Rotterdam: A.A. Balkema, pp. 191–207.

Hain, J. H. W., Hampp, J. D., McKenney, S. A., Albert, J. A., Kenney, R. D. 2013. Swim speed, behavior, and movement of North Atlantic right whales (*Eubalaena glacialis*) in coastal waters of northeastern Florida, USA. *PLoS ONE*. 8(1): e54340.

Hamilton, P. K., Davis, G. E. 2017. Right whale distribution: historical perspectives and recent shifts. North Atlantic Right Whale Consortium Meeting, Halifax, NS, 22 October 2017 (oral presentation).

Hamilton, P. K., Marx, M. K., Kraus, S. D. 1998. *Scarification Analysis of North Atlantic Right Whales (Eubalaena glacialis) as a Method of Assessing Human Impacts. A Report Submitted to The Island Foundation, Marion MA*. Boston, MA: The New England Aquarium.

Honma, Y., Chiba, A., Ushiki, T. 1997. Histological observations on a muscle mass from a large marine mammal struck by a jetfoil in the Sea of Japan. *Fisheries Science*. 63: 587–591.

IMO. 2006. *Routeing of Ships, Ship Reporting and Related Matters, Amendments of the Traffic Separation Scheme 'In the Approach to Boston, Massachusetts'. Submitted by the United States. Sub-committee of Safety of Navigation. 52nd Session, Agenda Item 3. NAV52/3/XX*. London: International Maritime Organization.

IMO. 2008. *Routeing of Ships, Ship Reporting and Related Matters, Amendments of the Traffic Separation Scheme 'In the Approach to Boston, Massachusetts'. Submitted by the United States. Sub-committee of Safety of Navigation. 54th Session, Agenda Item 3. NAV54/3/XX*. London: International Maritime Organization.

IMO. 2012. *International Shipping Facts and Figures – Information Resources on Trade, Safety, Security, and Environment*. London: International Maritime Organization, Maritime Knowledge Centre.

Jensen, A. S., Silber, G. K. 2003. *Large Whale Ship Strike Database. U.S. Department of Commerce, NOAA Technical Memorandum*. Silver Spring, MD: NMFS-OPR.

Klinck, H., Mellinger, D. K., Klinck, K. *et al.* 2012. Near-real-time acoustic monitoring of beaked whales and other cetaceans using a Seaglider™. *PLoS ONE*. 7: e36128.

Knowlton, A. R., Brown, M. W. 2007. Running the gauntlet: right whales and vessel strikes. In: S. D. Kraus, R. M. Rolland, eds., *The Urban Whale*. Cambridge, MA: Harvard University Press, pp. 409-435.

Knowlton, A. R., Kraus, S. D. 2001. Mortality and serious injury of northern right whales (*Eubalaena glacialis*) in the western North Atlantic Ocean. *Journal of Cetacean Research and Management*. 2: 193–208.

Knowlton, A. R., Hamilton, P. K., Marx, M. K., Pettis, H. M., Kraus, S. D. 2012. Monitoring North Atlantic right whale *Eubalaena glacialis* entanglement rates: a 30 yr retrospective. *Marine Ecology Progress Series*. 466: 293–302.

Kot, B. W., Sears, R., Zbinden, D., Borda, E., Gordon, M. S. 2014. Rorqual whale (Balaenopteridae) surface lunge-feeding behaviors: standardized classification, repertoire diversity, and evolutionary analyses. *Marine Mammal Science*. 30(4): 1335–1357.

Lagueux, K. M., Zani, M. A., Knowlton, A. R., Kraus, S. D., 2011. Response by vessel operators to protection measures for right whales *Eubalaena glacialis* in the southeast US calving ground. *Endangered Species Research*. 14(1): 69–77.

Laist, D. W., Knowlton, A. R., Mead, J. G., Collet, A. S., Podesta, M. 2001. Collisions between ships and whales. *Marine Mammal Science*. 17(1): 35–75.

Laist, D. W., Knowlton, A. R., Pendleton, D. 2014. Effectiveness of mandatory vessel speed limits for protecting North Atlantic right whales. *Endangered Species Research*. 23(2): 133–147.

Mascia, M. B., Brosius, J. P., Dobson, T. A. *et al.* 2003. Conservation and the social sciences. *Conservation Biology*. 17(3): 649–650.

Mayo, C. A., Letcher, B. H., Scott, S. 2001. Zooplankton filtering efficiency of the baleen of a North Atlantic right whale, *Eubalaena glacialis. Journal of Cetacean Research and Management*. 3: 245–250.

McKenna, M. F., Calambokidis, J., Oleson, E. M., Laist, D. W., Goldbogen, J. A. 2015. Simultaneous tracking of blue whales and large ships demonstrates limited behavioral responses for avoiding collision. *Endangered Species Research*. 27: 219–232.

McLellan, W. A., Berman, M., Cole, T. *et al.* 2013. Blunt force trauma induced by vessel collisions with large whales. *Diseases of Aquatic Organisms*. 103: 245–251.

Melbourne, B. A., Hastings, A. 2008. Extinction risk depends strongly on factors contributing to stochasticity. *Nature*. 454: 100–103.

Merrick, R. L. 2005. *Seasonal Management Areas to Reduce Ship Strikes of Northern Right Whales in the Gulf of Maine*. Northeast Fisheries Science Center Reference Document 05–19, 26. Washington, DC: US Department of Commerce, NOAA.

Merrick, R. L., Cole, T. V. N. 2007. *Evaluation of Northern Right Whale Ship Strike Reduction Measures in the Great South Channel of Massachusetts*. NOAA Technical Memorandum NMFS-NE, 202. Woods Hole, MA: US Department of Commerce.

Meyer-Gutbrod, E. L., Greene, C. H. 2017. Uncertain recovery of the North Atlantic right whale in a changing ocean. *Global Change Biology*. 1: 10.

Meyer-Gutbrod, E. L., Greene, C. H., Sullivan, P. J., Pershing, A. J. 2015. Climate-associated changes in prey availability drive reproductive dynamics of the North Atlantic right whale population. *Marine Ecology Progress Series*. 535: 243–258.

Mills, K. E., Pershing, A. J., Brown, C. J. *et al.* 2013. Fisheries management in a changing climate: lessons from the 2012 ocean heat wave in the northwest Atlantic. *Oceanography*. 26(2): 191–195.

Moore, M. J., McLellan, W. A., Daoust, P.-Y., Bonde, R. K., Knowlton, A. R. 2007. Right whale mortality: a message from the dead to the living. In: S. D. Kraus, R. M. Rolland, eds., *The Urban Whale*. Cambridge, MA: Harvard University Press, pp. 358–379.

Mullen, K. A., Peterson, M. L., Todd, S. K. 2013. Has designating and protecting critical habitat had an impact on endangered North Atlantic right whale ship strike mortality? *Marine Policy*. 42: 293–304.

Neilson, J. L., Gabriele, C. M., Jensen, A. S., Jackson, K., Straley, J. M. 2012. Summary of reported whale–vessel collisions in Alaskan waters. *Journal of Marine Biology*. 2012: 1–18.

NOAA. 2008. Endangered fish and wildlife; final rule to implement speed restrictions to reduce the threat of ship collisions with North Atlantic right whales. *Federal Register*. 73: 60173–60191.

NOAA. 2013. Endangered fish and wildlife; final rule to remove the sunset provision of the final rule implementing vessel speed restrictions to reduce the threat of ship collisions with North Atlantic right whales. *Federal Register*. 78: 73726–73736.

NOAA. 2017. 2017 North Atlantic right whale unusual mortality event. www.fisheries.noaa.gov/national/marine-life-distress/2017-2020-north-atlantic-right-whale-unusual-mortality-event

Nowacek, D. P., Johnson, M. P., Tyack, P. L. *et al.* 2001. Buoyant balaenids: the ups and downs of buoyancy in right whales. *Proceedings of the Royal Society B: Biological Sciences.* 268: 1811–1816.

Nowacek, D. P., Johnson, M. P., Tyack, P. L. 2004. North Atlantic right whales (*Eubalaena glacialis*) ignore ships but respond to alerting stimuli. *Proceedings of the Royal Society of London B: Biological Sciences.* 271(1536): 227–231.

NMFS 2017. *North Atlantic Right Whale (Eubalaena glacialis) 5-Year Review: Summary and Evaluation.* Gloucester, MA: National Marine Fisheries Service Greater Atlantic Regional Fisheries Office.

Pace, R. M., Corkeron, P. J., Kraus, S. D. 2017. State–space mark–recapture estimates reveal a recent decline in abundance of North Atlantic right whales. *Ecology and Evolution.* 7: 8730–8741.

Parks, S. E., Warren, J. D., Stamiezkin, K. M., Mayo, C. A., Wiley, D. 2011. Dangerous dining: Surface foraging of right whales increases risk for vessel collisions. *Biology Letters.* 8: 1–4.

PWC. 2011. A Game Changer for the Shipping Industry? An analysis of the future impact of carbon regulations on environment and industry. An analysis prepared for the ongoing discussions in IMO and other international for a regarding future global regulations of carbon emissions. www.scribd.com/document/68027866/Shipping-GHG-PwC-Final

Reeves, R. R., Smith, T. D., Josephson, E. A. 2007. Near-annihilation of a species: right whaling in the North Atlantic. In: S. D. Kraus, R. M. Rolland, eds., *The Urban Whale.* Cambridge, MA: Harvard University Press, pp. 39–74.

Reimer, J., Gravel, C., Brown, M. W., Taggart, C. T. 2016. Mitigating vessel strikes: the problem of the peripatetic whales and the peripatetic fleet. *Marine Policy.* 68: 91–99.

Ressurreição, A., Gibbons, J., Kaiser, M. *et al.* 2012. Different cultures, different values: the role of cultural variation in public's WTP for marine species conservation.- *Biological Conservation.* 145(1): 148–159.

Reygondeau, G., Beaugrand, G. 2011. Future climate-driven shifts in distribution of *Calanus finmarchicus. Global Change Biology.* 17(2): 756–766.

Ritter, F. 2012. Collisions of sailing vessels with cetaceans worldwide: first insights into a seemingly growing problem. *Journal of Cetacean Research and Management.* 12(1): 119–127.

Rolland, R. M., Parks, S. E., Hunt, K. E. *et al.* 2012. Evidence that ship noise increases stress in right whales. *Proceedings of the Royal Society of London B: Biological Sciences.* 279(1737): 2363–2368.

Rolland, R. M., Schick, R. S., Pettis, H. M. *et al.* 2016. Health of North Atlantic right whales *Eubalaena glacialis* over three decades: from individual health to demographic and population health trends. *Marine Ecology Progress Series.* 542: 265–282.

Sawchuk, J. H., Beaudreau, A. H., Tonnes, D., Fluharty, D. 2015. Using stakeholder engagement to inform endangered species management and improve conservation.- *Marine Policy.* 54: 98–107.

Silber, G. K., Bettridge, S. 2012. *An Assessment of the Final Rule to Implement Vessel Speed Restrictions to Reduce the Threat of Vessel Collisions with North Atlantic Right Whales.* US Department of Commerce, NOAA Technical Memorandum NMFS-OPR-48. Silver Spring, MD: NMFS-OPR.

Silber, G. K., Slutsky, J., Bettridge, S. 2010. Hydrodynamics of a ship/whale collision. *Journal of Experimental Marine Biology and Ecology*. 391(1): 10–19.

Silber, G. K., Adams, J. D., Bettridge, S. 2012a. Vessel operator response to a voluntary measure for reducing collisions with whales. *Endangered Species Research*. 17(3): 245–254.

Silber, G. K., Vanderlaan, A. S., Arceredillo, A. T. *et al.* 2012b. The role of the International Maritime Organization in reducing vessel threat to whales: process, options, action and effectiveness. *Marine Policy*. 36(6): 1221–1233.

Silber, G. K., Adams, J. D., Fonnesbeck, C. J. 2014. Compliance with vessel speed restrictions to protect North Atlantic right whales. *PeerJ*. 2: 399.

Silber, G. K., Adams, J. D., Asaro, M. J. *et al.* 2015. The right whale mandatory ship reporting system: a retrospective. *PeerJ*. 3: 866.

Stankey, G. H., Shindler, B. 2006. Formation of social acceptability judgments and their implications for management of rare and little-known species. *Conservation Biology*. 20(1): 28–37.

Taggart, C. T., Vanderlaan, A. S. M., Brown, M. W., McLeod, A., Fox, C. H. 2017. Reducing vessel-strike to large whales: science-driven policy, real-time whale-alerts to vessels, role and responsibilities of the shipping industry to make it happen. World Ocean Council, Sustainable Ocean Summit, Halifax, NS (oral presentation).

Tournadre, J. 2014. Anthropogenic pressure on the open ocean: the growth of ship traffic revealed by altimeter data analysis. *Geophysical Research Letters*. 41(22): 7924–7932.

Transport Canada. 2017. News release: Transport Canada issues Gulf of St. Lawrence speed restriction fine. www.canada.ca/en/transport-canada/news/2017/09/transport_canadaissuesgulfofstlawrencespeedrestrictionfine.html

UNCTAD. 2016. *Review of Maritime Transport*. New York and Geneva: United Nations Conference on Trade and Development.

van der Hoop, J. M., Vanderlaan, A. S., Taggart, C. T. 2012. Absolute probability estimates of lethal vessel strikes to North Atlantic right whales in Roseway Basin, Scotian Shelf. *Ecological Applications*. 22(7): 2021–2033.

van der Hoop, J. M., Moore, M. J., Barco, S. G. *et al.* 2013. Assessment of management to mitigate anthropogenic effects on large whales. *Conservation Biology*. 27(1): 121–133.

Van Waerebeek, K., Baker, A. N., Félix, F. *et al.* 2007. Vessel collisions with small cetaceans worldwide and with large whales in the Southern Hemisphere, an initial assessment. *Latin American Journal of Aquatic Mammals*. 6(1): 43–69.

Vanderlaan, A. S., Taggart, C. T. 2007. Vessel collisions with whales: the probability of lethal injury based on vessel speed. *Marine Mammal Science*. 23(1): 144–156.

Vanderlaan, A. S., Taggart, C. T., Serdynska, A. R., Kenney, R. D., Brown, M. W. 2008. Reducing the risk of lethal encounters: vessels and right whales in the Bay of Fundy and on the Scotian Shelf. *Endangered Species Research*. 4: 283–297.

Vanderlaan, A. S., Corbett, J. J., Green, S. L. *et al.* 2009. Probability and mitigation of vessel encounters with North Atlantic right whales. *Endangered Species Research*. 6: 273–285.

Wang, C., Lyons, S. B., Corbett, J. J., Firestone, J., Corbett, J. J. 2007. Using ship speed and mass to describe potential collision severity with whales: an application of ship traffic, energy and environment model (STEEM). In: *Compendium of Papers: Transportation Research Board 86th Annual Meeting*. Newark, DE: University of Delaware, pp. 1–18.

Ward-Geiger, L. I., Silber, G. K., Baumstark, R. D., Pulfer, T. L. 2005. Characterization of ship traffic in right whale critical habitat. *Coastal Management*. 33(3): 263–278.

Wiley, D. N., Thompson, M., Pace, R. M., Levenson, J. 2011. Modeling speed restrictions to mitigate lethal collisions between ships and whales in the Stellwagen Bank National Marine Sanctuary, USA. *Biological Conservation*. 144(9): 2377–2381.

Wiley, D. N., Mayo, C. A., Maloney, E. M., Moore, M. J. 2016. Vessel strike mitigation lessons from direct observations involving two collisions between noncommercial vessels and North Atlantic right whales (*Eubalaena glacialis*). *Marine Mammal Science*. 32(4): 1501–1509.

Zappes, C. A., de Sá Alves, L. C. P., da Silva, C. V. *et al.* 2013. Accidents between artisanal fisheries and cetaceans on the Brazilian coast and Central Amazon: proposals for integrated management. *Ocean and Coastal Management*. 85: 46–57.

11

Nuclear-Powered Vessels

RICHARD JOHN (JAN) PENTREATH

11.1 The Development of Nuclear-Powered Marine Propulsion

In spite of their mystique, nuclear reactors are relatively simple things. They convert the kinetic energy of the decay products of specific uranium and plutonium isotopes into thermal energy, and they do so in a controlled manner to produce superheated steam within a comparatively small volume and with great efficiency. The steam is then normally used to drive turbines and hence generate electricity. Such reactors were first developed in the USA as a result of initiatives that emerged in 1946 from the Manhattan Project, at what was then the Clinton Laboratory, now the Oak Ridge National Laboratory (ORNL), Tennessee. The US Navy had shown an immediate interest in the possibilities of such developments and, particularly through the vision and tenacity of an irascible naval electrical engineer, Hyman George Rickover, they not only built the first nuclear-powered marine vessel, but also produced the type of nuclear reactor that has subsequently been the most successful in the history of electricity generation, on either land or sea.

It is true to say that Rickover's vision was not initially appreciated by his immediate superiors. But in 1949, together with Alvin Weinberg, the ORNL's Director of Research, he set up and led a team that began the design of what was to become the pressurized water reactor (PWR) that could easily fit into the hull of a submarine. The first such reactor went critical in 1953. Key to its success was solving the problem of keeping it small enough. The basic requirements for a nuclear reactor were: a source of fuel for which, in theory, natural uranium containing about 0.7 per cent ^{235}U would suffice; a medium (the 'moderator') to slow the resultant neutrons down to increase their probability of being captured by other ^{235}U atoms and thus sustain a chain reaction; a means of extracting the resultant heat (the 'coolant') that is then used to create superheated steam; plus rods to capture neutrons and thus enable the reactor's rate of reaction to be controlled. To minimize the size of the reactor, it was evident that water could be used to act as both a moderator and a coolant within a high-pressure reactor vessel, but the problem then was that the materials needed to encase the fuel would absorb neutrons so that a chain reaction could not be sustained. Natural uranium could therefore not be used in this design and thus, in order to compensate for the neutron absorption of the fuel rod casing, it was necessary to alter (enrich) the normal ratio of ^{235}U to ^{238}U in the fuel pellets, which in itself was – and still is – a difficult and expensive process.

The effort required was considered very worthwhile, however, because the advantages of nuclear power reactors for submarines, or even for surface vessels likely to be at sea for extended periods of time, were obvious. Such reactors could be used to produce steam to turn a turbine that, in turn, could either be used directly to turn a propeller or to produce electricity that, indirectly through a turbo-electric drive system, could turn a propeller. They would require a minimum amount of fuel, and (of considerable relevance to submarines) no oxygen was necessary to burn it. And because nuclear reactors had a high power output, it was envisaged that such vessels would be able to attain considerable speeds underwater. Nevertheless, it was also acknowledged that there were considerable challenges that needed to be overcome in order to make such ambitions a reality. For a start, the reactors needed to be able to cope with much greater mechanical stress than they would experience on land, plus increased rates of metal corrosion. And being so compact, their components would also be liable to a greater risk of damage from irradiation than large, land-based reactors, and thus the entire assemblies would need to be extremely reliable and their operators and support teams highly self-sufficient.

All of these difficulties were nevertheless eventually overcome, and the first nuclear submarine, the 98-m-long USS *Nautilus* (SSN-571), was launched on 21 January 1954 into the Thames River at Gorton, Connecticut, powered by a PWR using enriched uranium as fuel, becoming a fully commissioned component of the US naval fleet on 30 September 1954. It was an extraordinary achievement. It not only heralded a completely new chapter in the power source of submarines, which could now remain submerged for months at a time while travelling at 25 knots and more, but also a new chapter in the military power – and thus influence – of those nation states that possessed them. Other US Navy vessels quickly followed, primarily by way of the 'Skate-class' submarines, also powered by single PWRs. Indeed, within less than a decade, the US Navy had 26 nuclear submarines operational and another 30 under construction.

After the Skate-class vessels, nuclear reactor development in the USA proceeded apace, and standardized designs were built by both Westinghouse and General Electric. The technology was eventually shared with Great Britain through the 1958 USA–UK Mutual Defence Agreement. This sharing of technology was essential because, although the UK had already been producing electricity on a commercial scale from nuclear power for its national grid since 1956, this was by way of a completely different type of reactor. And so in as short a time as possible, the UK's HMS *Dreadnought* was launched on 21 October, Trafalgar Day, 1960 and entered into service on 17 April 1963, powered by an American S5W nuclear reactor. Following upon the success of this vessel, Rolls Royce was then able to build these American-designed PWR units for the Royal Navy before subsequently further developing a basic design for its own use.

But the UK had already been beaten in this new arms race by the USSR, whose first nuclear-powered submarine, the *Leninskiy Komsomol*, built in Severodvinsk, had entered into service in 1958. It was powered (perhaps hardly surprisingly, in view of the espionage activity at that time) by a PWR of similar design to that of the USA's *Nautilus*, although the use of other reactor designs was soon being explored by both the USA and the USSR, albeit with mixed success. During 1957 and 1958, the USA experimented with a sodium-cooled reactor in the submarine USS *Seawolf*. This design produced higher temperatures than a PWR, and it was highly efficient, but it also had serious operational disadvantages. Large electric heaters were required to keep the entire system warm when the reactor was shut down in order to avoid the sodium 'freezing', and the sodium itself became highly radioactive (the ^{24}Na isotope formed by neutron capture being a γ-emitter with a half-life of 15 hours, decaying to ^{24}Mg) so that the reactor had to be more heavily shielded than a PWR. The reactor compartment of the submarine could thus not be entered for many days after shutdown. Meanwhile, the USSR's navy was also considering other forms of reactors, and its Alfa-class submarines had a highly enriched fuel and a lead-bismuth form of coolant. But this, too, had to be kept warm to prevent it from 'freezing' when the reactor was 'down', and only eight trouble-plagued vessels were ever built, all of which were eventually withdrawn from service. The early USSR submarines also had problems with their narrow pressure-vessel designs, which became brittle as a result of the high neutron bombardment.

The French joined the nuclear submarine club in 1971 with their commissioning of the *Redoutable*; the Chinese in 1983 with the *Xia*; and then India in 2009 with the launch of their own first nuclear submarine, the *Arihant*, although they had previously leased the Russian Akula-class attack submarine, the *Chakra*.

Other forms of reactors, such as the boiling water reactors (BWRs) that had been developed for civilian electricity power generation, were also considered for use in submarines, but these were deemed unsuitable because of the necessary circulation of radioactive water beyond the primary reactor vessel system. And it has to be said – and equally important for submarines – the BWRs were also unsuitable because, quite simply, they were too noisy! There were also differences in design for producing the method of propulsion for submarines. The US, British and Russian configurations relied on direct steam turbine propulsion to drive the props, whereas the French and later the Chinese were to use the steam turbines to generate electricity for propulsion. (The most recent submarine designs, incidentally, do not include propellers at all, but rely on a pump-jet system for propulsion.)

Large surface vessels were also obvious candidates for being nuclear powered, particularly if they were expected to be at sea for extended periods of time and operating at considerable distances from home or friendly ports. The first such vessel was the USS *Long Beach*, a guided missile-carrying cruiser, launched on 14 July 1959 and commissioned on 9 September 1961. She was decommissioned on 1 May 1995, having been refuelled three times. Eight other nuclear-powered cruisers followed, and the USSR built four similar vessels. The more obvious choice for large surface vessels, however, was for aircraft carriers, and thus the USS *Enterprise* was launched on 24 September 1960 and commissioned on 25 November 1961. Powered by eight Westinghouse nuclear reactor units, she was in service for over 50 years, being 'inactivated' on 1 December 2012.

The case for civilian nuclear-powered vessels was, however, much more difficult to make. The most obvious case, mirroring some of the needs of large surface military vessels, was for those operating for long periods of time in remote and hostile locations – such as the polar regions. And so in 1957 the USSR launched the first non-military nuclear-powered vessel (indeed, the first nuclear-powered surface vessel), the *Lenin*, an icebreaker that was to be deployed to keep open the northern sea route along the Siberian coast, where the ice can typically range from 1 to 2 m thick. The *Lenin* operated, although not without problems, until 1989, by which time its hull had apparently become considerably thinner as a result of ice abrasion. Such was the success of the vessel, however, that many more nuclear-powered icebreakers were to follow.

Unfortunately, the same could not be said for the *Savannah*, which was launched in 1959 by the USA as a passenger-cargo vessel and becoming fully operational in 1962, although in many ways she could be regarded as being successful. Unfortunately, her basic design (with a very limited hull capacity) and shape were not really conducive to the way in which cargo was then being handled, and her requirement for a large crew, plus other factors, made her uneconomical to run. She was therefore withdrawn from service in 1971. Another experiment was the (then) West German *Otto Hahn*, designed as a passenger and ore carrier, launched in 1964 and operated commercially from 1968 to 1978. She performed very well, covering over 250,000 nautical miles in her first 4 years alone. But commercially she suffered from a range of irrational restrictions on which ports she could and could not enter, and she was refused permission to traverse the Suez Canal. And so in 1979 her reactor was removed and replaced with diesel engines. From then on, and mirroring the ever-fickle and sometimes farcical attitudes of governments worldwide to the feasibility of nuclear power in general, she went through no less than seven name changes before finally being scrapped as the *Madre* at Alang, India, in 2009.

But at least she was technically successful, which was more than could be said of the Japanese vessel *Mutsu*. Launched in 1969, she suffered from reactor problems on her maiden voyage in 1974, resulting in a thorough redesign and thus subsequent reconstruction of her reactor shielding. She took to the seas again in 1990, but was withdrawn and decommissioned in 1992. In stark contrast, however, was the Russian *Sevmorput*. Launched in 1986, she was built to serve northern Siberian ports, particularly by way of carrying lighters to coastal ports in shallow-water locations (referred to as a LASH vessel) and capable of breaking ice of 1–2 m thickness. Despite the inevitable chequered history of such a vessel, she is still in service.

The real commercial success story, however, has been the Russian icebreakers. Indeed, their further development and deployment have completely changed the use of northern polar waters. Capable of sailing through 2 m of ice or more, the *Arktika* was the first surface vessel to reach the North Pole in 1977, and since then such icebreakers have enabled navigation in the Arctic generally to be increased from about 2 months to 10 months each year, and even all year round in the more western areas. There were six vessels built in the Arktika class, including the *Yamal* (Figure 11.1), which has been used for scientific research and for tourism. The Arktika class was followed by the Taimyr class, which were slightly smaller vessels and powered by a single nuclear reactor. More recently, the first of a new class of icebreaker was launched on 16 June 2016. Again named *Arktika*, this new vessel is powered by two 175-MW (thermal) reactors. Two similar vessels are under construction.

Figure 11.1 Russian icebreaker *Yamal* launched in 1992 and powered by two OK-900 170-MW reactors.
Source: Creative Commons Attribution – Share Alike Licence

11.2 Current Nuclear-Powered Fleets

In addition to the Russian icebreakers, nuclear-powered submarines proved to be a great success. As the Cold War drew to a close in 1989, there were at least 400 nuclear submarines in operation or being built. Such vessels are usually referred to as subsurface (SS) vessels and grouped as SSNs (attack), SSGNs (carrying cruise missiles) or SSBNs (carrying ballistic missiles). Being military vessels, the number deployed by any one country at any one time is not always clear, but as of 2017, these numbers are more or less as follows.

By far the biggest submarine fleet is that of the USA, currently operating some 40 Los Angeles-class SSNs, 3 Sea-Wolf-class SSNs, plus 12 Virginia-class SSNs, with about another 15 of this class being built or about to be built. And to these must be added 4 Ohio-class SSGNs and 14 Ohio-class SSBNs. Russia has some 16 SSNs (Victor III, Sierra, Typhoon and Yasen class), 7 SSGNs (Oscar class) and 12 SSBNs (Delta III, Delta IV and Borei class). Its latest addition to the fleet, the Yasen-class K561, was launched on 31 March 2017.

The UK is currently operating 3 Astute-class SSNs, 4 Trafalgar-class SSNs and 4 Vanguard-class SSBNs. The French have a similar sized fleet of 6 Rubis-class SSNs and 4 Triomphant-class SSBNs.

Elsewhere, China is believed to have some 10 SSNs and 6 SSBNs and India 1 SSN (on lease from Russia) and 1 SSBN Arihant class, with another being built.

Aircraft carriers, too, have been a success, with the USA currently operating 12 (Nimitz-class) carriers and France 1, the *Charles de Gaulle*. Russia still operates the *Pyotr Velikiy*, a nuclear-powered battle cruiser. The other success story – the icebreakers – has resulted in Russia currently operating six such vessels in Arctic waters, with another launched and two being built.

The current larger vessels – aircraft carriers, icebreakers and the Russian Typhoon-class submarines – have two reactors; all of the others have one. But from a potential environmental impact point of view, the relevant factors – apart from safety – are the type of fuel required and the frequency with which it has to be replaced. And these two factors are related. The reactors themselves generate from about 50 to 500 MW (thermal), depending on the vessel. The fuel, by and large, consists of a uranium/zirconium or a uranium/aluminium alloy, or is cast in a ceramic form and thus is quite different from that of land-based civil reactors. The uranium is enriched from the natural level of 0.7 per cent ^{235}U to 7 or 8 per cent in a few designs, and thus not much different from that used in land-based commercial reactors (which typically use 3–5 per cent), but it is usually enriched to >20 per cent (up to 45 per cent in some classes of submarines and surface vessels), and even to >90 per cent in some designs. The objective of such enrichment is, of course, not simply to maintain power output, but also to minimize the frequency of changing the fuel (and thus time spent out of commission). By and large, this is only required about every 10 years, but more recent designs only need refuelling every 25 years or more, and the most recent should not need refuelling throughout their intended 50-year lifespan. Such long 'burn-up' times are achieved not only by having highly enriched uranium fuel, but also by incorporating a 'burnable poison' such as boron or, more recently, gadolinium, which is progressively depleted, thus evening out the reactor power output over time.

11.3 Normal Operation, Refuelling and Discharges to Sea

As with any other ship, nuclear-powered vessels generate a variety of wastes. For surface nuclear ships, ordinary day-to-day wastes are dealt with in the same way as in any other ship, as discussed in Chapter 4. But submarines are, of necessity, somewhat different. As one would expect, the quantities of 'ordinary' waste are relatively small, compared with a surface vessel, because the provisions taken on board in the first place are relatively small. In (at least US) submarines, the day-to-day waste is separated into biodegradable and non-biodegradable categories: the former is compacted and ejected at sea through a 'trash' disposal unit; the latter is cleaned as necessary, compacted and stored for disposal ashore. Freshwater is made in a desalination plant and purified for drinking (as well as for the propulsion unit) using evaporators or reverse osmosis. Freshwater is used very sparingly. Oxygen is made (by electrolysis) and the hydrogen ejected. Carbon dioxide and other odours are 'scrubbed' from the submarine's controlled atmosphere using amine, which, in turn (as any submariner will readily attest!), produces its own odour.

Running the reactors on any nuclear vessel requires a high level of water quality control. Because of their high design criteria, none of the fuel rods are anticipated to release any fission product nuclides into the primary coolant – which is an accepted and anticipated occurrence in land-based reactors. Neutron activation product nuclides are nevertheless formed in the metals of the coolant pipework (such as those of cobalt, chromium, nickel and iron), which may therefore arise in trace amounts in the coolant water. Gaseous isotopes may also arise. Such impurities are therefore removed via ion exchange resins and so on, but the quantities of waste produced are very small. Liquid wastes may be discharged at sea, but any solid wastes, such as resins, are (at least on US and UK vessels) stored on-board and returned to land for disposal.

Radiation exposure data for naval personnel under normal operations in military vessels are not readily available. The effective dose rate limit recommended by the International Commission on Radiological Protection (ICRP) for occupational exposures with respect to planned exposure situations (which is how such exposures would be classified in this case) is 20 mSv/year, averaged over a defined 5-year period (ICRP, 2007). Bearing this in mind, the US ONR Report (2014) indicates that personnel of the US Office of Naval Reactors received an average annual dose of 0.06 mSv in 2013, and none have exceeded 20 mSv/year. The average occupational exposure since 1958 was 1.03 mSv/year.

Ashore, discharges from nuclear submarine bases are subject to regulations and emission controls, as required. These may change with time. As an example, discharges to the environment from all establishments in the UK where vessels are or have been repaired, refuelled and refitted are necessarily authorized by the relevant national authorities. Their surrounding environments are thus carefully monitored and any subsequent radiation doses to the public assessed and reported upon publically. Food materials are usually analysed for such radionuclides as ^{3}H, ^{14}C, ^{60}Co, ^{134}Cs, ^{137}Cs, ^{154}Eu, ^{155}Eu and ^{241}Am and external dose rates on soils, sediments and other materials measured directly. The results, as reported annually in the Radioactivity in Food and the Environment (RIFE) series (as in EA *et al.*, 2016), indicate that the discharges are typically only a few per cent of the authorized limits, and dose rates to the public are in the order of 0.005 mSv/year or less, and thus about 0.5 per cent or less of the internationally recommended (ICRP, 2007) effective dose limit of 1 mSv/year for members of the general public under planned exposure situations.

Variations in discharge rates over time from nuclear submarine bases are often related to the technical evolution of the vessels themselves. Thus, as shown in Figure 11.2, the discharge history of ^{3}H since 1990 with respect to Devonport (UK) reflects the increasing requirement to refit Vanguard-class submarines, which have a high ^{3}H inventory because they do not routinely discharge primary circuit coolant until they undergo refuelling at Devonport. In contrast, the overall decrease in ^{60}Co discharges over the same period of time is largely due to the improvement in submarine reactor design, so that less ^{60}Co is produced and thus less is released during routine maintenance operations. With regards to the nuclear fuel itself, this has to be manufactured, and the waste resulting from its 'burn-up' disposed of as part of the fuel cycle relating to all nuclear reactors. The quantities of uranium mined, milled and enriched for marine reactors varying in size from about 50 to 500 MW (thermal) is nevertheless of little consequence compared with the quantities dealt with to serve the 448 or so large land-based reactors worldwide that deliver some 390 GW

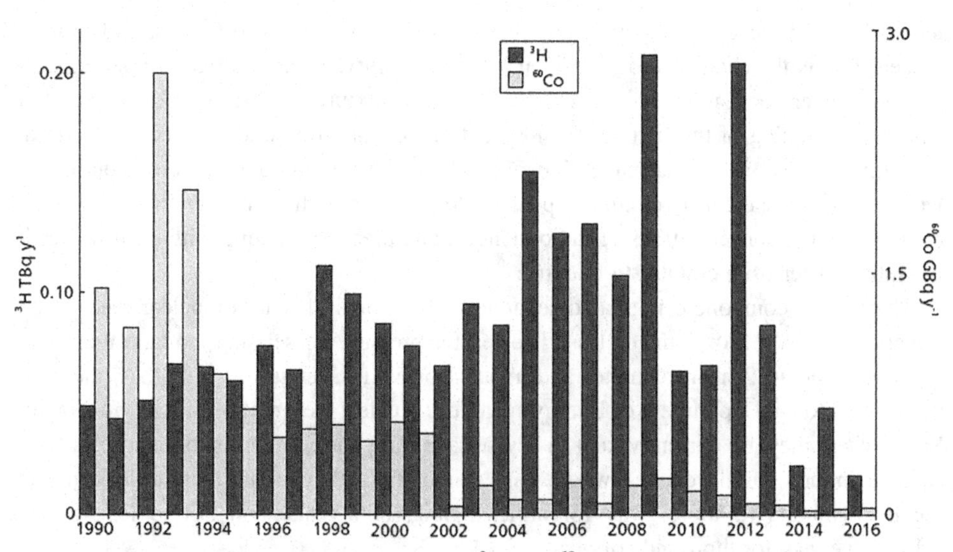

Figure 11.2 Trends in liquid discharges of ^{3}H and ^{60}Co from Devonport, Devon, UK, 1990–2015.
Source: From EA *et al.* (2017), reproduced with permission

(electrical) power into national grids. But the disposal of the reactor cores and their spent nuclear fuel is another somewhat different matter that has to be faced when the vessels are decommissioned.

11.4 Decommissioning and Disposal

The principal environmental concerns with respect to all reactor operations, apart from something going wrong, are the issues surrounding the decommissioning and ultimate disposal of the spent nuclear fuel, the reactor and other components that have become contaminated or made radioactive as a result of neutron activation.

As already noted, by the time the Cold War drew to a close in 1989, there were well over 400 nuclear-powered submarines operational or being built. Indeed, the USSR alone built some 248 nuclear-powered submarines, five naval surface vessels, plus 9 icebreakers between 1950 and 2003, collectively powered by about 468 reactors. Over the last 25 years, at least 300 nuclear vessels have been taken out of commission. Nevertheless, following defueling, not all of them have been broken up. The icebreaker *Lenin* is now a museum in Murmansk, although its reactors reside, somewhat controversially, elsewhere. And in the USA, and possibly in other countries, a small number of submarines have been taken out of service, but remain as moored training ships.

Not surprisingly, decommissioning and disposing of nuclear vessels comes at a significant cost. The US Navy has a nuclear Ship Recycling Program at its Puget Sound Naval Shipyard in Bremerton, although ships may be defueled elsewhere. Current practice is to remove and isolate all hazardous and toxic wastes (such as polychlorinated biphenyls) from the vessels and recycle as much as possible of the hull in order to offset costs. The spent nuclear fuel is then shipped to the Naval Reactor Facility, Idaho, and kept in storage. As with any other spent reactor fuel, this consists of a mixture of fission products resulting from the fissioning of the fuel itself, plus neutron activation products, all of which decay over time, but some of them also give rise to 'daughter' radionuclides, the quantities of which will rise and fall over time. Spent nuclear fuel is both thermally 'hot' as well as containing high concentrations of radionuclides, and plans for dealing with such wastes in the future differ from country to country.

The reactor component is treated separately. The cores of marine nuclear reactors are remarkably small: 2 or 3 m^3 or less. The reactor pressure vessels that contain them may perhaps be about 2 m in diameter and about 4 or 5 m in length. In the USA, these are transported to the Department of Energy's Hanford Nuclear Reservation site in Washington State, where they are currently kept in dry storage. The current plan is to dispose of these sections in land burial trenches, where it is estimated that they are unlikely to leak in even the most minute (pinhole) way for at least 500–1000 years, and not to leak in any substantive way for thousands of years.

The desert lands of central Washington State might seem to be a rather strange resting place for these devices that spent years of service driving nuclear vessels underwater around the oceans of the world. And, in fact, in the early days of the nuclear submarine programme, it was planned that such reactors from decommissioned vessels would indeed be returned to whence they came – the sea. Thus, in 1959, the reactor from the USS *Seawolf* (SSN 575) was dumped in the Atlantic Ocean, some 200 km east of Delaware at a depth of 2700 m. The disposal of radioactive waste into the sea was then a fairly common practice, having commenced in 1946, but such practices were to come to a halt, at least for most countries, with the introduction in 1972 of the London (Dumping) Convention (IMO, 1972) that placed various restrictions on the disposal of different categories of radioactive waste and banned the disposal of 'high-level' waste entirely. Indeed, as from 20 February 1994, all radioactive wastes were prohibited from being dumped (IMO, 1993). Such dumping practices of high-level wastes nevertheless continued well beyond 1972, but before examining the likely environmental impact of this legacy, it is first useful also to review the other incidents and accidents that have occurred at sea, including the loss of entire vessels, not just their reactors.

11.5 Incidents, Losses and Disposals at Sea

It was inevitable that accidents and incidents would occur with the development of nuclear vessels, particularly in the arms race that typified the early years of what was to become the 'Cold War'. In some cases, such vessels were involved in incidents or sank, but the events had no direct relationship with the fact that they were nuclear powered. And in some instances, accidents occurred to the reactors but the vessels survived; nevertheless, in some cases, the two were undoubtedly directly related. Known cases include the following.

- ❖ 1960 USSR: SSN *K-8* had a loss-of-coolant accident resulting in the loss of substantial quantities of radioactive material.
- ❖ 1961 USSR: SSN *K-19* had a loss-of-coolant accident resulting in 8 deaths and more than 30 other people being overexposed to radiation.
- ➢ 1963 USA: SSN *Thresher* sank while undergoing deep-sea diving trials, apparently due to a combination of mechanical problems, loss of propulsion and the formation of ice in valves that prevented the emergency ballast tanks from working normally, causing the submarine to exceed its 'crush depth'. All 129 personnel on board died, including shipyard personnel involved in the deep-sea diving tests. The incident occurred about 350 km east of Cape Cod.
- ❖ 1965 USSR: SSN *K-11* suffered from both reactors being damaged while refuelling, apparently during the process of removing the heads from the reactor vessels. The reactor compartments were subsequently scuttled off the east coast of Novaya Zemlya, in the Kara Sea.
- ❖ 1967 USSR: SSN *K-3*, the first Soviet nuclear submarine, developed a fire on board, associated with the hydraulic system, resulting in 39 deaths.
- ➢ 1968 USA: SSN *Scorpion* (SSN-589) was lost at sea, 740 km south-west of the Azores, with 99 men on board. The most likely cause was thought to be a malfunctioning battery attached to a torpedo, which exploded with sufficient force to ignite the torpedo's charge.
- ➢ 1968 USSR: SSN *K-27* experienced reactor core damage to one of its liquid metal (lead-bismuth)-cooled VT-1 reactors, resulting in 9 fatalities and 83 other injuries. The reactor compartment was then filled with a mixture of bitumen and other chemicals in order to seal it, and the vessel scuttled in shallow waters of the Kara Sea (Stepovoy Bay) in 1981.
- ❖ 1968 USSR: SSN *K-140* suffered reactor damage following an uncontrolled automatic increase in power during shipyard work.
- ❖ 1969 USA: *Guitarro* (SSN-665) sank while moored within a shipyard because of improper ballasting. It was then recovered.
- ➢ 1970 USSR: SSN *K-8* caught fire, killing eight members of the crew. The remaining crew were evacuated to a surface vessel, but then, as a result of an accident while under tow, the vessel sank 260 nautical miles north-west of Spain with the loss of a further 52 crewmen who had reboarded the vessel to help in its recovery.
- ❖ 1970 USSR: SSN *K-429* suffered a reactor fire due to an uncontrolled 'start-up', which resulted in a local release of radioactive material.
- ❖ 1970 USSR: SSN *K-116* suffered a loss-of-coolant accident in the port reactor, resulting in a substantial release of radioactive material.
- ❖ 1972 USSR: SSN *K-64*, the first of the Alfa-class submarines that had a liquid metal-cooled reactor, suffered reactor failure, and the vessel was then scrapped.
- ❖ 1973 USSR: SSN *K-56* collided with another Soviet vessel leading to flooding of the battery compartment, which resulted in the release of chlorine gas, killing an undisclosed number of crew.
- ❖ 1980 USSR: SSN *K-222* suffered a reactor accident during maintenance.

- ❖ 1982 USSR: SSN *K-123* Alfa-class submarine had a reactor core damaged by a liquid metal coolant leak; the vessel was forced out of commission for 8 years.
- ❖ 1983 USSR: SSN *K-429* sank due to flooding, killing 16 crewmen, but was subsequently raised, only to sink again at her moorings, and was then again raised in order to be scrapped.
- ❖ 1985 USSR: SSN *K-431* incurred a reactor accident while refuelling, resulting in 10 fatalities and 49 other people suffering radiation-related injuries.
- ➢ 1986 USSR: SSN *K-219* suffered an explosion and fire in a missile tube, eventually leading to a reactor accident. The submarine subsequently sank while under tow, 950 km east of Bermuda. Six crew members were killed.
- ❖ 1989 USSR: SSN *K-131* suffered a fire while in the Norwegian Sea, leading to a loss-of-coolant accident in the starboard reactor; a substantial but unknown quantity of radioactive water was discharged into the sea.
- ➢ 1989 USSR: SSN *K-278*, the *Komsomolets*, sank in the Barents Sea due to a fire, with the loss of 42 personnel, many of whom died either from smoke inhalation or from exposure (not from radiation) in the cold water.
- ➢ 2000 Russia: SSN *K-141 Kursk* sank dramatically in the Barents Sea as a result of an explosion in the torpedo compartment, killing all 118 seamen on board.
- ❖ 2001 USA: SSN *Greeneville* surfaced underneath the Japanese training vessel *Ehime Maru*. Nine Japanese were killed when their ship sank as a result of the collision.
- ➢ 2003 Russia: SSN *K-159*, some 14 years after being decommissioned, but not defueled, sank in the Barents Sea while being towed to be scrapped, killing 9 crewmen.
- ❖ 2005 USA: SSN *San Francisco* (SSN-711) collided with a seamount in the Pacific Ocean. A crew member was killed and 23 others were injured.
- ❖ 2012 USA: SSN *Miami* (SSN-755) was destroyed by a deliberately ignited fire while in a shipyard and was hence subsequently scrapped.

It might also be noted that, although not nuclear powered, the diesel–electric USSR submarine *K-129* was lost in the Pacific Ocean in 1968 under somewhat mysterious circumstances. She was armed with, among other weapons, three ballistic nuclear missiles and two nuclear torpedoes. She was subsequently located by the USA at a depth of 4.9 km north-west of Hawaii and parts of the vessel were recovered by the specially built *Hughes Glomar Explorer* in a clandestine operation called Project Azorian, for which this vessel was purportedly contracted at that time to 'mine' manganese nodules from the seabed.

With regards to the above list of nuclear-powered submarines, it can be concluded that eight entire vessels (indicated by arrows in the above list) now still reside at the bottom of the sea. The losses of the USA's *Thresher* (1963) and the *Scorpion* (1968) were the first, but the loss of neither could apparently be attributable to the fact that they were nuclear powered, although a question mark will always hang over the loss of the *Scorpion*. With regards to the USSR vessels, however, the situation is somewhat different. The damage incurred by *K-27* was directly the result of the malfunction of its novel reactor design, but it need not have been scuttled, and indeed was not scuttled until some 18 years later. It was not therefore 'lost' at sea, but deliberately disposed of there. The *K-8* sank due to an accident while under tow, and the loss of *K-219* was also the result of an accident while under tow, the original cause for her demise having nothing to do with the fact that she was nuclear powered. The same may be said of the loss of *K-278*, the *Komsomolets*. An electrical fire spread throughout the submarine, resulting in a reactor shutdown and hence loss of propulsion, so that she was then forced to surface. The major loss of life was primarily due to exposure in the freezing water of the Barents Sea while awaiting rescue.

More recently has been the dramatic loss of *K-141*, the *Kursk*, also in the Barents Sea. This, again, had nothing to do with her form of propulsion, but resulted from a failure in the firing of a torpedo during a naval exercise due to the leakage of its hydrogen peroxide propellant. The vessel sank in relatively shallow waters and much of it was subsequently recovered. And, yet again in the Barents Sea, the rusting hulk of *K-159* sank in a storm while under tow to be scrapped, some 14 years after being decommissioned.

More unfortunate still, in many ways, was the *K-429*. She is not included in the list of eight 'lost at sea' because, although she sank at sea during a test dive in 1983, she was subsequently recovered but then suffered the ignominy of sinking again 2 years later as a result of flooding, this time at her moorings. She was then, quite understandably, taken out of commission.

Another interesting inclusion in the above list is that of *K-11*; she did not sink, but it is known that her reactor was subsequently dumped at sea. This incident was by no means an isolated one. Notwithstanding the London (Dumping) Convention's prohibition on the sea disposal of such materials, and a 1983 moratorium on the disposal of all radioactive waste at sea, to which the USSR was also a signatory, it emerged in 1993 that a considerable number of marine (primarily submarine) vessel reactors had nevertheless been dumped at sea by that country. The known list of reactors (factory numbers are given, not the numbers of the vessels from which they came) includes the following.

- 1965 USSR: factory # 285, two reactors, one with fuel, one without, Abrosimov Fjord, Kara Sea
- 1965 USSR: factory # 254, two reactors, without fuel, Abrosimov Fjord, Kara Sea
- 1965 USSR: factory # 901, two reactors, with fuel, Abrosimov Fjord, Kara Sea
- 1966 USSR: factory # 260, two reactors, without fuel, Abrosimov Fjord, Kara Sea
- 1967 USSR: *Lenin*, three reactors, ostensibly without fuel, Tsivolka Fjord, Kara Sea
- 1971 USSR: factory # 143, two reactors, without fuel, Sea of Japan
- 1972 USSR: factory # 421, one reactor, with fuel, Novaya Zemlya Trough, Kara Sea
- 1979 USSR: factory # 172, two reactors, without fuel, Sea of Japan
- 1981 USSR: factory # 601 (*K-27*), two reactors, *in situ*, with fuel, Stepovoy Fjord, Kara Sea
- 1988 USSR: factory # 538, two reactors, without fuel, Techeniye Fjord, Kara Sea

To this list may be added large quantities of low- and intermediate-level radioactive wastes in both liquid and solid form associated with, primarily, the decommissioning of USSR submarine vessels in the seas around Russia. A number of vessels *containing* radioactive waste material have also been deliberately sunk in the deep sea by the USSR, but the origin of the waste is not always clear, and may or may not have been related to the operation of nuclear-powered vessels.

11.6 Human and Environmental Consequences of Reactors Lost and Dumped at Sea

As one might imagine, precise and accurate information concerning the radionuclide inventories of nuclear military vessels are not available because such data can be used to help provide sensitive information regarding the design of both reactors and any nuclear weapons being carried. Nevertheless, it is not too difficult to surmise what their content might be. With regards to the weapons themselves, these are likely to have a plutonium content that is at least 94 per cent ^{239}Pu, and the nuclear torpedoes, as carried by the *Komsomolets* (*K-278*), would probably have each contained about 3 kg of plutonium. Its reactor, having run for about 5 years, would contain in excess of 5×10^{15} Bq of radionuclides, but many of these would be short-lived, so that after about 100 years the principal radionuclides of concern would be some of the longer-lived fission products, some neutron activation products and the plutonium nuclides plus their decay products. The estimated quantities of the principal radionuclides within this reactor over time are given in Figure 11.3.

It is likely that similar results would be obtained for most of the other submarine reactors still residing at the bottom of the sea in terms of their radionuclide composition and relative abundance over time, and various estimates have indeed been made, particularly in relation to those Arctic waters where some of the materials lie in very shallow (a few tens of metres) depths. Thus, in the Kara Sea, which contains the best part of 16 reactors, of which 6 contained spent nuclear fuel, it was estimated that, as of 1994, the total inventory of radionuclides contained in these structures was in the order of 4.7×10^{15} Bq, made up of 4.1×10^{15} Bq of fission product nuclides (about 86 per cent of the total) and 5.7×10^{14} Bq of neutron activation products (about 12 per cent of the total), with the remaining 2 per cent consisting of 9.7×10^{13} Bq of actinides (IAEA, 1999). The principal fission product nuclides were estimated to be ^{137}Cs (1×10^{15} Bq) and ^{90}Sr (9×10^{14} Bq). The principal activation products were considered to be ^{63}Ni (3×10^{14} Bq) and ^{60}Co (1×10^{14} Bq) and the actinides dominated by ^{241}Pu (8×10^{13} Bq), which will of course decay into ^{241}Am.

As an example of the radionuclide composition within a submarine's reactor compartment, that of factory # 901 (actually twin reactors, containing spent nuclear fuel) was considered to be composed 3.6×10^{14} Bq of fission products, principally ^{137}Cs, ^{137}Ba, ^{90}Sr and ^{90}Y, as well as 3×10^{12} Bq of neutron activation products, principally ^{59}Ni, ^{63}Ni and ^{60}Co, plus 3.6×10^{14} Bq of actinides, primarily ^{241}Pu and ^{241}Am. The reactors identified as # 601 were somewhat different from the others, being liquid metal reactors that used a mixture of lead and bismuth as a coolant.

Sampling of environmental materials around these sites has yielded little in the way of elevated radionuclide concentrations, which is to be expected in view of the fact that the radionuclides are essentially an integral part of the structure of the vessels, their reactors and their reactor compartments plus, where relevant, their fuel rod assemblies

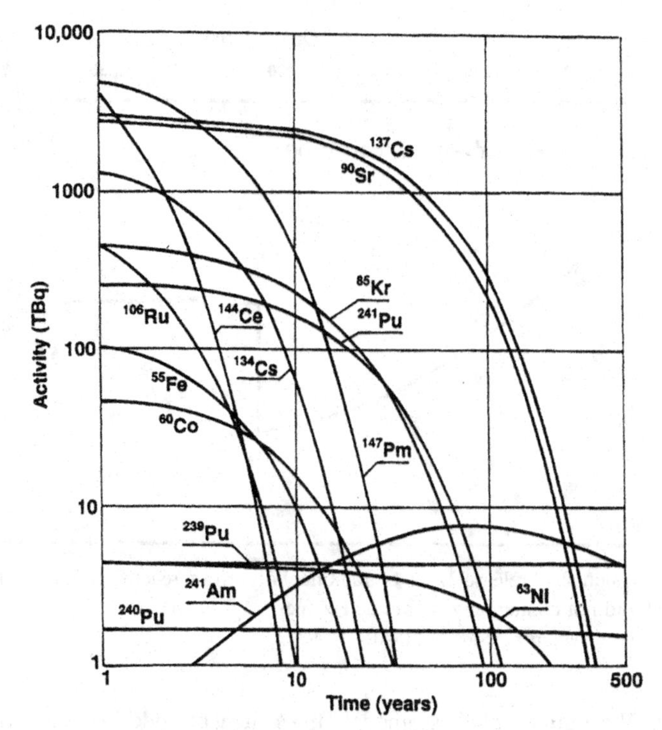

Figure 11.3 Computed changes of the principal radionuclides contained within the reactor of the submarine *K-278*, the *Komsomolets*, as a function of time. Such calculations produce different values for reactors that have been operational for different periods of time. In this case, the reactor was assumed to have been operational for only 5 years.
Source: Reproduced with permission from AMAP (1998)

(IAEA, 1998, 1999). The unique nature of the fuel itself would be expected to retain a large fraction of the actinide and fission product nuclides produced during the burn-up of the fuel, added to which the zircalloy fuel cladding, which is bonded to the fuel, provides an additional barrier. And the neutron activation product nuclides are integral parts of the fabric of the metals within which they are formed. Thus, the release of radionuclides into the surrounding media will only occur at a rate commensurate with the rate at which these components of the vessels corrode, plus the role of any materials used to cover or encase them prior to their disposal. Factors that need to be taken into account include the concentration of dissolved oxygen at the surface of the different materials, the ambient temperature, the nature of the surface of the materials (the extent of cracking and so on) and the presence or absence of any protective coatings that would affect the extent of electrochemical corrosion of, particularly, iron and titanium. And bearing in mind that the reactors themselves were designed to operate at high temperatures and pressures, their rate of corrosion might be expected to be extremely slow. During this process, the radionuclides will also be decaying.

Various modelling exercises have been undertaken using substantial and well-established databases relating to the rates of corrosion of the basic materials, including the different types of steel of which the reactors and their compartments were made

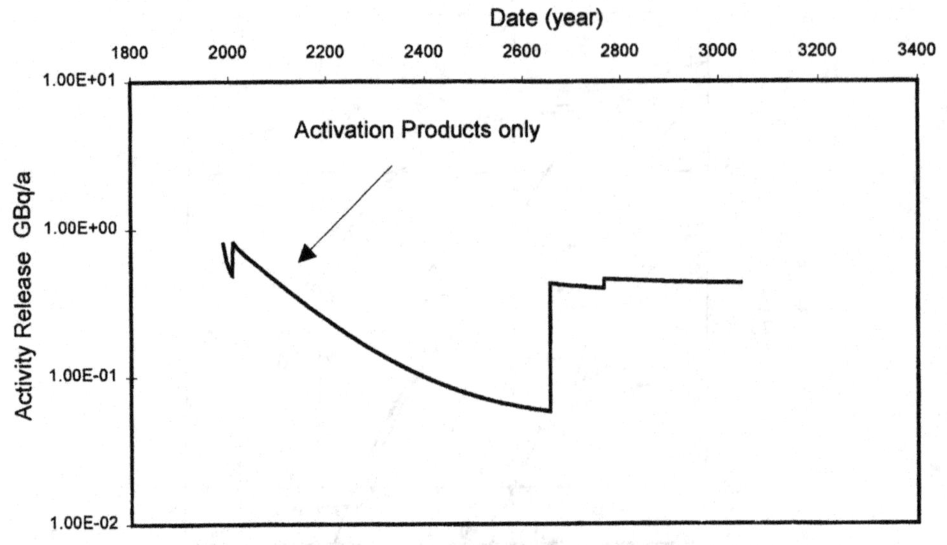

Figure 11.4 Calculated release rates of radionuclides from reactor # 538 in Techeniye Fjord, which did not contain any spent nuclear fuel.
Source: Reproduced with permission from IAEA (1997)

(IAEA, 1997). The outputs of these models have then provided source-term values for various environmental dispersion and food chain models to estimate when peak rates of release of different radionuclides are likely to occur in the future and what likely dose rates to the human population might then arise. The results are very encouraging.

For those reactors without spent nuclear fuel, such as # 538, dumped in 1988, only neutron activation products are of relevance. The predicted total release rate of these nuclides in GBq (10^9 Bq)/year, plotted at 5-year intervals, in calendar years into the future is shown in Figure 11.4. The modelled rate of release is initially dominated by the assumed corrosion of the pressure reactor vessels, falling from 0.8 GBq/year to 0.06 GBq/year by 2655, and subsequently by the corrosion of the thermal shields and cladding until these, too, have decayed away.

Where the reactors still contain spent nuclear fuel, as is the case with # 421, the situation is a bit more complicated (Figure 11.5). Again, the initial rate of radionuclide release is assumed to be dominated by the corrosion of the reactor pressure vessel, releasing neutron activation products. Then the fuel and interior stainless steel components are assumed to corrode so that, in this case, by the year 2035, both fission product and actinide nuclides are released, with peak rates of 370 GBq/year and 0.2 GBq/year, respectively. The more soluble fission products such as ^{137}Cs and ^{90}Sr are then assumed to be released fairly quickly, up to the year 2500, so that the release rate would then be dominated by the less soluble fission products and then by the actinides. From 2500 to 3360, the release rate is estimated to be about 0.2 GBq/year, dominated by the longer-lived activation product nuclides. Then, by about 3360, the reactor pressure vessel is assumed to have entirely disintegrated, resulting in a short peak of 2.5 GBq/year resulting from the release of further

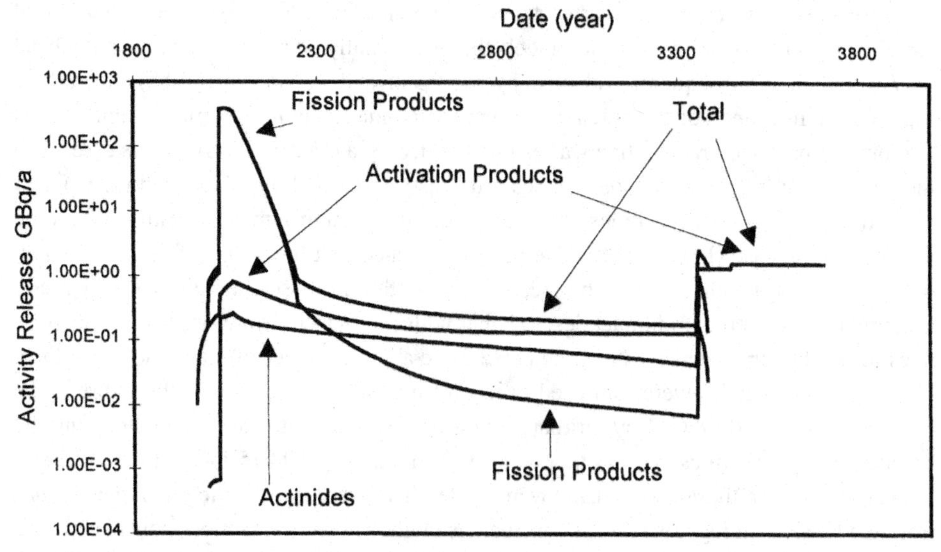

Figure 11.5 Calculated release rates of radionuclides from the single reactor # 421 in the Novaya Zimlaya Trough, Kara Sea, which did contain spent nuclear fuel.
Source: Reproduced with permission from IAEA (1997)

fission and activation products, with little being left of the fuel by about 3385. Finally, the reactor pressure vessel cladding and thermal shields are assumed to be fully corroded, the release rates then being about 1.6 GBq/year, again being dominated by the longer-lived activation product nuclides up until the year 3710.

The modelling scenario for reactor # 421 would probably equally apply to all of the other reactors disposed of, or contained within, sunken submarines. The one departure from these estimates is the submarine *K-27*, which was powered by twin liquid metal reactors (factory # 601) using lead and bismuth as a coolant. About 20 per cent of one reactor core suffered from uneven coolant flows, resulting in a release of fuel and fission product nuclides into the coolant. After it was decommissioned in 1979, the reactor compartment was filled with a mixture of furfuryl alcohol and bitumen and the entire vessel deliberately scuttled in the Kara Sea. The peak release rates are therefore somewhat lower than the estimates made for the reactors in isolation, but the entire process is estimated to continue for about 40,000 years, albeit at the slow rate of release of about 7 MBq/year.

In isolation, such release rates are not of any great consequence. But in the Kara Sea, there are 16 reactors in relative close proximity. Collectively, such release rates amount to peak values of about 1 TBq/year, but more generally about 10 GBq/year or less. When these release rates are used with various environmental dispersion models, together with food chain and radiological assessment models, the predicted future maximum dose rates to members of the public (IAEA, 1998) are estimated to be about 0.3 μSv/year or less, with the relevant current dose rate limit, as noted in Section 11.3, being 1 mSv/year. In other words, they would be trivial.

Dose rates experienced by marine fauna in the area have also been calculated, based largely on a range of solid ellipsoid models that had initially been developed by Woodhead (1979) and then subsequently adapted by Pentreath and Woodhead (1988) in order to evaluate the relevance of the potential impact on marine fauna in prescribing limits to the dumping of packaged radioactive wastes into the deep sea (IAEA, 1988). In this case, such models were augmented by those for seabirds, also developed by Woodhead (1986), and applied to the peak concentrations of radionuclides in the environment, as estimated for the area of peak release (IAEA 1998). The results indicated that the highest dose rates (about 0.0025 mGy/day) were likely to be received by benthic molluscs, largely from internally accumulated α-emitting radionuclides, together with external contributions from ^{60}Co; dose rates to benthic crustaceans were anticipated to be about an order of magnitude less. Fish, birds, seals and whales were estimated to receive maximum dose rates in the future in the order of 0.001–0.002 mGy/day, primarily from ^{137}Cs. Dose rates to fauna colonizing the surfaces of the structures were estimated to be in the order of 0.015 mGy/day.

At the time that these assessments were made, there were no specific recommendations from the ICRP with regards to what conclusions might be drawn or actions considered with respect to the exposure of fauna and flora under different exposure situations. Since then, however, recommendations have been made based on sets of reference fauna and flora that make use of the same general dose-modelling approaches as used in the Kara Sea assessments (ICRP, 2008, 2014). For the situation currently obtaining in the Kara Sea (regarded as an 'existing exposure situation'), the recommendations are that any action would only be contemplated if the dose rates were in excess of the relevant Derived Consideration Reference Levels, which, for the most exposed fauna in this situation, are 1–10 mGy/day for vertebrate animals and 10–100 mGy/day for the relevant invertebrates. As can be seen, the anticipated maximum dose rates for Kara Sea marine fauna are many orders of magnitude below these levels, and thus no measurable effects would be expected relating to the mortality, morbidity, reduced reproductive success or chromosomal damage or mutations of the relevant fauna, and thus it is unlikely that any remedial action would be considered on the grounds of environmental protection alone.

11.7 Overall Assessment

While it has been necessary to review and assess the various accidents and incidents relating to nuclear submarines and the disposal of some of their reactors, it must be born in mind that the deliberate disposals of the reactors by the USSR in the Kara Sea and the Sea of Japan was completely illegal and contrary to the international agreements to which the USSR was a signatory, the USA's dumping of the reactor from the *Seawolf* having taken place before any international agreements had been drawn up. And, apart from the loss of the *Thresher* and the *Scorpion* in the early days of the development of nuclear submarines, neither of which can be shown to be a direct consequence of their being powered by nuclear reactors, all of the other incidents resulting in the uncontrolled release of radionuclides into the environment relate to USSR vessels. And even in these cases,

Figure 11.6 USS *George Washington*.
Source: Photograph by M. Anderson, released by US Navy, 020928-N-3653A-001

because the radionuclides are present either as a result of becoming part of the fabric of the vessel (by neutron activation) or are contained within sections of the vessel that have been designed and engineered to withstand corrosion under normal operational conditions, their release rates are so slow as to constitute a very low hazard to either humans or marine fauna and flora.

Indeed, it is more useful to consider the operational history and current status of nuclear-powered vessels in their proper perspective. For many decades, over 700 reactors have been operational at sea. Current ships are quite impressive in their performance, such as the military craft operated by the USA (Figure 11.6). Modern reactors being designed or installed in marine vessels are now capable of generating up to some 700 MW (thermal) of energy for 50 or 60 years, with a single mid-life refuelling operation. Other reactors, of less power, are designed to operate for well over 30 years without any need to refuel at all. These reactors have cores that are only 2 or 3 m^3 in volume, and can yet power such enormous vessels as fully armed aircraft carriers that are over 1000 ft (>300 m) in length, with a displacement of over 100,000 t, at speeds in excess of 30 knots and with a complement of over 4000 personnel, supplying all of their and their vessel's energy needs for 30 years or more without refuelling.

It is in fact amazing to consider that over a period of more than 50 years, and thanks in large part to the stringent standards and powerful focus on personal integrity imposed by Hyman George Rickover as Head of Naval Reactors, the US Navy alone has accumulated well in excess of 6200 reactor-years of accident-free experience, involving 526 nuclear reactor cores, in vessels travelling over 240 million km, without a single radiological incident.

The future of nuclear-powered vessels is nevertheless far from clear, but it remains the subject of considerable exploration and debate. In the military field, the weakness of nuclear submarines lies in their continual need to cool the reactors, with about 70 per cent of the reactors' heat being dissipated into the ambient water. Being of lower density, this warm water rises to the surface and is then liable to detection by thermal imaging. Noise reduction is also a never-ending goal of submarine design, and reactors need pumps for the circulation of their cooling waters. Nevertheless, given the apparent commitment from many countries to continue or expand their current military programmes, nuclear-powered submarines will continue to ply the oceans for many decades to come.

The prospects for commercial vessels are less certain. It is inevitable that the burning of fossil fuels (particularly of the relatively 'dirty' bunker fuels) by marine commercial vessels will come under increasing scrutiny and attention. The world's merchant shipping fleet currently has a total power capacity in excess of 400 GW (thermal).

There are clearly considerable attractions of nuclear power for certain types of ships, apart from those operating in polar seas. Some large bulk carriers have similar requirements to aircraft carriers in that they are at sea for long periods of time and, in some cases, have considerable energy needs in addition to those necessary for propulsion – such as is the case for transporting liquid natural gas in large tankers. Cruise liners might also be candidates, although public reassurance would undoubtedly be an issue. From a purely commercial point of view, however, the economic realities of nuclear-powered commercial vessels are unlikely to improve until more effort, commitment and realism are invested in the use of nuclear power on land, thereby creating the infrastructure and facilities around the world necessary to serve them.

There is, however, no doubt that an interest in nuclear-powered ships remains, and designs for them continue to emerge (Hirdaris *et al.*, 2014). Should such projects come to fruition, then further consideration would also have to be given to the international framework within which their safety, in terms of impact on humans and the environment, would have to be assessed with respect to the ports and harbours that would receive them. This is an issue that has never been satisfactorily addressed. For military vessels, it is generally understood that warships of one state have no right of entry to the waters of another, and thus such visits are subject to the consent of the host, on the basis of whatever terms might be agreed. It is unlikely that sensitive technical information relating to the visiting vessel would be divulged in order to obtain entry and, indeed, the concerns of the visitor with regard to the adequacy of the host port (particularly bearing in mind the limited manoeuvrability of vessels such as single prop submarines in confined spaces) may well be paramount. But for civilian vessels, the situation is quite different.

The basic starting point for such vessels is the set of conventions drawn up regarding nuclear merchant ships (IGMCO, 1980; IMO, 1982) that essentially require that the host government of a port visited by a nuclear-powered vessel be given a detailed safety assessment and, potentially, any other information regarding the design, construction and operation of the vessel so that it is then able to make its own evaluation of the risks and consequences of receiving the vessel. A convention on the liability of operators of nuclear ships was drawn up a long time ago, but never came into effect (IMC, 1962). It aimed to ensure that the operator of any visiting nuclear ship would be absolutely liable for any damage caused by an incident involving the nuclear fuel or wastes arising from it. The failure to agree to and implement such conventions may well have helped limit the commercial viability of actual or intended nuclear-powered vessel projects in the past. If such conventions are to be dusted off and implemented in the future, then a more detailed and up-to-date incorporation of radiological protection criteria for both humans and the environment would be necessary, but this should not prove to be a barrier to the use of such vessels in the future.

References

AMAP (1998). *AMAP Assessment Report: Arctic Pollution Issues*. Oslo: Arctic Monitoring and Assessment Programme.

EA, FSA, FSS, NRW, NIEA & SEPA (2017). *Radioactivity in Food and the Environment, 2016. RIFE-22*. Lowestoft: CEFAS.

Hirdaris, S. E., Cheng, Y. F., Shallcross, P. *et al.* (2014). Considerations on the potential use of nuclear small modular reactor (SMR) technology for merchant marine propulsion, *Ocean Engineering* **79**, 101–130.

IAEA (1988). *Assessing the Impact of Deep Sea Disposal of Low Level Radioactive Waste on Living Marine Resources*. Technical Reports Series No. 288. Vienna: International Atomic Energy Agency.

IAEA (1997). *Predicted Radioactive Release from Marine Reactors Dumped in the Kara Sea*. IAEA-TECDOC-938. Vienna: International Atomic Energy Agency.

IAEA (1998). *Radiological Conditions of the Western Kara Sea*. Report of the International Arctic Seas Assessment Project (IASAP). Vienna: International Atomic Energy Agency.

IAEA (1999). *Radioactivity in the Arctic Seas*. IAEA-TECDOC-1075. Vienna: International Atomic Energy Agency.

ICRP (2007). *The 2007 Recommendations of the International Commission on Radiological Protection*. ICRP Publication 103, Ann. ICRP 37 (2–4). Ottawa: International Commission on Radiological Protection.

ICRP (2008). *Environmental Protection: The Concept and Use of Reference Animals and Plants*. ICRP Publication 108. Ann. ICRP 38 (4-6). Ottawa: International Commission on Radiological Protection.

ICRP (2014). *Protection of the Environment under Different Exposure Situations*. ICRP Publication 124. Ann. ICRP 43 (1). Ottawa: International Commission on Radiological Protection.

IGMCO (1980). *Safety Recommendations on the Use of Ports by Nuclear Merchant Ships*. London: Inter-Governmental Maritime Consultative Organisation and the International Atomic Energy Agency.

IMC (1962) *Brussels' Convention, 1962: Liability of Operators of Nuclear Ships. Convention on the Liability of Operators of Nuclear Ships and Additional Protocol*. Brussels: Diplomatic Conference on Maritime Law, International Maritime Committee.

IMO (1972). *International Conference on the Convention of Dumping Wastes at Sea, London, Final Act of the Conference with Attachment Including the Convention on the Prevention of Marine Pollution by Dumping of Wastes and Other Matter*. London: International Maritime Organization.

IMO (1982). *Code of Safety for Nuclear Merchant Ships*. London: International Maritime Organization.

IMO (1993). *Resolution LC.51(16): Report of the Sixteenth Consultative Meeting of the Contracting Parties to the Convention on the Prevention of Marine Pollution by Dumping of Wastes and Other Matter, LC 16/14*. London: International Maritime Organization.

Pentreath, R. J. & Woodhead, D. S. (1988). Towards the development of criteria for the protection of marine fauna in relation to the disposal of radioactive wastes into the sea. In: *Radiation Protection in Nuclear Energy*. Proc Conf, IAEA-CN-51, IAEA, Vienna, Vol. 2, 213–243. https://inis.iaea.org/search/search.aspx?orig_q=RN:20026319

US ONR (2014). *Occupational Radiation Exposure from Naval Eeactors' DOE Naval Nuclear Propulsion Programme*. Report NT-14-3 2014. Washington, DC: Office of Naval Reactors.

Woodhead, D. S. (1979). Methods of dosimetry for aquatic organisms. In: *Methodology for Assessing Impacts of Radioactivity on Aquatic Ecosystems*. Technical Reports Series No. 190. Vienna: International Atomic Energy Agency, pp. 43–96.

Woodhead, D. S. (1986). The radiation exposure of blackheaded gulls (*Larus ridibundus*) in the Ravenglass Estuary, Cumbria, UK: a preliminary assessment. *Science of the Total Environment* **58**, 273–281.

12

Environmental Impacts of Shipbreaking

M. MARUF HOSSAIN

12.1 Background

Ships are mobile structures of comprehensive size and consist mostly of steel. At the end of their active life (20–30 years of operation), they become a sought-after source of steel. As stated by the NGO Shipbreaking Platform (NGOSP) (2015), out of 768 recently dismantled ships, 469 ships were beached. Shipping companies sell old ships in return for a final profit: about 90 per cent of a ship's structure is made of steel, which is recovered during the demolition process and provides several millions USD of profit for the owner – the amount depending on the size and type of vessel (NGOSP, 2015). Obsolete vessels available for scrapping may also represent a useful source of supply for second-hand equipment and components. The very nature of vessels represents risks both to the environment and in terms of general safety aspects (OSHA, 2001) in the dismantling context. The considerable dimensions, their mobility and the presence of materials and substances, both those integrated in the structure as well as those required for operation, are all factors contributing to such risks. There are at present no international regulatory standards relating to shipbreaking. As a result of this inconsistency, practices and procedures for decommissioning and dismantling, which are in grave breach of basic environmental and human health protection norms, have been adopted in many countries.

Resource recovery of the material stream, sometimes referred to as the waste stream, generated from the dismantling process can be put to good use in a significant way. Usable equipment and components, electrical devices (radios, computers, televisions, etc.), life-saving equipment (life buoys, survival suits, rafts, etc.), sanitary equipment, compressors, pumps, motors, valves, generators and so on can all be reused for alternative applications and the scrap steel structures reprocessed. Steel production from scrap, in comparison to that from ore, offers considerable savings in energy consumption. From this perspective, shipbreaking may be claimed to comply with the principles of sustainability, even though there may be some discrepancy between areas of application (BCS, 2003).

12.2 Shipbreaking Activities around the World

After about 20–30 years of operation, a seagoing vessel reaches the end of its operational life. These ships represent a scrap volume of around 20.4 million gross tonnage (GT). The importance of shipbreaking around the world is well illustrated in Figure 12.1.

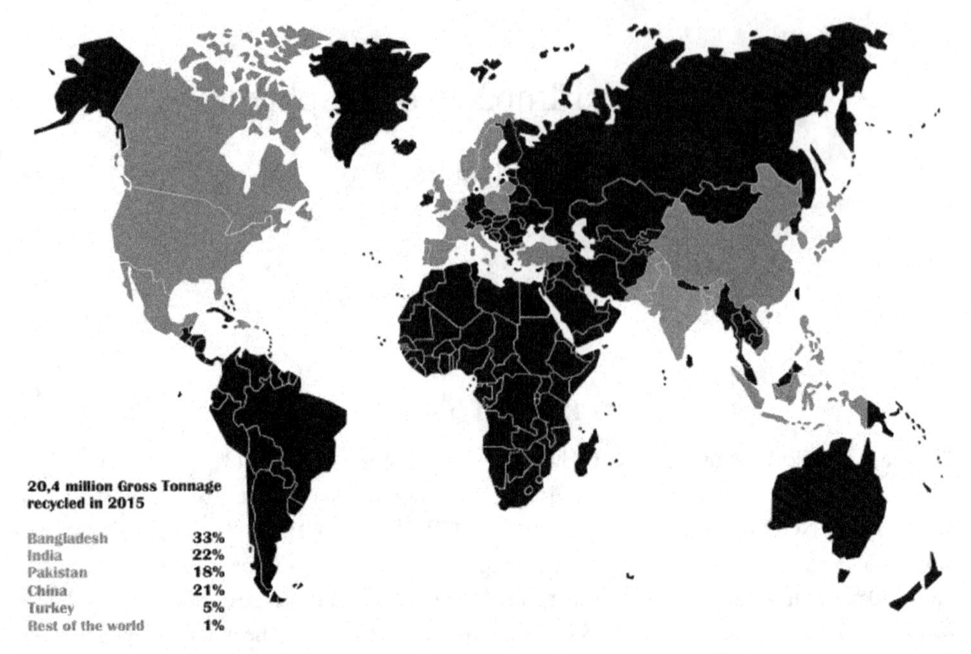

Figure 12.1 Shipbreaking activities around the world in 2015.
Source: NGOSP (2016)

After the 1980s, South Asia became the preferred dumping ground for shipowners looking to make the highest possible profit from end-of-life sales: out of the total 768 large commercial vessels dismantled around the world in 2015, 469 were sold to South Asian beaching yards. Therefore, beaching yards represented 73 per cent of the total tonnage scrapped. The rest of these old ships were recycled in facilities using more developed amenities, such as those applying the pier-side demolition method. For the first time in many years, Bangladesh once again became the preferred scrapping destination worldwide and overtook India in terms of tonnage dismantled, which previously represented the largest shipbreaking country. Thus, the country with the worst shipbreaking practices was able to strike the most end-of-life deals – this is very revealing of the shipowner community that has been promoting itself as environmentally friendly and concerned about the negative impacts of shipbreaking (NGOSP, 2015). The distribution channels of scrapped items as observed for a typical end-of-life ship can be seen in Figure 12.2.

Until the 1960s, shipbreaking activity was considered to be a highly mechanized operation that was concentrated in the industrialized countries – mainly the USA, the UK, Germany and Italy. The UK accounted for 50 per cent of the industry – Scotland ran the largest shipbreaking operation in the world. During the 1960s and 1970s, shipbreaking activity migrated to semi-industrialized countries, such as Spain, Turkey and Taiwan, mainly due to the availability of cheaper labour and the existence of a re-rolled steel market (Hossain & Islam, 2006). Tightening environmental regulations resulted in increased costs of hazardous waste disposal in industrialized and semi-industrialized

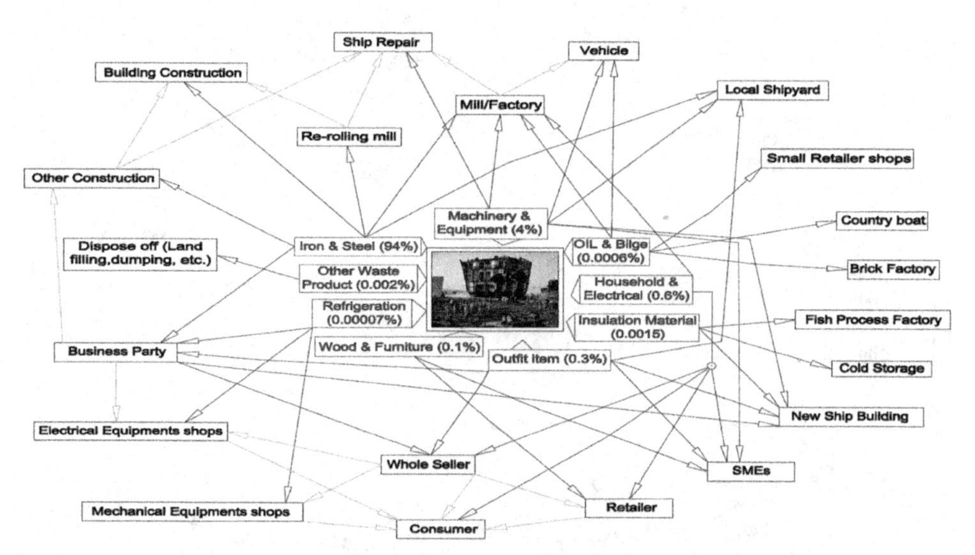

Figure 12.2 Distribution channel model of scrapped items from a typical end-of-life ship as observed fron ground observation at Chattogram Ship Recycling Yard, Bangladesh (2018). *Source:* Modification of Hossain (2010)

countries in the 1980s. Subsequently, this led to 'toxic traders' exporting hazardous waste to developing countries where environmental regulations were less stringent (BCS, 2003).

The global shipping industry depends heavily on ship scrapping and shipbreaking enterprises in the developing world to dispose of deep-sea vessels. For a developing country with relatively little industrial capacity to meet domestic demand for industrial products, deep-sea ships are almost entirely salvaged for materials ranging from steel to engines to toilets (Rousmaniere & Raj, 2007).

Most vessels scrapped in 2015 were bulk carriers, followed by general cargo and container ships, oil and gas tankers, roll-on/roll-offs (Ro-Ro), passenger vessels and oil platforms. Looking at the size of vessels scrapped on the beaches of South Asia, Pakistan received the largest vessels, followed by Bangladesh, while Indian yards scrapped more medium-sized ships. China and Turkey tend to recycle smaller vessels. Thus, the larger the vessel, the more likely it is to end up on a beach in Pakistan or Bangladesh, where conditions are worst in terms of environmental protection and provisions for health and safety of staff (NGOSP, 2015).

More than 70 per cent of the end-of-life ships sold for dismantling today end up in South Asia, the region that has served as the main destination for obsolete tonnage in the last two decades (NGOSP, 2016). The end-of-life vessels are run up on the tidal shores of India, Bangladesh and Pakistan, where they are dismantled mainly manually by a migrant workforce. The beaching method is at the source of severe coastal pollution and dangerous working conditions. Moreover, shipbreaking takes place in blatant violation of international hazardous waste management laws. These laws set out strict requirements for the transboundary movement and remediation of toxic substances.

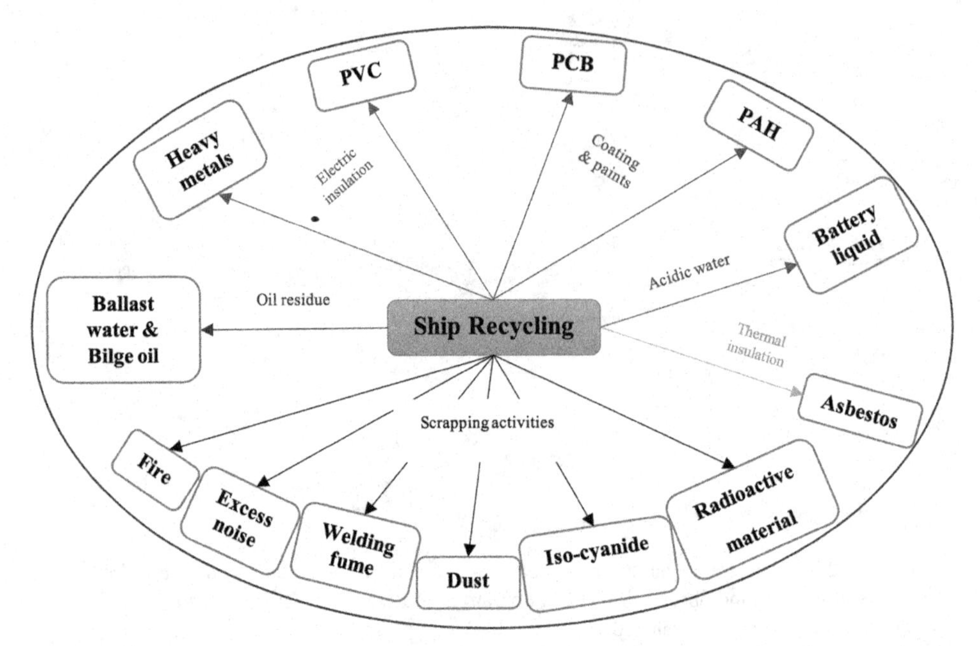

Figure 12.3 Major pollutants from ship recycling activities. PAH = polycyclic aromatic hydrocarbon; PCB = polychlorinated biphenyl; PVC = polyvinyl chloride.
Source: Local Environmental Consultants (2016)

China and Turkey are the two other main destinations for end-of-life ships. Concerns related to proper hazardous waste management, environmental protection and labour rights also exist in some yards located in these countries. Facilities operating in the rest of the world account for only 3 per cent of the GT scrapped globally every year (NGOSP, 2016).

Considering all of these issues, drawing a clear picture of the pollutants from ship recycling activities along would be difficult without a detailed pollutant profiling study of shipbreaking yards. However, the major pollutants from ship recycling activities can be seen in Figure 12.3.

12.2.1 Bangladesh

The shipbreaking yards are located in the south-east of the country, just outside the major port city of Chittagong. They stretch along the coastline of the Sitakunda area for approximately 15 km. During the 1960s and 1970s, locals started breaking down vessels that were wrecked on their shores. Shipbreaking first grew into an industry in the 1980s; however, it was not officially recognized as an industry by the Government of Bangladesh (and thus not regulated) until 2011. Some of the shipbreaking yards are set up on privately owned land, while the majority of companies lease government-owned land. Many of the shipbreaking workers come from the north-west of Bangladesh, a poverty-stricken area with very little industrial activity, while others hail from the greater Chittagong area and neighbouring areas in the south-east of the country. Dirty and dangerous shipbreaking practices in Bangladesh have been strongly criticized both by international and local non-governmental organizations (NGOs) for many years for the severe pollution caused to the marine environment, hazardous waste dumping, abysmal working conditions with many severe and fatal accidents, as well as for the widespread illegal exploitation of child workers. The import ban on hazardous materials on board of end-of-life vessels is simply circumvented by fake certificates declaring a vessel free of toxic materials, which in turn are accepted without contestation by local authorities. Some yards have set up waste storage facilities, although many of them are not used at all. Even if waste is properly stored, the Chittagong area is still devoid of a waste treatment facility and a landfill for hazardous wastes such as asbestos. Therefore, proper downstream waste management is absent, and hazardous materials are either dumped or resold.

The beaching method applied in Bangladesh raises serious concerns regarding the ineffective containment of pollutants and the contamination of the coast when dismantling a vessel on the tidal mudflat. Workers are usually not provided with proper and hygienic accommodation. They work long hours without extra pay or paid holidays and usually do not have any contractual arrangement with the yard management. In 2015, at least 16 workers died and more than 20 were severely injured. The yards do not employ the workers themselves, but work with contractors that bring in the workforce and deal with all related matters, including the payment of wages. Most workers are untrained migrant workers, unskilled to work safely in a hazardous industry that is a dangerous source of accidents. The Bangladesh Shipbreaking Association (BSBA) has set up a basic training facility; however, there is no proof of training being provided, and the workers do not receive certificates. According to IndustriALL Global Trade Union (2015), the workers earn between 1.50 and 4.20 euros a day, which amounts to a monthly wage of about 45–126 euros if they work every day. The living wage in Bangladesh is estimated at around 260 euros per month (NGOSP, 2016).

12.2.2 India

The shipbreaking yards in Alang are located in the state of Gujarat around 50 km by road from the port city of Bhavnagar. They were initially set up in 1983 and stretch along a 10-km-long tidal beach. Alang is the world's largest shipbreaking site. The Gujarat Maritime Board (GMB), a semi-public institution running all ports in Gujarat, is the main regulatory authority of the shipbreaking yards of Alang. GMB leases out shipbreaking plots to the shipbreakers on a 10-year lease basis. Most of the workers are migrant workers coming from poorer, less industrialized areas such as Uttar Pradesh, Orissa and Bihar. Shipbreaking practices in Alang have been under the spotlight of NGOs for more than 20 years. Appalling working conditions and pollution events were first documented by Greenpeace in 1998. Following cases filed by NGOs, the Indian Supreme Court has issued several rulings demanding improvement of the industry in line with national and international requirements for safe working conditions and environmental protection. Local trade unions reported that at least six workers died in accidents at shipbreaking yards in

India in 2015; however, the authorities do not publically disclose the accident record, and many more workers have suffered from injuries and occupational diseases. Yard owners have never been held responsible for the deaths of workers and have managed to put pressure on law enforcers to drop their charges. According to IndustriALL Global Trade Union (2015), the workers earn between 59 euros (unskilled workers) and 119 euros (skilled workers), while the living wage in India is around 195 euros per month. In the second half of 2015, the Indian shipbreaking industry in particular experienced a major downturn regarding the number of vessels dismantled due to depreciation of the rupee combined with cheap steel imports from China. The number of active yards in Alang fell to below 50 in 2015 from more than 100 in 2014 according to the Ship Recycling Industries Association of India (SRIA). India has thus currently lost its position as the number one shipbreaking destination and many workers were laid off (NGOSP, 2016).

12.2.3 Pakistan

The Pakistani shipbreaking yards are located on the Arabian Gulf, 50 km to the west of the country's largest city, Karachi. They stretch out along several kilometres of coast. Most of the plots are leased from private landowners, while a few others are operated on government-owned land. Most of the shipbreaking workers are migrant workers from the poorest parts of Pakistan, mainly from Khyber Pakhtunkhwa (the violence-stricken North-West Frontier Province). Most of the workers have to leave their families behind, as the Gadani yards do not offer appropriate housing and other social amenities for family members. For many years, shipbreaking practices in Pakistan have not been prominent on the radar of the international media. In recent years, both international and local NGOs, as well as trade unions, have been campaigning for better working conditions and environmental protection. The shipbreaking yards operate directly on the beach without any impermeable and drained working areas protecting the sea and sand from pollution. Gadani and the region do not have access to any kind of hazardous waste management system, resulting in hazardous waste, such as asbestos, simply being dumped behind the shipbreaking area. Local activists have filed a complaint under the Balochistan Environmental Protection Act, which clearly demands that shipbreaking must be in line with the requirements of environmentally sound management of hazardous waste as defined by the Basel Convention. The beaching method applied in Pakistan raises serious concerns regarding the containment of pollutants and the contamination of the coast. Moreover, the use of cranes to lift blocks directly onto drained working areas is not possible without further investment in infrastructure. Workers are killed in fires and explosions, by falling from great heights or when hit by large steel blocks. The yards do not employ the workers themselves, but work with contractors bringing in the workforce and dealing with all related matters, including the payment of wages. Most workers are untrained migrant workers, unskilled to work safely in a hazardous industry. According to IndustriALL Global Trade Union (2015), the workers currently earn between 2.70 and 6.00 euros per day; that is, between 80 and 180 euros per month if they work every day, whereas the living wage is estimated at around 218 euros per month (NGOSP, 2016).

12.2.4 China

The Chinese ship recycling yards are located in three different areas of the country. The two main ship recycling clusters are to be found along the Pearl River in Xinhui and around Shanghai (both along the Yangtze River and on Zhoushan Island). In addition, Dalian in the north of China hosts a ship recycling facility. Most of the companies are privately operated. The total workforce directly employed in Chinese ship recycling yards is not known. Parts of the workforce are made up of local workers, such as in the yards along the Yangtze River; other yards, such as those on Zhoushan Island, also use migrant workers. Shipbreaking practices in China first reached the radar of environmental organizations in the late 1990s. Shipbreaking yards had been active and had grown in China since the early 1980s, and by the mid-1990s around half of the world's obsolete tonnage was broken in China every year. Workers were not protected when removing asbestos, which was also found lying around in the yards. Moreover, the researchers found that explosions and fires regularly injured workers. In later years, the Chinese yards were not able to compete with the very competitive prices offered by shipbreaking yards in South Asia, where wages were even lower and environmental and social standards even weaker. The Chinese government introduced stricter environmental protection laws and banned the beaching method. In the early 2000s, Chinese companies invested in modern ship recycling facilities. In the last 10 years, China has developed into one of the preferred destinations for shipowners, mainly from Europe, seeking clean and safe recycling. Not all yards are adequately equipped and trained for the fully adequate removal of hazardous wastes, in particular asbestos. Asbestos is widely used in China, and yards do not necessarily guarantee that all asbestos-containing materials will be adequately disposed. In addition, the fate of other waste streams is not easy to track. Last but not least, there are no publicly available data on environmental impact assessments and environmental controls regarding, for instance, the possible contamination of the sea or river water, air or sediments. All of these concerns must be addressed for the yards to be accepted on the European Union (EU) List of Approved Ship Recycling Facilities (NGOSP, 2016).

12.2.5 Turkey

Ship recycling in Turkey is carried out in an organized industrial zone that is state-owned and leased out to private companies. The zone is located in Aliaga, around 50 km north of Izmir on the Aegean coast. The ship recycling facilities are located on a peninsula that hosts a large cluster of heavy industries. The ship recycling zone was first established by a government decree in 1976. Most of the workers are migrant workers from eastern Turkey around Tokat and Sivas. Since 2009, the Turkish ship recyclers and government have continued to improve practices in Aliaga regarding both environmental and social standards. The yards have opened their doors to independent researchers, consultants and experts. Moreover, the cooperation with European governments to dismantle obsolete navy vessels has further helped to improve practices. Some Turkish yards have joined the International Ship Recycler's Association (ISRA), which guarantees the quality of its members. Despite these improvements, several concerns persist that need to be addressed for the yards' application to the EU List of Ship Recycling Facilities. The Turkish yards have seen at least 11 fatal accidents in the last 5 years – an accident rate that needs to be brought down urgently. Moreover, there is no independent trade union to represent the workers. The Turkish ship recycling yards apply the so-called landing method. The bow of the vessel is grounded on the shore while the stern is still afloat. The blocks are then lifted by cranes onto a drained and impermeable working area. Therefore, the yard does not resort to the gravity method – that is, dropping blocks into the water or onto the beach; however, the environmental impact of the landing method has still to be scrutinized closely (NGOSP, 2016).

12.2.6 Rest of the World

Outside the five main ship recycling countries, dismantling facilities can also be found in Europe, North America and other Asian countries (i.e., Indonesia, Japan, the Philippines, South Korea and Vietnam). In the last 4 years, facilities located outside the main shipbreaking destinations scrapped approximately 115 vessels.

In Europe, there are several state-of-the-art yards that are either exclusively or partially involved with ship recycling. Currently, the most active dismantling operators can be found in Belgium, Denmark and the UK. Other facilities are located, for instance, in Spain, France, Italy, Portugal, Lithuania, Norway and Poland. European yards are commonly equipped with slipways and/or dry-docks and the peer-side breaking method is widely used. The majority of facilities recycle small- and medium-sized vessels, while some deal with navy ships only. Despite the current market trends, several European yards have scrapped large commercial vessels as well. Not all recycling operators located in Europe, however, offer the same high-quality service, and there have been cases of EU facilities that did not meet environmental standards. In this respect, the NGOSP is committed to monitoring the industry's performance and raising concerns whenever necessary.

In North America and the Caribbean, most of the recycling yards are located in the USA, Canada, the Dominican Republic, Mexico and Puerto Rico. There are inactive facilities in Ecuador and Curacao.

In Canada, the industry focuses predominantly on government-owned vessels and smaller vessels in the Great Lakes areas and has a significant underutilized capacity.

In Mexico, only in recent years has the industry been revived with the development of yards capable of dismantling medium or large seagoing vessels on both the Gulf Coast and the Pacific Coast; however, the environmental standards as well as downstream hazardous waste management in the Mexican yards need to be scrutinized closely by all shipowners seeking to cooperate with a yard.

In the USA, environmentally safe and sound practices have been developed and have been heavily influenced by the industry's primary client, the US Federal Government. The US ship recycling industry has focused primarily on the domestic inventory, but the current US Maritime Administration (MARAD) and Navy inventory will likely be disposed of soon. This opens up significant capacity for foreign flagged ships seeking environmentally sound and safe recycling. The yards are located on all three coastlines of the continental USA, with the facilities in Brownsville, Texas, being the most active ones. Ship dismantling operations are conducted in either graving dry-docks or alongside in slipways (NGOSP, 2016).

12.3 Shipbreaking Techniques

The decommissioning process is entirely different in developed and developing countries. Both start with an auction for which the highest bidder wins the contract (Chavdar, 2015). The shipbreaker then acquires the vessel from the international broker who deals in outdated ships (Peter, 2014). The price paid is normally about US$400 per ton (up to US$4–10 million), and the poorer the environmental legislation, the higher the price (Isabell *et al.*, 2013). The purchase of watercraft makes up 69 per cent of the income earned by the industry in Bangladesh, versus 2 per cent for labour costs (Sarraf *et al.*, 2010). The boat is taken to the decommissioning location either under its own power or with the use of tugs.

12.3.1 Developing Countries

In developing countries, chiefly the Indian subcontinent, ships are run ashore on gently sloping sand tidal beaches at high tide so that they can be accessed for dismantling. During the second phase, cutters and their helpers start cutting the vessel into parts. The breaking operation is undertaken based on the structural design of the vessel. The larger parts are dragged to the dry part of the shore with the help of motorized pulleys. A large number of workers are also engaged in this operation. Although the motor does the main job, workers need to help the pulley driver in dragging the part to the dry area of the shore.

Another group of cutters, helpers and workers start cutting the dragged parts of the ship into truck-able parts as per order of the purchasers. Heavy pieces of equipment such as boilers, motors, capstan stocking, etc., are carried to stack yards by moving crane. Unskilled workers carry metal plates, metal bars or pipes on their heads or shoulders and start walking in synchronized step to the rhythm of singers who call out a destination for the parts. The workers then pile up metal plates in stack yards or load them on trucks. The supervisors control the group of workers; the onlooker guides them and helps them in piling up the heavy metal plates in stacks.

The ship is cut down into different pieces and winched to the shore at high tide. Further large portions are cut into suitable pieces on the beach for easier loading and transportation. The valuable components (e.g., small motors and pumps, generators, navigation equipment, life-saving equipment, furniture, electrical cables, utensils, etc.) are dismantled and sold to second-hand markets. It takes 5–6 months to dismantle a typical cargo ship (Hossain & Islam, 2006). Stockpiled in Bangladesh, for example, are 79,000 tons of asbestos,

240,000 tons of polychlorinated biphenyls (PCBs) and 210,000 tons of ozone-depleting substances (Sarraf *et al.*, 2010).

12.3.2 Developed Countries

In developed countries, the dismantling process should mirror the technical guidelines for the environmentally sound management of the full and partial dismantling of ships, published by the Basel Convention (2003). Recycling rates of 98 per cent can be achieved in these facilities (BBC, 2010b).

Prior to dismantling, an inventory of dangerous substances should be compiled. All hazardous materials and liquids, such as bilge water, are removed before shipbreaking. Holes should be bored for ventilation and all flammable vapours extracted.

Vessels are initially taken to a dry-dock or a pier, although a dry-dock is considered more environmentally friendly because all spillage is contained and can easily be cleaned up. Floating is, however, cheaper than a dry-dock (Frey, 2015). Storm water discharge facilities will stop an overflow of toxic liquid into the waterways. The carrier is then secured to ensure its stability (Frey, 2015). Often the propeller is removed beforehand to allow the watercraft to be moved into shallower water (Frey, 2015).

Workers must completely strip the ship down to a bare hull, with objects cut free using saws, grinders, abrasive cutting wheels, handheld shears and plasma and gas torches (Frey, 2015). Anything of value, such as spare parts and electronic equipment, is sold for reuse, although labour costs mean that low-value items are not economical to sell. The Basel Convention demands that all yards separate hazardous and non-hazardous waste and have appropriate storage units, and this must be done before the hull is cut up. Asbestos, found in the engine room, is isolated and stored in custom-made plastic wrapping prior to being placed in secure steel containers, which are then landfilled (BBC, 2010b).

Many hazardous wastes can be recycled into new products. Examples include lead–acid batteries or electronic circuit boards. Another commonly used treatment is cement-based solidification and stabilization. Cement kilns are used because they can treat a range of hazardous wastes by improving their physical characteristics and decreasing the toxicity and transmission of contaminants. Hazardous waste may also be 'destroyed' by incinerating it at a high temperature; flammable wastes can sometimes be burned as energy sources. Some hazardous waste types may be eliminated using pyrolysis in a high-temperature electrical arc under inert conditions to avoid combustion. This treatment method may be preferable to high-temperature incineration in some circumstances, such as in the destruction of concentrated organic waste types, including PCBs, pesticides and other persistent organic pollutants (POPs). Dangerous chemicals can also be permanently stored in landfills as long as leaching is prevented.

Valuable metals, such as copper in electric cables, that are mixed with other materials may be recovered through the use of shredders and separators in the same fashion as e-waste recycling. The shredders cut the electronics into metallic and non-metallic pieces. Metals are extracted using magnetic separators, air flotation separator columns, shaker tables or eddy currents. The plastic almost always contains regulated hazardous waste (e.g., asbestos, PCBs, hydrocarbons) and cannot be melted down (Frey, 2015).

Large objects, such as engine parts, are extracted and sold as they become accessible (Frey, 2015). The hull is cut into 300-tonne sections, starting with the upper deck and working slowly downwards. While oxy-acetylene gas torches are most commonly used, detonation charges can quickly remove large sections of the hull. These sections are transported to a blast furnace to be melted down into new ferrous products, although toxic paint must be stripped prior to heating (BBC, 2010a).

12.4 Pollutants Discharged from Shipbreaking and Their Impacts

An ocean-going vessel is a miniature version of a city, and during scrapping discharges, every kind of pollutant that a metropolis can generate, such as liquid, metal, gaseous and solid pollutants, can be produced. So, shipbreaking activities became perilous in respect of the environment, human health and biodiversity.

Sarraf *et al.* (2010) investigated hazardous materials that were found on ships during ship dismantling on the basis of two categories of ships: naval and non-naval (merchant) vessels. Naval warships are built to different specifications than merchant vessels, particularly with regards to the management of risk of fire and explosions, and thus have a higher occurrence of hazardous materials. In contrast, the extent of hazardous materials in merchant vessels can vary as they come in many types, including liquefied gas carriers, dry bulk, reefers, oil tankers, Ro-Ro and containers. Such design types may affect which hazardous materials are typically found on-board. But due to the limited availability of data, it was not possible to consider such detail in this study. The results for merchant vessels are therefore presented on the basis of a measure of cargo capacity – that is, GT as recommended by the International Maritime Organization (IMO) – and represent 14 inventories, 3 of which were undertaken as part of this study. The results for naval vessels are based on displacement, which here equalled light displacement tons, a common measure of the steel weight of a vessel. No inspections were made of naval vessels during the study, and the Inventory of Hazardous Materials data sets include 13 vessels. The data are presented in Tables 12.1 and 12.2.

12.4.1 Fuel Oil Seepage

Oil is a highly volatile compound composed of mainly hydrocarbons (75 per cent) and sulfur-containing compounds. As a result of shipbreaking, oil residues and the other refuse are often spilled, where they mix with soil and water in the beach. Islam and Hossain (1986) detected between 10.800 and 9.280 mg/L of oil in water samples, and Khan (1994) detected between 239 and 248 μg/L of petroleum hydrocarbons (PHCs) in seawater from shipbreaking areas.

Oil may cause serious damage to a range of biota in different ways. Seabirds or diving birds spend much of their time on or near shore. They become oiled when feeding in

Table 12.1. *Amount of hazardous material per million GT on merchant and navy vessels (Sarraf* et al., *2010)*

		Merchant vessels[a]		Navy vessels[a]	
Hazardous materials	Unit	Material/ million GT	Panamax tanker 40,000 GT	Material/million light displacement tons	Destroyer class 5000 LDT
Asbestos	Ton	510	20	17,000	86
PCBs					
PCB liquids (transformers, etc.)	kg	0	0	No information available	
PCB solids (capacitors, ballasts, etc.)[b]	kg	1.7	0.07	5500	28
Hydraulic oils	Ton	110	5	1600	8
ODS					
ODS liquids (chloroflouro carbons, halons, etc.)	Ton	7	0.3	No information available	
ODS solids (e.g., polyurethane)	Ton	1800	70	No information available	
Paints					
Paints, no further information	Ton	420	17	39,000	200
Paints containing tributyltin	Ton	14	0.56	No information available	
Paints containing PCBs	Ton	No information available		No information available	
Paints containing metals	Ton	No information available		25,500	130
Heavy metals					
Cadmium (merchant), lead (naval)	Ton	1.9	0.08	34	0.17
Mercury	kg	44	1.8	75	0.38
Radioactive substances	kg	No information available		No information available	
Organic waste liquids	m^3	5650	230	1900	9
Reusable organic liquids (heavy fuel oil, diesel)	Ton	3200	130	23,000	110
Miscellaneous					
Ballast water (C-34)	Ton	60,000	2400	280,000	1400
Sewage (C-35)	m^3	660	26	No information available	
Garbage (C-42)	Ton	2.3	0.09	No information available	
Incinerator ash (C-41)	Ton	1.9	0.08	No information available	
Oily rags (C-45)	Ton	3.1	0.12	No information available	
Batteries, nickel/ cadmium	Units	170	7	No information available	

Table 12.1. (*cont.*)

Hazardous materials	Unit	Merchant vessels[a] Material/ million GT	Merchant vessels[a] Panamax tanker 40,000 GT	Navy vessels[a] Material/million light displacement tons	Navy vessels[a] Destroyer class 5000 LDT
Inorganic waste liquids (acids)	m^3	0.28	0.01	430	2
Reusable organic liquids (others)	m^3	620	25	1500	7
Equipment					
Batteries, lead (C-46)	Ton	2.2	0.09	34,000	170

[a] For both categories, an example is given for a typical vessel. The underlying Inventory of Hazardous Materials database includes 14 merchant and 13 navy vessels.
[b] Merchant vessels do not estimate PCBs in cables due to a lack of data.
ODS = ozone-depleting substances.

contaminated areas such as saltwater marshes, and consequently may die due to hypothermia (i.e., an abnormally low body temperature), drowning or poisoning. Other effects can include: damage to the liver, lungs, kidneys, intestines and other internal organs; destruction of red blood cells important for good immune response; pneumonia; reductions in the ability to reproduce; declines in the number of eggs laid; decreased fertility of eggs and thinner shell thicknesses; and disruption of normal breeding and incubating behaviours.

Regarding marine mammals, effects due to oil pollution include: hypothermia, resulting in metabolic shock; poisoning and toxic effects due to ingestion of oil; congested lungs and damaged airways; and gastrointestinal ulceration and haemorrhaging.

When oil seeps into shallow or confined waters, fish can be seriously affected and may even die. Fish eggs, larvae and juveniles are much more sensitive to damage. Eggs may not hatch or may be totally destroyed, especially when the water body is shallow. Fish absorb oil that is dissolved in water through their gills, accumulating it within the liver, stomach and gall bladder. Although they are able to cleanse themselves of contaminants within weeks of exposure, there may be a period when they are unfit for human consumption. Like fish, many molluscs, shrimp and worms have a natural ability to purify themselves of contaminants if the concentration is low or if the source has been removed. Experts are concerned about the possible poisoning effects of oil spills on sea turtles due to absorption of impurities through the skin or ingestion of contaminated food, leading to damage to the digestive tract and other internal organs, damage or irritation to airways, lungs and eyes and contamination of eggs, which may inhibit their development.

Kelp, marsh grass, mangroves and sea grasses are some of the types of marine vegetation that can be damaged by oil spills. Plants occupying the area between high- and low-tide marks can also be impacted by oil spills.

In freshwater environments, perhaps the most dangerous problem in terms of public health is the contamination of drinking water sources. Food sources, such as both marine and freshwater fish and crustaceans, may be tainted, and the consumption of tainted food may causes human health problems, as well as loss of export trade in foreign markets. Oil pollution may threaten the livelihoods of fishers.

Table 12.2. *Amounts of hazardous materials during ship dismantling around the world*

Hazardous materials	Unit	Bangladesh	India	China	Pakistan	Turkey	Rest of the world
Asbestos	Ton	3447.6	2305.2	2116.5	1902.3	550.8	66.3
PCBs							
PCB liquids (transformers, etc.)	kg	0	0	0	0	0	0
PCB solids (capacitors, ballasts, etc.)	kg	11.492	7.684	7.055	6.341	1.836	0.221
Hydraulic oils/lubricants	Ton	743.6	497.2	456.5	410.3	118.8	14.3
ODS							
ODS liquids (chloroflurocarbons, halons, etc.)	Ton	47.32	31.64	29.05	26.11	7.56	0.91
ODS solids (e.g., polyurethane)	Ton	12,168	8136	7470	6714	1944	234
Paints							
Paints, no further information	Ton	2839.2	1898.4	1743	1566.6	453.6	54.6
Paints containing tributyltin	Ton	94.64	63.28	58.1	52.22	15.12	1.82
Paints containing PCBs	Ton	No information available					
Paints containing metals	Ton	No information available					
Heavy metals							
Cadmium (merchant), lead (naval)	Ton	12.844	8.588	7.885	7.087	2.052	0.247
Mercury	kg	297.44	198.88	182.6	164.12	47.52	5.72
Radioactive substances	kg	No information available					
Organic waste liquids	m^3	38,194	25,538	23,447.5	21,074.5	6102	734.5
Reusable organic liquids (heavy fuel oil, diesel)	Ton	21,632	14,464	13,280	11,936	3456	416
Miscellaneous							
Ballast water (C-34)	Ton	405,600	271,200	249,000	223,800	64,800	7800
Sewage (C-35)	m^3	4461.6	2983.2	2739	2461.8	712.8	85.8
Garbage (C-42)	Ton	15.548	10.396	9.545	8.579	2.484	0.299
Incinerator ash (C-41)	Ton	12.844	8.588	7.885	7.087	2.052	0.247
Oily rags (C-45)	Ton	20.956	14.012	12.865	11.563	3.348	0.403
Batteries, nickel/cadmium	Units	1149.2	768.4	705.5	634.1	183.6	22.1
Inorganic waste liquids (acids)	m^3	1.8928	1.2656	1.162	1.0444	0.3024	0.0364
Reusable organic liquids (others)	m^3	4191.2	2802.4	2573	2312.6	669.6	80.6
Equipment							
Batteries, lead (C-46)	Ton	14.872	9.944	9.13	8.206	2.376	0.286

ODS = ozone-depleting substances.

Sources: Based on NGOSP (2015) and Saffar *et al.* (2010)

12.4.2 Metals

Heavy metals are found in many parts of ships, such as in paints, coatings, anodes and electrical equipment. The heavy metals described in the following subsections were discovered in old ships being scrapped by workers without the correct protective equipment. These parts are often dumped or burnt on beaches, causing widespread pollution of the area. Heavy metal concentrations in the soil samples taken by DNV (2001) from the shipbreaking area of Chittagong, except for iron (Fe), were all found to be far above the background level. The values are especially high for copper (Cu), lead (Pb) and zinc (Zn). The high concentrations were all found to reflect the heavy metals that are typically found in marine anti-fouling paints.

12.4.2.1 Mercury

Mercury, a highly toxic heavy metal, exists in various forms, including metallic mercury, as well as inorganic and organic mercury compounds. Research studies show that contamination brought about by natural and human-made activities is clearly a growing problem today. The level of mercury found in the soil of the shipbreaking area of Chittagong ranges from 0.8 to 3.0 mg/kg, where the background value is only 0.1 mg/kg in dry weight (dw) (DNV, 2001).

Impacts: This toxic heavy metal affects the nervous system. In 1956, the first recognized mercury poisoning outbreaks occurred, called 'Minamata disease'. This is a disorder of the central nervous system caused by the consumption of fish and shellfish contaminated with methylmercury. Clinical manifestation differs from inorganic mercury poisoning in which the kidneys and the renal system are damaged. Young children are most vulnerable. Long-term exposure to low levels can cause irreversible learning difficulties. Mercury can also cause mental retardation and delayed neurological and physical development (Greenpeace, 2005; Hossain, 1994; Rivera *et al.*, 2003).

12.4.2.2 Lead

Lead may enter the human system mainly through inhalation, but also through ingestion. Most of the lead that enters the human body accumulates in the bones. The concentration of lead in the soil found in the shipbreaking area of Chittagong ranges from 4232 to 5733 mg/kg, where the background value is 144 mg/kg, all in dw (DNV, 2001).

Impacts: The effects of lead upon human health have been known for a long time. Long-term exposure (greater than 14 days) to lead may cause brain and kidney damage and increased blood pressure. Young children are most vulnerable to its toxic effects. Studies on the effects of lead in children have demonstrated a relationship between exposure to lead and a variety of adverse health effects. These effects include impaired mental and physical development, decreased haem biosynthesis, elevated hearing thresholds and decreased IQ and serum levels of vitamin D. Long-term exposure to even low levels can cause irreversible learning difficulties, mental retardation and delayed neurological and physical development, and even death (ATSDR, 1998; Greenpeace, 2005).

12.4.2.3 Arsenic

The environmental problems associated with arsenic due to its toxicity are well known. Presently, arsenic is treated as a human carcinogen, and environmental agencies are pressing for stricter consumption standards. Acute exposure to inorganic arsenic has serious adverse health effects. Studies of acute toxicity show that inorganic arsenic compounds are more potent than organic ones.

Impacts: Exposure can result in lung, skin, intestine, kidney, liver or bladder cancers. It can also cause damage to blood vessels. Inflammation of nervous tissue caused by arsenic may result in paralysis. After exposure, disfiguring growths may appear on the skin (Greenpeace, 2005).

12.4.2.4 Chromium

The level of chromium found in the soil of shipbreaking area ranges from 507 to 568 mg/kg, where the background value is 144 mg/kg, all in dw (Hossain & Islam, 2006).

Impacts: A small amount of chromium (0.0007–0.003 mg/kg/day) is considered as an essential nutrient for humans that helps normal metabolism of cholesterol, glucose and fat. However, large amounts of chromium can be harmful. The hexavalent form of chromium is irritating and can cause acute adverse effects on the skin, gastrointestinal tract, liver and kidneys (EPA, 1998). Some chromium-based chemicals can cause eczema. When exposed to dust and fumes containing chromium, people may develop respiratory diseases such as lung cancer (Greenpeace, 2005).

12.4.2.5 Other Heavy Metals

Among the other heavy metals found in the shipbreaking area of Chittagong are Cu (ranges from 573 to 1211 mg/kg, background value 51 mg/kg), Hg (ranges from 0.076 to 0.266 mg/kg, background value 0.05–2.0 mg/kg), Mn (ranges from 1792 to 2321 mg/kg, background value 1363 mg/kg) and Zn (ranges from 2929 to 5888 mg/kg, background value 144 mg/kg, all in dw) (DNV, 2001).

12.4.3 Lubricants

The use of lubricants and lubricant parameters on ships is important, and their use is managed by the main engine lubricant oil system. This consists of a pump that pressurizes the oil, circulating it through a set of filters and a cooler before distribution to all of the moving components of the diesel engine (Latarche, 2017).

Lubrication is necessary to help overcome friction; lubricant oils can be from animals (whale oil), vegetables (linseed) or minerals (petroleum). All three types of lubricant have been used over the centuries to dissipate heat, reduce friction and combat wear and tear between the surfaces of two moving components.

Today, however, we have a choice of two main types of lube oil:

- Mineral lubricants – these are obtained from the processing of crude oil.
- Synthetic oils – a relatively new innovation involving the use of polyglycols or esters.

12.4.4 Asbestos

Asbestos is the name given to a group of six different fibrous minerals (amosite, chrysotile, crocidolite and the fibrous varieties of tremolite, actinolite and anthophyllite) that occur naturally in the environment. All forms of asbestos are hazardous and all can cause cancer, but amphibole forms of asbestos (with straight, needle-like fibres) are considered to be somewhat more hazardous to health than chrysotile. Such fibres cannot be identified or ruled out as being asbestos using either the naked eye or simply by looking at these fibres under a regular microscope. The most common methods of identifying asbestos fibres are by using polarized light microscopy (PLM) or transmission electron microscopy (TEM). PLM is less expensive, but TEM is more precise and can be used at lower concentrations of asbestos.

If asbestos abatement is performed, completion of the abatement is verified using visual confirmation and may also involve air sampling. Air samples are typically analysed using phase-contrast microscopy (PCM). PCM involves counting fibres on a filter using a microscope. Airborne occupational exposure limits for asbestos are based on using the PCM method.

The American Conference of Governmental Industrial Hygienists has a recommended threshold limit value for asbestos of 0.1 fibres/mL over an 8-hour shift. The Occupational Safety and Health Administration (OSHA) in the USA and occupational health and safety regulatory jurisdictions in Canada use 0.1 fibres/mL over an 8-hour shift as their exposure limits (ATSDR, 2008b).

On the shipbreaking beaches of Asia, asbestos fibres and flocks fly around in the open air. Workers take out asbestos insulation materials with their bare hands. They dry it in the sun so that they can sell it (DNV, 2001).

Asbestos fibres do not have any detectable odour or taste. They do not dissolve in water or evaporate and are resistant to heat, fire, chemical and biological degradation.

Impacts: Workers who repeatedly breathe in asbestos fibres with lengths greater than or equal to 5 μm may develop a slow build-up of scar-like tissue in the lungs and in the membrane that surrounds the lungs. This scar-like tissue does not expand and contract like normal lung tissue and so breathing becomes difficult, and it hampers the lungs' ability to exchange gases. Blood flow to the lungs may also be decreased, and this causes the heart to enlarge. This disease is called asbestosis. People with asbestosis have shortness of breath, often accompanied by a cough. This is a serious disease and can eventually lead to disability or death in people exposed to high amounts of asbestos over a long period.

Asbestos workers have increased chances of getting two principal types of cancer: cancer of the lung tissue itself and mesothelioma, a cancer of the thin membrane that surrounds the lungs and other internal organs. Mesothelioma is considered to be exclusively related to asbestos exposure. By the time it is diagnosed, it is almost always fatal. Similar to other asbestos-related diseases, mesothelioma has a long latency period of 30–40 years.

Changes in the membrane surrounding the lungs, called pleural plaques, are quite common in people occupationally exposed to asbestos and are sometimes found in people living in areas with high environmental levels of asbestos. There is also some evidence from studies of workers that breathing asbestos can increase the chances of getting cancer in other locations (e.g., the stomach, intestines, oesophagus, pancreas and kidneys).

Members of the public who are exposed to lower levels of asbestos may also have increased chances of getting cancer, but the risks are usually small and are difficult to measure directly (ATSDR, 1998). It should be noted that there is a synergistic effect between smoking and asbestos exposure, which creates an extreme susceptibility to lung cancer.

The most common diseases associated with chronic exposure to asbestos are asbestosis and mesothelioma (ATSDR, 2008a).

According to the OSHA (2014), 'There is no "safe" level of asbestos exposure for any type of asbestos fibre. Asbestos exposures as short in duration as a few days have caused mesothelioma in humans. Every occupational exposure to asbestos can cause injury or disease; every occupational exposure to asbestos contributes to the risk of getting an asbestos related disease.'

Diseases commonly associated with asbestos include:

- Asbestosis: progressive fibrosis of the lungs of varying severity, progressing to bilateral fibrosis and honeycombing of the lungs on radiological view, with symptoms including rales and wheezing. Individuals who have been exposed to asbestos at home, in the environment or at work should notify their doctors about their exposure history.
- Asbestos warts: caused when the sharp fibres lodge in the skin and are overgrown, causing benign, callus-like growths.
- Pleural plaques: discrete fibrous or partially calcified thickened areas that can be seen on X-rays of individuals exposed to asbestos. Although pleural plaques are themselves asymptomatic, in some patients they develop into pleural thickening.
- Diffuse pleural thickening: similar to above, and they can sometimes be associated with asbestosis. Usually no symptoms are shown, but if exposure is extensive, it can cause lung impairment.
- Pneumothorax: some reports have also linked the condition of pneumothorax to asbestos-related diseases.

It is important to consult a doctor particularly if the following symptoms develop: shortness of breath, wheezing or hoarseness, persistent cough that worsens over time, blood in fluid coughed up, pain or tightening in chest, difficulty swallowing, swelling of the neck or face, decreased appetite, weight loss, fatigue or anaemia (NCI, 2012).

12.4.5 Persistent Organic Pollutants

POPs are chemicals that are highly toxic, remain intact in the environment for long periods, become widely distributed geographically, bioaccumulate through the food web, accumulate in the fatty tissues of living organisms and pose a risk of causing adverse effects to human populations, wildlife and the environment. There has been a realization that these pollutants, upon exposure to the human population, can cause serious health effects ranging from increased incidence of cancers to the disruption of hormonal system. These effects have also been observed and recorded for various animal species. Developing countries are particularly vulnerable due to the often indiscriminate use and disposal of POPs. Shipbreaking activities are a potential source of lethal POPs. Exposure to POPs may cause acute, medium- or long-term impacts.

The early manifestations of acute poisoning (usually due to exposure to large quantities in a very short time) include an increase in sensitivity and a tingling sensation in the face and limbs, although giddiness, lack of coordination, tremors and mental confusion may also occur. In the case of ingestion, gastrointestinal irritation (vomiting and diarrhoea) is present. In the most severe cases of intoxication, there are muscular contractions, followed by generalized convulsions. High concentrations of these substances increase cardiac irritability and may produce cardiac arrhythmia. Coma and respiratory depression may also occur.

These effects can be observed after a single exposure to a large, non-lethal dose or after repeat exposures to doses that are usually low. The reproductive impact of POPs includes abortion, retardation of foetal growth, birth defects (teratogenesis) and an increase in mortality among the newborn of exposed mothers. Infertility and loss of libido have been reported in humans exposed to aldrin and dieldrin. Except for mirex and toxaphene, all POPs are mutagenic. All of these pesticides are considered potential carcinogens of greater or lesser strength. Chlordane, DDT, heptachlor, aldrin and dieldrin are associated with liver cancer. DDT is also associated with breast cancer and dieldrin with cancer of the adrenal glands.

Other effects that are observed after medium- or long-term exposure to POPs may include loss of appetite, weight loss, nausea, headaches, sleep alterations, signs that numerous peripheral nerves are affected, liver and renal damage, increased generation of liver enzymes (which can accelerate the metabolism of drugs and other products), cardiac arrhythmia, eye damage such as allergic conjunctivitis, blepharitis and retinal angiopathy. Changes in personality and difficulties in concentrating have been noted in people exposed to heptachlor.

Scrapped ships are infested with several POPs in particular, described in the following subsections.

12.4.5.1 Dioxins

Dioxins, which are known human carcinogens, are produced when chlorine products such as PCBs and polyvinyl chlorides (PVCs) are made or burned. Dioxins are the most toxic substances humans have ever released into the environment. Burned in open fires in shipbreaking yards, dioxins are constantly inhaled by workers. In most Western countries, the emissions of dioxins are strongly regulated. A United Nations (UN) treaty on POPs banned dioxins worldwide in 2001.

Impacts: Dioxins are carcinogenic and can suppress the immune system. Dioxins are suspected of pre-natal and post-natal effects on children's nervous systems. Among other things, dioxin has been linked to endocrine disruption, reproductive abnormalities, neurological problems and infertility in humans and animals. In addition, large amounts of chemicals called 'phthalates' are used to manufacture PVC products. A commonly used phthalate plasticizer called diethyl hexyl phthalate (DEHP) is probably a reproductive toxicant, as well as a toxicant of the liver and kidneys. Moreover, dioxin exposures to humans are associated with increased risk of severe skin lesions such as chloracne and hyperpigmentation, altered liver function and lipid metabolism, general weakness associated with drastic weight loss, changes in the activities of various liver enzymes and depression of the immune system.

12.4.5.2 Polyvinyl Chloride

Lots of equipment and materials in ships are made of PVC. PVC poses serious threats to environmental health at every stage of its existence (production, use and disposal). At the end of its life, PVC waste creates intractable disposal problems because it is expensive and unsafe to burn. It releases hazardous chemicals into groundwater and air when buried, and it is difficult and expensive to recycle. Degraded PVC releases volatile organic chemicals such as 3-ethyl-1-hexanol and 1-butanol into the air, which can cause asthma. Among the dangers when PVC burns in open fires are dioxin generation, the formation of hydrochloric acid mist and the generation of thick, choking smoke.

Impacts: PVC can have a negative impact on the environment and human health. PVC has been known to cause Raynaud's syndrome, scleroderma, cholangiocarcinoma, asthma, angiosarcoma, liver cancer, brain cancer, acroosteolysis, risks of impaired human reproduction, etc. (Steingraber, 2004).

12.4.5.3 Polycyclic Aromatic Hydrocarbons

PAHs are released during torch cutting and afterwards when paints continue to smoulder or when wastes are deliberately burnt. About 250 PAHs are known. Some harmful PAHs are naphthalene, phenanthrene, anthracene, fluoranthene, pyrene, benzo(a)anthracene, chrysene benzo(k)fluoranthene, benzo(a)pyrene, benzo(ghi)perylene and indenopyrene. The health hazards from PAHs come from directly inhaling fumes during torch cutting, smouldering of paints and burning of wastes. PAHs accumulate in dust and sediment and in the tissues of life forms. As a result, they are available for uptake through inhalation, dermal contact or via the food chain.

Impacts: Some PAHs have been shown to cause cancer in laboratory animals and also in humans following occupational exposure at high concentrations. Some 30 compounds and several hundreds of derivatives are carcinogenic. A number of PAHs have been shown to be genotoxic (i.e., they interact with the genetic material in cells). The EU Scientific Committee on Food (SCF) identified 15 PAHs that may be genotoxic and carcinogenic to humans (SCF, 2002). PAHs cause malignant tumours by interfering with enzymatic breakdown, affecting the lungs, stomach, intestines and skin.

12.4.5.4 Polychlorinated Biphenyl Compounds

PCBs are synthetic chemicals that are used as coolants and lubricants in transformers, capacitors and other electrical equipment because they do not burn easily and are good insulators. The manufacture of PCBs was stopped in the USA in 1977 because of evidence that they build up in the environment and can cause harmful effects to health. Products made before 1977 that may contain PCBs include old fluorescent lighting fixtures and electrical devices containing PCB capacitors and hydraulic oils (Latarche, 2017). During ship dismantling, some old vessels may still emit PCBs.

PCBs are found in solid and liquid forms. There is a lack of studies or reports on PCB contamination from Bangladesh, except for a study on marine shrimp and fin fish from Bangladesh by Hossain (1989). The study by Hossain (1989) on PCB contamination by $\sum$(PCB28–PCB180) in commercially important marine samples (three shrimp and three fin fish species) collected from Bangladesh ranges between undetected and 80 ng/g wet weight. Relatively higher values were observed in the shrimp *Penaeus indicus* (30 ng/g wet weight) and in the fin fish *Tenualosa ilisha* (80 ng/g wet weight). Higher concentrations were recorded from shrimps and fin fishes in the North Sea and the Mediterranean, but values similar to Hossain (1989) were found in subtropical Malaysian species (Hossain, 1989). Another analysis of soil from a steel plate reprocessing site in the Chittagong shipbreaking yards on $\sum$PCBs (seven toxic congeners: $\sum$(PCB28–PCB180)) found that the amount varied from 1.444 to 0.2 mg/kg in dw (DNV, 2001). However, over the last 10–15 years in Bangladesh, as in other developing countries, the usage of capacitors, transformers and other PCB-containing products from the dismantling of old ships and other sources has increased manyfold, and there is no legislation to control or manage the old stocks of PCB-containing products in Bangladesh, posing real danger or risk for PCB contamination of marine biodiversity (higher trophic levels) and human health in Bangladesh. Certainly, the contamination of PCBs will likely remain an issue in the years ahead (Hossain, 2002).

Impacts: PCBs are highly toxic and persistent pollutants. They bioaccumulate and are highly magnified in the fatty tissues, especially at higher trophic levels of the food chain. Exposure to PCBs has been associated with a variety of adverse health conditions. PCBs have been linked to cancer, liver damage, reproductive impairments, immune system damage and behavioural and neurological damage (Hossain, 1989, 2004; Tanabe, 1988).

12.4.6 Organotins

Organotins are nerve toxins that accumulate in the blood, liver, kidneys and brain. Some notorious organotins are tributyltin (TBT), triphenyltin, dibutyltin (DBT), dicyclohexyltin, diphenyltin and tricyclohexyltin. Analysis of soil from a steel plate reprocessing region in the Chittagong shipbreaking area found monobutyltin ranges from 1.9 to 0.72 mg/kg, DBT ranges from 2.4 to 1.38 mg/kg and TBT ranges from 25 to 17 mg/kg, all in dry weight (DNV, 1999). TBT is an aggressive biocide (kills living organisms) that has been used in anti-fouling paints since the 1970s. Concerns regarding the toxicity of these compounds (some reports describe biological effects to marine life at a concentration of 1 ng/L) have led to a worldwide ban by the IMO in 2002.

Impacts: TBT is considered to be one of the most toxic compounds in aquatic ecosystems; its impacts on marine organisms range from the subtle to the lethal. TBT is responsible for the disruption of the endocrine system of marine shellfish leading to the development of male characteristics in female marine snails. TBT also impairs the immune system of organisms. Shellfish are reported to have developed shell malformations after exposure to extremely low levels of TBT in seawater. High doses of organotins have been shown to damage the central nervous systems and reproductive mechanisms in mammals. The most widely used organotin, TBT, is an endocrine-disrupting chemical in mammals. There is evidence of organotins having an endocrine-disrupting effect in fish. Some recent studies appear to suggest that the consumption of contaminated fish poses a real threat to humans. Organotins also bioaccumulate in certain marine species, some of which are food species used by humans. Organotins have endocrine-disrupting effects in humans. Butyltins disrupt the critical functioning of human immune cells, particularly killer cells. As organotin compounds can damage human health even in small doses, in industrialized nations, legal regulations are in place to protect workers from exposure to anti-fouling paints containing TBT. Skin, eye and lung protections are mandatory for any contact work with TBT-containing paints (EPA, 2000; Greenpeace, 1999, 2001; ILO, 2003).

12.5 Fouling and Invasive Species

Invasive alien species are introduced species that become established in a new environment then proliferate and spread in ways that are destructive to human interests and natural systems. Invasive species have been identified as one of the four greatest threats to marine and coastal ecosystems, together with land-based sources of marine pollution, overexploitation of living marine resources and habitat degradation.

Biofouling algae and invertebrates may attach themselves to any boat surfaces that remain submerged, such as hulls and propellers, as well as those in intermittent or continuous contact with water, such as anchors and water pipes. Once the vessel has arrived at its destination, the biofoulers and their offspring may quickly 'jump ship' to colonize new areas, with nearby wharf pilings and breakwaters offering convenient settlement sites. Slow-moving barges and oil rigs are especially implicated in the transfer of invasive species, as they often spend lengthy periods in one location – increasing the risk of colonization by biofoulers – and are cleaned infrequently, since the increased drag and fuel consumption associated with biofouling are not critical to their operational profit margins.

Ballast water is used to provide stability to ships during transit. The ballast tanks are filled with water when the cargo holds are emptied before a journey and flushed out at the next port of call when new cargo is loaded. Suspended sediment is taken up with the water, and later settles out onto the floor of the ballast tank. It is estimated that up to 14 billion tons of ballast water are transferred globally each year, and that as many as 10,000 species of marine organisms may be present in ballast water at any given time. Most marine plants and animals have a planktonic stage in their life cycle that can easily be taken up with ballast water. The sediments commonly contain dinoflagellate cysts that can remain in a state of dormancy until they are deposited in a suitable environment, where they divide and multiply to form harmful algal blooms.

Apart from their ecological impacts, marine and coastal invasive species also have a range of economic and human health consequences.

Ecological impacts: They reduce biodiversity by displacing indigenous species through predation, competition, habitat modification and food web disturbance. They also disrupt ecological processes and compromise ecosystem services such as flood attenuation and shore protection.

Economic impacts: They cause losses due to reduced productivity and efficiency, and they also incur costs associated with their prevention and management. Biofouling poses a particular problem for shipping, mariculture infrastructure and industrial installations, while fisheries production may be hampered by operational impacts, such as gear damage, as well as a decline in target populations.

Health impacts: Ballast water is capable of transferring bacteria and viruses that may cause diseases, as well as toxic phytoplankton that form harmful algal blooms, often resulting in shellfish poisoning or allergic reactions in people. In addition, introduced animals may harbour diseases that could potentially spread to humans, and they may cause health problems such as high fever and skin irritations.

12.5.1 Some Invasive Species Introduced in New Territories through Discharged Ballast Water

The green mussel *Perna viridis* is indigenous to the tropical Indo-Pacific, but has been introduced to the Gulf of Mexico, probably either as adults attached to ships' hulls or as larvae in ballast water. First discovered in Trinidad in 1990, the mussel was later reported in Venezuela, Jamaica and Tampa Bay, Florida. Currents subsequently dispersed the larvae along Florida's Gulf Coast, while boat movements allowed it to spread to the neighbouring state of Georgia. The species has a broad salinity and temperature tolerance and quickly forms dense fouling populations on bridge pilings, jetties and buoys. Apart from concerns about its ecological impacts due to its potential to outcompete native filter-feeders for food and space, the species has a range of economic impacts. Its fouling clogs the cooling water intakes of industrial plants, reduces the fuel efficiency of boats and interferes with shell-fisheries and mariculture operations, all of which necessitate increased maintenance costs.

The cordgrass *Spartina alterniflora*, which is indigenous to the Atlantic coast of the USA, is advancing across south-eastern China after being introduced in 1979 for erosion control. It rapidly invades intertidal mudflats and channels, seriously impacting the birds, fish and invertebrates that depend on this vital habitat and inhibiting boat traffic. By increasing sedimentation rates and reducing seawater exchange, it causes further ecosystem degradation, threatening the region's important mangroves.

Management action against invasive species aims to:

- Prevent or minimize new introductions;
- Prevent or minimize their establishment and spread;
- Eradicate or control established populations of invasive species.

Since eradication is extremely difficult once a species has become established – and control is very expensive – prevention is considered the best defence against invasion. Prevention can be achieved through intervention at a number of different stages.

Pre-border interventions to prevent the introduction of invasive species include environmental impact assessments prior to the importation of alien species, permit systems for the trade or movement of listed species and appropriate ballast water management.

Border controls are designed to stop potentially invasive species at the point of entry, with officials responsible for inspecting cargo, checking compliance with permit systems and imposing quarantine periods or treatment regimes.

Post-border interventions occur once the potentially invasive species is present in the area, and they include early detection and rapid response, eradication and control and mitigation.

12.6 Conclusion

The global shipping industry heavily depends on ship scrapping and shipbreaking enterprises in the developing world. Specifically, more than 70 per cent of the total tonnage is scrapped in five major shipbreaking South Asian countries.

The decommission process is entirely different in developed and developing countries. Developed country dismantling processes follow the technical guidelines for the environmentally sound management of fully or partially dismantling ships following the Basel Convention (2003), but due to weak enforcement of environmental laws, cheap labour forces and huge demand for steel, iron, etc., as well as the absence of international regulatory standards relating to shipbreaking, Southeast Asia's shipping yards are sources of severe pollution and dangerous working conditions. Moreover, shipbreaking takes place in blatant violation of international hazardous waste management.

Therefore, in order to protect coastal and marine ecosystems, fishery resources and human health, technical and logistic support to major Southeast Asian shipbreaking countries is needed. Such support will help maintain the standards of eco-friendly shipbreaking and help us to achieve the vision and mission of sustainable global ocean management and Sustainable Development Goal 14, as set by UN to save the global ocean and its resources.

References

ATSDR. 1998. *The Nature and Extent of Lead Poisoning in Children in the United States: A Report to Congress, July 1988*. Atlanta, GA: Agency for Toxic Substances and Disease Registry.

ATSDR. 2008a. *Asbestos – Health Effects*. Atlanta, GA: Agency for Toxic Substances and Disease Registry.

ATSDR. 2008b. *Asbestos – Detecting Asbestos-related Health Problems*. Atlanta, GA: Agency for Toxic Substances and Disease Registry.

Basel Convention. 2003. *Technical Guidelines for the Environmentally Sound Management of the Full and Partial Dismantling of Ships*. Basel Convention series/SBC. Châtelaine: Secretariat of the Basel Convention.

BBC. 2010a. Praise for 'toxic' ship scrapping. http://news.bbc.co.uk/1/hi/england/tees/8439308.stm

BBC. 2010b. Able UK's TERRC yard on Teesside. http://news.bbc.co.uk/local/tees/hi/people_and_places/newsid_8147000/8147873.stm

Chavdar, C. 2015. Ship Breaking. www.cruisemapper.com/wiki/768-ship-breaking

DNV. 1999. *Decommissioning of Ships, Environmental Protection and Demolition Practices. Report No. 99-3065*. Oslo: DNV.

DNV. 2001. *Decommissioning of Ships – Environmental Standards Ship-Breaking Practices/On-Site Assessment Bangladesh – Chittagong*. Oslo: DNV.

EPA. 1998. *Chromium(III), Insoluble Salts, CASRN 16065-83-1*. Washington, DC: US Environmental Protection Agency.

EPA. 2000. *A Guide for Ship Scrappers: Tips for Regulatory Compliance*. Washington, DC: US Environmental Protection Agency.

Frey, R. S. 2015. Breaking Ships in the World System: An Analysis of two Ship Breaking Capitols, Alang -Sosiya, India and Chittagong, Bangladesh. *Journal of World System Research*, 21(1): 25–49.

Greenpeace. 1999. *Ships for Scrap II, Steel and Toxic Wastes for Asia. Worker Health & Safety and Environmental Problems at the Chang Jiang Shipbreaking Yard Operated by the China National Ship Breaking Corporation in Xiagang Near Jiangyin*. Amsterdam: Greenpeace.

Greenpeace. 2001. *Ships for Scrap III, Steel and Toxic Wastes for Asia. Findings of a Greenpeace Study on Workplace and Environmental Contamination in Alang-Sosya*. Amsterdam: Greenpeace.

Greenpeace. 2005. End of Life Ships: The Human Cost of Breaking Ships. www.fidh.org/IMG/pdf/shipbreaking2005a.pdf

Hossain, M. M. 1989. Contamination of some commercially important marine shrimp and fish species from Bangladesh by organochlorine pesticides, PCB's and total mercury. MSc thesis, Vrije University of Brussels.

Hossain, M. M. 1994. Concentration of total mercury in marine shrimps and fish species from the Bay of Bengal, Bangladesh. *'Marine Reserch' Journal of Marine Biology Pakistan*, 3(2): 27–32.

Hossain, M. M. 2002. PCB's Organic pollutants: Threatening Marine Marine Environments and Bangladesh Scenario. Presented at: 2nd International Seminar on Bangladesh Environment – 2002 (ICBEN, 2002). Dhaka, 19–21 December.

Hossain, M. M. 2004. *Sustainable Management of the Bay of Bengal Large Marine Ecosystem (BOBLME)*. National Report of BOBLME-Bangladesh. (GCP/RAS/179/WBG/179). Rome: Food and Agriculture Organization of the United Nations.

Hossain, M. M. 2010. *Ship Breaking Activities: Threat to Coastal Environment, Biodiversity and Fishermen Community in Chittagong, Bangladesh*. Chittagong: Young Power Social Action (YPSA).

Hossain, M. M. & Islam, M. M. 2006. *Ship Breaking Activities and Its Impact on the Coastal Zone of Chittagong, Bangladesh: Towards Sustainable Management*. Chittagong: Young Power Social Action (YPSA).

ILO. 2003. *Draft Guidelines on Safety and Health in Ship Breaking. Interregional Tripartite Meeting of Experts on Safety and Health in Ship Breaking for Selected Asian Countries and Turkey Bangkok, 20–27 May 2003*. Geneva: International Labour Office.

ILPI. 2016. *Shipbreaking Practices in Bangladesh, India and Pakistan: An Investor Perspective on the Human Rights and Environmental Impacts of Beaching*. Sørkedalsveien: International Law and Policy Institute.

IMO. 2002. *Recycling of Ships – Development of Guidelines on Recycling of Ships*. London: Marine Environment Protection Committee (MEPC).

IndustriALL Global Trade Union. 2015. *Global Worker*. 2. www.industriall-union.org/sites/default/files/uploads/documents/2015/GlobalWorker/global_worker_nov15_en_web.pdf

Isabell, H., Wieland, W. & Bernhard, Z. 2013. 'Booming Scrap Business: Ship-Breaking Lessons from the Exxon Valdez'. *Spiegel Online*.

Islam, K. L. and Hossain, M. M. 1986. Effect of ship scrapping on the soil and sea environment in the coastal area of Chittagong, Bangladesh. *Marine Pollution Bulletin*, 17(10): 462–463.

Khan, A. 1994. Study on oil pollution caused by ocean going vessels in Chittagong port and abandoned vessels in ship breaking area (Fauzdarhat) with reference to MARPOL 73/78. MSc dissertation (unpublished). University of Chittagong, Bangladesh.

Latarche, M. 2017. Other lubricants and greases used on a ship. https://shipinsight.com/articles/other-lubricants-and-greases-used-on-a-ship

Local Environmental Consultants. 2016. *Bangladesh Hazardous Waste Assessment Report – 2016*. London: International Maritime Organization.

NCI. 2012. *Asbestos Exposure and Cancer*. Bethesda, MD: National Cancer Institute.

NGOSP. 2015. *Annual Report 2015*. Brussels: NGO Shipbreaking Platform.

NGOSP. 2016. *Substandard Shipbreaking: A Global Challenge*. Brussels: NGO Shipbreaking Platform.

OSHA. 2001. *Ship Breaking Fact Sheet*. Washington, DC: United States Department of Labor, Occupational Safety & Health Administration.

OSHA. 2014. *Safety and Health Topics: Asbestos*. Washington, DC: United States Department of Labor, Occupational Safety & Health Administration.

Peter, G. 2014. *The Ship-Breakers*. Washington, DC: National Geographic Society.

Rivera, A. T. F., Cortes-Maramba, N. P. & Akagi, H. 2003. Health and environmental impact of mercury: past and present experience. *Journal de Physique IV (Proceedings)*, 107: 1139.

Rousmaniere, P. & Raj, N. 2007. Shipbreaking in the developing world: problems and prospects. *International Journal of Occupational and Environmental Health*, 13: 359–368.

Sarraf, M., Stuer-Lauridsen, L., Dyoulgerov, M. *et al.* 2010. *Ship Breaking and Recycling Industry in Bangladesh and Pakistan*. Report No. 58275-SAS. Washington, DC: World Bank.

SCF. 2002. Opinion of the Scientific Committee on Food on the risks to human health of polycyclic aromatic hydrocarbons in food. SCF/CS/CNTM/PAH/29 Final. Scientific Committee on Food. http://europa.eu.int/comm/food/fs/sc/scf/out153_en.pdf

Steingraber, S. 2004. Update on the Environmental Health Impacts of Polyvinyl Chloride (PVC) as a Building Material: Evidence from 2000–2004. A commentary for the U.S. Green Building Council, Healthy Building Network. https://pdfs.semanticscholar.org/b45f/a39cab5d491f5a5bd225ccc31e500ee41a2a.pdf

13

International Legislative Framework

GORANA JELIC MRCELIC, NIKOLA MANDIC AND RANKA PETRINOVIC

International environmental law is one of the most rapidly evolving branches of international law. It encompasses intergovernmental organizations (IGOs), non-governmental organizations (NGOs), international financial organizations (IFOs), associations of private-sector corporations and trade groups.

There are several sources of international law: bilateral and multilateral treaties (*conventional law*), practice of legal customs (*customary law*), general principles of law adopted in *national laws* and *judicial decisions* and *experts writings*. The new source of law comes from IGOs and other entities that produce resolutions, declarations, guidelines, etc., that have recommendatory power. It is so-called *soft law* because it is non-binding, but it can become binding over time.

Some soft law sources are ISO14000 – Environmental Management System[1] and ISO 26000 – Guidance for Social Responsibility,[2] the Institute of Environmental Management and Assessment's (IEMA) best practice standards,[3] etc.

Corporate social responsibility (CSR)[4] is a concept that signifies the integration of environmental, social and human rights awareness into the corporate business model. Today, corporations voluntarily adopt CSR policy because it brings competitive advantage. In spite of the fact that critics see it as a *greenwashing* instrument, corporations can promote environmental responsibility more efficiently than governments because they have financial and human resources at their disposal.[5]

13.1 The International Organizations Related to Shipping

13.1.1 Inter-Governmental Organizations

13.1.1.1 The International Maritime Organization

The International Maritime Organization (IMO) is a specialized agency of the United Nations (UN) responsible for maritime affairs. It was established under the Convention on the Intergovernmental Maritime Consultative Organization at the Maritime Conference of the UN in 1948 in Geneva, known then as the Inter-Governmental Maritime Consultative Organization (IMCO). Based in London, it started its activities in 1959. In 1982, it changed its name from the IMCO to the IMO, which still stands. Today, 171 countries are permanent Member States and three are associate Member States of the IMO.

The IMO's objectives are primarily technical, but also economic. The role of the organization is to consult and provide information, and its task is to discuss technical issues affecting the development of international marine merchant shipping, especially the safety of human life at sea, and to achieve the highest standards for unobstructed navigation. The IMO, among other things:

- Monitors and prevents pollution of the marine environment;
- Encourages the general acceptance of the highest applicable standards of maritime safety and efficiency of navigation;
- Provides extensive assistance to national governments, allowing the establishment of a system of cooperation between the governments in the area of regulation and practices relating to technical aspects important for shipping within international trade and the development of national fleets;
- Prevents any discrimination in international navigation.

Since it was established, the IMO has mostly been concerned with issues related to the protection of the marine environment and the prevention of marine pollution from ships. In addition, marine safety has been the focus of IMO activities since the early 2000s.

The IMO carries out a programme of technical assistance to developing countries aiming at the ratification of conventions and the achievement of the standards contained in international maritime conventions such as the International Convention for the Safety of Life at Sea (SOLAS) and the International Convention for the Prevention of Pollution from Ships (MARPOL). Within this programme, the IMO employs different advisers and consultants to advise the governments of developing countries. Every year, the IMO organizes or participates in numerous seminars, workshops and other events that are designed to assist in the implementation of their measures.

Nevertheless, the main task of the IMO is to convene international maritime conferences and to draw up international conventions. As a result of its activities so far, the IMO has adopted 50 conventions and more than 1000 resolutions and recommendations concerning maritime security and safety, pollution prevention and other related issues.[6] The IMO has played a leading role in drafting a number of very important conventions and protocols. The Comité Maritime International has consultative status with the IMO.

The IMO's governing body is the Assembly, which meets every 2 years in regular sessions when approving bi-annual budgets, technical resolutions and recommendations proposed by its other bodies. The Council is the executive body of the IMO and it is responsible, under the Assembly, for supervising the work of the organization. Between the sessions of the Assembly, the Council performs all of the functions of the Assembly, except the function of making recommendations to governments on maritime safety, which is exclusively reserved for the Assembly.

The Council consists of 40 Member States elected by the Assembly.[7] The Council's role is to: coordinate the activities of the organs of the organization; estimate budgets and draft the work programme of the Assembly; receive reports and proposals of the Committees and other organs and submit them to the Assembly and Member States, with comments and recommendations as appropriate; appoint the Secretary-General, subject to the approval of the Assembly; and enter into agreements or arrangements concerning the relationship of the organization with other organizations, subject to approval by the Assembly.

The IMO is a technical organization with most of its work done in a number of committees and subcommittees. The Maritime Safety Committee (MSC) is one of the main bodies of the IMO, together with the Assembly and the Council. The MSC is the highest technical body consisting of all Member States. The MSC deals with all matters related to the safety of ships and considers the issues of protection and maritime piracy and armed robbery on ships. More specifically, the MSC considers any matter focusing on aids to navigation, construction and equipment of vessels, manning from a safety standpoint, rules for the prevention of collisions, handling of dangerous cargoes, maritime safety procedures and requirements, hydrographic information, log books and navigational records, marine casualty investigations, salvage and rescue and any other matters directly affecting maritime safety.

The Marine Environment Protection Committee (MEPC) is responsible for coordinating the activities of the IMO in the prevention and control of pollution from ships. The MEPC consists of all Member States and deals with the prevention and control of pollution from ships, with a special emphasis on the adoption and amendment of conventions and other regulations, as well as measures to ensure their enforcement.

The MSC and MEPC are assisted in their work by subcommittees. As the incidence of marine incidents significantly differs from one country to another, certain subcommittees have been established such as those for flag state implementation or port state control.

The Legal Committee (LC) deals with legal matters within the scope of the IMO, and all Member States participate in its work. It was established in 1967 as a subsidiary body for the research of legal issues after the *Torrey Canyon* disaster, and subsequently it became a permanent committee. Since then, the IMO has largely assumed the unification of maritime law and has become an essential coordinator in coordinating and directing international instruments in the field of private international maritime law. The LC is responsible for considering any legal matter within the scope of the IMO.

The Technical Co-operation Committee (TCC) is concerned with technical cooperation projects and performs other activities in the technical cooperation field. The TCC is responsible for coordinating the work of the IMO in providing technical assistance in the field of maritime affairs and especially assisting developing countries.

The Facilitation Committee (FC) is responsible for the activities of the IMO and aims at reducing formalities and the simplification of documentation required by the procedures of ships entering and leaving ports or other terminals.

The Secretariat consists of the Secretary-General and personnel. The Secretary-General is assisted by the personnel of some 300 international civil servants. The Secretary-General is appointed by the Council, subject to the approval of the Assembly.

The IMO works through the seven subcommittees, as follows: the Sub-Committee on Human Element, Training and Watchkeeping (HTW), the Sub-Committee on Implementation of IMO Instruments (III), the Sub-Committee on Navigation, Communications and Search and Rescue (NCSR), the Sub-Committee on Pollution Prevention and Response (PPR), the Sub-Committee on Ship Design and Construction (SDC), the Sub-Committee on Ship Systems and Equipment (SSE) and the Sub-Committee on Carriage of Cargoes and Containers (CCC).

The process of a convention adoption usually begins in committees that meet more often than the Assembly. If there is consent of the committee, the proposal is referred to the Council and, if necessary, to the Assembly as well. If the Assembly or the Council agrees, the board considers the issue further and prepares the proposal text, with possible additional consideration of a subcommittee. During this process, the opinions of intergovernmental and non-governmental bodies are considered. In this way, the text of the draft convention is harmonized, and the Council and the Assembly are notified with a recommendation to convene a conference to consider the proposal for formal adoption. This is followed by invitations to the Member States and other members of the UN or one of its specialized agencies. Before the start of the conference, the draft of the convention is sent to be discussed at the conference with accompanying comments from all participants or with certain amendments so that the text is acceptable to the majority of Member States.

The conventions adopted by such procedures are deposited with the Secretary-General, who forwards the copies to governments, and the Convention is then ready for signature, usually for a period of 12 months. The signatories can ratify or accept the Convention, while the non-signatories may also have access. This is followed by the procedure of entry into force, when the Convention is formally adopted by governments and thus becomes mandatory.

Each Convention has its own procedure of entry into force, and it may include requirements that the Convention enters into force by ratification of a number of countries, sometimes with the additional criterion that these countries represent a certain percentage of the overall global fleet. In most cases, both signature and ratification are required, and accession is applied to the States that wish to become a member of the contract but have not signed it within the period open for signatures.

As to the question of amendments of individual Conventions, previously the Convention amendment process used to be slow, and thus the principle of tacit acceptance was agreed. This meant that an amendment would enter into force at a particular time unless before that date objections to the amendment are received from a specified number of Parties.

13.1.1.2 The International Labour Organization

The International Labour Organization (ILO) was established after the First World War at the Peace Conference in Paris in 1919 (Part XIII of the Versailles Peace Treaty[8]), with the task of supporting countries in the creation of employment laws and other regulations ensuring humane living and working conditions for workers, as well as the limitation of the power of employers in terms of preventing worker exploitation. Its purpose is to ensure employment by creating international agreements, better working conditions and the position of workers living in freedom, equality, security and respect for human dignity. The ILO promotes social justice and internationally recognized human and labour rights.[9] It began work on 11 April 1919 based on the ILO Constitution, which stipulates the institutional structure of the ILO and its objectives and tasks. During the Second World

War, the ILO continued its operations, and on 10 May 1944 in Philadelphia, the 26th session of the International Labour Conference was held, which resulted in the adoption of the Declaration of Philadelphia as an amendment to the Constitution. It formulated the tasks of the organization under new conditions. The treaty of 1946 enabled the union between the ILO and the UN. The ILO became the first specialized agency associated with the UN.

The ILO establishes international labour standards in the form of Conventions and Recommendations, establishing minimum standards of basic labour rights: freedom of association, the right to organize, collective bargaining, abolition of forced labour, equal opportunities and equal treatment, as well as other standards regulating conditions in a range of issues related to work. The ILO is devoted to advancing equal opportunities for women and men to obtain decent and productive work in conditions of freedom, equity, security and human dignity. It promotes rights at work and fair employment opportunities, enhances social protection and strengthens social dialogue. It can be said that the ILO was built on its fundamental constitutional principle that universal and lasting peace may be achieved only if it is based on social justice.

The ILO currently has 187 Member States.[10] In its tripartite structure, the ILO is unique among the UN organizations in which workers and employers make equal partners with governments in the creation of organization policies. In addition to the tripartite system that exists within the organization, the organization enhances its development in Member States, supporting social dialogue between trade unions and employers. In terms of the action of members of the national representations, each member is autonomous in relation to other members of the national representations in terms of commitment and vote.

The basis of the organizational structure of the ILO consists of three main permanent bodies that reflect the tripartite structure: the International Labour Conference, which represents the supreme body; the Governing Body, which has an executive office; and the International Labour Office, which is a permanent body in Geneva, headed by a Director-General.

From the very beginning, the ILO paid special attention to the maritime sector, whose activities are international by their very nature. In addition, seamen, as a special category of workers, have a place of work that at the same time is their place of living, and their living and working conditions are very different from the conditions of workers in other working environments on land. The seafarers' labour relations are unique within labour relations in general. The provisions of such specific rules (*lex specialis*) derogate the provisions of the general rules on labour relations (*lex generalis*). A number of Conventions and Recommendations on the regulation of seafarers' and stevedores' labour relations (working conditions) have been adopted through the Joint Maritime Department and the special maritime session of the International Labour Conference within the ILO. As mentioned earlier in this section, the draft conventions are proclaimed by the International Labour Conference, and the text obliges the state after it has been ratified and the ILO bodies duly notified.

Taking into account that the shipping industry is a global industry and that seafarers need special protection, in most of the laws governing the labour relations of seafarers, establishing the rights and obligations of seafarers is exempt from the exclusive autonomy of the parties in the individual employment contracts. To a large extent, these relations are regulated mostly by the imperative legal norms set out in legislation, and due to the international character of maritime industry, labour relations, and thus the contracts of employment, are regulated by a number of Conventions[11] and Recommendations.

The ILO does not have the means to compel the Member States to ratify the Convention. However, regular meetings and consultations of tripartite constituents from each of the Member States provide a forum for discussion, negotiation and demands, which can promote changes intended for a specific regulation or its acceptance by ratification.

Well-organized trade unions of traditional maritime countries play an important role in the global labour market. Many of them influence the market through their membership in the International Transport Workers' Federation (ITF) and the negotiation of the working conditions and wages of seafarers working on ships of other nationalities, especially ships flying so-called flags of convenience.

The greatest breakthrough in terms of employment relations in the maritime industry was made by adopting the Maritime Labour Convention, 2006 (hereinafter referred to as the MLC Convention). The MLC Convention was adopted at the 94th session of the ILO, held from 7 to 23 February 2006 in Geneva, as a result of a tripartite agreement: the government, the seamen's trade unions and shipowners' representatives. It was adopted in order to further improve the status (i.e., working, living and social rights) of seafarers. Because of its great importance for seafarers, the MLC Convention is called the fourth pillar of quality in the shipping industry, alongside the most important instruments of the IMO: the SOLAS, MARPOL and the International Convention on Standards of Training, Certification and Watchkeeping for Seafarers (STCW).[12] In accordance with the classification of the ILO, the MLC Convention is marked ILO 186.

The MLC Convention stipulates that it shall enter into force 12 months after being ratified by 30 states, provided that these states make up at least 33 per cent of the world's gross tonnage. Since these requirements were met in 2012, the MLC Convention entered into force in August 2013. In this way, it should ensure the full protection of the fundamental rights of all seafarers regardless of their nationality or the flag that their ships fly. So far, it has been ratified by 79 states,[13] including most European Union (EU) countries.[14]

The MLC Convention represents the consolidation and modernization of the standards defined in 68 existing Conventions (36)[15] and Recommendations (32) previously adopted (since 1920) under the auspices of the ILO. It does not introduce any new requirements, but through the obligations prescribed by the flag state and the port state inspections, the Convention allows for a more effective application of prescribed standards in relation to the previous conventions of the ILO. The Bodies of the Member States may also examine the working conditions on board ships of states that have not ratified the MLC Convention, inevitably speeding up the process of its entry into force.

The MLC Convention is structured similarly to the practice used by the IMO for its own instruments, and that means the integration of binding norms and recommendations into a single act. The MLC Convention encompasses the Preamble, Articles, Regulations and the Code.

The Articles and Rules contain the fundamental rights and principles, as well as the basic obligations of Parties to the MLC Convention, while the Code contains the particulars regarding the application of the Rules. Part A of the Code (Standards) is binding, and part B (Guidelines) contains recommendations that Member States should take into account when adopting the national regulations for MLC Convention implementation. The Regulations and the Code are divided into five chapters[16]: 'The minimum requirements for seafarers to work on a ship', 'Conditions of employment, accommodation and holidays', 'Accommodation and recreational facilities, food and catering', 'Health protection, medical care, welfare and social security' and 'Compliance and enforcement'.

Each Member shall ensure that ships that fly its flag carry a maritime labour certificate and the declaration of maritime labour compliance. The flag state is required to carry out surveillance of ships of their nationality, while the port state has the right of inspection and supervision in such a way that checks whether the ship has a certain certificate and whether the conditions on board are in accordance with the provisions of the MLC Convention. If the ship does not meet these requirements, the port state has the right to detain the ship. The port state inspections apply to all vessels over 500 gross tons, regardless of whether their country has ratified the MLC Convention.

The possibility of modifications and improvements to the provisions following MLC Convention adoption are incorporated into the structure of the MLC Convention because it is the first such act that regulates the overall work and rights in the maritime profession on a global level, and because it does not cover the rights and obligations of only one of the parties involved (seamen), but rather it is a tripartite contract that also includes employers and governments. Article XIII of the MLC Convention stipulates the establishment of a Special Tripartite Committee that deals with the permanent review of the provisions of the MLC Convention.

According to Article 19 of the Constitution of the ILO (and Article XIV of the MLC Convention), Articles and Rules may be amended only by the procedure prescribed for the adoption of conventions. However, the Code can be amended through the simplified procedure set out in Article XV of the MLC Convention. Since the Code relates to in-depth implementation, its amendments have to remain within the general scope of the Articles and Rules.

The Amendments to the Code of the MLC Convention adopted at the 103rd International Labour Conference on 11 June 2014 took effect in January 2017 (the latest 2 years and 6 months after adoption). For the first time in the history of shipping, these Amendments enabled the regulation of 'abandoned seafarers' and their financial claims through binding international legislation. Further amendments followed in 2016, from which it is clear that the process of the improving living and working conditions of seafarers has been active by way of amendments to the MCL Convention in the framework of a Special Tripartite Committee, and it will continue successfully in the future.

13.1.2 Non-Governmental Organizations

The International Chamber of Shipping (ICS), the Baltic and International Maritime Council (BIMCO), the International Association of Dry Cargo Shipowners

(INTERCARGO) and the International Association of Independent Tanker Owners (INTERTANKO) are the members of the Round Table of international shipping associations. Some other organizations related to shipping issues are: the International Association of Classification Societies (IACS), the International Association of Ports and Harbors (IAPH), the International Tanker Owners Pollution Federation (ITOPF), the International Union of Marine Insurance (IUMI), the International Ship Managers' Association (InterManager), etc.

The ICS, established in 1921 with headquarters in London, is the principal international trade association for the shipping industry, representing shipowners and operators in all sectors and trades (tanker operators, passenger ship operators, etc.). It cooperates with shipping industry organizations related to shipping, ports, insurance and classification societies (e.g., the International Group of P&I Clubs (IG) and the IACS). The ICS represents shipowners along with the various intergovernmental regulatory bodies that impact on shipping (as a consultative body), including the IMO, the ILO and the UN Division for Oceans Affairs and the Law of the Sea (DOALOS). The ICS consists of national shipowners' associations, representing over 80 per cent of the world's merchant tonnage.[17] The ICS helps national shipowners' associations and advises companies on international developments, while national shipowners' associations advise governments on ICS policy and participate in government delegations at international meetings.[18] The organization consists of the Board of Directors and five main committees: the Insurance Committee, the Maritime Law Committee, the Shipping Policy Committee, the Labour Affairs Committee and the Maritime Committee. Under the Marine Committee, there are four subcommittees (the Constructional and Equipment Sub-Committee, the Canals Sub-Committee, the Manning and Training Sub-Committee and the Environment Sub-Committee) and 14 panels (for different types of ships).

The ICS plays a significant role in the development, implementation and revision of important IMO Conventions (e.g., SOLAS, MARPOL and STCW) and Codes (the International Safety Management (ISM) and International Ship and Port Facility Security (ISPS) Codes), as well as ILO's MLC Convention.

13.2 The International Conventions Related to the Prevention of Pollution from Ships

Maritime law has been made uniform in international treaties. From the mid-nineteenth century onwards, a number of international treaties were adopted in order improve safety at sea. Today, the IMO plays the leading role in the development, adoption and revision of international treaties related to shipping.

13.2.1 The United Nations Convention on the Law of the Sea

After a long discussion and many meetings, the United Nations Convention on the Law of the Sea (UNCLOS) was signed in 1982 in Montego Bay (Jamaica), and it entered into force in 1994. It is considered one of the most important multilateral agreements in the international community after the Charter of the United Nations.

UNCLOS encompasses eight different legal regimes in different parts of the sea, the seabed and its subsoil and the air-space over the sea. The legal regimes of any of these areas depend on the rights to certain usage of the sea, such as sailing, fishing, laying of pipelines and cables, flyover aircraft, scientific research and forming artificial islands and installations within the limits of relevant areas and the usage within the exclusive rights of the coastal state or as may be the case of freedoms (rights) for the benefit of all countries, coastal and landlocked, and physical and legal persons worldwide. The listed usage on the high seas represents the rights that are to the benefit of all. By contrast, in internal waters, archipelagic waters and territorial seas, most of these uses are within the exclusive rights of the coastal state. Due to the increasing threat of pollution, coastal states have additional important competence in the protection and preservation of the marine environment. This is governed by a special Section XII of the Convention on Protection and Conservation of the Marine Environment, which contains norms that are of crucial importance not only for coastal states, but also for landlocked states, and, in fact, for the future and destiny of humankind in general.

UNCLOS consists of 17 parts that deal with all of the institutes of international law on the sea, such as the territorial sea and contiguous zone, straits used for international navigation, archipelagic states, the exclusive economic zone, continental shelf, open sea, the regimes of islands, closed or semi-closed seas, the right of access of landlocked countries to and from the sea and freedom of transit, protection and preservation of the marine environment, marine scientific research and the development and transfer of marine technology.

13.2.2 The International Convention for the Safety of Life at Sea

SOLAS was adopted in 1974 in London, and along with its numerous amendments, it is the most important and most comprehensive international instrument on the safety of navigation at sea. In 1978, the Protocol to the SOLAS Convention was adopted as part of the IMO at the International Conference on the Safety of Tankers and the Prevention of Pollution after several oil tanker accidents. The SOLAS Convention and the Protocol entered into force in 1980.

Permanent amendments to the SOLAS Convention have followed to date, occurring almost once a year. Technical innovations are rapidly increasing and they require prompt legal intervention to assist in the efficient protection of human lives. The main objective of the SOLAS Convention is to establish the minimum safety standards for the construction, equipment and operation of ships. It is the responsibility of the state of the ship that ships flying its flag comply with the requirements. Therefore, the SOLAS Convention lists the certificates to be issued as evidence of compliance with the terms of the Convention. The control or inspection provisions allow the contracting states to perform a review of compliance with the required conditions of ships of the other contracting parties in their ports if there are reasonable grounds to assume that the situation does not correspond to the prescribed requirements.

The SOLAS Convention is divided into 12 chapters, as follows: Chapter I – General Provisions, includes provisions on clearance for different types of ships and issuing certificates of compliance with the terms of the Convention, as well as provisions on the control of ships of other contracting states in the state inspection port; Chapter II – 1. Construction (rearrangement and stability, machinery and electrical appliances), 2. Design (fire protection, fire detection and firefighting); Chapter III – Means and Equipment for Salvage; Chapter IV –Radio; Chapter V – Safety of Navigation; Chapter VI – Cargo Transport; Chapter VII – Transport of Dangerous Goods; Chapter VIII – Nuclear Ships; Chapter IX – Safe Management of Ships (ISM Code); Chapter X – Safety Measures for High Speed Boats (International Code of Safety for High-Speed Craft); Chapter XI – 1. Specific Measures to Increase Safety at Sea, 2. Specific Measures to Increase the Security Protection at Sea; and Chapter XII – Additional Security Measures for the Transport of Bulk Cargo.

In addition, the SOLAS Convention contains an Appendix with certificates and annexes of Resolutions and Recommendations.

The ISM Code is the result of many years of effort at the international level to improve security throughout the whole process of the exploitation of ships and to protect the environment from pollution. The IMO Resolution of 1993 accepts the ISM Code and requires that the flag states begin to apply the new rules in accordance with these terms. In practice, the process of harmonization of existing security procedures and practices to prevent contamination with the ISM Regulations is done through shipping company developing special safety management systems.

The purpose of the ISM Code is not the creation of entirely new safety management systems for ships, but the unification and improvement of existing procedures for reaching the final goal: the safety of all operations related to the use of ships and the protection of the environment from pollution. The final result of this process is to obtain the appropriate certificates from a competent authority (i.e., a flag state). This takes the form a Document of Compliance issued to the company as evidence that the safety management system of the ship is in accordance with the ISM Code. The next step is the issue of a document called a safety management certificate, which proves that the ship carries out all of its operations in accordance with the safety management system of the ship.

13.2.3 The International Convention for the Prevention of Pollution from Ships

MARPOL was adopted in London in 1973. The aim of the MARPOL Convention was to completely eliminate intentional or accidental pollution of the marine environment from ships of all substances harmful to humans and other living things in any way by throwing, releasing, discharging, pouring out and leaking. The MARPOL Convention of 1973 was amended and extended by the Protocol in 1978, and later amended again; thus, it is more correct to specify it as MARPOL 1973/78. After the required number of ratifications, the Convention entered into force together with the Protocol in 1983.

MARPOL deals with the marine pollution from ships by oil (Annex I); the pollution by other harmful substances carried in bulk (Annex II); the pollution by harmful substances in special packaging, containers or portable tanks (Annex III); faecal contamination from ships (Annex IV); the pollution of the sea by debris from ships (Annex V); and air pollution from ships (Annex VI).

The provisions of the MARPOL Convention apply to all ships except for military and public vessels. The Convention establishes an absolute prohibition of any discharge of oil or oily mixture from a ship in all areas of the sea, including the areas of internal waters and territorial seas, as well as open sea area. In addition to this general prohibition, there are special areas that, due to their geo-climatic characteristics and special ecological exposure, require the most stringent measures and strict supervision. When it comes to oil pollution, such areas encompass the entire Mediterranean, the Baltic, the Black and the Red seas and the Persian Gulf.

13.2.4 The International Convention on Standards of Training, Certification and Watchkeeping for Seafarers

The STCW Convention was adopted in 1978 and entered into force in 1984.[19] The amendments to the Convention were adopted at the Diplomatic Conference held in 1995 in London, which came into force in 1997, and at the Diplomatic Conference held in Manila in 2010, which came into force in 2010. This is the first Convention at the international level that prescribes the basic requirements for training, certification and watchkeeping for seafarers.

The STCW Convention sets minimum requirements that oblige state parties and unifies the national regulations for seafarers' acquisition of knowledge in order to raise the general level of training of ships' crews. This is also the main purpose of the adoption of the STCW Convention due to the fact that so-called human factor is the most common cause of maritime accidents. The STCW Convention is considered, not without reason, to be one of the most important international instruments ever adopted by the IMO for the security of navigation and protection of the marine environment.

The STCW Convention consists of eight chapters, as follows: Chapter I – General Provisions; Chapter II – Master and Deck Department; Chapter III – Engine Department; Chapter IV – Radio Communication and Radio Operators; Chapter V – Special Training Requirements for Personnel on Certain Types of Ships; Chapter VI – Emergency, Occupational Safety, Medical Care and Survival Functions; Chapter VII – Alternative Certification; and Chapter VIII – Watchkeeping.

The STCW Convention adopted a certain number of Resolutions that support it. There are 23 Resolutions, which seem more like recommendations than obligations, and they expand on the issues of the Convention. In particular, the Resolutions deal with the requirements for specialized knowledge for certain types of ships, parts of the ship and the performance of duties on board.

The amendments to the STCW Convention were adopted at the Diplomatic Conference in 1995. The amendments to the Annex to the Convention and the adoption of the STCW

Code substantially modify the Convention in terms of stringency, greater precision and the obligations of its provisions. The STCW Convention and STCW Code constitute a whole in which the rules of the Annex to the Convention are expanded on in the sections of the Code. The Code consists of two parts: part A is mandatory,[20] while part B contains recommended guidance.

13.2.5 The International Convention Relating to Intervention on the High Seas in Cases of Oil Pollution Casualties

The International Convention Relating to Intervention on the High Seas in Cases of Oil Pollution Casualties (INTERVENTION) confirms the rights of coastal states to take measures on the high seas aimed at preventing, reducing or eliminating threats to the coast or other similar interests as a result of oil pollution or the risk of pollution caused by a marine incident. The INTERVENTION Convention, which was adopted in 1969 and entered into force in 1975, has undergone subsequent changes, first through the Protocol of 1973, when it expanded to substances other than oil, and then through modifications in 1991, 1996 and 2000, which complement and correct the list of substances covered by the Convention.

The leading idea of the Convention was that in order to protect the environment, there is a need to recognize coastal states' rights to take certain measures on the high seas, which are not under national jurisdiction, after due consultation with the flag state of the ship, the owner of the ship and cargo and, if circumstances permit, independent experts appointed for that purpose.

The Convention applies to all seagoing vessels other than warships and the ones owned or used by a state and intended for a non-economic purpose. The coastal state that takes measures beyond what is permitted by the Convention shall be responsible for damages caused by such measures.

13.2.6 The Convention on the Prevention of Marine Pollution by Dumping of Wastes and Other Matter

The Convention on the Prevention of Marine Pollution by Dumping of Wastes and Other Matter (London Convention) has a global character and contributes to the control and prevention of marine pollution. The London Convention was adopted in 1972 and entered into force in 1975. Since then, it has been amended several times. It prohibits the dumping of certain hazardous substances, requires the prior special permit for the dumping of certain other designated substances and previous general license for other wastes or substances.

The London Convention defines dumping as any deliberate discarding into the sea of wastes or other substances from ships, aircraft, platforms and other installations at sea. The provisions of the London Convention do not apply when it is necessary to save human life at sea or a ship due to *force majeure*. Among other requirements, a contracting state has to establish the authority that shall issue permits, keep records and monitor the state of the sea.

Furthermore, provisions encourage regional cooperation, especially in the area of control and scientific research.

Depending on the degree of harm, the London Convention classifies waste and other substances into three categories, each subject to different legal treatment.

(1) Annex I (black list) – most dangerous substances for the marine environment, whose disposal into the marine environment is completely prohibited.
(2) Annex II (grey list) – less harmful substances whose dumping requires special vigilance and each case requires a special permission for the disposal.
(3) All substances that were not listed in the Annexes can be disposed of in the marine environment, but only on the basis of prior general approval. The criteria for the issuance of general and special permits are contained in Annex III.

13.2.7 The International Convention on Oil Preparedness, Response and Cooperation

The International Convention on Oil Preparedness, Response and Cooperation (OPRC), adopted in 1990, requires Member States to establish the measures for pollution incidents at the national level or in cooperation with other countries. According to the OPRC Convention, ships should have their own contingency plans in case of oil pollution and they have to inform the coastal authorities about incidents of pollution.

While the ship is in a port, it is subject to the control of the coastal state. In addition, the state is required by the port authority to have action plans in place in case of oil pollution. The state parties to the OPRC Convention have to provide assistance to others in the event of an emergency. They undertake, individually or in cooperation with other countries, to take all appropriate measures in accordance with the provisions of the OPRC Convention and its Annex (charges of providing assistance) to prepare and act in the case of oil pollution incidents. The procedures regarding oil pollution and the receipt of reports of oil pollution are thoroughly dealt with in the OPRC Convention.

The requirements include international cooperation in measures against pollution, cooperation in the promotion and exchange of the results of research and development programmes regarding improvements in preparedness and actions in case of oil pollution and technical assistance (particularly in the training of personnel). The OPRC Convention states that the Member States shall enter into bilateral and multilateral agreements in the fields of preparedness and the implementation of actions in case of oil pollution. Copies of such agreements must be submitted to the IMO, which shall make them available to the parties upon request.

The Protocol on Convention on Oil Preparedness, Response and Co-operation in Pollution Incidents by Hazardous and Noxious Substances was adopted in 2000.

13.2.8 The International Convention for the Control and Management of Ships' Ballast Water and Sediments

Ballast water, which is one of the most dangerous pollutants today, is a common cause of pollution of the marine environment. The International Convention for the Control and

Management of Ships' Ballast Water and Sediments (BWC) was adopted on 13 February 2004 and entered into force on 8 September 2017.

The purpose of the BWC Convention is to prevent, reduce and ultimately eliminate the transfer of dangerous aquatic organisms and pathogens through monitoring and managing ballast water and sediments in accordance with the principles of international law.

The Member States shall ensure that ballast water does not cause damage to the environment, human health, property and resources of their own and other countries. The BWC Convention deals with the capacities for receiving ballast water, regulates the issues of research and control and provides a procedure for survey, certification and inspection procedures, as well as technical assistance. The Annexes to the BWC Convention, apart from general provisions, state the methods of management and control of ships and additional measures and standards for ballast water management, especially ballast water exchange with observance of the standards of treatment. Furthermore, the Annexes include: Regulation D-1 – Ballast Water Exchange Standard; Regulation D-2 – Ballast Water Performance Standard; Regulation D-4 – Prototype Ballast Water Treatment Technologies; and Regulation D-5 – Review of Ballast Water Performance Standards.

13.2.9 The International Convention on the Control of Harmful Anti-fouling Systems on Ships

In 1990, the MEPC adopted a resolution that recommended that governments adopt measures to eliminate the use of anti-fouling paint containing organotin compounds (tributyltin (TBT)) on non-aluminium-hulled vessels of less than 25 m in length and to eliminate the use of anti-fouling paints with a leaching rate of more than 4 μg TBT/day. Chapter 17 of Agenda 21 developed by the 1992 Rio Conference on Environment and Development called on states to take measures to reduce pollution caused by organotin compounds used in anti-fouling systems. In November 1999, the IMO adopted an Assembly resolution that called for a global prohibition on the application of organotin compounds in anti-fouling systems on ships by 1 January 2003, and a complete prohibition by 1 January 2008. On 5 October 2001, this instrument was adopted in London as the International Convention on the Control of Harmful Anti-fouling Systems on Ships (AFS Convention). The AFS Convention entered into force on 17 September 2008. It has been ratified by 73 states representing approximately 93 per cent of the gross tonnage of the world's merchant fleets.

Parties to the Convention are required to prohibit and/or restrict the use of harmful anti-fouling systems on ships flying their flag, as well as ships not entitled to fly their flag but that operate under their authority and all ships that enter a port, shipyard or offshore terminal of a party.[21] Anti-fouling systems to be prohibited or controlled are listed in Annex I. The process of an anti-fouling system evaluation is set in Article 6 of the AFS Convention. Article 12 states that a ship shall be entitled to compensation if it is unduly detained or delayed while undergoing inspection for possible violations of the AFS Convention.

Ships of above 400 gross tonnage and engaged in international voyages must carry an International Anti-fouling System Certificate. In order to get such a Certificate, they have to undergo an initial survey and a survey every time the anti-fouling systems are changed or replaced. Ships of 24 m or more in length but less than 400 gross tonnage engaged in international voyages must carry a Declaration on Anti-fouling Systems signed by the owner or authorized agent, accompanied by appropriate documentation (a paint receipt or contractor invoice).

Four Resolutions to the AFS Convention were adopted at the Diplomatic Conference: Resolution 1 on Early and Effective Application of the Convention; Resolution 2 on Future Work of the Organization Pertaining to the Convention; Resolution 3 on Approval and Test Methodologies for Anti-Fouling Systems on Ships; and Resolution 4 on Promotion of Technical Co-operation.

In order to ensure global and uniform application of the AFS Convention, the following documents have been developed and adopted: Guidelines for Survey and Certification of Anti-Fouling Systems on Ships – adopted by Resolution MEPC.102(48) and superseded by Resolution MEPC.195(61); Guidelines for Brief Sampling of Anti-Fouling Systems on Ships – adopted by Resolution MEPC.104(49); Guidelines for Inspection of Anti-Fouling Systems on Ships – adopted by Resolution MEPC.105(49) and superseded by Resolution MEPC.208(62); and Guidance on Best Management Practices for Removal of Anti-Fouling Coatings from Ships, including TBT Hull Paints (AFS.3/Circ.3).

13.2.10 The Hong Kong International Convention for the Safe and Environmentally Sound Recycling of Ships

The Hong Kong International Convention for the Safe and Environmentally Sound Recycling of Ships (Hong Kong Convention) was adopted at a diplomatic conference in May 2009. It was developed by the IMO, NGOs, the ILO and the parties to the Basel Convention on the Control of Transboundary Movements of Hazardous Wastes and their Disposal. It will enter into force 24 months after 15 states, representing 40 per cent of world's merchant shipping by gross tonnage, have either signed it without reservation as to ratification, acceptance or approval or have deposited instruments of ratification, acceptance, approval or accession with the Secretary-General. Furthermore, the combined maximum annual ship recycling volume of those states must, during the preceding 10 years, constitute not less than 3 per cent of their combined merchant shipping tonnage (Resolution MEPC.178(59) on the calculation of the recycling capacity for meeting the entry-into-force conditions of the Hong Kong Convention and document MEPC 64/INF.2/Rev.1).[22]

The Hong Kong Convention covers: the design, construction, operation and preparation of ships in order to facilitate safe and environmentally sound recycling; the operation of ship recycling facilities in a safe and environmentally sound manner in order to provide safe working and environmental conditions; and the establishment of an appropriate enforcement mechanism for ship recycling, incorporating certification and reporting requirements. The MEPC adopted guidelines in order to assist states in the implementation of the Hong Kong Convention,[23] as well as the Hong Kong Convention's technical standards[24] after it enters into force. Upon entry into force, ships designated for recycling will be required to carry an inventory of hazardous materials (Green Passport). A list of hazardous materials that are prohibited or restricted in shipyards, ship repair yards and ships of parties to the Hong Kong Convention is provided in an appendix to the Convention. Ships will be required to have an initial survey, renewal surveys during the life of the ship and a final survey prior to recycling in order to verify the inventory of hazardous materials. Ship recycling yards will be required to provide a ship recycling plan for each individual ship depending on its particulars and its inventory.

13.3 Flag State, Coastal State and Port State

The IMO has no powers to enforce conventions. The enforcement of IMO conventions depends upon the governments of Member Parties. Contracting governments enforce the provisions of IMO conventions as far as their own ships are concerned and also set the penalties for infringements, where these are applicable. They may also have certain limited powers in respect of the ships of other governments.[25] The IMO alone has the authority to vet the training, examination and certification procedures of Contracting Parties to the STCW. Governments have to provide relevant information to the IMO in order to decide whether or not the party meets the requirements of the Convention.

The effectiveness of marine environment protection depends on state practice in implementing the obligations. On 1 January 2016, the IMO Member State Audit Scheme became mandatory. An audited Member State should prove that it effectively administers and implements mandatory IMO instruments that are covered by the Scheme. A new module is developed within the IMO Global Integrated Shipping Information System (GISIS) through which governments can publish their reports.

The UNCLOS recognizes states competence to prescribe and apply anti-pollution laws and regulations and to set standards for vessels that meet the minimum accepted international standards. The most basic aspect of the enforcement provisions embodied in UNCLOS depends upon the relationship between the duties and responsibilities of the coastal state, flag state and port state authorities.[26]

13.3.1 Flag State

In most conventions, the flag state is primarily responsible for enforcing conventions as far as its own ships and their personnel are concerned. An inspection of ships take place within the jurisdiction of the port state, but in international waters the responsibility rests with the flag state. Many of the IMO's conventions contain provisions for ships to be inspected when they visit foreign ports to ensure that they meet IMO requirements. In addition, all ships must be surveyed and verified by officers of the flag state administrations or recognized organizations/recognized security organizations/nominated surveyors so that relevant certificates can be issued to establish that the ships are designed, constructed, maintained and managed in compliance with the requirements of IMO Conventions, Codes and other instruments.[27]

According to UNCLOS, flag states are competent to prescribe laws and regulations and to set standards for vessels flying their flags or of their registry, but it also obligates flag states to have their laws meet generally accepted international rules and standards established through the competent international organization (IMO) or general diplomatic conference (UNCLOS Art. 211(2)).[28] UNCLOS recognizes states' competence to prescribe and apply laws that shall be no less effective than the international rules, regulations and procedures (UNCLOS Art. 209(2)) for ships flaying the flag or operating under the authority or registry of state parties.

UNCLOS requires the flag state to take all measures necessary to implement international and national rules, including certification and inspection, prohibition of sailing until the ship complies with the requirements and mandatory institution of proceedings in respect of any violation of the relevant rules, regulations and standards.[29]

According to the MARPOL Convention, the flag state has a primary role in MARPOL's implementation. The coastal state and port state also have specific jurisdiction under customary law. According to Article 5 of MARPOL, the flag state shall ensure that vessels meet the requirements of the MARPOL Convention. Therefore, the flag state shall undertake periodic inspections of vessels flying its flag and supply them with appropriate documents as evidence that they comply with the relevant requirements of MARPOL. The MARPOL Convention obliges the flag state to institute proceedings if any violation of the Convention's provisions occurs on the high seas. The coastal state may institute proceedings for violation of the provisions of the MARPOL Convention committed on the territory under its jurisdiction. The port state may institute proceedings only if the violation occurs in an area under its jurisdiction.

In 1992, the IMO set up a special Sub-Committee on Flag State Implementation (FSI) to improve the performance of governments. In 2013, the Sub-Committee was renamed the Sub-Committee on Implementation of IMO Instruments (III).

The shipowner may be situated in a country other than the flag state because of the commercial advantages of some flags, but they have to be cautious as some flags do not meet international obligations. The ICS publishes the Shipping Industry Flag State Performance Table that assesses the performance of flag states using the following criteria: port state control records, ratification of Conventions, IMO Resolution A.739-recognized organizations, age of fleet, reporting requirements (STCW 95 white lists and completed ILO reports) and IMO meeting attendance. Port state control records include data from three principal regional port state control authorities: the Paris Memorandum of Understanding (MoU) white list, the Tokyo MoU white lists and the US Coast Guard (USCG) Qualship 21.[30] IMO Resolution A.739 requires flag states to establish controls over recognized organizations conducting survey work on their behalf.[31] The ICS Table is treated seriously by maritime administrations. The ICS Table shows that arbitrary distinctions between the open registers (so-called flags of convenience) and so-called traditional flag states are no longer helpful. Almost two-thirds of the world's fleet is registered with the eight largest open registers (Panama, Liberia, Marshall Islands, Singapore, Bahamas, Malta, Cyprus and Isle of Man), all of which show impressive levels of performance.[32]

13.3.2 Coastal State

UNCLOS recognizes coastal states' interest in controlling pollution in coastal waters. In territorial waters, UNCLOS authorizes coastal states to set standards for discharges but not for the construction, design, equipment and manning of ships unless such standards give effect to generally accepted international rules and standards (UNCLOS Art. 211(6)(c)).[33] The coastal states' enforcement competence is unlimited in territorial waters outside of straits used for international navigation, but it must not hamper innocent passage through territorial waters. There are certain limitations on punishment for activities in territorial seas. Imprisonment can only be imposed in the case of a wilful and serious act of pollution in territorial seas (UNCLOS Art. 230(2)). The coastal state is authorized to detain vessels for violation resulting in a discharge causing major damage or threat of major damage to the coastline, the territorial sea or the exclusive economic zone (UNCLOS Art. 220(6)).

In the exclusive economic zone, UNCLOS does not authorize the coastal state to set different standards from the international standards established by competent international organizations or general diplomatic conferences (UNCLOS Art. 21(1)(d)). UNCLOS ensures that coastal states do not abuse the power and authority given to them, causing unnecessary delay by investigations and proceedings.

13.3.3 Port State

The port states' jurisdiction in the implementation of international regulations for the prevention of pollution is covered by UNCLOS and MARPOL, global international agreements and the MoUs.

UNCLOS introduces an innovative concept of port state jurisdiction. The port state is a state in which a ship has entered voluntarily after polluting the sea area outside the state's sovereignty. Under the provisions of UNCLOS, the port state is authorized to adopt national regulations. The specific requirements that the port state has established for the entry of foreign vessels into their ports or internal waters or for a call at their offshore terminals shall be given due publicity (UNCLOS Art. 211 (3)). Pursuant to Article 218, the port state has the authority to investigate and institute proceedings in respect of any discharge from vessels that sail into its port that occurred anywhere: on the high seas, internal waters, territorial sea or exclusive economic zone of another state. The action of the port state, in the case of areas under national jurisdiction, has to be at the request of the coastal state, the flag state or the state that is damaged or threatened by the discharge. A port state may also inspect vessels in its port, and if a vessel is unseaworthy, the port state may refuse to release it or have the release made conditional upon proceeding to the nearest appropriate repair yard (UNCLOS Art. 226(1)(c)). When the port state or the coastal state institutes proceedings against a foreign vessel for violation of pollution from ships, then: it is not allowed to delay a foreign vessel longer than is essential; it shall notify the flag state immediately about the proceedings against its vessel; and the legal proceedings must be suspended if the flag state institutes proceedings for the same offense. The port state or coastal state does not have to suspend the proceedings if the offense that has caused marine pollution from ships occurred in the territorial sea, or if it caused major damage to the coastal state, or if the flag state has not instituted proceedings within 6 months of the institution of proceedings in the port state or the coastal state, or if the flag state has repeatedly disregarded its obligation to enforce effectively the applicable international rules and standards in respect of violations committed by its vessels (UNCLOS Art. 228 (1)).

The flag state has the primary role in implementing MARPOL, but it is also possible for the port state to undertake the inspection of vessels. The inspectors of the port state may inspect vessels, particularly those that have no MARPOL certificates, if there are clear grounds for believing that the condition of the vessel or its equipment does not correspond substantially to the particulars of those certificates. In this case, the port state should not allow such a vessel to continue further sailing, but it should be instructed to proceed for repairs (Art. 5). In accordance with Article 6, paragraph 5 of the MARPOL Convention, the port state may inspect a foreign vessel that enters its port or terminal to check whether the vessel has discharged any harmful substances contrary to the provisions of the Convention, if it indicates that the discharge has occurred in its internal waters or territorial sea or at the request of another state if they submit clear evidence that the ship has discharged harmful substances anywhere, even on the high seas. The port state may institute proceedings for breach of the provisions of the MARPOL Convention that took place in the area under its sovereignty, but under MARPOL and customary law, not for violation of the provisions that have occurred outside the territory under its sovereignty. In that case, the port state shall notify the flag state, which shall institute proceedings if it finds sufficient evidence for that purpose. Violation of the provisions of the MARPOL Convention is penalized under the laws of the flag state and the provisions of the state in which the violation has occurred.[34]

Port state control is the inspection of foreign ships in national ports to verify that the ships comply with international regulations. The intention of these inspections is to provide a backup to flag state implementation. The practice of some regional port state control authorities is to request information from flag states as to whether the voluntary IMO audits have been conducted in their criteria for targeting inspections.

During inspection of a foreign ship, inspectors check: whether the ship has valid documents in accordance with the provisions of international conventions; that it does not pose a risk to safety, health and the marine environment; that it has safety protection; and that the crew has a satisfactory level of knowledge for the safe operation of the ship. If the inspector has clear grounds for believing that the vessel does not observe required standards, thorough inspections may be undertaken. The port state inspection shall take appropriate measures, including detention of the vessel in case of irregularities for which the ship poses threats to the safety, health and the marine environment.

It has already been pointed out that the port state may inspect a foreign vessel that enters into its port. The inspection and control of the port state is more effective if implemented in a systematic, consistent and coordinated manner. Therefore, the IMO adopted Resolution A.682(17) on regional co-operation in the control of ships and discharges promoting the conclusion of regional agreements. In 1982, the Western European states adopted the Paris MoU on port state control. The MoU was adopted by 14 European states, and up to the present time, they have been joined by 13 other states, making a total of 27 Member States.[35] The Paris MoU aims to unify and coordinate inspection procedures at the ports of Member States of the MoU. The states parties of the MoU have committed themselves to organize an effective and coordinated control system of foreign vessels, ensuring that these vessels meet international safety at sea, security and environmental standards and that crew members have adequate living and working conditions. The MoU does not impose new standards, but aims at controlling existing standards laid down in international conventions adopted in the frameworks of the IMO and ILO, which are listed in the MoU. Port states only apply conventions to which they are state parties, regardless of whether the flag state is a party of the relevant contracts.

On 1 January 2011, the Member States of the Paris MoU entered into force the New Inspection Regime, which has the purpose of removing substandard vessels with the introduction of stricter criteria to vessels and companies that do not fulfil the requirements of international maritime conventions. One of the new measures being introduced to the new inspection regime is a permanent removal of substandard ships from the ports of the Paris MoU. Evaluation of Paris MoU statistics on inspections enables the identification of flag states whose vessels are most frequently detained, as well as the establishment of classification societies responsible for classing these vessels. Port states may prohibit such vessels from entering their ports. Currently, there are nine different agreements on port state control: Europe and the North Atlantic (Paris MoU); Asia and the Pacific (Tokyo MoU); Latin America (Acuerdo de Viña del Mar); the Caribbean (Caribbean MoU); West and Central Africa (Abuja MoU); the Black Sea region (Black Sea MoU); the Mediterranean (Mediterranean MoU); the Indian Ocean (Indian Ocean MoU); and the Gulf Region (Riyadh MoU). The USCG maintains the tenth port state control regime.

International environmental law is a system that protects common interests in the maintenance of a healthy environment. In the international environmental law regime, the weakest link is state practice implementation, and therefore many international treaties contain the obligation of states to report on filling treaty obligations, but this remains to be improved.

Notes

1 www.iso.org/iso/iso_14000_essentials
2 www.iso.org/iso/social_responsibility
3 www.iema.net
4 US Department of Commerce: www.commerce.gov
5 Nanda, VP, Pring, GW, 2013. *International Environmental Law and Policy for the 21st Century*. Leiden/Boston: Martinus Nijhoff Publishers.
6 Key IMO Conventions are: the International Convention for the Safety of Life at Sea (SOLAS), 1974, as amended; the International Convention for the Prevention of Pollution from Ships, 1973, as modified by the Protocol of 1978 relating thereto and by the Protocol of 1997 (MARPOL); and the International Convention on Standards of Training, Certification and Watchkeeping for Seafarers (STCW) as amended, including the 1995 and 2010 Manila Amendments. Other conventions relating to maritime safety and security and ship/port interface are: the Convention on the International Regulations for Preventing Collisions at Sea (COLREG), 1972; the Convention on Facilitation of International Maritime Traffic (FAL), 1965; the International

Convention on Load Lines (LL), 1966; the International Convention on Maritime Search and Rescue (SAR), 1979; the Convention for the Suppression of Unlawful Acts Against the Safety of Maritime Navigation (SUA), 1988, and the Protocol for the Suppression of Unlawful Acts Against the Safety of Fixed Platforms located on the Continental Shelf (and the 2005 Protocols); the International Convention for Safe Containers (CSC), 1972; the Convention on the International Maritime Satellite Organization (IMSO C), 1976; the Torremolinos International Convention for the Safety of Fishing Vessels (SFV), 1977, superseded by the 1993 Torremolinos Protocol; the Cape Town Agreement of 2012 on the Implementation of the Provisions of the 1993 Protocol Relating to the Torremolinos International Convention for the Safety of Fishing Vessels; the International Convention on Standards of Training, Certification and Watchkeeping for Fishing Vessel Personnel (STCW-F), 1995; the Special Trade Passenger Ships Agreement (STP), 1971; and the Protocol on Space Requirements for Special Trade Passenger Ships, 1973. Other conventions relating to the prevention of marine pollution are: the International Convention Relating to Intervention on the High Seas in Cases of Oil Pollution Casualties (INTERVENTION), 1969; the Convention on the Prevention of Marine Pollution by Dumping of Wastes and Other Matter (LC), 1972 (and the 1996 London Protocol); the International Convention on Oil Pollution Preparedness, Response and Co-operation (OPRC), 1990; the Protocol on Preparedness, Response and Co-operation to Pollution Incidents by Hazardous and Noxious Substances, 2000 (OPRC-HNS Protocol); the International Convention on the Control of Harmful Anti-fouling Systems on Ships (AFS), 2001; the International Convention for the Control and Management of Ships′ Ballast Water and Sediments, 2004; and the Hong Kong International Convention for the Safe and Environmentally Sound Recycling of Ships, 2009. Conventions covering liability and compensation are: the International Convention on Civil Liability for Oil Pollution Damage (CLC), 1969; the 1992 Protocol to the International Convention on the Establishment of an International Fund for Compensation for Oil Pollution Damage (FUND 1992); the Convention Relating to Civil Liability in the Field of Maritime Carriage of Nuclear Material (NUCLEAR), 1971; the Athens Convention Relating to the Carriage of Passengers and Their Luggage by Sea (PAL), 1974; the Convention on Limitation of Liability for Maritime Claims (LLMC), 1976; the International Convention on Liability and Compensation for Damage in Connection with the Carriage of Hazardous and Noxious Substances by Sea (HNS), 1996 (and its 2010 Protocol); the International Convention on Civil Liability for Bunker Oil Pollution Damage, 2001; and the Nairobi International Convention on the Removal of Wrecks, 2007. Other subjects are: the International Convention on Tonnage Measurement of Ships (TONNAGE), 1969; and the International Convention on Salvage (SALVAGE), 1989. The status of the Conventions adopted by the IMO is available at its official website: www.imo.org/en/About/Conventions/StatusOfConventions/Pages/Default.aspx.

7 In electing the members of the Council, the Assembly respects the criterion that it consists of 10 states with the largest interests in providing international maritime services and 10 other states with the largest interests in international seaborne trade, as well as 20 other states that have special interests in maritime transport or navigation and whose appointment has ensured the representation of all major geographical areas.

8 The Commission on International Labour Legislation was established at the first session of the Versailles Peace Conference, held on 25 January 1919. It comprised 15 members representing the states, including workers' representatives. This was the first case in diplomatic history that a diplomatic delegation of a state included workers' representatives. The task of the Commission was to propose the measures necessary to improve workers' conditions and to propose a form of a permanent organization that, in cooperation with the League of Nations and under its authority, would continue relevant research and evaluation. In the first part of the report, the Commission proposed the establishment of the ILO as a permanent international body with a task to promote and regulate the international protection of workers and working conditions. The principles for future international labour legislation were set forth in the second part of the report. The report was adopted by the Versailles Peace Conference on 11 April 1919, and the report was built into Chapter XIII of the Peace Agreement.

9 These characteristics were of great importance for its later, uninterrupted universal presence and action in international relations. Its real universality varied depending on the political and

international situation in the world. The highest degree of universality was achieved when the USA and the Soviet Union accepted its membership.

10 www.ilo.org/global/about-the-ilo

11 Minimum Age (Sea) Convention, 1920 (No. 7); Unemployment Indemnity (Shipwreck) Convention, 1920 (No. 8); Placing of Seamen Convention, 1920 (No. 9); Medical Examination of Young Persons (Sea) Convention, 1921 (No. 16); Seamen's Articles of Agreement Convention, 1926 (No. 22); Repatriation of Seamen Convention, 1926 (No. 23); Officers' Competency Certificates Convention, 1936 (No. 53); Holidays with Pay (Sea) Convention, 1936 (No. 54); Shipowners' Liability (Sick and Injured Seamen) Convention, 1936 (No. 55); Sickness Insurance (Sea) Convention, 1936 (No. 56); Hours of Work and Manning (Sea) Convention, 1936 (No. 57); Minimum Age (Sea) Convention (Revised), 1936 (No. 58); Food and Catering (Ships' Crews) Convention, 1946 (No. 68); Certification of Ships' Cooks Convention, 1946 (No. 69); Social Security (Seafarers) Convention, 1946 (No. 70); Seafarers' Pensions Convention, 1946 (No. 71); Paid Vacations (Seafarers) Convention, 1946 (No. 72); Medical Examination (Seafarers) Convention, 1946 (No. 73); Certification of Able Seamen Convention, 1946 (No. 74); Accommodation of Crews Convention, 1946 (No. 75); Wages, Hours of Work and Manning (Sea) Convention, 1946 (No. 76); Paid Vacations (Seafarers) Convention (Revised), 1949 (No. 91); Accommodation of Crews Convention (Revised), 1949 (No. 92); Wages, Hours of Work and Manning (Sea) Convention (Revised), 1949 (No.93); Seafarers' Identity Documents, 1958 (No. 108); Wages, Hours of Work and Manning (Sea) Convention (Revised), 1958 (No. 109); Accommodation of Crews (Supplementary Provisions) Convention, 1970 (No. 133); Prevention of Accidents (Seafarers) Convention, 1970 (No. 134); Continuity of Employment (Seafarers) Convention, 1976 (No. 145); Seafarers' Annual Leave with Pay Convention, 1976 (No. 146); Merchant Shipping (Minimum Standards) Convention, 1976 (No. 147); Protocol of 1996 to the Merchant Shipping (Minimum Standards) Convention, 1976 (No. 147); Seafarers' Welfare Convention, 1987 (No. 163); Health Protection and Medical Care (Seafarers) Convention, 1987 (No. 164); Social Security (Seafarers) Convention (Revised), 1987 (No. 165); Repatriation of Seafarers Convention (Revised), 1987 (No. 166); Labour Inspection (Seafarers) Convention, 1996 (No. 178); Recruitment and Placement of Seafarers Convention, 1996 (No. 179); Seafarers' Hours of Work and the Manning of Ships Convention, 1996 (No. 180); Seafarers' Identity Documents (Revised), 2003 (No. 186); Maritime Labour Convention, 2006 (No. 186); Work in Fishing Convention, 2007 (No. 188).

12 Apart from the ILO, in the drafting and adoption of the MLC Convention, the IMO was also actively involved.

13 www.ilo.org/global/about-the-ilo

14 The EU issued a recommendation (Council Decision 2997/431/EC of 7 June 2010) to its Member States to ratify the MLC Convention by the end of 2010, which speeded up its entry into force (a sufficient number of ratifications needed for entry into force was obtained).

15 Convention Nos. 7, 8, 9, 16, 22, 23, 53, 54, 55, 56, 57, 58, 68, 69, 70, 72, 73, 74, 75, 76, 91, 92, 93, 109, 133, 134, 145, 146, 147, 163, 164, 165, 166, 178, 179 and 180.

16 Each chapter contains groups of provisions relating to a particular right or principle (or enforcement measure in Chapter 5), connected by numbers. The first group in Paragraph 1, for example, consists of Regulation 1.1., Standard A1.1 and Guideline B1.1 relating to minimum age.

17 www.ics-shipping.org

18 www.ics-shipping.org/docs/default-source/about-ics/the-international-chamber-of-shipping-ics-representing-the-global-shipping-industry.pdf?sfvrsn=18

19 According to the IMO, as of 21 January 2016, the STCW Convention has been ratified by 160 countries, which represent 98.55 per cent of the gross tonnage of the world's merchant fleet.

20 Part A of the Code contains mandatory provisions particularly referred to by the Appendix to the Convention, which give the details of the minimum standards that state members (contracting states) have to secure in order to achieve its full and complete effect. It further comprises a standard on qualifications, which has to be proven by the applicants for the issuance or renewal of the certificate of competence in accordance with the provisions of the Convention.

21 www.imo.org/en/About/Conventions/ListOfConventions/Pages/International-Convention-on-the-Control-of-Harmful-Anti-fouling-Systems-on-Ships-%28AFS%29.aspx

22 www.imo.org/en/OurWork/Environment/ShipRecycling/Pages/Default.aspx
23 2012 Guidelines for the survey and certification of ships under the Hong Kong Convention, adopted by Resolution MEPC.222(64); and 2012 Guidelines for the inspection of ships under the Hong Kong Convention, adopted by Resolution MEPC.223(64).
24 2011 Guidelines for the Development of the Inventory of Hazardous Materials, adopted by Resolution MEPC.197(62); 2011 Guidelines for the Development of the Ship Recycling Plan, adopted by Resolution MEPC.196(62); 2012 Guidelines for Safe and Environmentally Sound Ship Recycling, adopted by Resolution MEPC.210(63); and 2012 Guidelines for the Authorization of Ship Recycling Facilities, adopted by Resolution MEPC.211(63).
25 www.imo.org/en/About/Conventions/Pages/Home.aspx
26 Nanda, VP, Pring, GW, 2013: *International Environmental Law and Policy for the 21st Century*, Martinus Nijhoff Publishers, Leiden/Boston, pp 665.
27 www.imo.org/en/OurWork/MSAS/Pages/PortStateControl.aspx
28 Nanda, VP, Pring, GW, 2013. *International Environmental Law and Policy for the 21st Century.* Leiden/Boston: Martinus Nijhoff Publishers.
29 Sersic, M, 2003. *Medunarodno – pravna zastita morskog okolisa.* Zagreb: Pravni fakultet Sveucilista u Zagrebu.
30 Initiative to identify high-quality ships – Preliminary Qualified Flag Administrations and USCG List of Targeted Flag Administrations.
31 www.ics-shipping.org/free-resources/flag-state-performance-table
32 www.ics-shipping.org/docs/default-source/ICS-Annual-Review-2016/ics-annual-review-2016 .pdf
33 Nanda, VP, Pring, GW, 2013. *International Environmental Law and Policy for the 21st Century.* Leiden/Boston: Martinus Nijhoff Publishers.
34 Coric, D, 2009. *Oneciscenje mora s brodova.* Rijeka: Pravni fakultet Sveucilista u Rijeci.
35 These are: Belgium, Bulgaria, Croatia, Cyprus, Denmark, Estonia, Finland, France, Germany, Greece, Iceland, Ireland, Italy, Canada, Latvia, Lithuania, Malta, Netherlands, Norway, Poland, Portugal, Romania, Russia, Slovenia, Spain, Sweden and the UK.

14

Shipping Industry's Perspective

PETER HINCHLIFFE

14.1 Shipping: Indispensable to the World

In 2016, the International Maritime Organization (IMO) chose 'Shipping: Indispensable to the World' as its annual World Maritime Day theme. That the IMO should select this is a reflection of IMO Member States' confidence in shipping's environmental performance and the need for society to reflect upon the contribution it makes to the sustainability of modern life and the functioning of the global economy. The IMO proudly displays its strap line 'safe, secure and efficient shipping on clean oceans' because it has confidence in the 'sustainability of shipping' in the true sense of the phrase.

The United Nations (UN) defines sustainable development through three fundamental pillars – environmental, social and economic. None of these pillars is paramount, and if one should fail, then global sustainability fails.

This is a time of economic crisis for the business of international shipping. In 2017, the shipping industry was struggling with the lengthy fallout from the Global Recession of 2008 and from the 'own goal' of the purchase of new ships leading to an enduring overcapacity problem at a time of slowing growth in physical demand, and so the shipping industry is perhaps at a watershed. The possible delinking of gross domestic product (GDP) growth from trade in favour of service industries suggests that shipping – already changing its character – has yet more changes to make. For example, Figure 14.1 illustrates services as a proportion of GDP in China. The services industries, which generate less demand for shipping, account for a rising share of the economy. Even in 2019 these problems have not been resolved, and shipping still struggles with very low freight rates, leading to poor profitability in the sector.

Notwithstanding the impact of these global events, the commitment of the shipping industry to the protection of the environment is unshaken. The global shipping industry fully respects the role of the IMO as its global regulator and participates actively in the development of new and revised regulation to protect the environment. The industry's focus is not on delaying regulation, but on ensuring that regulation is effective and fit for purpose.

A few key facts highlight the global significance of shipping. About 90 per cent of global trade is seaborne. Shipping transports about 10 billion tonnes of cargo a year. Shipping is the world's most environmentally efficient form of commercial transport. A large containership of 18,000 twenty-foot equivalent unit (TEU) burns just 1 g of fuel

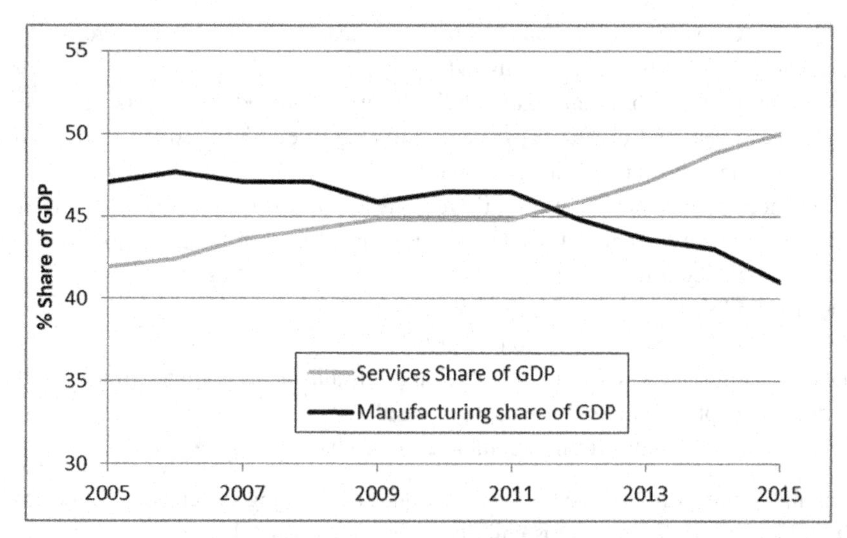

Figure 14.1 Services as a proportion of China's GDP.

per tonne of cargo carried 1 km. There are around 80,000 ships trading internationally, owned by around 10,000 companies. The sustainability of international shipping directly affects the employment of over 1.6 million seafarers[1] and of millions more working ashore in the service of maritime transport.

14.2 Sustainable Shipping

In his 2016 World Maritime Day message, UN Secretary-General Ban Ki-moon said:

> The importance of shipping in supporting and sustaining today's global society makes it indispensable to the world, and to meeting the challenge of the 2030 Agenda for Sustainable Development.

The 1987 'Report of the World Commission on Environment and Development: Our Common Future' (also known as the Brundtland Report) provided a definition for sustainable development:

> Development that meets the needs of the present without compromising the ability of future generations to meet their own needs.[2]

Such development is currently limited by technological capability, social organization of environmental resources and the ability of the biosphere to absorb greenhouse emissions. The report went on to note that the first two of these factors – technology and social organization of environmental resources – are manageable by humans when the goal is economic growth in the interest of preventing widespread poverty. This underlines that sustainable development is only possible where there is political will.

In 2015, the UN adopted 17 Sustainable Development Goals (SDGs) to be achieved in the next 15 years, and the shipping industry respects that it has an important role to play in many of these, but in particular in:

- SDG 14 – Conserve and sustainably use the oceans, seas and marine resources for sustainable development. Targets include:
  - ☐ Prevent marine pollution (particularly from land-based activities), manage and protect marine and coastal ecosystems and conserve at least 10 per cent of coastal and marine areas through international law;
  - ☐ Enhance the sustainable use of the oceans by implementing international law as reflected in the United Nations Convention on the Law of the Sea (UNCLOS) – the appropriate legal framework.
- SDG 1 – End poverty.
- SDG 2 – End hunger through food security.
- SDG 8 – Promote sustained, inclusive and sustainable economic growth, full and productive employment and decent work for all.
- SDG 13 – Take urgent action to combat climate change.

Shipping is indeed indispensable to the world, but shipping depends upon the world's need to trade, and in that respect is entirely a service industry. It provides the function of carrying cargo where there is a need to do so. For example, it serves the most remote and poorest areas of the world such as small island developing states, where maritime transport is often the only cost-effective option.

However, in carrying out this service for society and for the world economy on demand, the industry recognizes that it also has social and environmental responsibilities, as depicted in Figure 14.2.

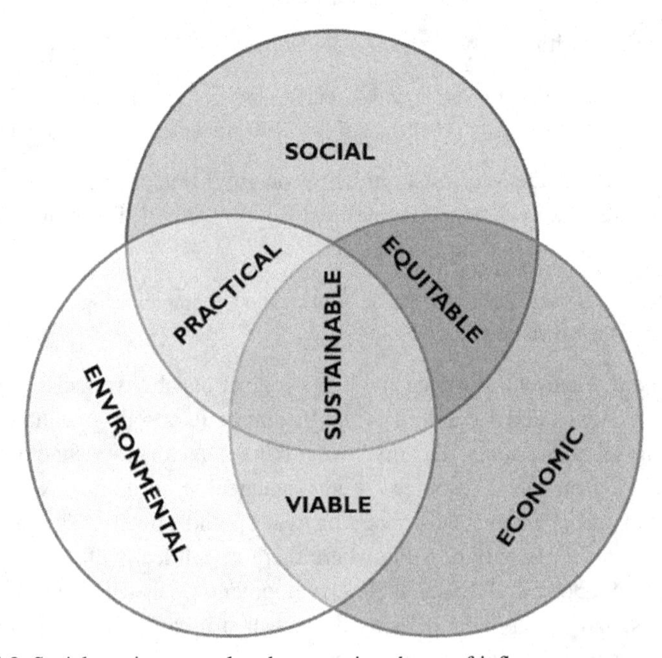

Figure 14.2 Social, environmental and economic spheres of influence.

The majority of maritime trade now serves developing countries. According to United Nations Conference on Trade and Development (UNCTAD) statistics,[3] developing countries have become major importers and exporters over the last few years, and their contribution to international seaborne trade was estimated at 60 per cent for goods loaded and 61 per cent for goods unloaded in 2014. The equivalent for developed economies was about 34 per cent for goods loaded and 38 per cent for goods unloaded. It is obvious that this means that maintaining a healthy shipping industry with low freight rates is crucial for combatting poverty and hunger by providing market access to those that need it the most. The contribution of shipping in the achievement of these goals is through facilitating the transport of goods to provide food security and well-being to those who need them the most. The impact of shipping as a global employer has already been emphasized.

It is little recognized outside the industry that virtually every potential impact that a ship may have on its environment is already regulated to some degree under international law. It is not disputed that some of the thresholds of this regulation may need to be revised, but the basic premise is sound. As a result of efficient regulation at the IMO and industry compliance and initiative, oil spills have dramatically dropped since the 1970s (Figure 14.3), and in more recent years, carbon emissions have significantly reduced per tonne-mile of cargo carried.

14.3 Shipping's View on CO_2 Emissions Regulation

The agreement reached in 2015 within the United Nations Framework Convention on Climate Change (UNFCCC) at the 21st Annual Conference of the Parties (COP21) in Paris, widely known as the 'Paris Agreement', sets out to limit global warming to less than 2°C (and aspires to less than 1.5°C) compared to pre-industrial levels. This will be done via State Party commitments known as Intended Nationally Determined Contributions (INDCs), which will be reviewed and updated every 5 years.

Shipping is still referred to in the Kyoto Protocol, which set the IMO the task of reducing greenhouse gas (GHG) emissions from the sector. Some commentators find it odd that shipping is not explicitly referred to in the Paris Agreement, but this omission reflects much more on the political drive to achieve an agreement than it implies any lessening of the obligation on shipping. In reality, the IMO is already addressing the challenge of reducing CO_2 emissions from shipping with the full support of the global industry.

It would make no sense to include shipping in INDCs committed by governments when most of the industry's emissions are produced beyond national jurisdiction and cannot be attributed to a particular country. The IMO has already adopted a mandatory package of technical and operational measures – the CO_2 amendments to Annex VI of the MARPOL Convention – that entered into force worldwide in 2013. This is the first ever global agreement outside of the UNFCCC to address the CO_2 emissions of an entire industrial sector.

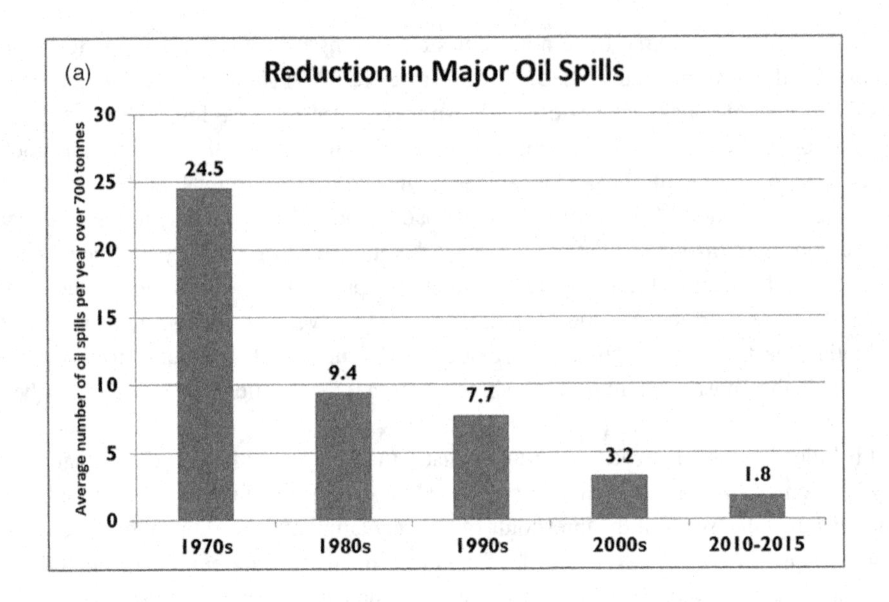

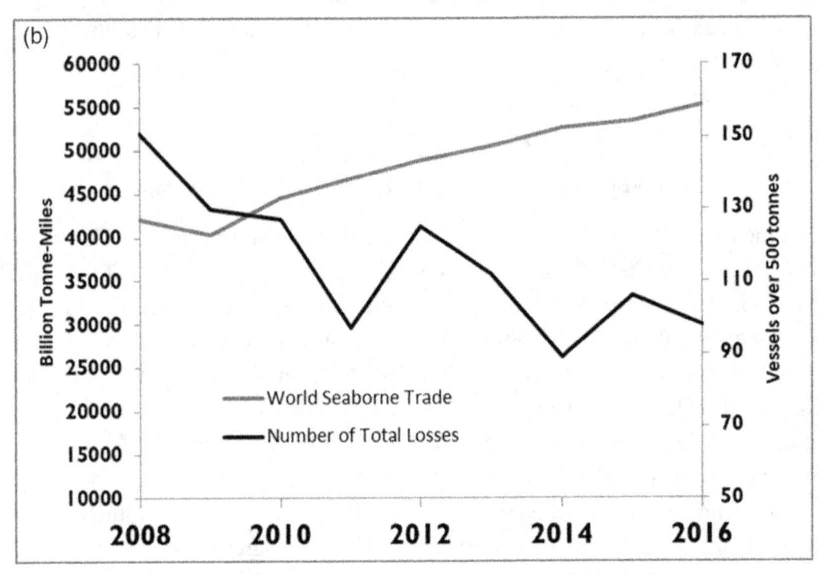

Figure 14.3a and b Improvements in the safety of shipping as evident from a decrease in major oil spills and fewer maritime causalities over time.

According to the latest available data, international shipping contributes about 2.2 per cent of the world's total CO_2 emissions.[4] The size of this figure does not indicate that shipping is environmentally hostile, but merely that the shipping industry is a large and vital industry, moving about 90 per cent of global trade. Shipping takes responsibility for its CO_2 emissions and is encouraging the IMO to establish additional measures to further limit CO_2 emissions on a global basis (Figure 14.4). In addition, the search is on for

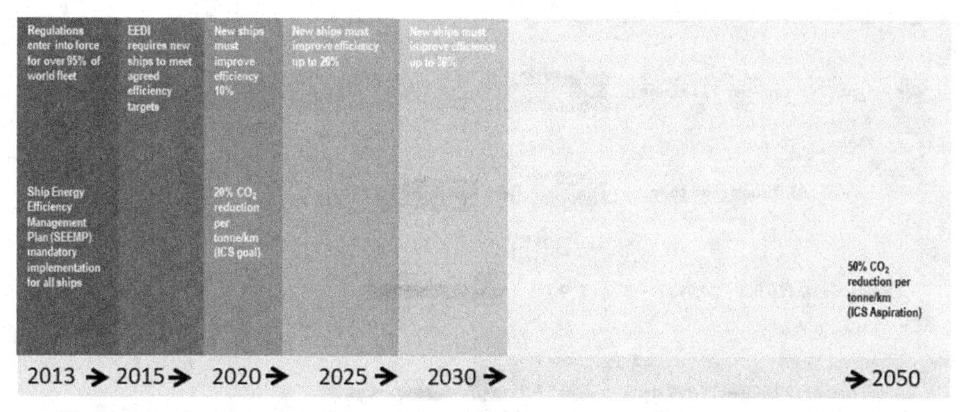

Figure 14.4 Reduction of CO_2 emissions through existing IMO agreements (MARPOL Annex VI, Chapter 4 adopted July 2011). EEDI = Energy Efficiency Design Index; ICS = International Chamber of Shipping.

advanced technologies and new fuels, as well as to improve operational efficiency through continuous research.

It may seem strange to the outsider that an industry should actively support new regulatory requirements but, put simply, global regulations are in the best interests of those that work across almost all of the world's jurisdictions. Equality of application of regulation is the leveller that makes the shipping industry competitive and cost-effective.

It is true that the shipping industry would like 'smarter' regulation that takes into account tensions between the sustainability pillars and between existing regulations and is also sustainable, but that is another story!

In terms of transport carbon efficiency, shipping is typically 30 times more efficient than air cargo and 5 times more efficient than trucks (Figure 14.5). For this reason, any regulation that would cause a shift of cargo from sea to land can readily be seen as ultimately self-defeating. There is a real need to avoid unwittingly creating an appetite for such modal shifts.

It is worth examining in a little more detail the regulations already in force: the Energy Efficiency Design Index (EEDI) and its partner regulation, the Ship Energy Efficiency Management Plan (SEEMP), which have both been in force since 1 January 2013. The EEDI is just one of the internationally enforceable measures on CO_2 reduction and is still being adjusted and expanded to tighten the requirements for new construction ships.

The EEDI requires a minimum energy efficiency level per tonne-mile to be set at the design stage and verified on initial post-construction sea trials. The level is to be tightened incrementally every 5 years and covers ship types responsible for approximately 85 per cent of the CO_2 emissions from international shipping by holistically utilizing ship design, ship equipment and main and auxiliary engines. At the same time, the SEEMP is a document that must be carried on every ship to assist the crew to monitor and improve energy efficiency in a cost-effective manner, incorporating, among others, best practices for fuel-efficient ship operation.

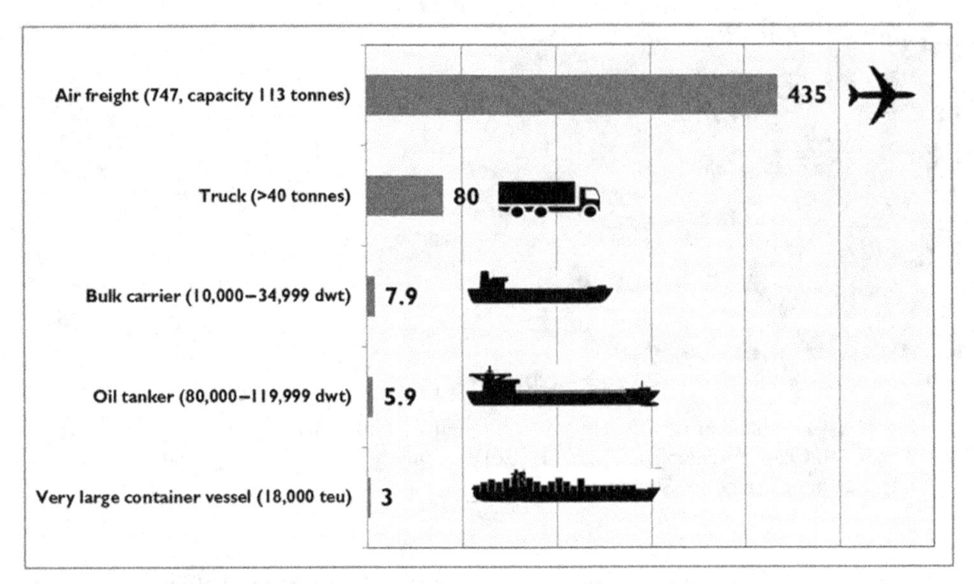

Figure 14.5 Comparison of CO_2 emissions between modes of transport.

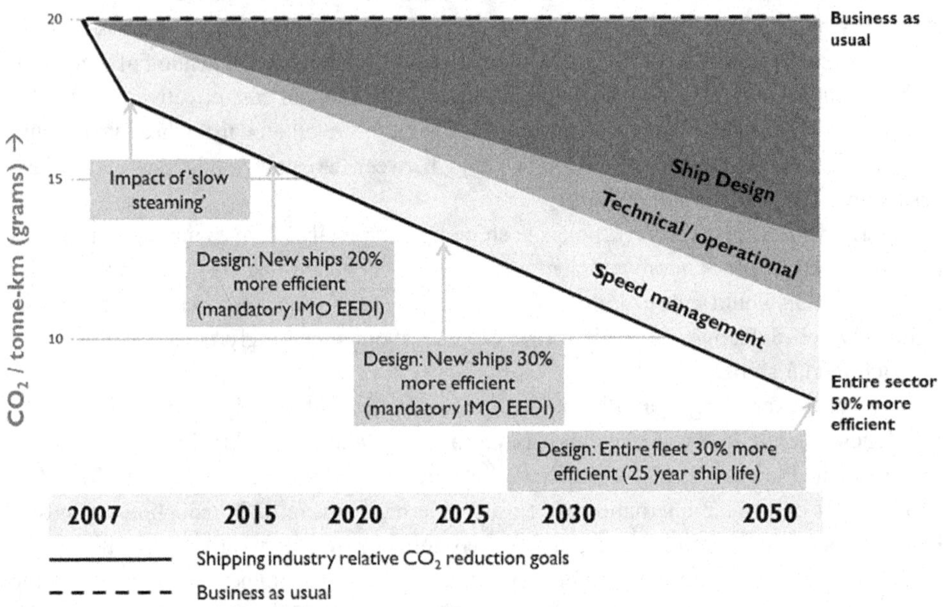

Figure 14.6 The shipping industry's CO_2 reduction goals.

As a result of the EEDI and SEEMP, ships built in 2025 will be at least 30 per cent more efficient than those constructed in the 2000s. With the introduction of additional technical and operational measures, the International Chamber of Shipping (ICS) is confident that shipping can reduce its CO_2 emissions per tonne-km by at least 50 per cent by 2050 compared to 2007 (Figure 14.6).

However, the story is not yet over, and more efficiency-based regulation is already emerging. In October 2016, the IMO adopted a mandatory system for collecting ships' fuel consumption data. Following its adoption, amendments to MARPOL Annex VI on the data collection system for the fuel oil consumption of ships entered into force on 1 March 2018. This data collection system will be the first step in a 'three-step' approach to determine shipping's contribution towards the reduction of global GHG emissions and to decide how best to limit these.

➢ **Stage 1: Data collection system**

Ships of 5000 gross tonnage and above will have to collect, *inter alia*, consumption data of each type of fuel they use and send them to their flag states at the end of each calendar year. Flag states will issue a Statement of Compliance for ships if they deem their report to be in accordance with the requirements and will subsequently transfer these data to the IMO's Fuel Consumption Database.

➢ **Stage 2: Data analysis**

The next step is for the IMO to summarize the data and produce an annual report for its Marine Environment Protection Committee (MEPC) to consider. This will provide the basis for objective and transparent information ahead of any policy-making in the IMO.

➢ **Stage 3: Decision on measures to be taken**

Member States would then be able to decide on updating the international measures or taking further measures in order to improve the energy efficiency of the sector with a view to reducing GHG emissions.

This approach is again fully supported by the international shipping industry:

(1) It reaffirms the IMO's role at the centre of international shipping regulation. The MEPC adopted this system after extensive debate, with valid points coming from many different governments. The system demonstrably enjoys widespread political support, which promises effective implementation.
(2) The IMO's 'no more favourable treatment' principle means that all ships have to be compliant with the environmental rules set at the IMO when calling at ports of ratifying states across the globe, regardless of whether the ship's flag state has or has not ratified these rules. This is a unique feature of the IMO regulation and provides for effective policing of the requirements.

Whilst these measures set a trajectory in the right direction, it was always recognized that they would not lead to industry de-carbonization. Therefore, the IMO has now adopted an Interim Strategy on GHG Reduction from Shipping; a strategy that will be reassessed and amended in 2023. The Strategy focuses on real CO_2 reduction from the sector, and investment in and trials of new fuels and propulsion systems are already underway. There is real optimism that shipping will be able to transition to very-low-carbon and onto zero-carbon systems in line with the expectations of the Strategy. The IMO is now working on solutions using short-, medium- and long-term timelines. The industry itself is fully supportive of the strategy, and the focus has shifted from paying for carbon emissions to finding ways to remove the carbon using research and development that may well be partly funded by the industry. The situation is dynamic and the work on carbon reduction in particular develops with each new day.

However, not all is straightforward for the industry. In an environment that thrives on international regulation, regional and national regulation is an anathema. But it exists, and there is a constant threat of more.

The European Union (EU) has created a regional requirement for fuel data collection in the form of a regulation known as Monitoring, Reporting and Verification (MRV),[5] which was adopted in 2015. This establishes an EU-wide framework for the monitoring, reporting and verification of CO_2 emissions from large ships, and it is to be implemented from 1 January 2018. The parameters differ from the IMO data collection system and in effect require ships to report similar data twice.

There is no doubt that shipping can do more to reduce fuel consumption and hence emissions. But there is currently an absolute physical limit to what can be achieved, as shipping is dependent on fossil fuels, at least for the foreseeable future. Renewables that do anything more than supplement auxiliary power demands at sea are probably several decades away. Battery-powered ships are presently limited to very short voyages, and nuclear energy can probably be ruled out on cost and political grounds. Shipping is therefore currently left with oil and gas as its main fuel sources, although the latter generates less CO_2 and may become an interim fuel for many ships until alternative low-carbon fuels, such as fuel cells using renewable energy, are eventually developed.

Therefore, if shipping is to match global calls for emission reductions that exceed what is currently technically feasible, it is possible that the way forward that will be demanded by governments is for the industry to pay to fund CO_2 emission reductions in other sectors or in developing nations. Many consider that some form of politically driven market-based measure (MBM), such as a fuel levy, is therefore probably inevitable.

While much of the shipping industry remains sceptical about MBMs, questioning whether or not they deliver a genuine environmental benefit, the industry is not shying away from this call and is looking at how a global fuel levy might be workable in practice, should this be what IMO Member States eventually decide.

14.4 Shipping's View on the Global Sulfur Cap

Like every mode of transport, shipping also has local impacts in terms of air quality, primarily through sulfur oxides (SO_x). The amount of SO_x emitted is directly proportional to the amount of sulfur in the fuel, and in shipping this has been limited by global regulation for many years, but at a higher level than that applied to road transport, for example. In 2012, the global sulfur limit was reduced from 4.5 per cent to 3.5 per cent mass/mass (m/m). In addition, Emission Control Areas (ECAs) have been established in the Baltic and North seas and off both coasts of North America, with the sulfur limits in these ECAs being set at just 0.1 per cent since 2015.

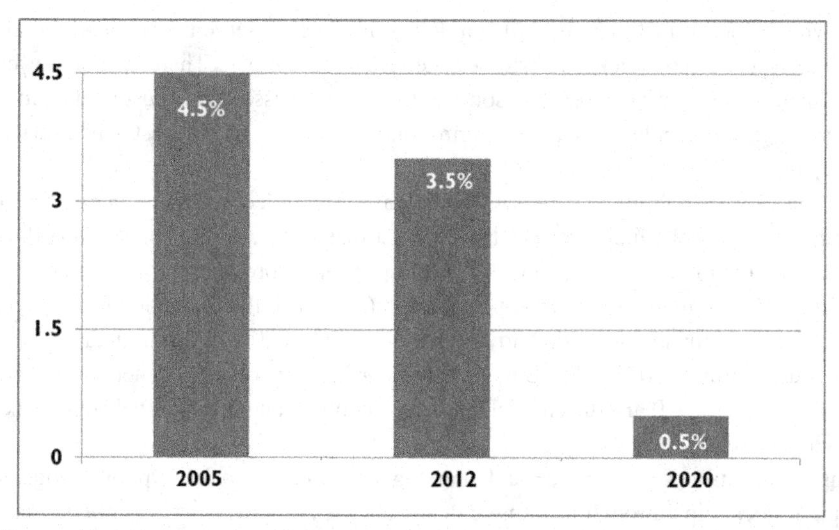

Figure 14.7 The global sulfur cap, showing the sulfur content of fuel permitted outside ECAs.

The next phase is to implement a reduction in the global sulfur content of fuel to 0.5 per cent m/m (outside ECAs). This sounds simple, but in fact it requires a complete shift in the fuel used by ships from the current viscous heavy fuel oil to something more akin to diesel. This is not a matter of merely changing the supply, but requires significant modifications to engines and fuel systems. There is also a need to ensure that sufficient fuel is in fact available in the market without having an unexpected impact on domestic and road fuels.

In October 2016, IMO Member States decided on the implementation of the 0.5 per cent global sulfur cap on marine fuel operative from 1 January 2020 (as required by MARPOL Annex VI and shown in Figure 14.7). The ICS took no view on the date of application, but it did urge IMO Member States to remove the uncertainty in the market at the time and to make a firm decision as soon as possible.

However, once again, the problem of regional regulation in Europe may have an unwanted impact. The EU has already adopted its own regulation requiring 0.5 per cent sulfur fuel inside the exclusive economic zones of its Member States (i.e., out to 200 nautical miles from the territorial sea baseline) from 2020, regardless of what the IMO decides. If the IMO were to postpone implementation, a narrow corridor would be created along the coast of North Africa in which the use of higher-sulfur fuel would still be acceptable, while elsewhere in the Mediterranean it would not. It is unknown whether EU states will also apply this regulation to their territories outside Europe. If the IMO global cap is postponed until 2025, the EU measure will create a different market for fuels, possibly making Europe less attractive for bunkering.

The industry stands behind the IMO regulation and asks only that IMO Member States consider the availability of compliant fuel, the possibility that any vacuum created by excess demand might be filled with substandard or unsuitable fuels and that the global cap is uniformly implemented to ensure fair competition and a level playing field. It is still unknown whether in fact there is sufficient low-sulfur fuel to serve the needs of the global fleet, but in just a few weeks we will see the transition happen. There will surely be an impact on the cost of marine fuel, and some of this will be passed on to customers, but in an industry largely dependent upon long-term contracts, much of the cost will impact the industry itself.

In terms of sustainability, it is impossible to leave this subject without touching upon the economic impact of the fuel switch. It has been estimated that, on implementation, the cost of marine fuel is likely to go up to well over 50 per cent more than the cost of heavy fuel oil. If fuel prices remain at the current low levels (i.e., since the dramatic fall in oil prices during 2015), the mandatory switch to low-sulfur fuel in 2020 will mean that bunker costs would return to their 2014 peak. But, on the other hand, if oil prices increase to around US$70 a barrel by 2020, the differential between compliant and residual fuel could spike by as much as US$400/tonne.

Shipowners have some serious and pressing decisions to make. Options range from scrapping older ships through to fitting exhaust gas cleaning systems (scrubbers) that will allow continued use of less expensive residual fuel, or looking at alternative low-sulfur fuels such as liquefied natural gas.

14.5 Shipping's View on Ballast Water Management Regulation

When ships discharge their cargo, they may then undertake a ballast voyage. This is a voyage where there is no cargo or perhaps less cargo than previously, and seawater has to be loaded to maintain the stability and seagoing capability of the ship. When the ship reaches the next port where cargo is to be loaded, this seawater will be discharged into the local environment. This movement of water between one ecosystem and another is a recognized vector in the promotion of unwanted and sometimes damaging invasive species. Ballast water can unavoidably contain microscopic aquatic organisms, plants and animal species. Some of these may be able to thrive in the new environment on discharge and can cause considerable local environmental and economic damage.

In 1991, the IMO's MEPC adopted the 'International Guidelines for Preventing the Introduction of Unwanted Aquatic Organisms and Pathogens from Ships' Ballast Water and Sediment Discharges',[6] urging countries to apply the appropriate provisions within the guidelines to tackle the problem, while in the meantime more research was encouraged. While work was being done at the IMO regarding the drafting and adoption of an international convention for ballast water management, the IMO Assembly (the highest governing body of the organization) adopted 'Guidelines for the Control and Management of Ships' Ballast Water to Minimize the Transfer of Harmful Aquatic Organisms and Pathogens'[7] in 1997 and, at the same time, work still continued at the MEPC towards the final adoption of a legally binding ballast water instrument.

Adopted in 2004, the International Convention for the Control and Management of Ships' Ballast Water and Sediments (BWM Convention) came into force on 8 September 2017 following the latest ratification by Finland. It has been a long and frustrating wait to reach this point, and the delay itself is the cause of many of the problems now associated with the implementation of this Convention.

It was a much-applauded feature of this Convention in 2004 that it was 'aspirational'. It aspired to mandate a requirement for which no suitable technology was then available. It is a classic case of regulation aspiring to lead technological development. In that respect, it is now clear that it expected far too much and that the parameters for ensuring that innovative equipment was fit for purpose were far too weak.

There is currently some flexibility in the timescale to which this equipment must be fitted and operated, but in the coming years there will probably be a shortage of available treatment systems and the dry-dock capacity necessary to carry out the work on perhaps 70,000 ships. This situation is made worse by unilateral regulation in the USA, which is unlikely ever to be a party to the BWM Convention.

Once again, the spectre of unilateral regulation serves not only to damage the sustainability of the shipping industry, but also to create confusion and inefficiency. By US unilateral regulation, all ships discharging ballast water in US waters will have to use a treatment system approved by the US Coast Guard (USCG). Although in practice AMSs are type-approved in accordance with the original IMO guidelines, they are only approved for 5 years, after which time a fully USCG-approved system must be installed. The USA is most unlikely to ratify the BWM Convention or accept the IMO methodology for approving ultraviolet (UV) treatment systems. This is a serious problem, as almost half of the treatment systems already installed on ships or ordered by shipowners employ UV technology. The US regulation will create market distortion and is already causing extreme uncertainty in the market, especially as the installation of ballast water treatment systems is estimated to cost the entire world fleet about US$100 billion. With the imminent enforcement of the BWM Convention, shipowners wishing to trade to the USA face an impossible dilemma when deciding what equipment to buy.

There is no doubt that the BWM Convention is needed – few would dispute that. But, equally, the shipping industry has a right to demand 'smarter' regulation that avoids unwanted confusion such as that which is now so very apparent in this case. The carriage of invasive species will be limited as ships start to use treatment equipment, and any improvement in the situation should be welcomed. We must not allow 'the best to be the enemy of the good'.

14.6 Shipping's View on Ship Recycling Regulation

Ships are huge investments, with many new ones costing in excess of US$100 million, and owners expect them to have a trading lifetime of at least 25 years. Indeed, safety regulations are predicated on ships being used for around this time in the weather conditions experienced in the North Atlantic in winter. However, uniquely, when a ship reaches the end of its serviceable life, around 90 per cent by weight of its structure is recycled: not just the steel, but many component parts find new lives in other industrial areas.

The main ship-dismantling countries are Bangladesh, India, China, Pakistan and Turkey. Ship recycling is the most environmentally sustainable method of disposing of end-of-life ships. It creates much-needed jobs in those countries, and in particular recycled steel is often a significant part of these countries' own requirements.

But, rightly, there are concerns about health, safety and the environment in these recycling countries, and consequently the process has been under the microscope for many years. The industry itself has produced guidance on the selection of the most environmentally friendly yards and the preservation of the highest possible standards in working conditions and safety. But, as is so often the case, an effective solution can only be found through regulation.

In 2009, the Hong Kong International Convention for the Safe and Environmentally Sound Recycling of Ships, 2009 (Hong Kong Convention) was adopted. The whole premise of the Hong Kong Convention is that it must be ratified by a significant proportion of the so-called recycling states to have any effect. Unfortunately, this seems a long way off.

The ICS (in conjunction with other industry organizations) has issued 'Transitional Measures for Shipowners Selling Ships for Recycling' to fill the regulatory gap and to allow shipowners to adhere to the Hong Kong Convention's requirements as far as practically possible, in advance of full implementation of a legally binding global regime. *Inter alia*, these measures set out detailed advice on the preparation and maintenance of inventories of hazardous materials as required by the Hong Kong Convention and (yet again) by a separate EU regulation that started to take effect in 2016 with implications on EU and non-EU ships alike calling at EU ports. Additionally, these guidelines provide recommended measures that shipping companies should take now when selling end-of-life ships for recycling.

Once again, Europe has stepped in with a regional requirement. According to the EU regulation,[8] a list of approved ship recycling yards was published in 2016, which EU shipowners will be required to use when disposing of redundant ships.

In the face of this regional obligation, it is important that the EU Commission acknowledge the efforts of non-EU shipyards to comply with the Hong Kong Convention's standards, such as yards in India, several of which have gained certification from classification societies. There is a danger that the EU regulation could undermine the process of Hong Kong Convention's ratification if yards that have demonstrated compliance with the Hong Kong Convention do not end up on the EU list.

The main point of controversy is whether beaching as a method of recycling (which the Hong Kong Convention does not expressly ban) is environmentally friendly or not. The shipping industry believes that under carefully controlled conditions beaching can be environmentally friendly. But some environmental non-governmental organizations take a negative stance on this matter, and the EU's position is unclear. If those shipyards that have already voluntarily complied with Hong Kong Convention standards (and use beaching in an environmentally friendly way) are not included on the EU list of approved shipyards, the remaining available shipyard capacity will be insufficient for current needs.

This may have unexpected environmental outcomes if ships have to be laid up afloat for years waiting for recycling capacity to become available.

14.7 Conclusion: Global Rules for a Global Industry

It is hoped that the foregoing has reflected the will of the shipping industry to take full responsibility for its environmental impact and its desire for truly effective global regulation. However, many obstacles stand in the way of that desire.

The first and most obvious of these is to combat the trend towards increasing national and regional regulation for all of the reasons already given, but not least to preserve the role of the IMO as the unique specialist agency for regulating shipping.

There is a need for regulation to be 'smarter'. This means developing regulation with an eye to factors that have not hitherto been acknowledged. A new regulation should be developed in the context of regulations that are already in place and to ensure that regulations do not become mutually damaging. For example, the requirement to fit ballast water treatment equipment means that ships burn more fuel to operate the equipment; this is very contradictory at a time when the need to reduce fuel consumption is paramount.

The ICS would like to see a requirement to conduct an impact assessment and cost–benefit analysis at the earliest stage of regulatory development at the IMO, with the intent of:

- Avoiding cross-cutting of regulations and harmonizing new ones with existing requirements;
- Preventing an increased administrative burden for seafarers and promoting an environment where procedures can be simplified;
- Helping understand cost efficiency in regulation;
- Avoiding unintended consequences;
- Allowing faster ratification by governments due to greater confidence in the efficiency of a regulation;
- Creating uniformity in rules for transparency; and
- Removing uncertainty for the shipping industry and its suppliers.

In short, the ICS is promoting 'smarter' regulation that is compliant with the three pillars of sustainability. It is only through this mechanism that the contribution of shipping to the global economy and to improved prosperity and well-being for the growing population can be sustained.

Notes

1 For further information, see International Chamber of Shipping (ICS) and BIMCO Manpower Report 2015.
2 The report is available online at www.un-documents.net/our-common-future.pdf.
3 See UNCTAD Review of Maritime Transport 2015, p. 10.
4 See Third IMO Greenhouse Gas Study 2014, p. 1.

5 Regulation (EU) 2015/757 of the European Parliament and of the Council of 29 April 2015 on the monitoring, reporting and verification of carbon dioxide emissions from maritime transport, and amending Directive 2009/16/EC.
6 See IMO resolution MEPC.50(31).
7 See IMO resolution A.868(20).
8 Regulation (EU) No. 1257/2013 of the European Parliament and of the Council of 20 November 2013 on ship recycling and amending Regulation (EC) No. 1013/2006 and Directive 2009/16/EC.

15

Environmental Impacts of Shipping

Can We Learn?

STEPHEN DE MORA, TIMOTHY FILEMAN AND THOMAS VANCE

15.1 Introduction

Environmental impacts of shipping arise through port development and ships, only the latter of which are considered here. Shipping is a vital industry supporting global trade, with over 90 per cent of goods transported by sea. While most of the worldwide fleet comprises cargo vessels of one type or another, ships are diverse in type, function and region of operation. Given that cruise liners, research ships and naval vessels travel throughout the world's oceans, ships impact the marine environment everywhere they can travel. These effects can be felt in inland waters that are navigable and/or are connected directly or indirectly to the seas. Well-known examples include the invasions of zebra mussels (*Dreissena polymorpha*) in the Great Lakes (Oliver, 2014) and of comb jellies (*Mnemiopsis leidyi*) in the Caspian Sea (Shiganova *et al.*, 2001).

15.2 The DPSIR Framework

The DPSIR (drivers, pressures, state, impact and response) framework provides a useful conceptual model for assessing and managing problems arising from the interactions between ships and the environment. The DPSIR framework comprises: *drivers* – the causes of the environments problem (e.g., ship operations); *pressures* – the effects of the activity (e.g., ship emissions); *state* – the environmental parameters and components that are affected (e.g., marine ecosystems); *impact* – the effect exerted on environmental and biological reservoirs (e.g., invasive species, habitat modification); and *response* – mechanisms put into effect to prevent and/or mitigate the environmental impacts (e.g., environmental policies, international conventions). The environmental impact of ships is illustrated using the DPSIR framework in Figure 15.1. Accordingly, a well-established cycle of processes characterizes the continuous means of protecting the marine environment from the deleterious effects of ships. Both end-of-life events, either by shipwreck or shipbreaking, and routine operational performance of vessels exert, respectively, acute and chronic environmental impacts.

The case history of tributyltin (TBT) exemplifies the cyclical nature of societal control of a contaminant in the marine environment (de Mora, 1996). TBT was introduced into

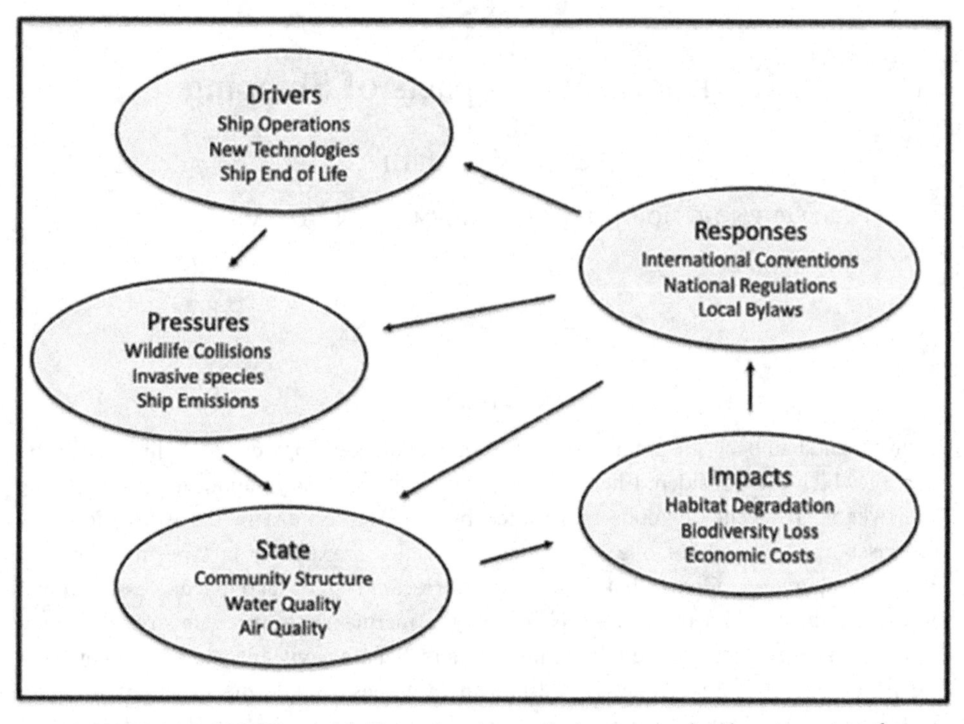

Figure 15.1 The DPSIR model illustrating the cyclical processes in place to protect the marine environment from ships.

marine anti-fouling paints in the 1960s. Owing to its effectiveness at preventing biofouling, TBT use become popular on marine vessels of all sizes, from yachts to supertankers, and as a result, the TBT leached from vessels' hulls was distributed throughout the world's oceans and coastal waters. Numerous harmful effects on non-target organisms were observed, notably chambering in oyster shells and imposex (the development of a penis in females) in marine gastropods. France first introduced local regulations banning the application of TBT on small vessels (<25 m in length) in its south-west region in order to protect and remediate the oyster shellfisheries industry. The quick re-establishment of the oyster industry admirably demonstrated the efficacy of the ban. In time, other nations adopted national legislation limiting the use of TBT. On a regional scale, the European Union (EU) imposed controls, and eventually the International Maritime Organization (IMO) established the International Convention on the Control of Harmful Anti-fouling Systems on Ships (AFS), which was signed in 2001 and came into force in 2008. Removing TBT from the marketplace stimulated research into alternative anti-foulant compounds. Several contenders became available for use, including the algaecides Irgarol- and Diuron, which act as photosynthetic inhibitors. However, environmental research demonstrated the damaging effects on the marine environment of these persistent pollutants as well (Gatidou *et al.*, 2007), leading to their use in marine anti-fouling paints being banned in several countries.

15.2.1 DPSIR: Drivers

The main driving force influencing the impact of ships on the environment is ship operations. Such activities are dominated by global trade and fishing. Other undertakings that result in ships operating worldwide in all marine environments include defence, tourism, research and offshore industries. A second key driver relates to new technologies, the introduction of which is generally either to achieve better economic efficiency or in response to legal requirements, including IMO Conventions. Finally, the end of life of vessels has an effect on the environment. The worst-case scenarios are accidents at sea, particularly shipwrecks. However, shipbreaking activities also contribute to deleterious consequences for the local environment.

15.2.2 DPSIR: Pressures

Numerous pressures can be identified. The most obvious ones pertain to emissions from ships. The contaminants here include various atmospheric emissions (especially carbon dioxide, sulfur oxides and nitrogen oxides), oils and lubricants, bilge and ballast waters and leachates from anti-fouling coatings. Invasive species can be released through ballast water operations, but also due to biofouling on the external surfaces of vessels. Less obvious pressures relate to noise pollution and wildlife collisions that especially affect charismatic marine organisms.

15.2.3 DPSIR: State

Clearly, atmospheric emissions can influence air quality. Similarly, the numerous discharges to the sea can affect water quality. Invasive species and, to a certain extent, wildlife collisions can modify community structures. Changing environmental conditions play a role, as marine biomes move northwards and can allow invasive species to become established in locations that previously were not conducive to their recruitment.

15.2.4 DPSIR: Impact

Consequences for marine ecosystems can involve habitat degradation and biodiversity loss. An important impact is economic costs. On the one hand, some such expenditure is borne by shipping stakeholders. This outcome is especially true in instances whereby shipboard modifications (e.g., retrofitting of ballast water treatment system or fitting of sewage treatment systems) or changes to operational activities (e.g., use of fuels with low sulfur contents) become necessary for legal compliance. There can also be economic costs relating to the environment, which are spectacularly visible in the case of shipwrecks, but much less evident with respect to noise pollution, air pollution and other physical disturbances.

15.2.5 DPSIR: Response

The most important responses involve legal instruments to mitigate, remediate and/or prevent the impact of ships on the environment. Mechanisms encompass local by-laws, national regulations and international conventions, notably those of the IMO.

The IMO has led worldwide efforts to regulate shipping and ships (Tan, 2012). Numerous conventions relating to the safety of lives and sea, as well as environmental protection, have been negotiated, signed and entered into force. In order to come into force, IMO conventions generally require a certain number of countries representing a given percentage of the global fleet tonnage to ratify these conventions. Table 15.1 lists the key conventions and indicates those that have not yet come into force. Notably, there may be a considerable time lag between a convention being signed and it coming into force. The International Convention for the Control and Management of Ships' Ballast Water and Sediments (BWM) is a case in point – having been signed in 2004, BWM did not come into force until 2017.

Despite the good intentions behind these legal instruments, they are unlikely to achieve the desired environmental outcome in isolation. They need to be combined with effective education, training and support for vessel operators. These secondary steps are vital to enable vessel operators to understand why the controls are required, and then to provide practical support to enable them to comply. Failure to communicate the requirements and benefits of environmental regulation and the lack of practical support to help implement changes to operation are likely to result in pushback, delay and non-compliance from industry.

Lastly, at an appropriate time, effective enforcement and review of environmental legislation are also required. Although enforcement of environmental legislation on a global and mobile industry is challenging, technological developments, such as remote sensing approaches, together with experience gained through the implementation of previous conventions, are making this aspect more achievable than ever before.

15.3 Conclusions

Ships impact the marine environment in multifarious ways. The integrated effect can be considerable given the large number of vessels involved, particularly in trade and fishing, which have access throughout the ocean and coastal ecosystems. The key driver for the shipping industry is cost. Accordingly, new technologies have been developed to improve efficiencies, notably via propulsion systems or through limiting biofouling. Such developments have often been introduced in order to comply with legal requirements. However, the introduction of new technologies has, on occasion, resulted in unexpected consequences for marine ecosystems, as is well exemplified by the case history of TBT (de Mora, 1996).

Research into marine pollution (including noise) and other impacts, such as wildlife collisions, has led to numerous requirements to alter behaviour and practices at sea. This has come about chiefly through the work of the IMO's Marine Environment Division. The IMO's senior technical body, the Marine Environment Protection Committee, aided by the

Table 15.1. *Key IMO conventions relating to ships, indicating the date of original approval and whether the convention is in force*

Convention	Original convention date	In force
International Convention for the Safety of Life at Sea (SOLAS) plus SOLAS PROT (1978 and 1988) and SOLAS AGR (1996)	1974	✓
International Convention for the Prevention of Pollution from Ships – as modified by the Protocol of 1978 relating thereto and by the Protocol of 1997 (MARPOL) plus Annexes III, IV and V and MARPOL PROT 1997	1973	✓
International Convention on Standards of Training, Certification and Watchkeeping for Seafarers (STCW) as amended including the 1995 and 2010 Manila Amendments	1978	✓
Other conventions relating to maritime safety and security and the ship–port interface		
Convention on the International Regulations for Preventing Collisions at Sea (COLREG) plus CLC PROT (1976 and 1992)	1972	✓
Convention on Facilitation of International Maritime Traffic (FAL)	1965	✓
International Convention on Load Lines (LL) plus LL PROT (1988)	1966	✓
International Convention on Maritime Search and Rescue (SAR)	1979	✓
Convention for the Suppression of Unlawful Acts Against the Safety of Maritime Navigation (SUA) and Protocol for the Suppression of Unlawful Acts Against the Safety of Fixed Platforms located on the Continental Shelf (and the 2005 Protocols)	1988	✓
International Convention for Safe Containers (CSC)	1972	✓
CSC AMEND 1993	1993	X
Convention on the International Maritime Satellite Organization (IMSOC)	1976	✓
The Torremolinos International Convention for the Safety of Fishing Vessels (SFV) superseded by the 1993 Torremolinos Protocol; Cape Town Agreement of 2012 on the Implementation of the Provisions of the 1993 Protocol Relating to the Torremolinos International Convention for the Safety of Fishing Vessels	1977	X
International Convention on Standards of Training, Certification and Watchkeeping for Fishing Vessel Personnel (STCW-F)	1995	✓
Special Trade Passenger Ships Agreement (STP) 1971 and Protocol on Space Requirements for Special Trade Passenger Ships	1973	✓
Other conventions relating to the prevention of marine pollution		
International Convention Relating to Intervention on the High Seas in Cases of Oil Pollution Casualties (INTERVENTION) plus INTERVATION PROT 1973	1969	✓
Convention on the Prevention of Marine Pollution by Dumping of Wastes and Other Matter (LC) (and the 1996 London Protocol)	1972 1978	✓ X
LC AMEND-78	1996	✓
LC PROT 1996		
International Convention on Oil Pollution Preparedness, Response and Co-operation (OPRC)	1990	✓
Protocol on Preparedness, Response and Co-operation to Pollution Incidents by Hazardous and Noxious Substances (OPRC-HNS Protocol)	2000	✓

Table 15.1. (*cont.*)

Convention	Original convention date	In force
International Convention on the Control of Harmful Anti-fouling Systems on Ships (AFS)	2001	√
International Convention for the Control and Management of Ships' Ballast Water and Sediments (BWM)	2004	√
The Hong Kong International Convention for the Safe and Environmentally Sound Recycling of Ships	2009	X
Conventions covering liability and compensation		
International Convention on Civil Liability for Oil Pollution Damage (CLC) plus CLC PROT (1976 and 1992)	1969	√
Protocol to the International Convention on the Establishment of an International Fund for Compensation for Oil Pollution Damage (FUND) plus FUND PROT (1976, 1992, 2003)	1992	√
FUND PROT 2000	2000	X
Convention relating to Civil Liability in the Field of Maritime Carriage of Nuclear Material (NUCLEAR)	1971	√
Athens Convention Relating to the Carriage of Passengers and Their Luggage by Sea (PAL) plus PAL PROT (1976 and 2002)	1974	√
PAL PROT 1990	1990	X
Convention on Limitation of Liability for Maritime Claims (LLMC) plus LLMC PROT 1996	1976	√
International Convention on Liability and Compensation for Damage in Connection with the Carriage of Hazardous and Noxious Substances by Sea (HNS) (and its 2010 Protocol)	1996	X
International Convention on Civil Liability for Bunker Oil Pollution Damage	2001	√
Nairobi International Convention on the Removal of Wrecks	2007	√
Other subjects		
International Convention on Tonnage Measurement of Ships (TONNAGE)	1969	√
International Convention on Salvage (SALVAGE)	1989	√
Convention on the International Maritime Organization	1948	√

Sub-Committee on Pollution Prevention and Response, reviews the available evidence and develops guidelines and procedures that facilitate the drafting, implementation and maintenance of the IMO's conventions (particularly MARPOL, the International Convention for the Prevention of Pollution from Ships). Notably, IMO conventions have led to a marked decrease in oil pollution at sea. Some conventions, such as the Ballast Water Convention, have stimulated research from the shipping industry in order to comply with strict water quality limits on discharges. Accordingly, the DPSIR framework provides a useful conceptual model for assessing and managing the problems arising from the interactions between ships and the environment.

References

de Mora, S. J. (1996). The tributyltin debate: ocean transportation versus seafood harvesting In: S. J. de Mora, ed., *Tributyltin: Case Study of an Environmental Contaminant.* Cambridge: Cambridge University Press, pp. 1–20.

Gatidou, G., Thomaidis, N. S. and Zhou, J. L. (2007). Fate of Irgarol 1051, Diuron and their main metabolites in two UK marine systems after restrictions in antifouling paints. *Environment International*, **33**, 70–77.

Oliver, M. (2014). Linking zebra mussel invasion and waterborne commerce in the U.S. *Water Policy*, **16**, 536–556.

Shiganova, T., Kamakin, A. M. and Zhukova, O. P. *et al.* (2001) The invader into the Caspian Sea ctenophore *Mnemiopsis* and its initial effect on the pelagic ecosystem. *Oceanology*, **41**, 517–524.

Tan, A. K.-J. (2012). *Vessel-Source Marine Pollution.* Cambridge: Cambridge University Press.

Index

Printed in the United States
By Bookmasters